Graphene Field-Effect Transistors

Graphene Field-Effect Transistors

Advanced Bioelectronic Devices for Sensing Applications

Edited by Omar Azzaroni and Wolfgang Knoll

WILEY VCH

Editors

Prof. Omar Azzaroni
INIFTA-CONICET-UNLP
Diagonal 113 y 64
1900 La Plata
Argentina

Prof. Wolfgang Knoll
Laboratory for Life Sciences and Technology
Faculty of Medicine and Dentistry
Danube Private University
3500 Krems
Austria

Cover Image: Surface engineering of graphene-based FET biosensors for virus detection. Credit: Gonzalo E. Fenoy & M. Lorena Cortez

Library of Congress Card No.: applied for

British Library Cataloguing-in-Publication Data:
A catalogue record for this book is available from the British Library.

Bibliographic information published by the Deutsche Nationalbibliothek
The Deutsche Nationalbibliothek lists this publication in the Deutsche Nationalbibliografie; detailed bibliographic data are available on the Internet at <http://dnb.d-nb.de>.

Print ISBN: 978-3-527-34990-6
ePDF ISBN: 978-3-527-84339-8
ePub ISBN: 978-3-527-84338-1
oBook ISBN: 978-3-527-84337-4

Typesetting Straive, Chennai, India

This book is dedicated to Elias Drake, Ivo, and Dante.

Contents

Foreword *xv*
Preface *xvii*

1 **2D Electronic Circuits for Sensing Applications** *1*
Diogo Baptista, Ivo Colmiais, Vitor Silva, Pedro Alpuim, and Paulo M. Mendes
1.1 Introduction *1*
1.2 Graphene Inductors *3*
1.2.1 Modeling of Graphene Inductors *4*
1.3 Graphene Capacitors *5*
1.3.1 Modeling Graphene Capacitors *8*
1.4 2D Material Transistors *9*
1.4.1 Most Common Topologies for Transistors *10*
1.4.2 Modeling of 2D Materials-Based Transistors *11*
1.5 2D Material Diodes *15*
1.5.1 Most Common Topologies *16*
1.5.2 Modeling of 2D Materials-Based Diodes *17*
1.6 Graphene Devices *18*
1.6.1 Graphene Frequency Multipliers *18*
1.6.2 Graphene Mixers *18*
1.6.3 Graphene Oscillators *19*
1.6.3.1 Ring Oscillators *19*
1.6.3.2 LC Tank Oscillators *19*
1.7 Conclusion *19*
References *20*

2 **Large Graphene Oxide for Sensing Applications** *25*
Jingfeng Huang, J. Amanda Ong, and I.Y. Alfred Tok
2.1 Graphene Oxide (GO) *25*
2.2 GO as Biosensors *25*
2.3 Large GO *26*
2.4 Mechanism of Large GO via Modified Hummers Method *27*
2.5 Large GO (Modified Hummers Method) Biosensors *28*

2.6 Mechanism of Large GO via Reduced GO Growth *29*
2.7 Large GO (Reduced GO Growth) Biosensors *34*
2.8 Conclusion *38*
2.9 Further Developments *38*
References *39*

3 Solution-Gated Reduced Graphene Oxide FETs: Device Fabrication and Biosensors Applications *43*
Nirton C. S. Vieira, Bianca C. S. Ribeiro, Rodrigo V. Blasques, Bruno C. Janegitz, Fabrício A. dos Santos, and Valtencir Zucolotto
3.1 Introduction *43*
3.2 Graphene, Graphene Oxide, and Reduced Graphene Oxide *45*
3.2.1 Chemical Reduction *48*
3.2.2 Thermal Reduction *49*
3.2.3 Electrochemical Reduction *51*
3.3 rGO-Based Solution-Gated FETs *52*
3.3.1 Manufacturing Strategies *53*
3.4 Applications of rGO SG-FETs as Biosensors *57*
3.4.1 rGO Functionalization *59*
3.4.2 Enzymatic Biosensors *60*
3.4.3 Affinity Biosensors *61*
3.4.4 Debye Length Screening and How to Overcome It *63*
3.5 Final Remarks and Challenges *64*
Acknowledgments *65*
References *65*

4 Graphene-Based Electronic Biosensors for Disease Diagnostics *71*
Ahmar Hasnain and Alexey Tarasov
4.1 Introduction *71*
4.1.1 A Promise for Diagnostics *71*
4.1.2 Principle of Graphene FET Sensor *72*
4.2 Device Fabrication Process *75*
4.2.1 Graphene Synthesis *75*
4.2.2 Graphene Transfer Over Substrates *76*
4.2.3 Fabrication of GFET *77*
4.2.4 New Developments *78*
4.3 Functionalization and Passivation *78*
4.3.1 Probe Molecules *79*
4.3.2 Immobilization of Probe Molecules *80*
4.3.3 Debye Length *81*
4.3.4 Passivation *82*
4.4 CVD GFETs for Diagnostics *83*
4.4.1 Graphene-Based FET Biosensors for Nucleic Acids *83*
4.4.2 Graphene-Based FET Biosensors for Antibody–Antigen Interactions *85*

4.4.3 Graphene-Based FET Biosensors for Enzymatic Biosensors *87*
4.4.4 Graphene-Based FET Biosensors for Sensing of Small Ions *90*
4.5 Discussion *92*
4.5.1 Summary *92*
4.5.2 Challenges *92*
4.5.3 Future Perspectives *93*
References *93*

5 Graphene Field-Effect Transistors: Advanced Bioelectronic Devices for Sensing Applications *103*
Kyung Ho Kim, Hyun Seok Song, Oh Seok Kwon, and Tai Hyun Park
5.1 Introduction *103*
5.1.1 Bioelectronic Nose Using Olfactory Receptor-Conjugated Graphene *106*
5.1.2 Bioelectronics for Diagnosis Using Bioprobe-Modified Graphene *112*
5.1.3 Biosensors for Environmental Component Monitoring Using Graphene *116*
5.2 Conclusion *120*
Acknowledgments *120*
References *120*

6 Thin-Film Transistors Based on Reduced Graphene Oxide for Biosensing *125*
Kai Bao, Ye Chen, Qiyuan He, and Hua Zhang
6.1 Introduction *125*
6.2 Working Principle of TFT-Based Biosensing *126*
6.3 TFTs Based on rGO for Biosensing *128*
6.3.1 Protein Detection *128*
6.3.2 Metal-Ion Detection *131*
6.3.3 Nucleic Acid Detection *134*
6.3.4 Small Biomolecular Biosensor *135*
6.3.5 Living-Cell Biosensor *137*
6.3.6 Gas Detection *138*
6.4 Conclusion *140*
References *142*

7 Towards Graphene-FET Health Sensors: Hardware and Implementation Considerations *149*
Nicholas V. Apollo and Hualin Zhan
7.1 Introduction to Health Sensing *149*
7.2 Graphene-FET in Liquid for Sensing *151*
7.2.1 Graphene Transistors *153*
7.2.2 Graphene Hall Structures in Liquid *156*
7.2.3 Graphene Membrane Transistors *159*
7.3 Device Implementation Considerations *160*

7.3.1 Hardware and Instrumentation *160*
7.3.2 Biostability and Biocompatibility *162*
7.3.3 Medical Imaging Compatibility *163*
References *164*

8 Quadratic Fit Analysis of the Nonlinear Transconductance of Disordered Bilayer Graphene Field-Effect Biosensors Functionalized with Pyrene Derivatives *169*
Sung Oh Woo, Sakurako Tani, and Yongki Choi
8.1 Introduction *169*
8.2 Fabrication of Graphene-Based Field-Effect Biosensors *170*
8.3 Fundamental Sensing Parameters of Graphene-Based Field-Effect Biosensors *173*
8.4 Disordered Bilayer Graphene Field-Effect Biosensors Functionalized with Pyrene Derivatives *174*
8.5 Quadratic Fit Analysis of the Nonlinear Transconductance of Disordered Bilayer Graphene Field-Effect Biosensors *177*
8.6 Conclusion *181*
Acknowledgment *181*
References *182*

9 Theoretical and Experimental Characterization of Molecular Self-Assembly on Graphene Films *185*
Kishan Thodkar, Pierre Cazade, and Damien Thompson
9.1 Introduction *185*
9.2 Experimental Tools to Characterize Molecular Functionalization of Graphene *186*
9.2.1 Considering the Three Distinct Techniques Available for Functionalizing Graphene Are the Outcomes of the Three Functionalization Techniques Consistent, Similar, Reproducible Across all Three Techniques? *187*
9.2.2 What Tools and Methods Are Available to Perform Such a Characterization of Molecular Self-Assembly Across the Nano to Macro Scale? *188*
9.3 Atomistic Insights to Guide Molecular Functionalization of Graphene *196*
References *203*

10 The Holy Grail of Surface Chemistry of CVD Graphene: Effect on Sensing of cTNI as Model Analyte *207*
Adrien Hugo, Teresa Rodrigues, Marie-Helen Polte, Yann R. Leroux, Rabah Boukherroub, Wolfgang Knoll, and Sabine Szunerits
10.1 Introduction *207*

10.2 General Overview of CVD Graphene Production, Substrate Transfer and Characterization *210*
10.3 Evaluation of Graphene Topographical Quality *212*
10.4 CVD Graphene for FET-Based Sensing *214*
10.4.1 Diazonium Chemistry on CVD Graphene *217*
10.4.2 Pyrene Chemistry on CVD Graphene *220*
10.5 Conclusion *225*
References *226*

11 Sensing Mechanisms in Graphene Field-Effect Transistors Operating in Liquid *231*
Tilmann J. Neubert and Kannan Balasubramanian *231*
11.1 Introduction *231*
11.2 Field-Effect Operation in Liquid Compared to Operation in Air *232*
11.3 Caveats When Operating FETs in Liquid *234*
11.4 Graphene FETs in Liquid *235*
11.5 Measurement Modes *236*
11.6 Using FETs for Sensing in Liquid – Sensing Mechanisms *238*
11.7 The Electrochemical Perspective *241*
11.8 The GLI and pH Sensing *245*
11.9 Detection of Nucleic Acids *246*
11.10 Other Examples *247*
11.11 Concluding Remarks *248*
References *248*

12 Surface Modification Strategies to Increase the Sensing Length in Electrolyte-Gated Graphene Field-Effect Transistors *251*
Juliana Scotto, Wolfgang Knoll, Waldemar A. Marmisollé, and Omar Azzaroni
12.1 Introduction *251*
12.2 Ion-Exclusion and Donnan Potential *253*
12.3 Surface Modification with Polymer Films *255*
12.4 Surface Modification with Lipid Layers *258*
12.5 Surface Modification with Mesoporous Materials *260*
12.6 Kinetic Cost of Extending the Sensing Length *262*
12.7 Conclusions *265*
References *266*

13 Hybridized Graphene Field-Effect Transistors for Gas Sensing Applications *271*
Radha Bhardwaj and Arnab Hazra *271*
13.1 Introduction *271*
13.2 Graphene *272*

13.3 Graphene FET *272*
13.4 Graphene in Gas Sensing *274*
13.5 Graphene FET for Gas Sensing *275*
13.6 Hybrid Graphene FET for Gas Sensing *277*
13.7 Conclusion *281*
Acknowledgments *281*
References *281*

14 Polyelectrolyte-Enzyme Assemblies Integrated into Graphene Field-Effect Transistors for Biosensing Applications *285*
Esteban Piccinini, Gonzalo E. Fenoy, Wolfgang Knoll, Waldemar A. Marmisollé, and Omar Azzaroni
14.1 Introduction *285*
14.2 Field-Effect Transistors Based on Reduced Graphene Oxide *286*
14.3 Enzyme-Based GFET Sensors Fabricated via Layer-by-Layer Assembly *287*
14.3.1 Layer-by-Layer (LbL) Assemblies of Polyethylenimine and Urease onto Reduced Graphene-Oxide-Based Field-Effect Transistors (rGO FETs) for the Detection of Urea *288*
14.3.2 Ultrasensitive Sensing Through Enzymatic Cascade Reactions on Graphene-Based FETs *292*
14.4 Conclusions *296*
References *297*

15 Graphene Field-Effect Transistor Biosensor for Detection of Heart Failure-Related Biomarker in Whole Blood *301*
Jiahao Li, Yongmin Lei, Zhi-Yong Zhang, and Guo-Jun Zhang
15.1 Introduction *301*
15.2 Experimental Systems and Procedures *304*
15.2.1 Fabrication of GFET Sensor *304*
15.2.2 Decoration of Platinum Nanoparticles *304*
15.2.3 Surface Functionalization *305*
15.2.4 Immunodetection in Whole Blood *305*
15.2.5 Electrical Measurements *305*
15.3 Sensing Principle of GFET for BNP Detection *306*
15.4 Device Characterization *306*
15.5 Sensing Performance *308*
15.5.1 Stability and Reproducibility *308*
15.5.2 Selectivity *309*
15.5.3 Sensitivity *309*
15.6 Clinical Application Prospects *311*
15.7 Advantages, Limitations, and Outlook of the FET-Based BNP Assay *311*
References *313*

16 Enzymatic Biosensors Based on the Electrochemical Functionalization of Graphene Field-Effect Transistors with Conducting Polymers *317*
Gonzalo E. Fenoy, Esteban Piccinini, Wolfgang Knoll, Waldemar A. Marmisollé, and Omar Azzaroni
16.1 Introduction *317*
16.2 Functionalization of Graphene Transistors with Poly(3-amino-benzylamine-*co*-aniline) Nanofilms *318*
16.3 Construction of Acetylcholine Biosensors Based on GFET Devices Functionalized with Electropolymerized Poly(3-amino-benzylamine-*co*-aniline) Nanofilms *322*
16.4 Glucose Detection by Graphene Field-Effect Transistors Functionalized with Electropolymerized Poly(3-amino-benzylamine-*co*-aniline) Nanofilms *327*
16.5 Conclusions *332*
References *333*

17 Graphene Field-Effect Transistors for Sensing Stress and Fatigue Biomarkers *339*
Biddut K. Sarker, Cheri M. Hampton, and Lawrence F. Drummy
17.1 Introduction *339*
17.2 Molecular Biomarkers *341*
17.3 Graphene Field-Effect Transistor Based Biosensors *343*
17.3.1 Graphene *343*
17.3.2 Structure of Graphene Field-Effect Transistors *345*
17.3.3 Sensing Mechanism of GFET Biosensors *346*
17.4 GFET Biosensor Fabrication *348*
17.4.1 Graphene Production *348*
17.4.2 Device Fabrication *349*
17.4.3 Graphene Functionalization *350*
17.5 GFET-Based Stress and Fatigue Biosensors *353*
17.6 Flexible, Wearable GFET Biosensors, and Biosensor Systems *358*
17.7 Current Challenges and Future Perspective *362*
17.7.1 Debye Length Screening *362*
17.7.2 Device-to-Device Variability *366*
17.7.3 Short Lifetime and Reusability Issue *366*
17.8 Conclusion *367*
References *367*

18 Highly Sensitive Pathogen Detection by Graphene Field-Effect Transistor Biosensors Toward Point-of-Care-Testing *373*
Shota Ushiba, Takao Ono, Yasushi Kanai, Naruto Miyakawa, Shinsuke Tani, Hiroshi Ueda, Masahiko Kimura, and Kazuhiko Matsumoto
18.1 Introduction *373*
18.2 Toward Detection of Pathogens by Mimicking Cell Surfaces *374*

18.2.1 Introduction 374
18.2.2 Fabrication of Sialoglycan-Functionalized GFETs 375
18.2.3 Evaluation of Sialoglycan-Functionalized GFETs 375
18.3 Signal Enhancement in GFETs 377
18.3.1 Sensitivity Enhancement by Increasing Receptor Density 377
18.3.1.1 Case of Linkers 377
18.3.1.2 Basis for Evaluation of Linker-Based Performance Enhancement 378
18.3.1.3 Evaluation of Performance Enhancement by Linkers 378
18.3.2 Ultrasensitive Detection of Small Antigens by Open-Sandwich Immunoassay on GFETs 380
18.3.2.1 Principle of Open-Sandwich (OS) Immunoassay 380
18.3.2.2 Advantages of OS-IAs with GFETs 380
18.3.2.3 Antibody Fragments and Device Fabrication 381
18.3.2.4 OS-IAs on GFETs 382
18.3.2.5 OS-IAs on GFETs in Human Serum 382
18.3.3 Real-Time Measurement of Enzyme Reaction in Microdroplets Using GFETs and Its Application to Pathogen Detection 384
18.3.3.1 Introduction 384
18.3.3.2 Measurement Mechanism and Model Measurement System 385
18.4 Practical Issues: Baseline Drift and Inspection Methods 387
18.4.1 Drift Suppression and Compensation of GFET Biosensors 388
18.4.1.1 Drift Suppression in GFETs by Cation Doping 388
18.4.1.2 Drift Compensation by State-Space Modeling 390
18.4.2 Deep-Learning-Based Optical Inspection of GFETs 393
18.5 Conclusion 398
References 398

19 High-Performance Detection of Extracellular Vesicles Using Graphene Field-Effect Transistor Biosensor *405*
Ding Wu, Yi Yu, Zhi-Yong Zhang, and Guo-Jun Zhang
19.1 What is Extracellular Vesicles 405
19.2 The Clinical Significance of Extracellular Vesicles 406
19.3 Introduction to Graphene Field-Effect Transistor Biosensor 406
19.4 GFET Biosensor for High-Performance Detection of Extracellular Vesicles 407
19.4.1 Detection of the Overall Level of Microvesicles Using GFET Biosensor 408
19.4.2 Specific Detection of Hepatocellular Carcinoma-Derived Microvesicles Using Dual-Aptamer Modified GFET Biosensor 409
19.4.3 Label-Free Detection of Cancerous Exosomes Using GFET Biosensor 410
19.5 Some Prospects for Graphene Field-Effect Transistor Biosensor 411
References 412

Index *417*

Foreword

Sensors are ubiquitous in modern life. They surround us, monitor our body function and our environment, protect us, help us communicate with each other, and make work more efficient and less strenuous. It is not at all surprising that the market demand for better and more efficient sensors is only increasing. For instance, in 2019, the sensor market size was of approximately US$ 167 billion, and it is expected to reach US$ 346 billion by 2028. However, progress in sensor's development passes unequivocally through the evolution of advanced materials that can convert different types of environmental energy into energy that can be processed either digitally or analogically with high accuracy, short response time, long-term thermal and noise stability, linearity, low power consumption, and cost. These characteristics impose stringent constraints on the type of materials that can be used in sensor development. Since its isolation in 2004, graphene has been lauded as the ultimate material for sensing applications because of its unique physical properties, namely, high electrical and thermal conductivity, large Young modulus, high sensitivity to chemicals and strain, and, most of all, its large surface area to volume ratio. It is famously said that the interface is the device. We can also say that *the surface is the sensor.* Detection usually occurs at the surface of a material and, hence, the larger the ratio of surface area to the volume of the device, the higher the ability of the device to detect. Graphene sensors have high accuracy because graphene is a pure surface (no bulk) and can detect minute amounts of chemical, electrical, magnetic, and stress signals and convert them into signals that can be processed and communicated to other devices. The literature on graphene sensors has grown exponentially in the last decade, and it is quite hard to follow this growth and even separate "tares from wheat." The book by Omar Azzaroni and Wolfgang Knoll on advanced bioelectronic devices for sensing applications using graphene field-effect transistors is a breath of fresh air in this crowded research space. I believe it will become a fundamental reference for any researcher, engineer, or industrialist who is interested in learning why graphene is the ultimate sensor material.

Singapore, November 2022

Prof. Antonio H. Castro Neto
Director, Centre for Advanced 2D Materials (CA2DM)
Co-Director, Institute for Functional Intelligent Materials (I-FIM)
Distinguished Professor, Department of Material Science Engineering
Distinguished Professor, Department of Physics
Professor, Department of Electrical and Computer Engineering
National University of Singapore

Preface

The discovery of a layer of graphene, one atom-thin, by Andre Geim and Konstantin Novoselov (Nobel Prize in Physics 2010) at the University of Manchester (UK) in 2004 created a whole new scientific field and resulted in robust technologies that strongly contributed to the development of advanced devices.

Among different graphene-based innovations, the graphene field-effect transistor (GFET) represents a remarkable example of an advanced electronic device. GFETs employ the typical configuration of a field-effect transistor (FET) device using a graphene channel – constituted of a lattice of carbon atoms that is only one atom thick! – between source and drain. Keeping this in mind, we can understand why GFETs exhibit unprecedented sensitivity, which can be exploited in a wide variety of sensing and biosensing applications. Contrary to the conventional view of bulk semiconductors, such as silicon, used in traditional FETs, the use of graphene to create FETs marks a profound departure from long-established notions and approaches to create semiconductor devices and puts them into practice.

From a historical perspective, traditional transistor sensors are three-dimensional semiconductor devices in which changes in electric charge at the surface of the channel do not always translate into a device response. As the reader can probably imagine, this fact can dramatically limit the sensitivity of the device.

In stark contrast, in GFETs, the channel is made from a two-dimensional material, which directly exposes the channel to any molecules in the surroundings. In this scenario, the local gating effect is much more effective than that in conventional devices because the species modulating the electric field can be directly attached to the "entire" transistor channel.

While the use of GFETs in biosensing and bioelectronics has been in the experimental phase for years, it is only recently that this technology has gradually shifted to a commercial stage [1–3]. For instance, a report from McKinsey estimates that graphene-related technologies will become a US$ 70B market by 2030 [4].

By nature, the subject of bioelectronics is diverse and interdisciplinary, and, as such, chemists, biologists, physicists, materials scientists, and engineers can make valuable contributions to the field. It is now clear within the scientific community that the successful translation of research in graphene transistors to commercial reality requires the harmonization of four universes: electronics, biology, physics, and chemistry. Indeed, the journey from basic science to technological applications has been compellingly described by Herbert Kroemer (Nobel Prize in Physics 2000) [5]: *"Even if the process from science and technology to applications is opportunistic rather than deterministic, we can speed up this process by better cross-discipline communication between scientists, technologists, and application engineers."* In this regard, a non-negligible merit of engineers, physicists, biologists, and chemists has been their willingness to overcome the fences in which they have been traditionally confined and to promote collaboration with "apparently" unrelated research areas.

In this book, our aim is to show the wide potential of GFETs as advanced bioelectronic platforms and also to promote the potential of the technology among scientists, students, postdoctoral fellows, engineers, and industrial researchers. In addition, we also hope that this book will help convince decision-makers from academia, government, and industry to cooperate in developing a comprehensive GFET roadmap to accelerate the manufacturing and commercialization of this promising technology.

We are pleased to have edited this book, and we are honored that so many contributors from all over the world (see map below) have accepted our invitation and taken time to write significant contributions. We are grateful to the authors who have always been very responsive and enthusiastic about the idea of the book. The invaluable efforts of these authors from 14 different countries with 38 different affiliations have helped build a comprehensive book that may be used as an advanced textbook by graduate students and young scientists, as well as a valuable reference for academic and industrial professionals performing research and development in the specific area of biosensing and bioelectronics.

Last, but not least, we thank Antonio H. Castro Neto, a pioneer in the field of graphene as an electronic material, for his Foreword, and we thank you so much for taking the time to read this book.

Wolfgang Knoll
Vienna
March 2023

Omar Azzaroni
La Plata
March 2023

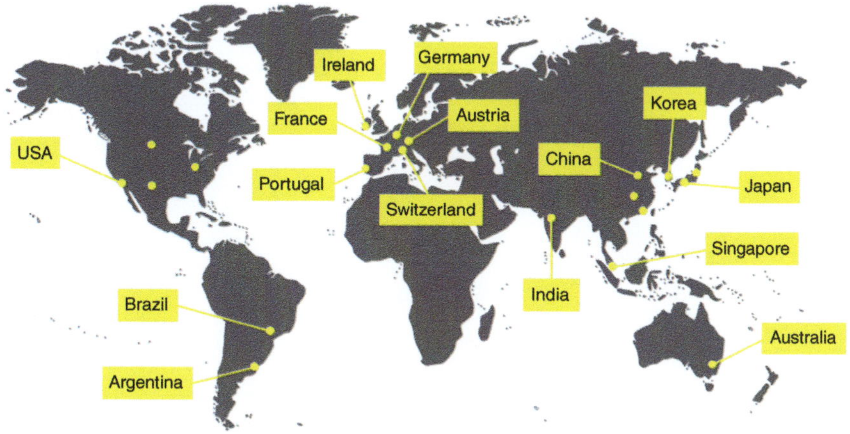

References

1 Graphenea. https://www.graphenea.com/ (accessed 30 March 2023).
2 Cardea. https://www.cardeabio.com/ (accessed 30 March 2023).
3 Gisens Biotech. https://www.gisensbiotech.com/ (accessed 30 March 2023).
4 Batra, G., Santhanam, N., Surana, K. (2018). Graphene: the next S-curve for semiconductors? McKinsey Insights. https://www.mckinsey.com/industries/semiconductors/our-insights/graphene-the-next-s-curve-for-semiconductors?utm_campaign=1millionth-sensor&utm_source=cision&utm_medium=cision&utm_term=mckinsey&utm_content=mckinsey (accessed 30 March 2023).
5 Kroemer, H. (2005). *Phys. Status Solidi A* 202: 957–964.

1

2D Electronic Circuits for Sensing Applications

Diogo Baptista[1,2,3], Ivo Colmiais[1,2,3], Vitor Silva[1,2,3], Pedro Alpuim[1,4], and Paulo M. Mendes[2,3]

[1]*INL – International Iberian Nanotechnology Laboratory, Av. Mestre José Veiga, 4715-330, Braga, Portugal*
[2]*CMEMS – Center for Microelectromechanical Systems, University of Minho, Campus de Azurém, 4800-058, Guimarães, Portugal*
[3]*LABBELS – Associate Laboratory, University of Minho, Campus de Gualtar, 4710-057, Braga, Portugal*
[4]*Center of Physics, University of Minho, Campus de Gualtar, 4710-057, Braga, Portugal*

1.1 Introduction

Technology has always been around, be it the most straightforward wooden wheel or a more complex system like a modern car. Technology's and science's evolution pave the way for society's evolution. One of the most relevant breakthroughs in the past years was the transistor's discovery in 1947. This discovery allowed for replacing the vacuum tube, also known as the valve, with a device with a smaller footprint and higher efficiency. This discovery led to a massive bump in industrial society's evolution. Technology fully integrates people's lives, either directly in the devices they use (e.g. smartphones) or indirectly through those who provide them with a service (e.g. mobile operators' signal coverage).

This revolution was possible due to the miniaturization of the electronic components that integrate transistors while maintaining or improving their performance. According to Moore's Law, the number of transistors in the same chip area doubles every two years. This law has been confirmed for over 50 years, but it is getting to a point where reducing component size is reaching its physical limitations due to short-channel effects and interconnect's heating [1, 2]. Further research on new materials is required to replace silicon before a new paradigm in nanoelectronics – more than Moore – can be reached.

In recent years, two-dimensional (2D) materials have become the focus of many investigations to replace silicon to continue downscaling electronic devices [3–6]. Many of these materials show excellent electronic, photoelectronic, and mechanical properties. Therefore, investigating the implementation of a technology based on

Graphene Field-Effect Transistors: Advanced Bioelectronic Devices for Sensing Applications,
First Edition. Edited by Omar Azzaroni and Wolfgang Knoll.

such materials is increasingly important as the Internet of Things (IoTs) becomes more prevalent and requires a massive number of devices with a small footprint to be integrated without much notice. Although these are promising materials, their integration with standard manufacturing processes and replicability outside of the laboratory is yet to be achieved [7]. Many other problems associated with interfacing 2D and other materials continue to be challenging since these interfaces usually degrade their electronic properties, especially the carrier mobility, reducing their overall performance [8, 9].

The first 2D material discovered was graphene by Andre Geim and Konstantin Novoselov in 2004, which led them to win the Nobel Prize in Physics in 2010 [10]. The graphene was obtained using tape to exfoliate graphite until it reached a single atomic layer. The discovery of graphene launched curiosity and investigations toward graphene and other 2D materials. These days, the investigation of 2D materials has progressed immensely, to the point where we can already separate 2D materials into families according to their elements' chemical composition, unit cell, electronic, optical, or structural properties [11]. The most known families are X-enes and transition-metal dichalcogenides (TMDs). X-enes group single element materials with atoms organized in a hexagonal lattice, which is the case of graphene, silicene, germanene, and others. TMDs group 2D materials of the form MX_2, where M is a transition metal from the 4th, 5th or 6th group, and X is a chalcogen from group 16th. The most known TMDs are molybdenum disulfide, tungsten disulfide, and molybdenum diselenide. Since some 2D materials were discovered recently, their science and technology are not sufficiently mature to place them as candidates for next-generation electronic materials. Therefore, limit our discussion to graphene and MoS_2 since, as of today, they are by far the most studied materials.

Graphene consists of a single graphite layer of carbon atoms arranged in a hexagonal lattice. Graphene can be grown at a large scale by chemical vapor deposition (CVD) [12, 13] or liquid-phase exfoliation (LPE). One of the problems with graphene fabrication is that it cannot be grown directly on most substrates. It is usually deposited on a transition metal catalyst foil – often copper or nickel – and then transferred using a wet or dry transfer process to the desired substrate [14]. Because of the need to transfer to the final substrate, graphene's performance is affected by the degradation of its carrier mobility during this transfer process [15]. Graphene has remarkable properties, such as extremely high carrier mobility ($2000\,cm^2\,V^{-1}\,s^{-1}$ for any mechanically transferred graphene [16] and $200\,000\,cm^2\,V^{-1}\,s^{-1}$ for suspended graphene [17]) when compared to silicon ($1400\,cm^2\,V^{-1}\,s^{-1}$ for electrons and $450\,cm^2\,V^{-1}\,s^{-1}$ for holes), good electrical conductivity ($\approx 10^4\,\Omega^{-1}\,cm^{-1}$ [17]), high thermal conductivity ($5300\,Wm^{-1}\,K^{-1}$ [18]), and high Young's modulus (0.5–1.0 TPa [16]). This material is a gapless semiconductor, meaning it has a 0 eV energy gap. Because of this intrinsic property, this material cannot be used in devices where the off state is needed since the material always conducts electricity by holes or electrons. These properties make graphene a possible solution to overcome silicon limitations in certain applications and can be implemented in devices with a broad range of uses, from high-speed electronics to sensing applications.

MoS_2 belongs to the TMDs family and consists of a molybdenum layer sandwiched between two sulfur layers. This material appears in nature as molybdenite, and, like graphene, can be fabricated using CVD or exfoliation techniques. Unlike graphene, MoS_2 properties are not all well-defined, but its carrier mobility has been shown to have values up to $200\,cm^2\,V^{-1}\,s^{-1}$ at room temperature and Young's modulus of 0.33 TPa [19]. Different from graphene, MoS_2 has a direct bandgap of 1.8 eV [20], which means it can be used in devices that need to have an off state.

Due to the remarkable properties of these materials, their implementation on capacitors, inductors, and field-effect transistors (FETs) has already been reported in numerous papers. In Sections 1.2, 1.3, and 1.4, the literature on these components will be explored, regarding their physical implementation and the models that try to predict these components' behavior.

1.2 Graphene Inductors

On-chip inductors revolutionized RF electronics in the 1990s, but not everything is excellent. These inductors are planar and must have a large area, as dictated by electromagnetic laws, which means they cannot be downsized alongside standard transistors while maintaining high inductance density. In some cases, it is reported that planar inductors occupy up to 50% of an integrated circuit area. Thus, they hinder further miniaturization and integration. Finding new approaches to making these devices is imperative.

It is well known that the inductance is shape- and size-dependent, but in graphene a third factor can be explored, known as *kinetic inductance*. This material property arises from the inertia of charge carriers moving in alternating electric fields. Like all mass particles, charge carriers preserve their momentum, so when in an alternating electric field, it takes a finite time to change their momentum according to the field, which manifests as kinetic inductance. It is not very important in conventional metals because their conductance is associated with higher carrier concentration and macroscopic thickness. The kinetic inductance manifests as an equivalent series inductance, adding to the geometric inductance associated with the shape/size. Therefore, materials with high kinetic inductance must be used to reduce inductor size while maintaining high inductance density. Graphene is being exploited as a possible solution to the inductance component miniaturization issue due to its atomic thickness and relatively high conductivity, based on high carrier mobility and low carrier concentration. Consequently, graphene has high kinetic inductance and a small footprint.

A multilayer graphene (MLG) inductor is proposed in [21], shown in Figure 1.1. The authors' choice of using MLG is to ensure a lower quantum contact resistance (resistance associated with the interface between graphene and metal contact). This approach raises two problems: when compared with metals, graphene has a much lower conductivity; compared to SLG, the MLG exhibits reduced charge carrier inertia due to interlayer coupling. Bromine intercalation is used to overcome these

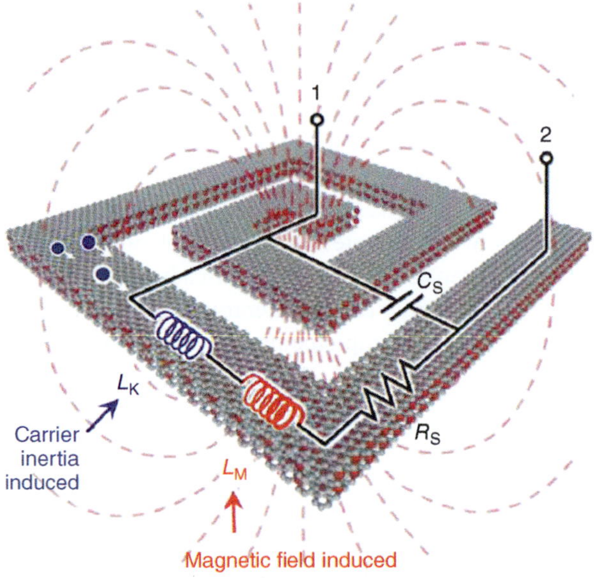

Figure 1.1 Schematic of a spiral inductor and its simplified equivalent circuit. L_M and L_K are the magnetic and kinetic inductance, respectively. R_S and C_S are the series resistance and the inter-turn capacitance, respectively. Source: Reproduced with permission from Kang et al. [21]/Springer Nature.

issues by increasing conductivity and reducing interlayer coupling. When compared to copper inductors, this method can achieve Q-factors of up to 12, and 1.5-times higher inductance in a two-turn inductor when compared to copper ones. The authors also claim that it is possible to achieve better results by improving intercalation technology and increasing contact quality.

1.2.1 Modeling of Graphene Inductors

The modeling of graphene inductors is not perfectly established, and most works are based on extracting parameters from or simulating typical metallic structures, adding some parameters to better match graphene characteristics.

In [21], the authors affirm that current simulation tools cannot capture the physics of graphene in modeling impedance/inductance. To try and predict the inductor behavior, the authors start by analyzing the performance of the inductors through finite element method (FEM) simulations in ANSYS HFSS and then modeled bulk coils with electrical conductivities considering grain boundary and surface scattering effects at the micro- and nanoscale for metals, and graphene conductivities extracted from DC analysis.

Although modeling graphene inductors remains a challenge, a simple model of a graphene spiral inductor can be seen in Figure 1.2. In this model, it can be corroborated that L_k appears in series with L_M, the substrate and dielectric need to be considered, and an inter-turn coupling capacitance appears, resulting from the inductor

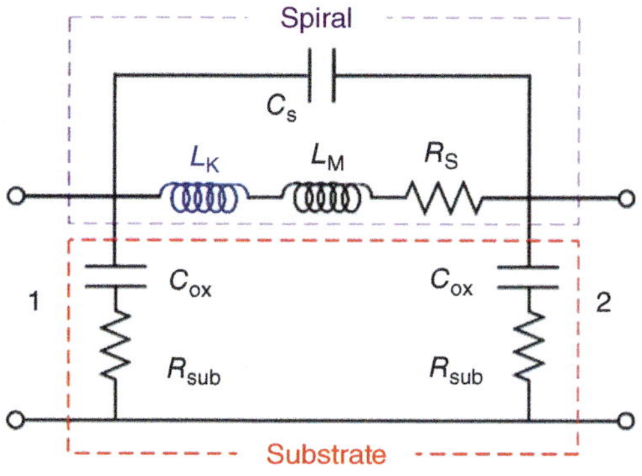

Figure 1.2 Simplified circuit model for a two-turn inductor. L_K, L_M, and R_S are the kinetic inductance, magnetic inductance, and series resistance of graphene inductor, respectively. C_s is the inter-turn coupling capacitance. C_{ox} and R_{sub} are substrate dielectric capacitance and substrate resistance, respectively. Source: Adapted from [21].

design. This model predicts that the substrate and the inter-turn coupling degrade the inductor's behavior at higher frequencies.

By using Launder's approach, the kinetic inductance per unit length is given by:

$$L_k = \frac{2\pi\hbar}{e^2 v_F M} \approx \frac{81\,\text{nH}}{M} / \mu\text{m} \tag{1.1}$$

where M is the number of quantum modes $\left(M = \frac{\Delta E_F}{\pi\hbar v_F}\right)$, ΔE_F is the difference between Dirac point energies and Fermi level, W is the width of graphene, e is the electron charge, $\hbar$ is the reduced Planck's constant, and v_F is the Fermi velocity.

The figure of merit (FOM) of an inductor is the quality factor (Q-factor) and is expressed as follows:

$$Q_{\text{factor}} = -\frac{\text{Imag}(Y_{12})}{\text{Real}(Y_{12})} = \frac{Z_L}{R} = \frac{wL}{R} \tag{1.2}$$

where Y_{12} is an admittance parameter $\left(Y_{12} = -\frac{2S_{12}}{(1+S_{11})(1+S_{22}) - S_{12}S_{21}}\right)$, L and R are the inductance and resistance of the inductor, and w is the angular frequency.

1.3 Graphene Capacitors

A capacitor is a passive device that stores electrical energy and adds capacitance to a circuit. The main usages of this device are to serve as a signal filter, for example, in a ladder design or as a temporary battery. The simplest capacitor consists of two parallel metal plates separated by a dielectric material, and the energy stored depends on the area of the plates, the distance between plates, the permittivity, and the dielectric function. Capacitors based on 2D materials have been explored, and the most common material used is graphene. Most of the work is done in biosensing, but research in the RF branch is emerging.

Graphene is used in the biosensing scene due to the possibility of functionalizing the graphene's surface. By immobilizing molecular probes on graphene using a linker, it is possible to change the graphene's surface charge density whenever there is a biorecognition event. This change in accumulation or depletion happens due to the charged or polar target molecules' local gating, which modulates the graphene channel conductance. The effect is capacitive, where the capacitance is that of the electrical double layer (EDL) forming at the graphene-solution interface. Consequently, different target molecule concentrations induce different amounts of charge in the EDL capacitor, which will be mirrored on the opposite plate of the capacitor, i.e the graphene surface.

There are no well-defined characteristics of graphene capacitors on the RF branch, but due to the high quantum capacitance and tuning possibility, some works claim that graphene is an excellent candidate. Moreover, the small footprint and low control voltage allow the development of compact systems like voltage-controlled oscillators, tunable filters, and phase shifters.

In [22], a variable capacitor based on graphene was implemented as a glucose sensor. This device takes advantage of the carrier density change in functionalized graphene when the adsorbed molecules' concentration changes, leading to the modulation of the channel conductance. As shown in Figure 1.3, aside from the capacitance dependence on adsorbed molecule concentration, it also depends on the gate voltage and ranges from 90 to 140 pF.

Another capacitor with tunable control based on graphene was studied for RF applications, with capacitance between 3.8 and 2.9 pF, a gate voltage of 1.25 V, and frequencies ranging from 1 to 10 GHz [23]. The best device performance was achieved with 1.25 V gate voltage and 0.4 GHz frequency, obtaining a *Q*-factor of up

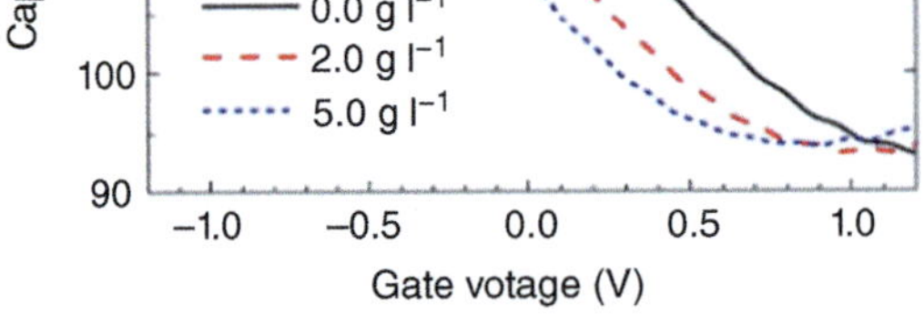

Figure 1.3 Capacitance-voltage curves. Source: Reproduced with permission from Zhang et al. [22]/American Chemical Society.

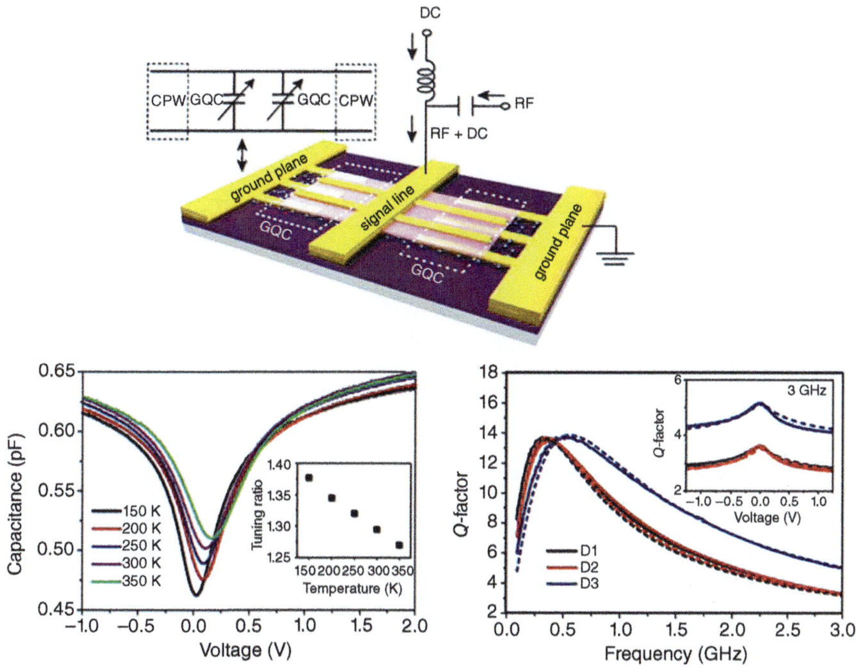

Figure 1.4 Schematic of the graphene quantum capacitor (left), capacitance dependence on frequency and DC bias (center) and *Q*-factor dependence on frequency (right). Source: Reproduced with permission from Moldovan et al. [23]/American Chemical Society.

to 14.5. The device consists of two symmetrically placed capacitors in a parallel configuration so that they can be characterized at microwave frequencies. This design also uses a multi-finger approach to increase the capacitance while reducing graphene's parasitic resistance. From Figure 1.4, it is possible to see that by changing the DC bias, the capacitance also changes, allowing for low-power ICs. With this design, the maximum *Q*-factor obtained is 15 at 0.4 GHz, but the authors say it is possible to increase it with simple design changes.

In [24], the same authors explored the design described above. By changing the fingers' number and size, the authors concluded the *Q*-factor increases when reducing the finger length and that the number of fingers does not have much impact on the *Q*-factor. It was also concluded that changing the number of fingers makes it possible to scale the capacitors to any capacitance while maintaining similar *Q*-factors. The best device achieved a *Q*-factor of 12 at 1 GHz, an improvement from the first study, where at 1 GHz the *Q*-factor was about 9.

Another tunable graphene capacitor is explored in [25]. This design (Figure 1.5) is more straightforward than the previous one discussed and consists of a parallel capacitor where the bottom plate is graphene, the dielectric is hexagonal boron nitride (hBN), and the top plate is chromium (Cr) and gold (Au). From the graph, it can be proved that the capacitor is tunable and that it has a minimum capacitance of around 3.5 pF.

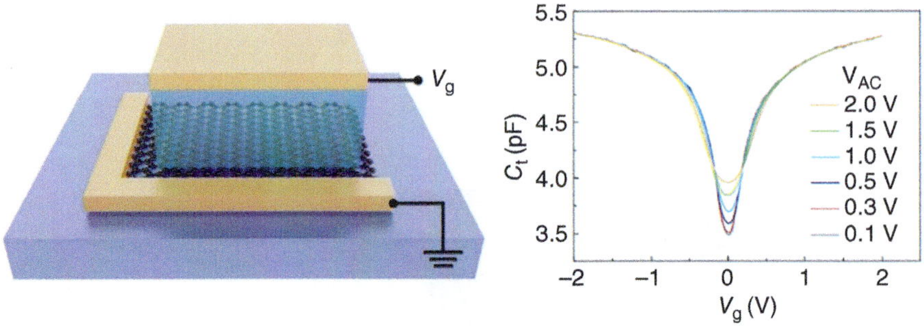

Figure 1.5 Schematic of graphene tunable capacitor (left) and capacitance dependence on bias voltage and AC signal voltage (right). Source: Reproduced with permission from Zhang et al. [25]/John Wiley & Sons.

1.3.1 Modeling Graphene Capacitors

As for graphene inductors, state-of-the-art in graphene capacitor modeling is not well established, and due to the differences between designs, the models available may not suit all designs.

In [24], the authors used the circuit model in Figure 1.6 to extract the device parameters for the proposed design. In this model, a series resistance ($R_C + R_M + R_{GS}$) is observed along with an oxide capacitance (C_{ox}) that raises from the material's intrinsic properties. The graphene layer raises variable resistance (R_G) and quantum capacitance C_q. The quantum capacitance of graphene is reported as:

$$C_q = \frac{2e^2 k_B T}{\pi(\hbar v_F)^2} \ln\left[2\left(1+\cosh\left(\frac{eV_{ch}}{k_B T}\right)\right)\right] LWN \tag{1.3}$$

where e is the electron charge, k_B is the Boltzmann's constant, T is the temperature, $\hbar$ is the reduced Plank's constant, v_F is the Fermi velocity, V_{ch} is the graphene potential, N is the number of fingers, and L and W are the length and width of the fingers.

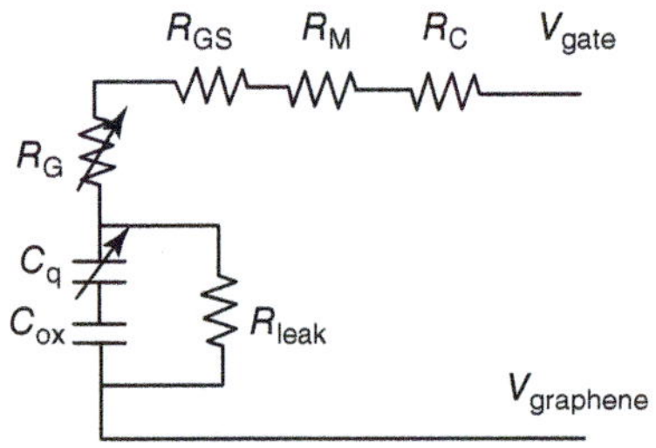

Figure 1.6 Circuit model used to extract the device parameters. R_C is the contact resistance. R_M is the metal fingers resistance. R_{GS} is the graphene resistance in the interspace. R_G is the graphene resistance. C_q is the graphene quantum capacitance. C_{ox} is the fixed oxide capacitance. Source: Reproduced with permission from Moldovan et al. [24]/American Chemical Society.

For a more straightforward design [25], the authors propose an equivalent circuit composed of two capacitors in series. One is the geometrical capacitance which depends on the hBN thickness and dielectric constant and is defined as $C_{\text{is}} = \frac{\varepsilon}{t_{\text{hBN}}}$. The other is the quantum capacitance of the graphene and is defined as $C_q = dQ/dV_{ch}$, where Q is the charge induced on the graphene and V_{ch} the graphene potential. Far from the Dirac point, C_q is much larger than C_{is}. So, in that region, the quantum capacitance can be neglected, but it must be considered when close to the Dirac point.

Similar to the inductors, the FOM of a capacitor is the Q-factor and is expressed as:

$$Q_{\text{factor}} = -\frac{\text{Imag}(Y_{12})}{\text{Real}(Y_{12})} = \frac{Z_C}{R} = \frac{1}{wCR} \tag{1.4}$$

where Y_{21} is an admittance parameter, C and R are the capacitance and resistance of the inductor, and w is the angular frequency.

1.4 2D Material Transistors

Transistors are one of the essential components of modern electronics and can be found in almost every electronic system, to amplify or switch electrical signals. There are several types of transistors, the most relevant for micro- and nanoelectronics being the FET. These devices are based on channel conductance modulation by applying a voltage to the gate. It means we can control the current that flows between the drain and source terminals, I_{DS}, using a gate voltage applied to a third contact, which is electrically insulated from the other two. The gate contact and the FET channel coupling are capacitive and have been discussed to some extent in Section 1.3.

The most important part of the transistor is the channel, which forms in the semiconductor material at the interface with the gate dielectric. Most common chips use transistors with silicon channels and rely on reducing the channel's size to improve their overall performance. In simple terms, reducing transistor size allows for more integration in the same chip's area, increasing the chip performance and speed. However, due to short-channel effects and transistor Cu interconnects heating due to increased speed; their downscaling is becoming a considerable challenge. To overcome this issue and ensure technological advancement, further research on new materials to replace silicon must be undertaken. As a result of the effort to find new materials to replace silicon, 2D materials appeared as a possible solution because of their high saturation velocity and high carrier mobility, being the most promising graphene and MoS_2.

These materials can be used in transistors for RF applications, such as oscillators, frequency multipliers, transceivers, or mixers.

1.4.1 Most Common Topologies for Transistors

The common topologies of graphene field-effect transistors (GFETs) are those found in other technologies, namely top-gated, back-gated, and top/back-gated GFETs, Figure 1.7. What gives the name to these topologies is the position of the gate. Although the difference between these topologies may look simple, changing the position of the gate has implications on the fabrication methodology, expected performance, modeling considerations, and applications.

Top-gated GFETs are reported in the literature with their implementation in RF applications, biosensors, and liquid-gate GFETs. This approach is excellent when it is necessary to have greater control over the channel modulation while using lower gate bias by using a thin gate oxide layer. The major problem with this topology is the need to grow an oxide layer on top of the graphene without damaging its structure and consequently degrading its carrier mobility. To overcome this issue, researchers are trying new approaches, such as the physical transfer of a nanowire to function as a gate electrode or the use of boron nitride (h-BN), or the refinement of standard fabrication techniques like atomic layer deposition (ALD) or thermal growth.

Back-gated structures expose the channel, allowing the graphene's surface to interact with light or molecules. Both interactions produce changes in graphene's properties and, therefore, changes in the transfer characteristic, allowing the evaluation of the element that caused the change. This behavior makes back-gated GFETs suitable for biosensors and photodetectors. The major problem with this approach is the necessity of high voltage to control the device, which is a drawback for most common applications.

Top/back-gated structures are used when it is advantageous to split the DC and AC parts of the gate voltage and apply them to different contacts. In this way, a constant quiescent bias can be applied to the back gate – setting the transistor functioning point – while a signal is applied to the top gate in this case, modulating the transfer characteristics around the GFET quiescent point. Since GFETs' technology is still early, researchers use this topology to tune all their devices equally. Another

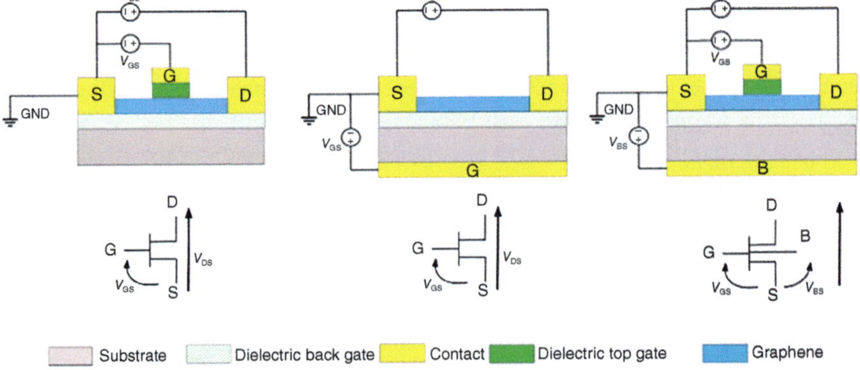

Figure 1.7 Top-gated (left), back-gated (center) and top/back-gated (right) graphene transistors. Source: Jmai et al. [26]/MDPI/CC BY 4.0.

architecture reported in [27] uses receded gate geometry. It is adequate for liquid-gate transistors working as chemical sensors since, like in the bottom gate case; it leaves the channel accessible for the molecules in the solution. Moreover, it uses the EDLs formed at the solid–liquid interfaces as the gate dielectric, providing a super-capacitor that allows operation at very low voltage, which is critical when dealing with biomolecules, cells, and microorganisms.

Graphene transistors (GFETs) have some unique characteristics. The first is that they cannot be turned off. Conventional transistors have a threshold gate voltage below which no current flows between drain and source and they are turned off. This property allows conventional transistors to be used in digital systems. On the contrary, GFETs do not have a minimum gate voltage to turn on; they have a specific voltage at which they exhibit the minimum I_{DS}, which is called the Dirac voltage. The second unique property of GFETs is their ambipolar character. Whereas, for example, silicon FETs are either n- or p-type, but not both simultaneously, because their doping is achieved by impurity doping, which acts as donors (n-doping) or acceptors (p-doping), but not both, GFETs can be seen as p- and n-type transistors in the same device, whereby adjusting the gate voltage to the left or right of the Dirac voltage switches from p- to n-type transistors. Although this is an obstacle for digital applications, it is possible to implement them in analog systems. These analog systems can be biosensors, flexible electronics, or radiofrequency circuits. The third unique property of GFETs stems from graphene's very high carrier mobility, which is essential for developing transistors with high cut-off (f_T) frequencies or biosensors with exceedingly high sensitivity. GFETs with $f_T = 100$ GHz were reported in [28], and many others and the purpose of their investigation can be seen in Table 1.1. GFETs found in the literature are not easy to replicate, so further research is still needed to integrate these devices into a system.

Unlike GFETs, MoS_2 transistors (MoS_2 FETs) work like conventional FETs; they can be turned off. Therefore, MoS_2 FETs can be used in digital systems, which makes them a possible replacement for silicon-based transistors. MoS_2 single-layer FETs with f_T of 6.7 GHz and f_{max} of 5.3 GHz were reported in [44]. Although the design frequencies and the carrier mobility in MoS_2 are lower than those in graphene, the presence of a bandgap enables more significant voltage gain compared to GFETs. In another publication, by using a few-layer MoS_2, it was achieved a f_T of 42 GHz and f_{max} of 50 GHz [20]. Although MoS_2 FETs may look great, the low mobility of MoS_2 can be a limitation for their application in the higher frequency domain. The lack of models makes it difficult to predict the behavior of these devices, and the state-of-the-art of such devices is still too poor compared to graphene FETs. Further research is needed to understand the true potential of these devices.

1.4.2 Modeling of 2D Materials-Based Transistors

Simulating a device's performance is a key success factor of modern electronics, and because of that, modeling GFETs plays an important role in helping researchers achieve GFETs' best performance and understanding if their implementation in more complex devices is reliable, allowing for the substitution of silicon transistors.

Table 1.1 Graphene FETs found in the literature.

	Purpose	Graphene type	Differentiating approach	Performance	References
Top gated	Achieve high transconductance and drain-current saturation	CVD	Self-aligned source/drain electrodes	Maximum $g_m = 250\ \mu S\ \mu m^{-1}$	[29]
	RF applications	Mechanical exfoliation	Nanowire as the gate and self-aligned source/drain	Maximum $g_m = 1.27\ mS\ \mu m^{-1}$ and intrinsic $f_T = 300$ GHz	[30]
	Improve the drain-current saturation	CVD	Thin Al_2O_3 gate oxide dielectric (~4 nm)	$\frac{f_{max}}{f_T} > 3$ $A_v > 30$ dB	[31]
	RF applications	Mechanical exfoliation	Dual top gate	Maximum $g_m = 550\ \mu S\ \mu m^{-1}$ and intrinsic $f_T = 14.7$ GHz	[32]
	RF applications	CVD	T-shaped gate and drain/source	Intrinsic $f_{max} = 200$ GHz and extrinsic $f_{max} = 106$ GHz	[33]
	DNA biosensor	CVD	Liquid gate and PDMS well to isolate source/drain electrodes	Detection of full hybridization of the complementary strand down to 15 aM	[34]
Back gated	Study of velocity saturation: design and performance	Pulsed CVD	Use the h-BN as a gate oxide. Dual-gate device	$\frac{f_{max}}{f_T} > 5$	[35]
	Achieve high f_{max}	CVD	Buried gates to reduce gate resistance	Intrinsic $f_T = 35$ GHz and $f_{max} = 50$ GHz	[36]
	Achieve high f_{max}	CVD	T-gate structure to reduce the gate resistance	Extrinsic $f_T = 11.4$ GHz and $f_{max} = 15$ GHz	[37]
	Improvement of the process-induced mobility degradation of graphene	CVD	Development of buried gates	$I_{on}/I_{off} = 5.31$ Maximum $g_m = 6.85\ \mu S\ \mu m^{-1}$ Intrinsic $f_T = 2$ GHz and $f_{max} = 13$ GHz	[38]
	High-sensitivity label-free DNA biosensor	CVD	Electrolysis bubbling method for graphene transfer	The detection limit depends on the length of the DNA	[39]
	Graphene FET biosensor for the label-free sensing of exosome	CVD	Back gate contact made with silver paint	Exosome detection of at least $0.1\ \mu g\ ml^{-1}$	[40]
Top/back gated	GFET	Mechanical exfoliation	Gate oxide (Al_2O_3) deposited by ALD	Preservation of graphene mobility after gate dielectric deposition ($8000\ cm^2\ V^{-1}\ s^{-1}$)	[41]
	Frequency doubler	Mechanical exfoliation	Yttrium oxide as gate dielectric	Able to work with 200 kHz input frequencies	[42]
	RF applications	Mechanical exfoliation	h-BN used as top and back gate dielectric	Current density of $1.2\ A\ mm^{-1}$ Extrinsic $f_T = 33$ GHz	[43]

Source: Adapted from Jmai et al. [26].

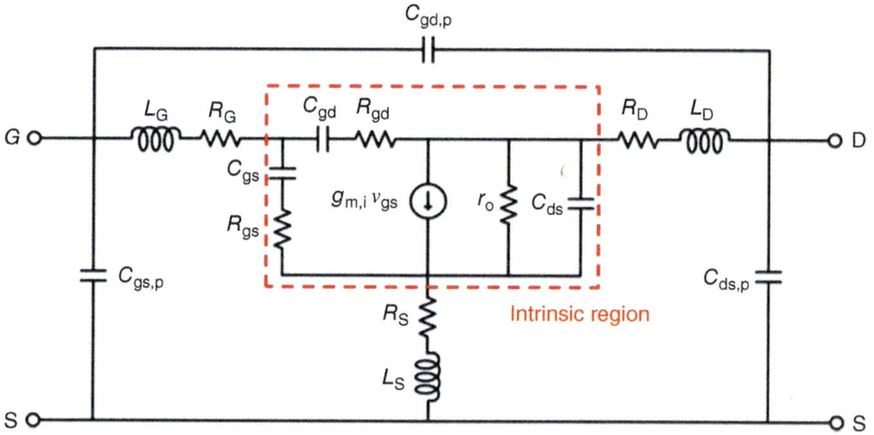

Figure 1.8 Small-signal equivalent circuit model of FETs.

Bearing this in mind, many papers try to achieve the closest and most reliable GFET model possible. Like all transistors, GFETs can be modeled using the small-signal model, shown in Figure 1.8, and that is what is done in [45]. This is an empirical model and is composed of a current source, I_{ds}, an output resistance, r_o, gate resistance and inductance, R_g and L_g, drain resistance and inductance, R_d and L_g, source resistance and inductance, R_s and L_s, drain to source capacitance, C_{ds}, gate to drain capacitance and resistance, C_{gd} and R_{gd}, and gate to source capacitance and resistance, C_{gs} and R_{gs}. The resistances and inductances associated with the terminals alone raise from their geometrical design. The capacitances and resistances between the GFET terminals appear due to the devices' small size and must be considered for the high-frequency domain. It is well known that the transconductance, g_m, is the slope of I_{ds} against $V_{gs}\left(g_m = \frac{dI_{ds}}{dV_{gs}}\right)$. Therefore, most of the time I_{ds} appears as the product of g_m and V_{gs}. Because this is an empirical model all parameters can be extracted from the measurement of the S parameters and DC analysis. The DC analysis consists in measuring I_{ds} while changing V_{gs}. This allows for a draft of the DC characteristic of the GFET and obtaining the corresponding g_m. The other parameters can be obtained from the Y parameters (standard conversion from the S parameters) using the following expressions:

$$D = 1 + w^2 C_{gd}^2 R_{gd}^2 \tag{1.5}$$

$$C_{gd} = -\frac{\mathrm{Im}(Y_{12})}{w} \cdot \left[1 + \left(\frac{\mathrm{Re}(Y_{12})}{\mathrm{Im}(Y_{12})}\right)^2\right] \tag{1.6}$$

$$R_{gd} = \frac{\mathrm{Re}(Y_{12})}{\mathrm{Im}(Y_{12})} \cdot \frac{1}{wC_{gd}} \tag{1.7}$$

$$R_{gs} = \frac{\mathrm{Re}(Y_{11}) - \frac{w^2 C_{gd}^2 R_{gd}}{D}}{\mathrm{Im}(Y_{11}) - \frac{wC_{gd}}{D}} \cdot \frac{1}{wC_{gd}} \tag{1.8}$$

$$C_{ds} = \frac{\mathrm{Im}(Y_{22})}{w} - \frac{C_{gd}}{D} \tag{1.9}$$

$$C_{gs} = \frac{1}{w} \left[\mathrm{Re}(Y_{11}) - \frac{w^2 C_{gd}^2 R_{gd}}{D} \right] \cdot \left[\frac{\mathrm{Im}(Y_{11}) - \frac{wC_{gd}}{D}}{\mathrm{Re}(Y_{11}) - \frac{w^2 C_{gd}^2 R_{gd}}{D}} + \frac{\mathrm{Re}(Y_{11}) - \frac{w^2 C_{gd}^2 R_{gd}}{D}}{\mathrm{Im}(Y_{11}) - \frac{wC_{gd}}{D}} \right] \tag{1.10}$$

In this chapter, the authors use a fixed g_m to simulate the S parameters, and because of that the generalization of this model becomes difficult since operating the GFET at the Dirac point is especially important to some applications, like the ring oscillator, and close to this point g_m changes a lot.

In [46], the authors take a different approach. Instead of relying only on measured data, they try to predict the transistor behavior using analytical expressions as well as some tabulated values of the materials' properties. With their work, the authors were able to implement a compact equivalent circuit that evaluates the value of I_{ds} in the three working regions and verified the model against experimental DC data. The major difference between both models here presented, is that one relies on measured data to analyze RF performance, and the other using only theoretical data to predict the DC behavior of the transistor.

Reported in the literature are several models that try to predict the behavior of the GFET, and by taking different approaches, they can predict its behavior in a closed operation zone. To replace silicon, the GFET model needs to be standardized in all operation zones, allowing researchers and chip manufacturers to design and predict the device performance accurately.

The FOMs of GFETs are the cut-off frequency (f_T) and the maximum oscillation frequency (f_{max}). The f_T is defined as the frequency at which the magnitude of the small-signal current gain is unitary ($h_{21} = 0$ dB). This FOM is usually extracted from the h_{21} parameter, which is obtained by the measurement and conversion of the S parameters of the device, using the following expression:

$$h_{21} = \frac{-2S_{21}}{(1 - S_{11})(1 + S_{22}) + S_{12}S_{21}} \tag{1.11}$$

Also, to predict the cut-off frequency, it is generally used the following expression:

$$f_T = \frac{g_m}{2\pi\left[(C_{gs} + C_{gd})(1 + (R_d + R_s)g_{ds}) + C_{gd}g_m(R_d + R_s) + C_{pg}\right]} \tag{1.12}$$

$$k = \frac{1 + |S_{11}S_{22} - S_{12}S_{21}|^2 - |S_{11}|^2 - |S_{22}|^2}{2|S_{12}||S_{21}|} \equiv \text{Stability Factor} \tag{1.13}$$

The f_{max} is described as the frequency when the maximum available power gain (MAG/MSG) becomes unitary (MAG/MSG = 0 dB). This gain is not directly calculated, it must satisfy some conditions. By using the S parameters, the first thing to evaluate is the stability factor for all frequencies using the following expression:

$$U = \frac{\left|\frac{S_{12}}{S_{12}} - 1\right|^2}{2k\left|\frac{S_{21}}{S_{12}}\right| - 2\,\mathrm{Re}\left(\frac{S_{21}}{S12}\right)} \equiv \text{Mason's Gain} \tag{1.14}$$

Afterward, comes the evaluation of all k's. If all k's are less than one ($k < 1$ for all frequencies), the MAG/MSG corresponds to the Mason's Gain, and it can be calculated using the following expression:

If all k's are not less than one, for each frequency must be evaluated if k is less or greater than one. If k is less than one ($k < 1$), the MAG/MSG corresponds to the maximum stable gain, and can be calculated using the following expression:

$$\text{MSG} = \frac{|S_{21}|}{|S_{12}|} \equiv \text{Maximum Stable Gain} \tag{1.15}$$

If k is greater than one ($k > 1$), the MAG/MSG corresponds to the maximum available gain, and can be calculated using the following expression:

$$\text{MAG} = \text{MSG}\left(k - \sqrt{k^2 - 1}\right) \equiv \text{Maximum Available Gain} \tag{1.16}$$

Finally, the MAG/MSG can be converted to dB by the evaluation of 10 times the logarithmic of each value of MAG/MSG (MAG/MSG (dB) = $10\log_{10}$(MAG/MSG)).

Like for f_T, there is a general expression used to try to predict f_{max}, and it is the following:

$$f_{max} = \frac{f_T}{2\sqrt{g_{ds}\left(R_g + R_s\right) + 2\pi f_T R_g C_{gd}}} \tag{1.17}$$

1.5 2D Material Diodes

A diode is an electronic component that allows current to flow in one direction while it blocks transport in the reverse direction, thus rectifying the electric signal. The most common type of semiconductor diode is a p–n junction. The p–n junction induces an electric field in a space-charge carrier with a depleted volume, which enables current rectification. There are homo- and heterojunctions, depending on whether both sides of the junction are made of the same or different materials. 2D materials junctions can be made 2D or 1D. Certain authors add a gate to the junction diode, to tune the chemical potential on one or both sides of the junction, thus improving the device performance.

1.5.1 Most Common Topologies

P–n junctions based on 2D materials can be made of only one material (*2D homostructures*), two different materials (*2D heterostructures*), or different dimensions materials (*mixed dimensional*) [47]. 2D homostructures can be obtained using different methods, shown in Figure 1.9, which are as follows:

- *Thickness based*: p and n regions are formed by regions with different thicknesses.
- *Electrostatically doped*: p and n regions are obtained using local gates.
- *Chemical doping*: p and n regions obtained by the surface adsorption of molecules, nanoparticles, or quantum dots.
- *Elemental doping*: two flakes with different doping are stacked.

2D heterostructures can be:

- *Vertical*: stacking two different 2D materials on top of each other.
- *Lateral*: combining two different 2D materials on the same plane.

Mixed dimensional can be:

- *2D–0D and 2D–1D*: 2D–0D and 2D–1D material junctions.
- *2D and 3D*: stacking of 2D and 3D material on top of each other.

A thickness-based diode using MoS_2 is reported in [48], with a rectification ratio of $\approx 10^3$ and a small ideality factor (a value that compares the diode with the ideal diode) of 1.95. Besides the good electronic properties, it also has good photoresponsivity of $10\,A\,W^{-1}$ and high photosensitivity of 10^5.

In [49], it is reported that two different mixed dimension-based diodes are similar to metal-insulator-metal (MIM) diodes, but one of the metals is replaced by graphene, creating a metal-insulator-graphene interface (MIG), as shown in Figure 1.10. The difference between both approaches is the type of interface between graphene and the insulator, being a 2D or 1D interface, shown in Figure 1.11. From the 2D to 1D interface, the capacitance and series resistance decrease, allowing to fully exploit the high mobility of graphene, which increases the device cut-off frequency (predicted to be up to 2.4 THz) and current density (from 7.5 [50] to $7.5\times10^6\,A\,cm^{-2}$ [51]).

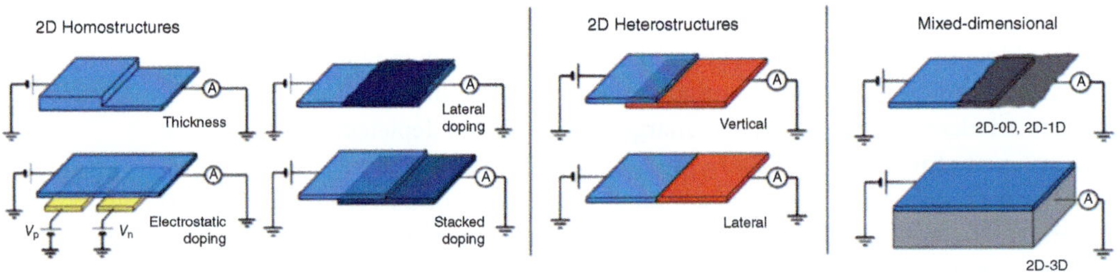

Figure 1.9 Schematic of different p–n junctions. Source: Adapted from [42].

Figure 1.10 Schematic of the 2D (left) and 1D (right) MIG diodes. Source: Wang et al. [49]/ John Wiley & Sons.

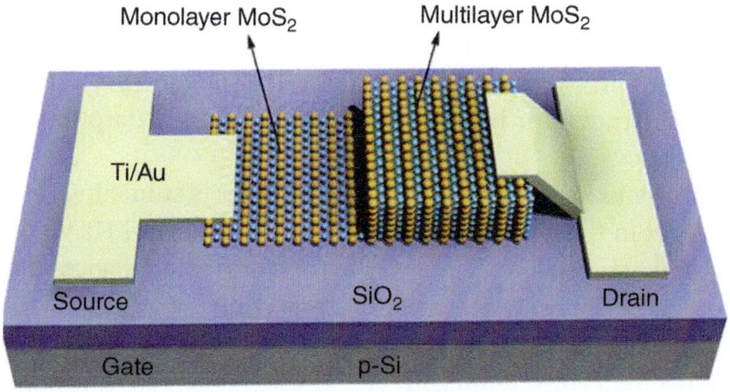

Figure 1.11 Schematic of the thickness-based diode. Source: Sun et al. [48]/Springer Nature/CC BY 4.0.

1.5.2 Modeling of 2D Materials-Based Diodes

The modeling of 2D material diodes has not been well explored, so modeling such devices relies on parameter extraction. In [52], it is described as a small signal equivalent, Figure 1.12, that is composed of a linear capacitance, C_1, and nonlinear bias-dependent capacitance, C_2, two leakages variable resistances, R_1 and R_2, and the

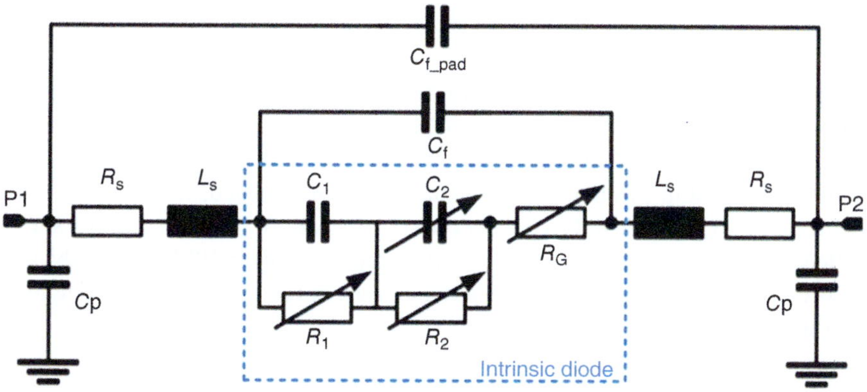

Figure 1.12 Small-signal equivalent circuit model of diodes.

graphene sheet resistance, R_G. In addition to the intrinsic region, the model also includes extrinsic parasitics.

1.6 Graphene Devices

Due to the high conductivity of graphene, GFETS can have large values of f_T and f_{max}. Therefore, GFETs are becoming the focus of much research for RF devices like frequency multipliers, mixers, and oscillators.

1.6.1 Graphene Frequency Multipliers

Due to the V-shaped transfer curve of GFETs, it is possible to obtain a frequency doubler when operating at the Dirac point. In simple terms, if a signal with DC bias equal to the Dirac point of the transistor is applied to the gate, V_{gs}, the output current I_{ds} has double the frequency.

In [42], a top/back-gated against back-gated frequency doubler is studied, showing a significant improvement in the operating frequency from 10 to 200 kHz when the top gate is added to the back-gated device. For the top/back-gated device, the output power is concentrated at 400 kHz with a relative power of 75% Another frequency doubler on a flexible substrate is reported [53], which achieved a spectral purity higher than 97% and a high conversion gain of −13 dB.

A W-shaped transfer curve is obtained when two GFETs with different Dirac points are combined in series. When operating at different points of the W-shape it is possible to obtain a frequency tripler or quadrupler. A frequency tripler is reported in [54] with spectral purity higher than 70% at an output frequency of 600 Hz.

A different approach is implemented in [55, 56], where the W-shaped transfer curve is achieved with a single GFET by biasing the back and top gates of top/back-gated transistors. In [55], a frequency tripler is studied, and a device with spectral purity higher than 90% was achieved at an output frequency of 3 kHz. In [56], a frequency quadrupler with spectral purity of 50% at 800 kHz was reported.

1.6.2 Graphene Mixers

It is reported in [57] that it is possible to implement an RF mixer with operating frequencies up to 10 GHz while having a high conversion loss of 30 dB at 1 GHz using a single GFET. Another graphene RF mixer is studied in [58], where frequencies up to 10 GHz and excellent thermal stability were achieved, with its peak performance around 4.5 GHz and a conversion loss of 27 dB. In [59], the authors study the effects of reducing channel length on the graphene mixer. With this study, the authors concluded that the conversion loss increases by reducing the channel length while the IIP3 increases.

1.6.3 Graphene Oscillators

1.6.3.1 Ring Oscillators

Another implementation of graphene transistors is reported in [60]. In this chapter, the authors propose the implementation of GFETs as a ring oscillator. A ring oscillator is a circuit built of an odd number of cascaded logic inverters in a loop. This loop induces instability and therefore induces oscillations at high frequencies. Each inverter must be identical, and over-unity voltage gain $\left(A = \frac{g_m}{g_{ds}} > 1\right)$ is required. The FOM used for this kind of device is the maximum oscillation frequency, f_O, since it is smaller than f_T. Although the positive voltage of the drain induces a shift on the Dirac point [61], the complementary GFETs of the inverters were obtained by using a back gate voltage to ensure a proper shift of the Dirac point. In this study, the authors made three types of devices: large (L = 3 μm and W = 20 μm), medium (L = 2 μm and W = 10 μm), and small (L = 1 μm and W = 10 μm), obtaining 284 MHz $< f_O <$ 350 MHz, 504 MHz $< f_O <$ 750 MHz, and 1 GHz $< f_O <$ 1.28 GHz for each device, respectively.

A similar ring oscillator is presented in [62], in which the authors also studied the effects of changing the transistors' channel size, access length, and source and drain contact thickness. The best device achieved a f_O = 4.3 GHz.

1.6.3.2 LC Tank Oscillators

Although full graphene-based LC tank oscillators have not been accomplished yet, the implementation of graphene inductors, capacitors, and transistors alone to study their performance in LC tank oscillators has been reported.

The capacitor developed in [25] was implemented in an LC tank by adding a 2 mH inductor in series, achieving a tunable resonant tank from 1.45 to 1.73 MHz and a Q-factor ranging from 65 to 25.

In [63], the performance of a graphene LC tank used in an oscillator was assessed through simulation. This simulation relied on graphene capacitor and inductor values found in the literature, and an oscillation frequency of 1.5 GHz was achieved with a phase noise of $-134\,\text{dB}\,\text{Hz}^{-1}$.

1.7 Conclusion

From the gathered information in this writing, it is possible to conclude that the research on 2D materials has a long way to go. Although graphene is the most researched 2D material and shows excellent electronic properties to outperform silicon devices in certain applications, state-of-the-art in devices' fabrication and modeling predictions have yet to be well established. Moreover, the devices' performance reported in the literature shows excellent potential to improve or substitute RF circuit designs, but more research on this matter needs to be conducted to assess the true potential of such devices.

References

1 Theis, T.N. and Philip Wong, H.S. (2017). The end of Moore's law: a new beginning for information technology. *Comput. Sci. Eng.* 19 (2): 41–50.

2 Schmidt, M., Lemme, M.C., Gottlob, H.D.B. et al. (2009). Mobility extraction in SOI MOSFETs with sub 1 nm body thickness. *Solid-State Electron.* 53 (12): 1246–1251.

3 Das, S., Sebastian, A., Pop, E. et al. (2021). Transistors based on two-dimensional materials for future integrated circuits. *Nat. Electron.* 4 (11): 786–799.

4 Lemme, M.C., Wagner, S., Lee, K. et al. (2020). Nanoelectromechanical sensors based on suspended 2D materials. *Research* 2020: 1–25.

5 Romagnoli, M., Sorianello, V., Midrio, M. et al. (2018). Graphene-based integrated photonics for next-generation datacom and telecom. *Nat. Rev. Mater.* 3 (10): 392–414.

6 Liu, C., Chen, H., Wang, S. et al. (2020). Two-dimensional materials for next-generation computing technologies. *Nat. Nanotechnol.* 15 (7): 545–557.

7 Lupina, G., Kitzmann, J., Costina, I. et al. (2015). Residual metallic contamination of transferred chemical vapor deposited graphene. *ACS Nano* 9 (5): 4776–4785.

8 Illarionov, Y.Y., Knobloch, T., Jech, M. et al. (2020). Insulators for 2D nanoelectronics: the gap to bridge. *Nat. Commun.* 11 (1): 3385. https://doi.org/10.1038/s41467-020-16640-8.

9 Passi, V., Gahoi, A., Marin, E.G. et al. (2019). Ultralow specific contact resistivity in metal–graphene junctions via contact engineering. *Adv. Mater. Interfaces* 6 (1): 1801285. https://doi.org/10.1002/admi.201801285.

10 Novoselov, K.S., Geim, A.K., and Morozov, S. v, Jiang, D., Zhang, Y., Dubonos, S. v, Grigorieva, I. v, and Firsov, A.A. (2004). Electric field effect in atomically thin carbon films. *Science (1979)* 306 (5696): 666–669.

11 Dragoman, M. and Dragoman, D. (2017). *2D Nanoelectronics: Physics and Devices of Atomically Thin Materials*. Cham: Springer.

12 Lee, Y., Bae, S., Jang, H. et al. (2010). Wafer-scale synthesis and transfer of graphene films. *Nano Lett.* 10 (2): 490–493.

13 Bae, S., Kim, H., Lee, Y. et al. (2010). Roll-to-roll production of 30-inch graphene films for transparent electrodes. *Nat. Nanotechnol.* 5 (8): 574–578.

14 Ullah, S., Yang, X., Ta, H.Q. et al. (2021). Graphene transfer methods: a review. *Nano Res.* 14 (11): 3756–3772.

15 Qing, F., Zhang, Y., Niu, Y. et al. (2020). Towards large-scale graphene transfer. *Nanoscale* 12 (20): 10890–10911.

16 Soldano, C., Mahmood, A., and Dujardin, E. (2010). Production, properties and potential of graphene. *Carbon N Y* 48 (8): 2127–2150.

17 Allen, M.J., Tung, V.C., and Kaner, R.B. (2010). Honeycomb carbon: a review of graphene. *Chem. Rev.* 110 (1): 132–145.

18 Balandin, A.A., Ghosh, S., Bao, W. et al. (2008). Superior thermal conductivity of single-layer graphene. *Nano Lett.* 8 (3): 902–907.

19 Li, X. and Zhu, H. (2015). Two-dimensional MoS_2: properties, preparation, and applications. *J. Materiomics* 1 (1): 33–44.

20 Cheng, R., Jiang, S., Chen, Y. et al. (2014). Few-layer molybdenum disulfide transistors and circuits for high-speed flexible electronics. *Nat. Commun.* 5: 5143. https://doi.org/10.1038/ncomms6143.

21 Kang, J., Matsumoto, Y., Li, X. et al. (2018). On-chip intercalated-graphene inductors for next-generation radio frequency electronics. *Nat. Electron.* 1 (1): 46–51.

22 Zhang, Y., Ma, R., Zhen, X.v. et al. (2017). Capacitive sensing of glucose in electrolytes using graphene quantum capacitance varactors. *ACS Appl. Mater. Interfaces* 9 (44): 38863–38869.

23 C. F. Moldovan, W. A. Vitale, M. Tamagnone, J. R. Mosig and A. M. Ionescu (2016). "Graphene quantum capacitors for high-Q tunable LC-tanks for RF ICs," 2016 46th European Solid-State Device Research Conference (ESSDERC), Lausanne, Switzerland, pp. 345–348, https://doi.org/10.1109/ESSDERC.2016.7599657.

24 Moldovan, C.F., Vitale, W.A., Sharma, P. et al. (2016). Graphene quantum capacitors for high frequency tunable analog applications. *Nano Lett.* 16 (8): 4746–4753.

25 Zhang, D., Zhang, L., Feng, X. et al. (2021). A tunable resonant circuit based on graphene quantum capacitor. *Adv. Electron. Mater.* 7 (4): 2001009. https://doi.org/10.1002/aelm.202001009.

26 Jmai, B., Silva, V., and Mendes, P.M. (2021). 2D electronics based on graphene field effect transistors: tutorial for modelling and simulation. *Micromachines* 12 (8): 979. https://doi.org/10.3390/mi12080979.

27 Vieira, N.C.S., Borme, J., MacHado, G. et al. (2016). Graphene field-effect transistor array with integrated electrolytic gates scaled to 200 mm. *J. Phys. Condens. Matter* 28 (8): 085302. https://doi.org/10.1088/0953-8984/28/8/085302.

28 Lin, Y.M., Dimitrakopoulos, C., Jenkins, K.A. et al. (2010). 100-GHz transistors from wafer-scale epitaxial graphene. *Science (1979)* 327 (5966): 662.

29 Bai, J., Liao, L., Zhou, H. et al. (2011). Top-gated chemical vapor deposition grown graphene transistors with current saturation. *Nano Lett.* 11 (6): 2555–2559.

30 Liao, L., Lin, Y.C., Bao, M. et al. (2010). High-speed graphene transistors with a self-aligned nanowire gate. *Nature* 467 (7313): 305–308.

31 Guerriero, E., Pedrinazzi, P., Mansouri, A. et al. (2017). High-gain graphene transistors with a thin AlO_x top-gate oxide. *Sci. Rep.* 7 (1): 2419. https://doi.org/10.1038/s41598-017-02541-2.

32 I. Meric, N. Baklitskaya, P. Kim and K. L. Shepard (2008), "RF performance of top-gated, zero-bandgap graphene field-effect transistors," 2008 IEEE International Electron Devices Meeting, San Francisco, CA, USA, pp. 1–4, https://doi.org/10.1109/IEDM.2008.4796738.

33 Yu, C., He, Z.Z., Liu, Q.B. et al. (2016). Graphene amplifier MMIC on SiC substrate. *IEEE Electron Device Lett.* 37 (5): 684–687.

34 Danielson, E., Sontakke, V.A., Porkovich, A.J. et al. (2020). Graphene based field-effect transistor biosensors functionalized using gas-phase synthesized gold nanoparticles. *Sens. Actuators, B* 320 (March): 128432.

35 Wilmart, Q., Boukhicha, M., Graef, H. et al. (2020). High-frequency limits of graphene field-effect transistors with velocity saturation. *Appl. Sci. (Switzerland)* 10 (2): 446. https://doi.org/10.3390/app10020446.

36 Han, S.J., Oida, S., Jenkins, K.A. et al. (2013). Multifinger embedded T-shaped gate graphene RF transistors with high ratio. *IEEE Electron Device Lett.* 34 (10): 1340–1342.

37 Lyu, H., Lu, Q., Liu, J. et al. (2016). Deep-submicron graphene field-effect transistors with state-of-art f_{max}. *Sci. Rep.* 6: 35717. https://doi.org/10.1038/srep35717.

38 Voronin, K.v., Ermolaev, G.A., Stebunov, Y.v. et al. (2020). Substrate effects in graphene field-effect transistor photodetectors. *J. Phys. Conf. Ser.* 1461 (1): 012188. https://doi.org/10.1088/1742-6596/1461/1/012188.

39 Ping, J., Vishnubhotla, R., Vrudhula, A., and Johnson, A.T.C. (2016). Scalable production of high-sensitivity, label-free DNA biosensors based on back-gated graphene field effect transistors. *ACS Nano* 10 (9): 8700–8704.

40 Kwong Hong Tsang, D., Lieberthal, T.J., Watts, C. et al. (2019). Chemically functionalised graphene FET biosensor for the label-free sensing of exosomes. *Sci. Rep.* 9 (1): 2–11.

41 Kim, S., Nah, J., Jo, I. et al. (2009). Realization of a high mobility dual-gated graphene field-effect transistor with Al_2O_3 dielectric. *Appl. Phys. Lett.* 94 (6): 062107. https://doi.org/10.1063/1.3077021.

42 Wang, Z., Zhang, Z., Xu, H. et al. (2010). A high-performance top-gate graphene field-effect transistor based frequency doubler. *Appl. Phys. Lett.* 96 (17): 173104. https://doi.org/10.1063/1.3413959.

43 Wang, H., Taychatanapat, T., Hsu, A. et al. (2011). BN/graphene/BN transistors for RF applications. *IEEE Electron Device Lett.* 32 (9): 1209–1211.

44 Sanne, A., Ghosh, R., Rai, A. et al. (2015). Radio frequency transistors and circuits based on CVD MoS_2. *Nano Lett.* 15 (8): 5039–5045.

45 Yeh, C.H., Teng, P.Y., Chiu, Y.C. et al. (2019). Gigahertz field-effect transistors with CMOS-compatible transfer-free graphene. *ACS Appl. Mater. Interfaces* 11 (6): 6336–6343.

46 Umoh, I.J., Kazmierski, T.J., and Al-Hashimi, B.M. (2013). A dual-gate graphene FET model for circuit simulation - SPICE implementation. *IEEE Trans. Nanotechnol.* 12 (3): 427–435.

47 Frisenda, R., Molina-Mendoza, A.J., Mueller, T. et al. (2018). Atomically thin p-n junctions based on two-dimensional materials. *Chem. Soc. Rev.* 47 (9): 3339–3358.

48 Sun, M., Xie, D., Sun, Y. et al. (2017). Lateral multilayer/monolayer MoS2 heterojunction for high performance photodetector applications. *Sci. Rep.* 7 (1): 1–7.

49 Wang, Z., Hemmetter, A., Uzlu, B. et al. (2021). Graphene in 2D/3D heterostructure diodes for high performance electronics and optoelectronics. *Adv. Electron. Mater.* 7 (7): 2001210. https://doi.org/10.1002/aelm.202001210.

50 Shaygan, M., Wang, Z., Elsayed, M.S. et al. (2017). High performance metal-insulator-graphene diodes for radio frequency power detection application. *Nanoscale* 9 (33): 11944–11950.

51 Wang, Z., Uzlu, B., Shaygan, M. et al. (2019). Flexible one-dimensional metal-insulator-graphene diode. *ACS Appl. Electron. Mater.* 1 (6): 945–950.

52 Saeed, M., Hamed, A., Wang, Z. et al. (2018). Metal-insulator-graphene diode mixer based on CVD Graphene-on-Glass. *IEEE Electron Device Lett.* 39 (7): 1104–1107.

53 Liang, Y., Liang, X., Zhang, Z. et al. (2015). High mobility flexible graphene field-effect transistors and ambipolar radio-frequency circuits. *Nanoscale* 7 (25): 10954–10962.

54 Chen, H.Y. and Appenzeller, J. (2012). Graphene-based frequency tripler. *Nano Lett.* 12 (4): 2067–2070.

55 Cheng, C., Huang, B., Liu, J. et al. (2016). A pure frequency tripler based on CVD graphene. *IEEE Electron Device Lett.* 37 (6): 785–788.

56 Cheng, C., Huang, B., Mao, X., Zhang, Z., Zhang, Z., Geng, Z., Xue, P., and Chen, H. (2017) A graphene based frequency quadrupler. *Sci. Rep.*, 7, 46605. https://doi.org/10.1038/srep46605.

57 Liao, L. and Duan, X. (2012). Graphene for radio frequency electronics. *Mater. Today* 15 (7–8): 328–338. https://doi.org/10.1016/S1369-7021(12)70138-4.

58 Lin, Y.-M., Valdes-Garcia, A., Han, S.-J. et al. (2011). Wafer-scale graphene integrated circuit. *Science (1979)* 332 (6035): 1294–1297.

59 Gao, Q., Li, X., Tian, M. et al. (2017). Short-channel graphene mixer with high linearity. *IEEE Electron Device Lett.* 38 (8): 1168–1171.

60 Guerriero, E., Polloni, L., Bianchi, M. et al. (2013). Gigahertz integrated graphene ring oscillators. *ACS Nano* 7 (6): 5588–5594.

61 Li, S.L., Miyazaki, H., Kumatani, A. et al. (2010). Low operating bias and matched input-output characteristics in graphene logic inverters. *Nano Lett.* 10 (7): 2357–2362.

62 Bianchi, M., Guerriero, E., Fiocco, M. et al. (2015). Scaling of graphene integrated circuits. *Nanoscale* 7 (17): 8076–8083.

63 Colmiais, I., Borme, J., Alpuim, P., and Mendes, P.M. (2021). Graphene LC oscillator for biosensing applications. In: *2021 XXXVI Conference on Design of Circuits and Integrated Systems (DCIS), Vila do Conde, Portugal*, pp. 1–5, https://doi.org/10.1109/DCIS53048.2021.9666173.

2

Large Graphene Oxide for Sensing Applications

Jingfeng Huang, J. Amanda Ong, and I.Y. Alfred Tok

Nanyang Technological University, School of Materials Science and Engineering, 50 Nanyang Avenue, Singapore, 639798, Singapore

2.1 Graphene Oxide (GO)

Carbon has pure allotropes with properties ranging from hard as diamond too soft as graphite. A two-dimensional (2D) allotrope of carbon, graphene, can combine with different moieties of oxygen to form graphene oxide (GO), which exhibits properties ranging from electrical insulator to electrical conductor. Similar to graphene, GO has a very high-surface area for a molecule. Thus, it is very sensitive to any disturbances in the environment and highly suitable for ultrasensitive biosensors.

GO contains variable amounts of oxygen atoms on the surface. These can act as catalytic active centers for both covalent and non-covalent functionalization on the surface. Finally, the large detection area of the GO is biocompatible [1, 2].

In contrast, the mechanical cleavage of graphene has low productivity and is unsuitable for large-scale use. The synthesis of graphene by high temperature chemical vapor deposition (CVD) is complicated, expensive, and often time-consuming. CVD technique is also not compatible with plastic or cellulose substrates [3].

2.2 GO as Biosensors

GO has been utilized in biological sensor devices in many different transduction modes [4, 5]. Biological sensors from GO field-effect transistors (FETs) had also attracted great interest. The architecture of GO FETs enables real-time, label-free, yet sensitive measurements from readable electrical signals which arise between the interactions of the target biomolecule and the FET transducer surface [4, 6–10]. In conjunction with facile on-chip integration, high transconductance, and the cost-effective fabrication, FET-based biosensors can become an excellent alternative to existing biosensor technologies [4, 6].

Graphene Field-Effect Transistors: Advanced Bioelectronic Devices for Sensing Applications,
First Edition. Edited by Omar Azzaroni and Wolfgang Knoll.

Agarwal et al. [11] demonstrated the biocompatibility of reduced graphene oxide (rGO) by functionalizing the FET device with proteins in order to detect various metal ions [12]. He et al. [3] fabricated a reduced-GO biosensor for detecting the dynamic secretion of hormonal catecholamine from live cells.

Mao and coworkers reported the fabrication and operation of a highly sensitive and selective thermally rGO FET decorated with gold nanoparticle-antibody conjugates for detection of protein-binding events [13]. This approach of modifying the graphene surface with nanoparticles to increase the number of probe molecules at the FET gate was beneficial in terms of a linear and enhanced sensor response [5].

2.3 Large GO

Synthesis methods and characterization were of increasing importance for molecular electronics, sensing, and battery applications in which GO of better quality could improve the reported performance results with all else held constant. Not all GO were created equal. With larger flakes, the success rate of conductive FET device fabrication increased from 20% (40 μm graphite) to 100%. The lower density of inter-sheet junctions in large GO also induces stronger mechanical performance, higher electrical/thermal conductivity, and stronger electromagnetic interference shielding performance [14].

Improved Hummer's method can yield large graphene flake sizes. This subchapter will bring the reader through in a systematic manner, the role and mechanism of preoxidation and *in situ* size fractionation in an improved Hummer's method to obtain large mono-layered GO.

The GO synthesis method described by Su et al. [15] was modified by taking only large floating flakes after both the ultrasonication and oxidation steps for further processing. According to [15], a graphite preoxidation procedure was needed to avoid incompletely oxidized graphite-core/GO-shell particles in the final product.

Due to the large size of the natural graphite flakes (2–5 mm), it was expected that the floating flakes were not fully oxidized, so GO sheets were weakly attached to the graphite core (Figure 2.1). Large floating flakes were collected after the main

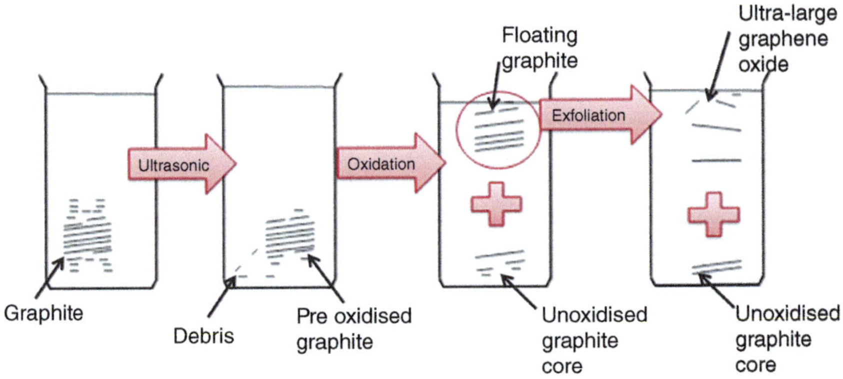

Figure 2.1 Schematic of improved GO synthesis.

oxidation step and resuspended in 500 ml DI water by stirring overnight in the modified synthesis. Due to the mechanical movement, large GO flakes that were weakly adhered to the core peeled off.

Since GO was soluble in water because of its carboxyl and hydroxyl groups, exfoliation was achieved by this dilution step. In contrast to similar considerations [15, 16], no ultra-sonication was used. With this facile but crucial resuspension step addition, the total yield of GO sheets was increased to 25% [17, 18].

2.4 Mechanism of Large GO via Modified Hummers Method

It was noted that although preoxidation steps had been commonly used in synthesis of GO or aqueous graphene, the role and mechanism of preoxidation on graphite had yet to be determined. For smaller graphite flakes, the first step of preoxidation was used to obtain a higher yield of GO from micron-sized graphite [19]. Then in 2009, Su et al. [15] used ultra-sonication on large graphite flakes (3–5 mm across) to substitute chemical preoxidation step and this resulted in large flakes with higher mobility values up to 12 $cm^2V^{-1}s^{-1}$. Adding a preoxidation step to the Hummer's method, was reported to enlarge the size of aqueous graphene [15, 20]. With larger flakes, the carrier mobility and the electrical conductivity [21] could be increased while maintaining high optical transparencies [18].

The scanning electron microscopy (SEM) image of an ultra-large hydrazine vapor rGO flake, which was approximately 60 μm in lateral size when deposited on SiO_2/Si substrate, is shown in Figure 2.2a. Smaller GO fragments were also observed beside the ultra-large flakes. Figure 2.2b illustrates a size histogram of GO flakes observed using optical microscopy and SEM. The average size of the GO flakes obtained by the reported method was 700 μm^2, hence the GO flakes were seven times larger in area than those produced using previously reported method [22].

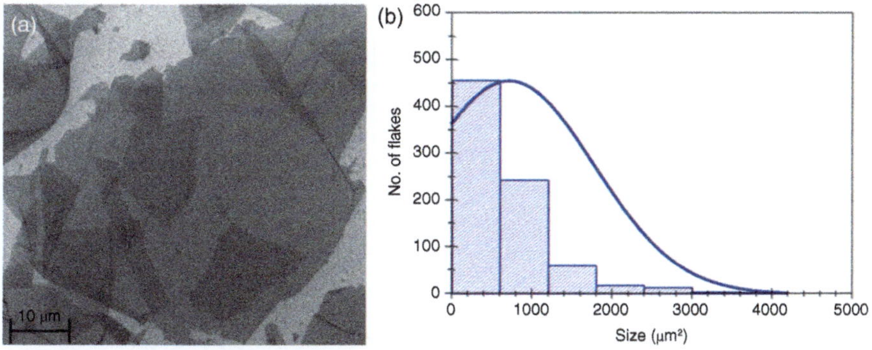

Figure 2.2 (a) SEM image of a large GO sheet on SiO_2 substrate produced from the modified method. (b) Size distribution of GO flakes. Source: Reproduced with permission from Larisika et al. [18]/Elsevier.

2.5 Large GO (Modified Hummers Method) Biosensors

Figure 2.3a shows an optical micrograph of a typical graphene FET sensor device. It was frequently reported that hydrazine reduction was not sufficient to achieve high conductivities of the flakes, so further reduction by annealing in Ar/H_2 at high temperatures was necessary [15, 16]. In this context, the transfer curves were measured to evaluate the efficiency of the reduction process. Figure 2.3b shows the output characteristics (drain current I_D vs. drain voltage V_D) of the SiO_2-gated graphene FETs at eleven different gate voltages (V_G).

The device showed a clear increase in conductance as the gate voltage changed from +100 to −100V, indicating that the RGO films behaved as p-type semiconducting materials [23]. Figure 2.3c demonstrated the typical transfer characteristics

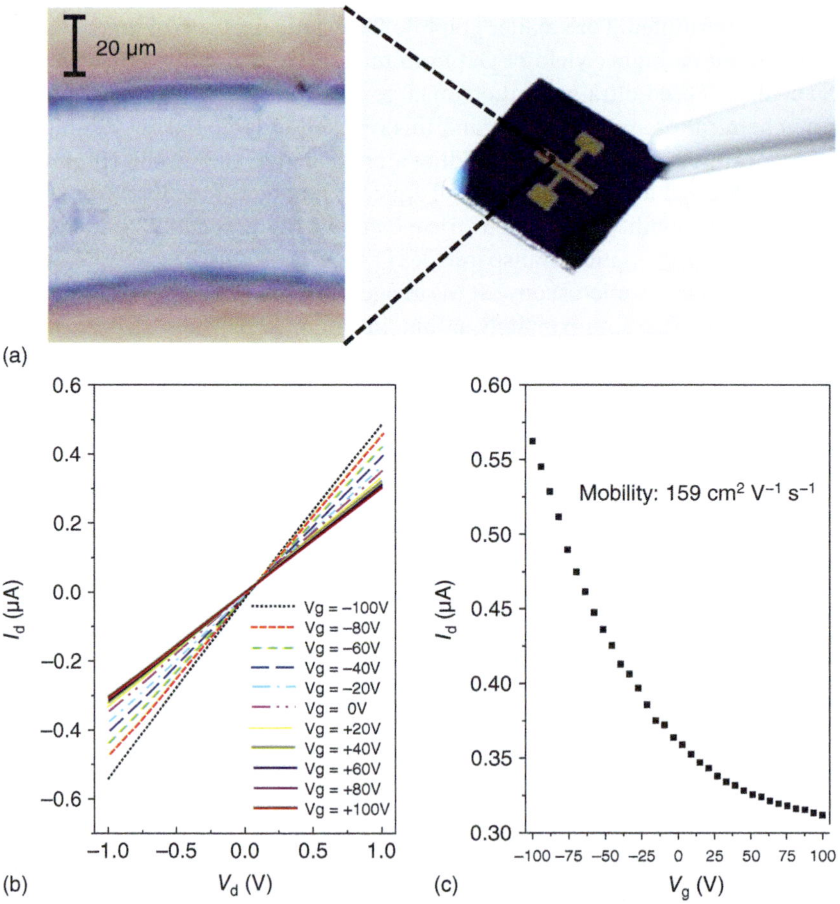

Figure 2.3 (a) Optical micrograph of the fabricated graphene device. (b) Output characteristics of the transistor device under different applied gate voltages. (c) Current–voltage transfer curve of bottom-gated graphene FET at a drain–source bias of V_D = 0.1 V. Source: Reproduced with permission from Larisika et al. [18]/Elsevier.

(I_D–V_G) of the graphene FET measured under ambient conditions (source–drain bias voltage V_{DS} = 100 mV) and its effective field effect hole mobility was 159 $cm^2 V^{-1} s^{-1}$.

In order to extract the field-effect mobility of holes, only selected GO sheets with suitable size (20 μm × 50 μm) were used for the transistor measurements [15]. Due to the indicated flake area, a graphene coverage corresponding to a channel width of 500 μm was assumed. The channel width (W) and length (L) were 4000 and 60 μm, respectively. The hole mobility was calculated from the linear regime of the transfer characteristics using the equation $\mu = (L/WC_{ox}V_d)(\Delta I_d/\Delta V_g)$, where L and W were the channel length and width and C_{ox} the gate capacitance.

2.6 Mechanism of Large GO via Reduced GO Growth

Similar to graphene and carbon nanotubes, GO has the ability to grow in CVD to form a complete substrate of rGO with drastically reduced electrical resistance and electrical resistance homogeneity, thus enabling its use in practical electronics and biosensors [24, 25].

In theory, any hydrocarbon source such as methane, ethylene, and ethanol could be used to extend the growth of GO. However, liquid carbon sources such as ethanol are a safer alternative than gaseous hydrocarbons, which are extremely flammable and may form explosive mixtures with air. Thus, ethanol was used as the carbon source.

During the deposition phase, the gaseous hydrocarbon was thermally pyrolyzed, and each of the C—C and C—H bonds could break to form two free radicals, with each fragment keeping one electron. The presence of radicals in the molecule allowed rapid rearrangement of carbon bonds. Though the reaction was more complicated for alcohols, the possible cracking of carbon radical species to form carbon deposition in basic form could be expressed as Eq. (2.1), and the catalytic hydrogenation [26] process of solid carbon in which H_2 formed CH_4 as Eq. (2.2).

Cracking of carbon radical species

$$C_xH_y(g) \leftrightarrow xC(s) + (y/2)H_2(g) \tag{2.1}$$

Catalytic hydrogenation of solid carbon

$$C(s) + 2H_2(g) \rightarrow CH_4(g) \tag{2.2}$$

According to Le Chatelier's principle and Eq. (2.1), the growth rate of carbon will decrease with decreasing hydrocarbon concentrations and increasing hydrogen gas. From Eq. (2.2), increasing the ratio of hydrogen gas increased the dissociation rate of carbon and hydrogen into gaseous hydrocarbons. Thus, the addition of hydrogen gas could help to etch away amorphous or excess carbon formed [27]. In the experiment, by increasing the hydrogen to argon gas ratio from 0.17 to 0.34 while maintaining the total flow volume of gas mixture, no deposition of carbon occurred even after one hour. With the ethanol precursor removed from the system for only one hour, no carbon Raman signature was detected on the substrate; the entire GO

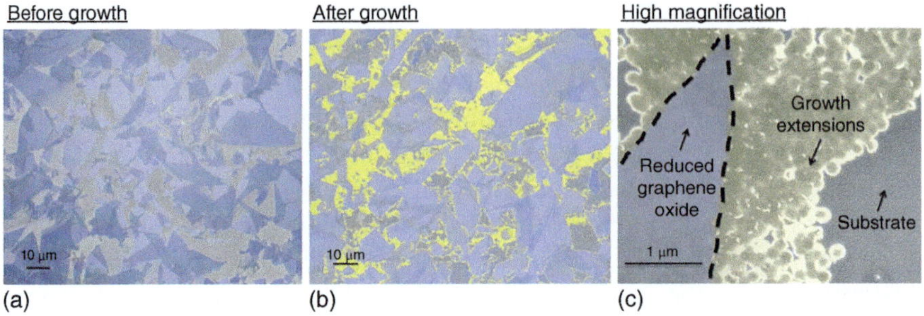

Figure 2.4 Scanning electron microscope image of (a) before ethanol CVD treatment GO flakes with c. 60% coverage and (b) after 30-minutes ethanol CVD treatment rGO flakes with c. 80% coverage (GO flakes highlighted in purple and new growth highlighted in yellow, please refer to online version for color) (c) High-magnification image on the new growths. Source: Reproduced with permission from Huang et al. [25]/Royal Society of Chemistry.

template was etched away. Thus, the deposition of rGO in the experiment obeyed Le Chatelier's principle as depicted in Eq. (2.1) and occurs in direct competition with the catalytic hydrogenation process as depicted in Eq. (2.2).

Also, according to Lerf–Klinowski's GO model, carboxyl (–COOH) groups were found at the edges of GO [28–30]. Thus, an etching action by H_2 was added to activate the carbon sites, followed by possible carbon deposition at the edges.

In Figure 2.4, existing rGO flakes were highlighted in purple and the new growths were highlighted in yellow. Figure 2.4a showed the GO coverage of c. 60% after a typical APTES-functionalization and GO incubation on SiO_2. However, after a partial ethanol CVD treatment of 30 minutes, the coverage increased to c. 80% as shown in Figure 2.4b. A high magnification of the new growth was shown in Figure 2.4c.

It was shown in the high-resolution SEM image that the rGO new growth continued to act as a catalyst for subsequent growth, subsequently covering more of the gap between existing rGO flakes. To prove the role of GO as an effective catalyst, substrates in the same conditions (clean and APTES-functionalized) were processed under the same ethanol CVD conditions for up to two hours. Neither carbon deposits nor carbon spectra were detected by SEM nor Raman, respectively [31]. The initial GO flakes acted as crystal nuclei for the subsequent growth of these flakes [24].

Since rGO could be a template for subsequent growth of rGO, this work supported the free radical condensate concept in which metal catalysts only help to initiate carbon precipitation and rearrangement at the nucleation stage: once the carbon structure is created, e.g. 1D carbon nanotube structure [32], the metal catalyst becomes nonfunctional. The carbon radicals from the pyrolysis of ethanol could then catenate into longer chains by forming covalent bonds directly on the existing edges of rGO template, resulting in the growth observed [33]. It had been reported that the energetics of the primary particles dictated the new growth's phase; if the precursor was metastable, the growth would follow a metastable phase [34]. This could explain the growth of rGO during an extended ethanol CVD posttreatment on GO.

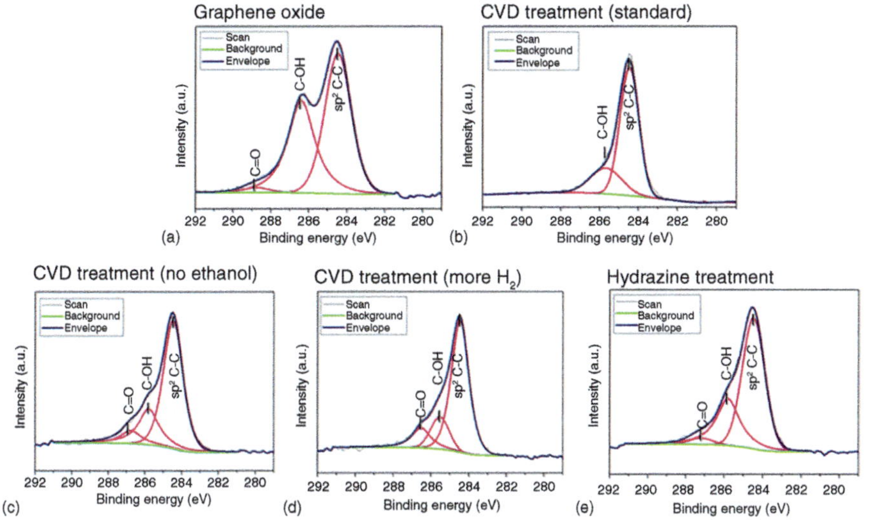

Figure 2.5 C1s XPS spectra (hv = 1486.6 eV) collected on (a) GO; and reduced GO reduced in similar CVD setup with (b) one hour standard CVD treatment (c) 15 minutes CVD treatment with no ethanol (d) one hour CVD treatment with more H_2 gas (e) 12-hour hydrazine reduced sample. The different components related to various chemical shifts of carbon bonds were indicated. Source: Reproduced with permission from Gilje et al. [24]/ Royal Society of Chemistry.

To elucidate more on the different growing conditions of the rGO, the ethanol and hydrogen gas input were varied. The C1s spectral of rGO shown in Figure 2.5 could be deconvoluted using Lorentzian with Gaussian broadening curve fitting into C=O (287.2 eV), C–OH (286.3 eV) and C–C (284.5 eV). These values were in agreement with previous reports [15, 24, 35]. The C–C spectra positions for all five samples were detected to be at 284.5 eV. However, there were variations in C=O and C—OH spectra positions for different samples. The sample treated with more hydrogen gas had both C=O and C–OH down shifted, followed by sample with standard CVD treatment, CVD treatment with no ethanol, hydrazine treatment, and GO. The CVD treatment with more hydrogen had chemical shifts more toward that of highly oriented pyrolytic graphite (HOPG) which indicated that the bonds present in the compound was more similar to that of HOPG. This showed that the presence of hydrogen gas in the ethanol system acted as an etchant which could anisotropically etch amorphous carbon to produce rGO closer to that of HOPG, similar to what had been reported for graphene systems [36]. Hydrogen also preferentially etch away stacked carbon (16 kJ mol^{-1}) and some oxygen moieties due to the weaker bond compared to in-plane carbon (345 J mol^{-1}) thus leading to more graphitized rGO instead of amorphous carbon [37]. The atomic percentages of the bonds present in the XPS spectra was then calculated and presented in Table 2.1.

In all the rGO samples, the atomic percentage of C—OH bonds decreased while the percentage of C—C bonds increased. For the sample with standard CVD

Table 2.1 Atomic percentage of different carbon bonds identified by XPS on mono-layered GO and rGO reduced under different conditions.

Samples	C=O (at%)	C—OH (at%)	C—C (at%)
GO	2.4	38.7	58.9
CVD treatment – Standard	0	17.2	82.8
CVD treatment – No ethanol	9.8	21.2	69.0
CVD treatment – More hydrogen	12.5	18.2	69.3
Hydrazine treatment	5.4	25.8	68.8

treatment, the C=O bond peak (initially at 2.4%) was completely removed indicating an efficient graphitization. The C=O bond in the CVD treatment had been reduced to C—O bonds, a reduced species of C=O [24]. However, in other reduced samples, the atomic percentage of C=O bonds increased. This was not in agreement with existing literature [35, 38, 39]; this was due to the initial large GO flakes used. With larger GO flakes, the atomic percentages of both C=O and C—OH, which were predominantly found at the edges of the flakes, were much lower in the reported GO compared to the GO synthesized in existing literature [15, 17, 40].

The remaining C=O bonds and higher atomic percentage of the C—OH bonds indicated that the presence of ethanol, and the right flow rate of hydrogen gas, were critical in the complete and efficient graphitization of rGO. The CVD procedures, gas flow parameters, and ethanol setup used in this study had been optimized for the growth of carbon nanotubes cleanly without amorphous carbon. Deviations from the optimal condition resulted in residual C=O bonds. Using CVD treatment with no ethanol, more hydrogen, or hydrazine treatment resulted in rGO products with similar amount of C–C atomic percentage (69%), but with varying C—OH and C=O depending on condition used.

From 0 to 15 minutes of ethanol CVD treatment (Table 2.2), the AFM height profile indicated that the thickness of preexisting GO template had negligible increase in height and no growth from the edges was observed [27]. However, as the process time increased to 30 minutes, new growths appeared and the thickness of the GO template also increased slightly by 0.5 nm. With increasing process time to two hours, the new growths continued to extend outward and completely cover the gaps in between the preexisting rGO flakes. The new growth thickness after two hours of growth was 2 nm and this corresponded to —two to three layers of rGO [18].

From four-point probe measurements, the average electrical resistance of the chips decreased from 1.99E+6 to 4.68E+3 Ω/□ (99.76% reduction) while the relative standard deviation decreased from 1.60E+06 to 7.72E+02 Ω/□ (99.95% reduction) after a 2-hour ethanol CVD treatment as compared to 12-hour hydrazine vapor reduction. The decrease in electrical resistivity in the ethanol CVD treated samples was attributed to the increased percolation pathways between flakes due to the new growth completely filling up gaps between pre-existing rGO flakes. Due to the more uniform coverage and increased coverage of the rGO on the substrate,

Table 2.2 D/G Raman spectrum peak intensity, AFM height profile and four-point probe sheet resistance for GO and rGO reduced under different conditions.

		GO	15 min CVD	30 min CVD	1 h CVD	2 h CVD	12 h hydrazine [18]
UV Raman spectroscopy	Peak intensity of D/G	0.42	0.35	0.32	0.17	0.09	NA
AFM height profile	Flake thickness (nm)	1.4	1.5	2	4	4	NA
	New growth thickness (nm)	0	0	0.8	1	2	NA
Four-point probe	Average resistance (Ω/□)	NA	6.10E+4	2.87E+4	2.23E+4	4.68E+3	1.99E+6
Sheet resistance ($n = 5$)	Standard deviation (Ω/□))	NA	4.43E+4	1.25E+4	7.16E+3	7.72E+2	1.60E+6
	Relative standard deviation (%)	NA	72.6	43.6	32.1	16.5	80.5

Source: Reproduced from Huang et al. [25] with permission from the Royal Society of Chemistry.

the reported growth method decreased the standard deviations between chips, thus solving a critical problem for the practical use of rGO in practical electronic devices.

Having considered the data from the experiment, the growth sequence of rGO schematic was described (Figure 2.6). Ethanol CVD treatment of up to 15 minutes on the GO lead to an improved graphitization from the free radicals produced from pyrolysis of gaseous ethanol. After 30 minutes, a thin graphitic shell was formed around the GO catalyst. Due to the lower growth activation energy at the edge of graphene sheets for sequential addition of carbon hexagons [41], rGO preferentially grew out from the edges. Finally, rGO completely covered the substrate after two hours of the CVD ethanol growth process.

The presented data suggested that rGO could be grown via the free radical condensate mechanism. The mechanism of carbon nanotube growth had been debated, and two different mechanisms had been proposed: the dissolution of elemental carbon into metal nanoclusters followed by precipitation [42] or the free radical condensate that carbon could act as reaction medium for production of nanotubes [26]. In the experiment, no metal catalyst was used, but rGO growth was observed, thus supporting the free radical condensate growth theory. After the pyrolysis of ethanol, free radicals preferentially condensed at the edges of the preexisting GO flakes and thus rearranged itself into a new carbon skeleton until a local minimum had been reached.

This subchapter reported that the complete coverage of rGO on SiO_2 substrate was achieved using GO template under ethanol CVD at ambient pressure without any metal catalysts. GO edges preferentially acted as a catalyst for further growth of rGO. The growth sequence supported the free radical condensate growth theory that carbon could nucleate and grow without a metal catalyst. After a two-hour ethanol

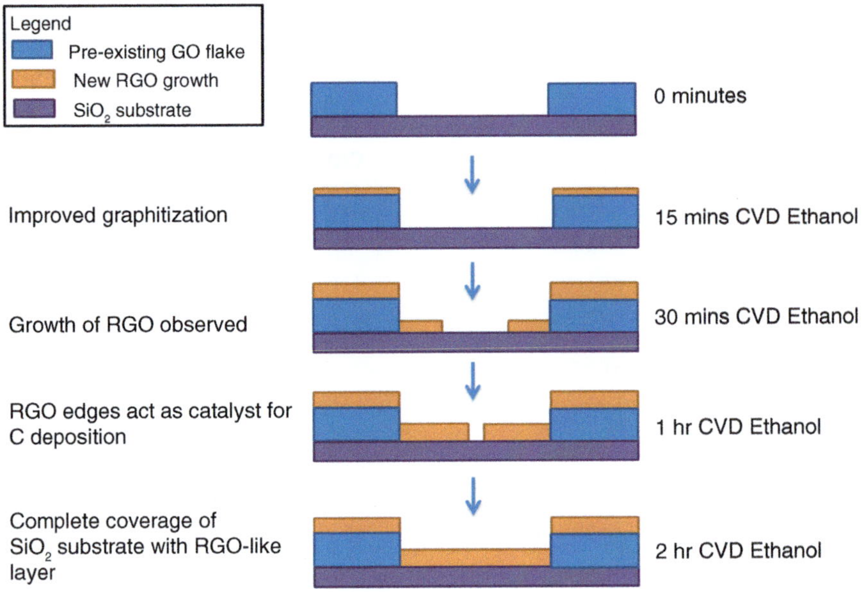

Figure 2.6 Schematic showing the growth sequence of rGO from a GO template during ethanol CVD treatment. Source: Reproduced with permission from Huang et al. [25]/Royal Society of Chemistry.

CVD treatment, the electrical resistivity decreased by 99.8% and its relative standard deviation decreased by 79.5% compared to hydrazine treated samples. The results and proposed mechanism show that the use of two-hours CVD treatment on GO was a promising way toward a better and homogenous rGO for practical, high-yielding preparation of novel functional biosensors, electronics, batteries or crystal growth templates.

2.7 Large GO (Reduced GO Growth) Biosensors

The ethanol CVD-treated rGO chips were then fabricated into liquid-gated FETs (Figure 2.7a,b) in order to test their biosensing capability. Bottom-gate operated transistors were fabricated, without using elaborate electron-beam lithographic techniques which could leave polymer residue, by evaporating Au electrodes (100 nm thick) directly on top of the selected rGO films that were previously deposited on 300 nm SiO_2/Si substrates. Silicone rubber well was used to hold the liquid analyte in place during sensing. Silicone rubber has good electrical insulating and sealing properties. A sensing fluidic channel was fabricated using polydimethylsiloxane (PDMS) mold and attached for continuous monitoring.

Shadow masks were specially designed and fabricated on stainless steel template with extended windings and enlarged contact pads for robust and easy connection to external measurement equipment. Autodesk Autocad was used for the design shown in full in Figure 2.8a, and the magnified design as Figure 2.8b.

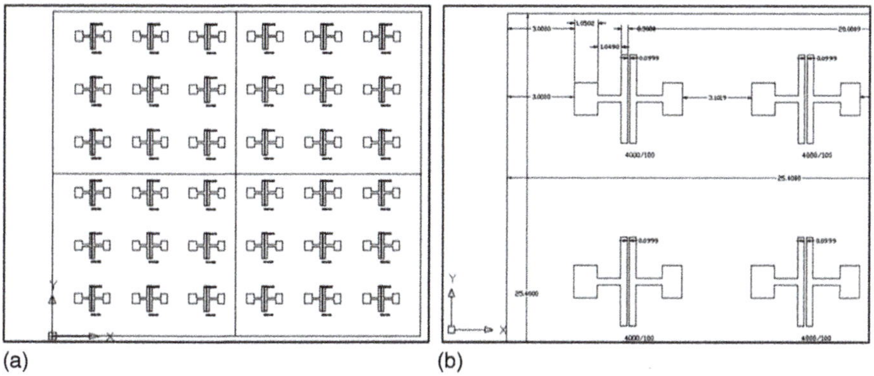

Figure 2.7 (a) Full and (b) magnified, image of Shadow mask design for FET fabrication.

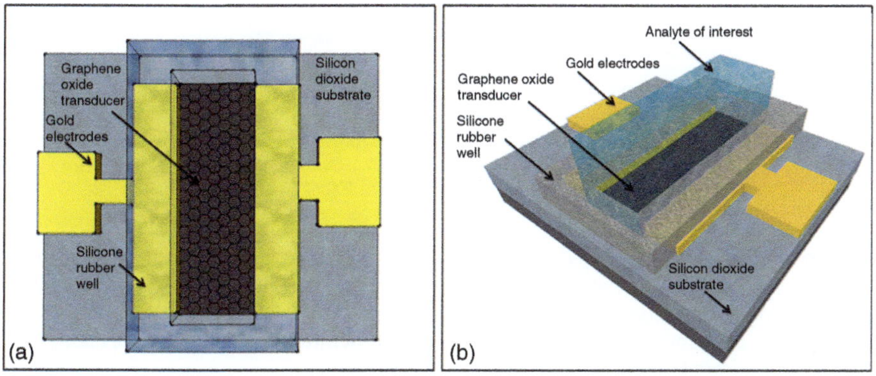

Figure 2.8 Schematic showing (a) top-view and (b) side-view of fabricated FET devices with silicon well suitable for analyte testing.

The change in I_{SD} could then be utilized to be a sensor for the IL-6 concentration by fixing the gate voltage value. The general trend of the I_{SD} decreasing with increasing concentration of IL-6 was evident in the calibration curve (Figure 2.9b). The linear dynamic range (LDR) of the sensor was from 4.7 to 18.8 pg ml^{-1} of IL-6. Although at higher concentrations, the change in the sensor electrical signal was not much higher than at 18.8 pg ml^{-1}, but the sensor still acted as an active material that gave a reliable reading against IL-6 concentrations (Figure 2.9b insert). The slightly nonlinearity relationship between interleukin-6 concentration and the decrease of measured source–drain current could be attributed to the carrier injection during the *I–V* measurement [13]. It had also been commonly observed for carbon-based FET biosensors [3, 13, 44–46]. Further studies were needed to reveal the details of the sensing mechanisms. Figure 2.9c shows no change in the I_{SD} when the sensor was exposed to interference protein, BSA. This showed the sensor resistance (after IL-6 attachment and BSA protection) in preventing interferences and unspecific binding.

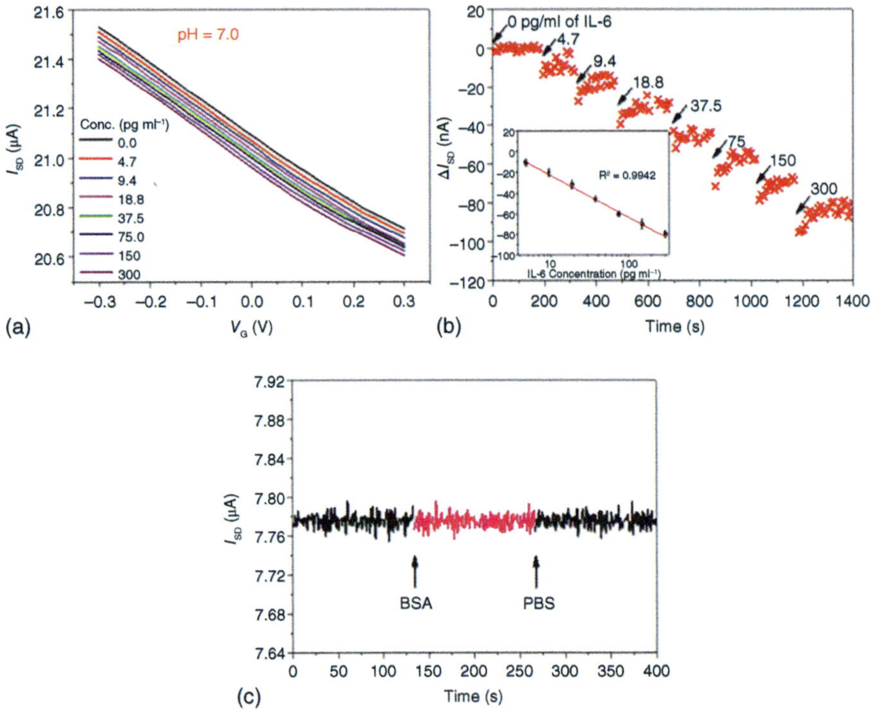

Figure 2.9 Detection characteristics of rGO FET biosensors. (a) Source–drain current (I_{SD}) vs. gate voltage (V_G) plot with different concentration of IL-6 protein, showing decreasing I_{SD} at increasing IL-6 addition; (b) I_{SD} vs. time with addition of 0–300 pg ml^{-1} of IL-6; and (c) real-time I_{SD} plot, showing negligible effect when BSA interference protein was added. Source: Adapted from Larisika et al. [43].

The shift in source–drain current (I_{SD}) of the anti-IL-6 immobilized rGO transistor was sensitive to increasing concentrations of IL-6 protein due to electronic n-doping of the device. Figure 2.9a shows that binding of increasing amounts of IL-6 protein to the anti-IL-6 antibody resulted in a down-shift of the I_{SD}–V_G curve, from 21.09 (0 pg ml^{-1}) to 20.96 μA (300 pg ml^{-1}) at V_G = 0 V.

Low-gate voltages were used to minimize possible disturbances to the biological sample and prevent instability of the biosensor due to electrochemical and charging effect [3]. The sensor was operating at the p-type region at V_G = 0 V and the negative shifts were tentatively attributed to the electrostatic gating effect on the rGO surface. The increasing positive charge from the immune-complex formation could induce negatively charges atom in the rGO thus n-doping (down-shifting) the I_{SD}–V_G curve [47, 48]. It could also be due to a decreased ionic screening effect after the protein–antibody interaction [49, 50]. The specific doping mechanism during biomolecular interactions on 2D carbon liquid-gated FET sensing surface was still not fully understood and required further study [48, 51].

The decrease in I_{SD} corresponded to a sensitivity of 1.5 pA per pg ml^{-1} of IL-6. The slope of the I_{SD}–V_G curve did not change thus indicating that the charge carrier mobility in the rGO transducer remained the same after the protein-antibody reaction. It was noted that the I_{SD} downshift (electrostatic gating effect) exhibited in the reported biosensor was more reproducible and reliable than other effects using Schottky barrier and changes in gate coupling and carrier mobility [47].

The LOD of the reported IL-6 sensor based on 3σ was 2.9 pA or 1.53 pg ml^{-1}. Plasma IL-6 levels in physically active individuals had been reported to be as low as ≤1 pg ml^{-1} [52, 53] while levels in sedentary middle-aged men 11 pg ml^{-1} and healthy premenopausal women had been reported in the range of 10–11 pg ml^{-1} [54]. Following prolonged exercise such as cycling, IL-6 could be elevated as much as 38-fold, and as much as 128-fold following completion of a marathon [53]; thus, the detection range of the reported sensor was within the physiological range of IL-6. However, further work was still needed to improve the detection limit of the sensor.

The eventual endpoint application would be for point-of-care, whole blood analysis. Thus, the ability of the rGO sensor to sensitively and selectively detect the biomarker in serum demonstrates its relevance to potential usage in medical diagnostics. In this experiment, fetal calf serum was used to simulate the assay conditions.

Table 2.3 compared the sensitivity range of different types of biosensors to the levels of IL-6 reported recently. Using the reported label-free GO sensor, the

Table 2.3 Comparison of reported biosensors to IL-6.

References	Transducer	Platform	Sensitivity range (pg ml^{-1})	Note
Reported	Post-processed GO	FET immunoassay	4.7–300	Label-free
[46]	SWCNTs/AuNPs	Electrochemical impedance immunoassay	1e^{-5}–0.1	Label-free
[55]	Functionalized Biacore sensor chip	Surface plasmon resonance (SPR) immunoassay	0.78–12.5	Label-free
[56]	Microfabricated silicon flexural plate wave (FPW) resonators	Micro-electromechanical (MEM) immunoassay	1.2–unknown	Label-free
[57]	Fiber-optic dip-probe	Optical immunoassay	120–12 000	Label-free
[58]	Polymeric nanoparticles on calcium carbonate	Electrochemical immunoassay	0.002–20	Porous polyelectrolyte nanoparticles label used
[59]	Modified carbon electrodes	Electrochemical immunoassay	1–1000	PS@PDA-metal label used
[60]	Functionalized fluorescent core-shell nanoparticles	Fluoro-immunoassay	20–1250	Fluorescent core–shell nanoparticles label used
[61]	Carbon nanotube forest electrodes	Electrochemical immunoassay	0.5–5	Ab2-MWNT-HRP label used

Source: Reproduced from Larisika et al. [43] with permission from the Royal Society of Chemistry.

sensitivity range was within the physiological range of 10 pg ml^{-1} in sweat samples from healthy women [54, 62], which were not to be significantly different from plasma levels [54]. A sensing fluidic channel was then fabricated using PDMS mold and attached for continuous monitoring.

2.8 Conclusion

Nonetheless, research of GO as FET transducer is in its infancy stage. We are exploring the different synthesis methods, functionalization, and testing the LOD limits. Only when we are able to control the variability in GO processing, and understand what each shape, size, and functionalization do, will we be able to fully utilize GO to its fullest potential. The author guesses that there will be specific names given to define the shape, size, and functionality of GO. Just like diamond, graphite, cholesterol, and fats.

The applications of graphene-based nanomaterials in biosensor platforms were manifold and had been proven by various studies. However, a large amount of work was needed to expand and adapt this biosensing technology to different areas.

GO had shown promising sensitivity and selectivity in FET biosensors. The future of GO was highly promising, although more research was needed to fully explore and control the benefits of this remarkable material and convert these works into practical biosensors,

2.9 Further Developments

The two distinct goals of successful biosensing are the determination of small quantity of specific protein biomarker and the validation or discovery of biomarkers, with the latter expected to accelerate early screening of high-risk populations at point-of-care.

With such a small size and countless molarity possibilities, it may be hard to manufacture GO, especially using the top-down method. Top-down methods are sometimes preferred as the quality of graphite and size could be better than bottom-up methods. The molarity, size, and shape of GO differ from batch to batch. Ground-up approach had been used to supplement and homogenize top-down approach. Using the ground-up method with dip-pen templating could be a good way to tailor the size and shape of the GO or graphene. Then, an electron gun could be used to functionalize the sheet with necessary molecules.

Even after successfully doing so, we would have only gotten the 2D part of the GO. GO could also be folded, crumbled, or rolled into different 3D configurations. These configurations could not only be disturbed electrically but magnetically, optically, and mechanically as well. These could be used for ultrasensitive sensors. The potential of GO for sensing has yet to be fully explored.

References

1 Shen, H., Zhang, L.M., Liu, M., and Zhang, Z.J. (2012). Biomedical applications of graphene. *Theranostics* 2 (3): 283–294.

2 Kim, Y.K., Kim, M.H., and Min, D.H. (2011). Biocompatible reduced graphene oxide prepared by using dextran as a multifunctional reducing agent. *Chem. Commun. (Cambridge).* 47 (11): 3195–3197.

3 He, Q., Sudibya, H.G., Yin, Z. et al. (2010). Centimeter-long and large-scale micropatterns of reduced graphene oxide films: fabrication and sensing applications. *ACS Nano* 4 (6): 3201–3208.

4 Jiang, H.J. (2011). Chemical preparation of graphene-based nanomaterials and their applications in chemical and biological sensors. *Small* 7 (17): 2413–2427.

5 Pumera, M. (2011). Graphene in biosensing. *Mater. Today* 14 (7, 8): 308–315.

6 Liu, S. and Guo, X.F. (2012). Carbon nanomaterials field-effect-transistor-based biosensors. *NPG Asia Mater.* 4: 1–10.

7 Huang, J., Harvey, J., Fam, W.H.D. et al. (2013). Novel biosensor for interleukin-6 detection. *Proc. Eng.* 60: 195–200.

8 Huang, J., Harvey, J., Chen, H. et al. (2013). Fully organic graphene oxide-based sensor with integrated pump for sodium detection. In: *Conference Proceeding of icSports*. Vilamoura, Portugal, SciTePress.

9 Huang, J., Chen, H., Fam, D. et al. (2014). Growth of reduced graphene oxide. *MRS Online Proceedings Library (OPL)* 1702: 1–6.

10 Huang, J., Chen, H., Jing, L. et al. (2016). Improved synthesis and growth of graphene oxide for field effect transistor biosensors. *Biomed. Microdevices* 18 (4): 1–10.

11 Agarwal, S., Zhou, X.Z., Ye, F. et al. (2010). Interfacing live cells with nanocarbon substrates. *Langmuir* 26 (4): 2244–2247.

12 Sudibya, H.G., He, Q.Y., Zhang, H., and Chen, P. (2011). Electrical detection of metal ions using field-effect transistors based on micropatterned reduced graphene oxide films. *ACS Nano* 5 (3): 1990–1994. Huang J, Harvey J, Chen H et al. (2013). Fully organic graphene oxide-based sensor with integrated pump for sodium detection. In: *International Congress on Sports Science Research and Technology Support (icSPORTS) 2013*. Portugal: INSTICC.

13 Mao, S., Lu, G.H., Yu, K.H. et al. (2010). Specific protein detection using thermally reduced graphene oxide sheet decorated with gold nanoparticle-antibody conjugates. *Adv. Mater.* 22 (32): 3521–3526.

14 Dong, L., Yang, J., Chhowalla, M., and Loh, K.P. (2017). Synthesis and reduction of large sized graphene oxide sheets. *Chem. Soc. Rev.* 46 (23): 7306–7316.

15 Su, C.-Y., Xu, Y., Zhang, W. et al. (2009). Electrical and spectroscopic characterizations of ultra-large reduced graphene oxide monolayers. *Chem. Mater.* 21 (23): 5674–5680.

16 Marcano, D.C., Kosynkin, D.V., Berlin, J.M. et al. (2010). Improved synthesis of graphene oxide. *ACS Nano* 4 (8): 4806–4814.

17 Huang, J., Larisika, M., Nowak, C., and Tok, I.Y.A. (2015). *New Methods in Aqueous Graphene (Graphene Oxide) Synthesis for Biosensor Devices*. Taylor & Francis Group.

18 Larisika, M., Huang, J., Tok, A. et al. (2012). An improved synthesis route to graphene for molecular sensor applications. *Mater. Chem. Phys.* 136 (2): 304–308.

19 Xu, Y., Bai, H., Lu, G. et al. (2008). Flexible graphene films via the filtration of water-soluble noncovalent functionalized graphene sheets. *J. Am. Chem. Soc.* 130 (18): 5856–5857.

20 Kovtyukhova, N.I., Ollivier, P.J., Martin, B.R. et al. (1999). Layer-by-layer assembly of ultrathin composite films from micron-sized graphite oxide sheets and polycations. *Chem. Mater.* 11 (3): 771–778.

21 Kholmanov, I.N., Stoller, M.D., Edgeworth, J. et al. (2012). Nanostructured hybrid transparent conductive films with antibacterial properties. *ACS Nano* 6 (6): 5157–5163.

22 Luo, Z., Lu, Y., Somers, L.A., and Johnson, A.T.C. (2009). High yield preparation of macroscopic graphene oxide membranes. *J. Am. Chem. Soc.* 131 (3): 898–899.

23 Gilje, S., Han, S., Wang, M. et al. (2007). A chemical route to graphene for device applications. *Nano Lett.* 7 (11): 3394–3398.

24 Huang, J., Larisika, M., Fam, W.H.D. et al. (2013). The extended growth of graphene oxide flakes using ethanol CVD. *Nanoscale* 5 (7): 2945–2951.

25 Huang, J., Fam, D., He, Q. et al. (2014). The mechanism of graphene oxide as a growth template for complete reduced graphene oxide coverage on an SiO_2 substrate. *J. Mater. Chem. C* 2 (1): 109–114.

26 Reilly, P.T.A. and Whitten, W.B. (2006). The role of free radical condensates in the production of carbon nanotubes during the hydrocarbon CVD process. *Carbon* 44 (9): 1653–1660.

27 Su, C.-Y., Xu, Y., Zhang, W. et al. (2010). Highly efficient restoration of graphitic structure in graphene oxide using alcohol vapors. *ACS Nano* 4 (9): 5285–5292.

28 Szabó, T., Berkesi, O., Forgó, P. et al. (2006). Evolution of surface functional groups in a series of progressively oxidized graphite oxides. *Chem. Mater.* 18 (11): 2740–2749.

29 Cai, W., Piner, R.D., Stadermann, F.J. et al. (2008). Synthesis and solid-state NMR structural characterization of ^{13}C-labeled graphite oxide. *Science* 321 (5897): 1815–1817.

30 Boukhvalov, D.W. and Katsnelson, M.I. (2008). Modeling of graphite oxide. *J. Am. Chem. Soc.* 130 (32): 10697–10701.

31 Ferrari, A.C. (2007). Raman spectroscopy of graphene and graphite: disorder, electron–phonon coupling, doping and nonadiabatic effects. *Solid State Commun.* 143 (1–2): 47–57.

32 Yao, Y., Feng, C., Zhang, J., and Liu, Z. (2009). "Cloning" of single-walled carbon nanotubes via open-end growth mechanism. *Nano Lett.* 9 (4): 1673–1677.

33 Dutta, S. and Pati, S.K. (2010). Novel properties of graphene nanoribbons: a review. *J. Mater. Chem.* 20 (38): 8207–8223.

34 Baumgartner, J., Dey, A., Bomans, P.H.H. et al. (2013). Nucleation and growth of magnetite from solution. *Nat. Mater.* 12 (4): 310–314.

35 Yang, D., Velamakanni, A., Bozoklu, G. et al. (2009). Chemical analysis of graphene oxide films after heat and chemical treatments by X-ray photoelectron and micro-Raman spectroscopy. *Carbon* 47 (1): 145–152.

36 Zhang, Y., Li, Z., Kim, P. et al. (2011). Anisotropic hydrogen etching of chemical vapor deposited graphene. *ACS Nano* 6 (1): 126–132.

37 Yuan, G.D., Zhang, W.J., Yang, Y. et al. (2009). Graphene sheets via microwave chemical vapor deposition. *Chem. Phys. Lett.* 467 (4–6): 361–364.

38 Becerril, H.A., Mao, J., Liu, Z. et al. (2008). Evaluation of solution-processed reduced graphene oxide films as transparent conductors. *ACS Nano* 2 (3): 463–470.

39 Eda, G., Fanchini, G., and Chhowalla, M. (2008). Large-area ultrathin films of reduced graphene oxide as a transparent and flexible electronic material. *Nat. Nanotechnol.* 3 (5): 270–274.

40 Mattevi, C., Eda, G., Agnoli, S. et al. (2009). Evolution of electrical, chemical, and structural properties of transparent and conducting chemically derived graphene thin films. *Adv. Funct. Mater.* 19 (16): 2577–2583.

41 Haghighatpanah, S., Börjesson, A., Amara, H. et al. (2012). Computational studies of graphene growth mechanisms. *Phys. Rev.B* 85 (20): 205448.

42 Gavillet, J., Loiseau, A., Ducastelle, F. et al. (2002). Microscopic mechanisms for the catalyst assisted growth of single-wall carbon nanotubes. *Carbon* 40 (10): 1649–1663.

43 Huang, J., Chen, H., Niu, W. et al. (2015). Highly manufacturable graphene oxide biosensor for sensitive interleukin-6 detection. *RSC Adv.* 5 (49): 39245–39251.

44 Kim, J.P., Lee, B.Y., Lee, J. et al. (2009). Enhancement of sensitivity and specificity by surface modification of carbon nanotubes in diagnosis of prostate cancer based on carbon nanotube field effect transistors. *Biosens. Bioelectron.* 24 (11): 3372–3378.

45 Abe, M., Murata, K., Kojima, A. et al. (2007). Quantitative detection of protein using a top-gate carbon nanotube field effect transistor. *J. Phys. Chem. C* 111 (24): 8667–8670.

46 Yang, T., Wang, S., Jin, H. et al. (2013). An electrochemical impedance sensor for the label-free ultrasensitive detection of interleukin-6 antigen. *Sens. Actuators, B* 178: 310–315.

47 Heller, I., Janssens, A.M., Männik, J. et al. (2007). Identifying the mechanism of biosensing with carbon nanotube transistors. *Nano Lett.* 8 (2): 591–595.

48 Sohn, I.-Y., Kim, D.-J., Jung, J.-H. et al. (2013). pH sensing characteristics and biosensing application of solution-gated reduced graphene oxide field-effect transistors. *Biosens. Bioelectron.* 45: 70–76.

49 Chen, F., Xia, J., and Tao, N. (2009). Ionic screening of charged-impurity scattering in graphene. *Nano Lett.* 9 (4): 1621–1625.

50 Chen, F., Qing, Q., Xia, J. et al. (2009). Electrochemical gate-controlled charge transport in graphene in ionic liquid and aqueous solution. *J. Am. Chem. Soc.* 131 (29): 9908–9909.

51 Kim, D.-J., Sohn, I.Y., Jung, J.-H. et al. (2013). Reduced graphene oxide field-effect transistor for label-free femtomolar protein detection. *Biosens. Bioelectron.* 41: 621–626.

52 Bruunsgaard, H., Galbo, H., Halkjaer-Kristensen, J. et al. (1997). Exercise-induced increase in serum interleukin-6 in humans is related to muscle damage. *J. Physiol.* 499 (Pt 3): 833–841.

53 Ostrowski, K., Rohde, T., Asp, S. et al. (1999). Pro- and anti-inflammatory cytokine balance in strenuous exercise in humans. *J. Physiol.* 515 (Pt 1): 287–291.

54 Marques-Deak, A., Cizza, G., Eskandari, F. et al. (2006). Measurement of cytokines in sweat patches and plasma in healthy women: validation in a controlled study. *J. Immunol. Methods* 315 (1, 2): 99–109.

55 Chou, T.-H., Chuang, C.-Y., and Wu, C.-M. (2010). Quantification of interleukin-6 in cell culture medium using surface plasmon resonance biosensors. *Cytokine* 51 (1): 107–111.

56 Pepper, J., Noring, R., Klempner, M. et al. (2003). Detection of proteins and intact microorganisms using microfabricated flexural plate silicon resonator arrays. *Sens. Actuators, B* 96 (3): 565–575.

57 Kapoor, R. and Wang, C.-W. (2009). Highly specific detection of interleukin-6 (IL-6) protein using combination tapered fiber-optic biosensor dip-probe. *Biosens. Bioelectron.* 24 (8): 2696–2701.

58 Li, T. and Yang, M. (2011). Electrochemical sensor utilizing ferrocene loaded porous polyelectrolyte nanoparticles as label for the detection of protein biomarker IL-6. *Sens. Actuators, B* 158 (1): 361–365.

59 Shi, J.-J., He, T.-T., Jiang, F. et al. (2014). Ultrasensitive multi-analyte electrochemical immunoassay based on GNR-modified heated screen-printed carbon electrodes and PS@PDA-metal labels for rapid detection of MMP-9 and IL-6. *Biosens. Bioelectron.* 55: 51–56.

60 Hun, X. and Zhang, Z. (2007). Functionalized fluorescent core-shell nanoparticles used as a fluorescent labels in fluoroimmunoassay for IL-6. *Biosens. Bioelectron.* 22 (11): 2743–2748.

61 Malhotra, R., Patel, V., Vaqué, J.P. et al. (2010). Ultrasensitive electrochemical immunosensor for oral cancer biomarker il-6 using carbon nanotube forest electrodes and multilabel amplification. *Anal. Chem.* 82 (8): 3118–3123.

62 Cizza, G., Marques, A.H., Eskandari, F. et al. (2008). Elevated neuroimmune biomarkers in sweat patches and plasma of premenopausal women with major depressive disorder in remission: the POWER study. *Biol. Psychiatr.* 64 (10): 907–911.

3

Solution-Gated Reduced Graphene Oxide FETs: Device Fabrication and Biosensors Applications

Nirton C. S. Vieira[1], *Bianca C. S. Ribeiro*[1], *Rodrigo V. Blasques*[2], *Bruno C. Janegitz*[2], *Fabrício A. dos Santos*[3], *and Valtencir Zucolotto*[3]

[1]*Federal University of São Paulo, Institute of Science and Technology, Department of Science and Technology, Talim St., São José dos Campos, São Paulo, 12231-280, Brazil*
[2]*Federal University of São Carlos, Center of Agrarian Sciences, Department of Natural Sciences, Mathematics and Education, Anhanguera Hwy., Araras, São Paulo, 13600-970, Brazil*
[3]*University of São Paulo, Physics Institute of São Carlos, Nanomedicine and Nanotoxicology Group, Department of Physics and Materials Science, Trabalhador São Carlense Av., São Carlos, São Carlos, São Paulo, 13560-970, Brazil*

3.1 Introduction

Graphene is an allotrope of carbon consisting of an atomic layer of sp^2 carbon arranged in a hexagonal two-dimensional (2D) crystal lattice. The study of pristine graphene by Novoselov and coworkers opened the way for the development of new electronic devices, especially field-effect transistors (FETs), where graphene acts as the material of the channel in replacement of silicon [1]. As a zero-bandgap semimetal, graphene FETs never shut down. The transfer curve of a graphene FET behaves differently from conventional transistors. Depending on the voltage applied to the gate, there is conduction by positive or negative charges, around a point of minimum conductivity, where the concentration of charge carriers equals. This fact can compromise the switching function of graphene FETs. On the other hand, applying these devices as electronic sensors is exceptionally advantageous because graphene is extremely sensitive to the electric field in its vicinity [2].

Graphene can be obtained mainly from two distinct routes: mechanical exfoliation, as proposed by Novoselov et al. [1], or chemical vapor deposition (CVD) [3]. Mechanical exfoliation produces the highest quality graphene and is suitable for studies of the fundamental properties of the material [1]. However, the latter method is not an appropriate technique for the mass production of devices since it has low yield and low reproducibility. CVD is a good technique for the mass production of graphene [3]. However, this technique is still costly and needs various steps. Briefly, CVD encompasses the growth of graphene on a precursor substrate (usually

Graphene Field-Effect Transistors: Advanced Bioelectronic Devices for Sensing Applications,
First Edition. Edited by Omar Azzaroni and Wolfgang Knoll.

copper), graphene transfer from the original substrate to a polymeric intermediate substrate, and hence graphene transfer to the device substrate [3].

The chemical exfoliation of graphite is an alternative method for producing a graphene derivative known as graphene oxide (GO). GO is a single-atomic layer of a mixture of sp^2 and sp^3 hybridized carbons containing oxygenated functional groups on its basal planes and at its edges [4]. The Hummers' method (including its modifications) has been used to generate graphite oxide, being very well established for obtaining large amounts of GO [4]. GO is not a conductive material because the oxygen groups break the double bonds that hold the carbon atoms together. This fact is not desired in many electronic devices, especially in FETs. In addition, the reduction of GO to obtain reduced graphene oxide (rGO) produces another derivative of graphene, where much of the π-conjugated structure is recovered by removing the majority of oxygen groups [5]. This reduction gives rGO a high conductivity, making it an excellent alternative to pristine graphene in many optoelectronic devices [6, 7]. Another highlight is that functional groups in GO and rGO may be beneficial for many applications. Functional groups enable better dispersion of these materials in various solvents, including water. Water dispersions of GO and rGO allow the regular deposition of these graphene derivatives on several surfaces. In addition, functional groups can be functionalized with receptor molecules (enzymes, antibodies, and DNA) to develop biosensors to detect several analytes.

Most biosensors operate in a liquid environment. For this reason, the configuration where the FET is polarized via an electrolyte solution using a reference electrode is more suitable in biosensors than the conventional back-gate transistor configuration [8]. Furthermore, solution-gated field-effect transistors (SG-FETs), also referred to as electrolyte-gated FETs, operate at lower voltages compared to traditional ones. Figure 3.1 shows the structure of the SG-FET. Graphene or rGO is deposited on two metallic electrodes (source S and drain D) [8]. The V_{DS} voltage is responsible for establishing a current in the channel. The gate-source voltage (V_{GS}), indicated by V_G in Figure 3.1, is transmitted through an electrolyte solution via the polarization of a metallic wire or a reference electrode. In this case, an electrical double layer (EDL1) is formed at the graphene/solution interface and acts similarly to the gate oxide in traditional transistors [8]. There is also the formation of an EDL2 at the reference electrode/solution interface, and the system can be modeled as a parallel plate capacitor. The total capacitance of the system consists of the series association of these two EDL capacitors with a third one that comes from the graphene quantum capacitance (C_Q) (Figure 3.1b). Unlike conventional semiconductor transistors, the transfer curve ($I_{DS} \times V_{GS}$ in a constant V_{DS}) of graphene or rGO SG-FET has a V-shaped around a minimum point of conductivity (Figure 3.1c) [8].

The choice of rGO instead of graphene in SG-FETs is advantageous in terms of cost and ease of processing, especially when considering the large-scale production of disposable biosensors. This chapter encompasses the use of rGO in SG-FETs, including some fabrication strategies for fabricating the devices and their application as biosensors. We first discuss the reduction of GO in Section 3.2. Section 3.3 discusses rGO-based SG-FETs, including manufacturing strategies and essential electrical characterization in the context of biosensing. Section 3.4 deals with the

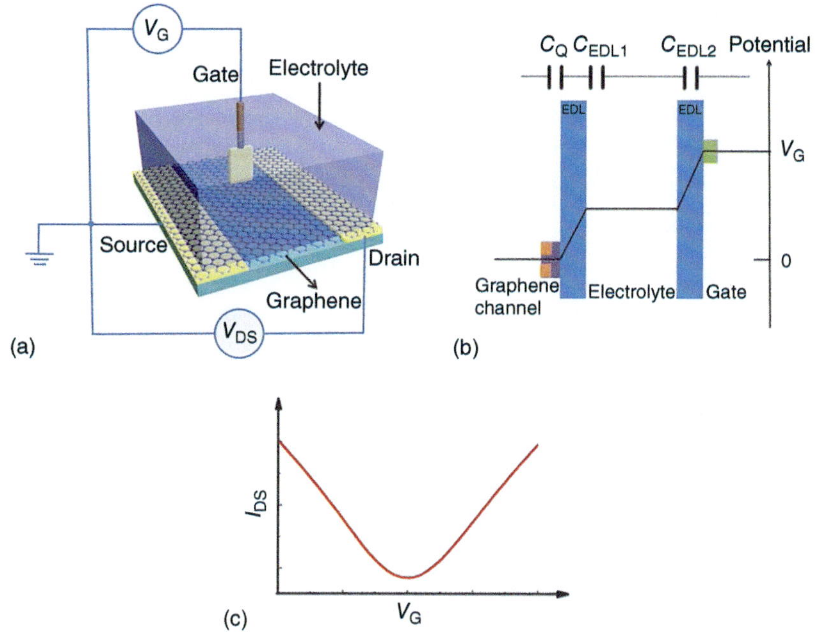

Figure 3.1 (a) Typical configuration of the SG-FET. (b) The potential drop in the SG-FET device under a gate voltage V_G. The blue regions represent the electric double layer (EDL). (c) A typical transfer curve of the SG-FET is characterized at a constant drain–source voltage V_{DS}. Source: Reproduced with permission from Yan et al. [8]/John Wiley & Sons.

application of SG-FETs as biosensors, emphasizing catalytic, and affinity biosensors, from the functionalization of the rGO, application to real samples, the charge screening limitations, and how to overcome it. In Section 3.5, we provide the challenges and perspectives regarding the application of rGO as FET biosensors.

3.2 Graphene, Graphene Oxide, and Reduced Graphene Oxide

GO can be produced through the oxidation of graphite. Its properties are different from those presented for pristine graphene. The most apparent difference between the two materials is the electrical conductivity: graphene has very high conductivity, while GO is an insulator. For most electronics applications, especially in the development of devices such as FETs, the transistor channel material needs to be semiconductor. Thus, it is necessary to transform GO into rGO. rGO presents lower electrical conductivity than pristine graphene. However, this material is quite advantageous in terms of cost and processing. In this section, some GO reduction methods are described.

GO can be produced from the oxidation of graphite in the presence of strong acids, and this process has been used for more than a century, using the methods

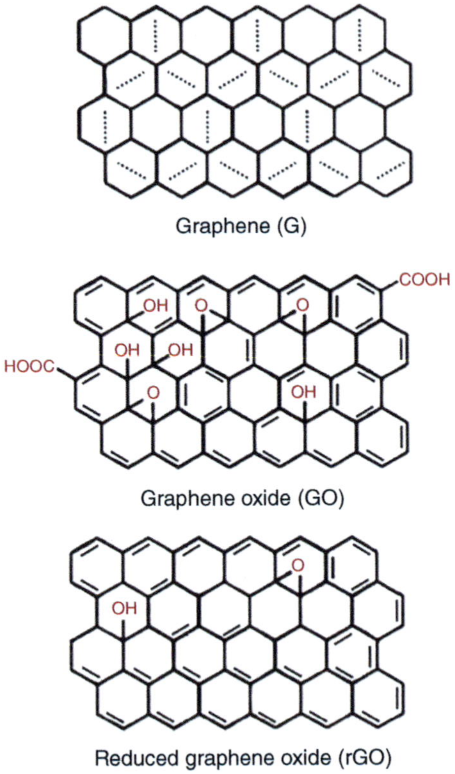

Figure 3.2 Structures of graphene (G), graphene oxide (GO), and reduced graphene oxide (rGO). Source: Tadyszak et al. [15]/MDPI/CC BY 4.0.

obtained by Brodie, Staudenmaier, and Hummers [9–11]. By using simple sonication, it is possible to convert all graphite oxide into GO (producing aqueous colloidal suspensions) [12]. The GO dispersion in an aqueous medium is allowed due to its functionalization with hydroxyl, epoxide, carbonyl, and carboxyl groups appearing from the oxidation process. sp^3 bonds are mostly found in GO sheets, causing a rupture in the crystal structure of graphene, thus changing its properties, being the most evident the optical and electrical ones, with strong absorption at 230 nm and low electronic conduction, with a gap around 3.5 eV [13]. This electrical behavior differs from that found in pristine graphene (bandgap of 0 eV) [14]. Figure 3.2 shows the structures of graphene (G), GO, and rGO.

Finding a way to make GO with properties approaching those of pristine graphene has been the subject of several studies, where the objective is to reduce as many oxygen-containing groups as possible, thus restoring π-conjugated structure, producing the rGO material [5]. Note that rGO must be considered a chemically modified "graphene" or a graphene derivative since its properties are different from those of pristine graphene.

Although it is difficult or unlikely to eliminate all oxygen groups from GO, the inherent properties of GO and rGO can be controlled, aiming at other functionalities where pristine graphene is not viable. For this reason, the use of GO and rGO is of great relevance, such as in applications in biosensors, gas sensors, chemical sensors, supercapacitors, and biological and medical applications [16–19]. It is possible to control the number of sheets in monolayers, with lateral sizes ranging from a few nanometers to a few micrometers [20]. The production cost of GO/rGO is lower compared to graphene production, including CVD and epitaxial growth. There is also the advantage of handling GO/rGO in aqueous solutions, or in some cases, in dispersed solutions containing surfactants, polymers, and organic solvents [21], thus allowing the production of thin films, devices, and composites, which represent the main reasons why GO/rGO remain the protagonists in several electroanalytical devices.

Figure 3.3 shows the conversion of graphite into graphite oxide, graphite oxide into GO and finally, rGO after chemical reduction. The synthesis of GO commonly uses potent oxidizing agents such as $K_2S_2O_8$, P_2O_5, and $KMnO_4$ in the presence of oxidizing acids, e.g. (HNO_3 or H_2SO_4). After oxidation, the dispersion is sonicated in an ultrasonic tip, transforming graphite oxide into GO [22].

GO and rGO differ in several points, visually, GO forms a brown/yellow dispersion, while rGO forms a black dispersion. This difference is due to the decrease in oxygen groups [23]. Another striking feature is the increase in conductivity. Dispersivity is also an important factor. In this context, several methodologies are used to stabilize rGO in aqueous dispersions, such as the use of surfactants, or the possibility that, after being reduced, it be functionalized with other groups, thus maintaining its dispersibility [24].

Figure 3.3 Synthesis of graphite oxide from graphite (indicated for green arrow). Graphite oxide is converted to GO through sonication (yellow arrow). The black dispersion contains rGO after chemical reduction.

There are different methodologies for the reduction of GO, of which chemical, thermal, and electrochemical reduction prevail as the most common. We can also highlight less common methods, such as photoreduction and biological reduction [25]. These different processes produce different materials, which in turn result in different properties. Therefore, the reduction is a key point when working with rGO. The most commonly used methods for reducing GO are described.

3.2.1 Chemical Reduction

The chemical reduction of GO is carried out using a suitable reducing agent at room temperature or under moderate heating conditions. Chemical reduction is generally the technique of choice for producing rGO in large quantities. In 2007, Stankovich et al. proposed the reduction of GO using hydrazine hydrate [26]. Since then, several types of reduction agents have been used to produce rGO. Besides hydrazine monohydrate, the following reagents stand out: dimethylhydrazine, anhydrous hydrazine, sodium borohydride, sodium hydride, lithium aluminum hydride, hydroquinone, ammonium hydroxide, ascorbic acid, and many others [27, 28]. The choice of the best reducer will depend on the application of rGO. One relevant characteristic of chemical reduction beyond the large-scale production and the elimination of most oxygen from GO is the possibility of incorporating other elements during the reduction process, such as nitrogen, boron, and sulfur, among others [29].

The chemical reduction using hydrazine is well known, and different reduction protocols have been extensively tested. Parameters like pH, temperature, GO concentration, hydrazine concentration, and time of reaction are studied. For example, Li and coworkers suggested that residual electrolytes from graphite exfoliation, such as metals, salts, and acids, are critical in the rGO stabilization process, since these residues neutralize the surface charges of the rGO sheets resulting in aggregation [23]. An important factor is the pH of the dispersion. In a highly alkaline medium (pH 10), –COOH side chains remain ionized due to GO reduction with hydrazine. It is suggested that these groups are responsible for keeping the rGO sheets negatively charged, preventing them from agglomerating [30, 31].

Due to the high toxicity and dangers associated with hydrazine reactivity, metal hydrides play an essential role as an alternative for GO reduction. Metal hydrides like sodium hydride, sodium borohydride, and lithium aluminum hydride have been widely used in the literature [32]. Sodium borohydride ($NaBH_4$) is one of those most commonly used reagents. $NaBH_4$ is a salt containing a tetrahedral BH_4^- anion, which readily solubilizes in the aqueous and alcoholic medium. Aldehyde and ketone groups are reduced by $NaBH_4$ to alcohol groups, while carboxylic groups are not [27]. Lee and coworkers first showed that rGO films reduced via $NaBH_4$ presented lower sheet resistance than when reduced with hydrazine, while the films showed similar C–O composition. The higher sheet resistance observed in hydrazine-reduced films is due to the formation of C–N groups that may act as donors, compensating the hole carriers in rGO [33]. Other reagents are used as reducer agents, which can be used at room temperature, concomitantly with other molecules and/ or in more than one processing step [5]. For example, to increase the rGO conductivity, Moon et al. showed that the hydroiodic (HI) acid/acetic acid system has an excellent reduction path because of the high degree of deoxygenation and high graphitization without nitrogen incorporation [34]. The authors showed that air-dried rGO powder pellets presented a conductivity of 30 400 S m^{-1}. This high conductivity was due to the lesser structural defect density and closer restacking between the rGO sheets because of van der Waals forces [34].

The chemical reduction of GO is a simple, fast, and low-cost process capable of providing high yields, at temperatures below 100 °C [25]. Despite the advantages, it is important to highlight some limitations of chemical reduction techniques. It is not possible, for example, to deoxygenate GO completely since the reduction mechanisms are different for different functional groups and reduction agents [5]. Still, there are problems concerning rGO and the reducer itself, which can incorporate into the crystal lattice, limiting or decreasing some properties required of rGO for a given application. With the occurrence of aggregation with the decreased functional groups, alternatives are sought to minimize these effects, in this case, many researchers have been looking for new functionalization methodologies, mitigating this problem, without compromising the electronic properties of the material [24]. As an alternative to maintaining stability in the dispersion, Santos et al. proposed the functionalization of GO with (3-Aminopropyl)triethoxysilane (APTES) as an alternative for stabilizing the sheets in acidic pH and basic medium, thus enabling the deposition of rGO in the form of self-assembled films, without the need for a polymer or surfactant [35].

3.2.2 Thermal Reduction

Thermal reduction provides a safer and easier GO reduction strategy, compared to other techniques, and is very effective in the production of optoelectronic devices [36]. Upon increasing the annealing temperature, the degree of crystallinity increases, consequently increasing the conductivity of the rGO, with mobility near to that of graphene grown by CVD [37]. Liang and colleagues, produced rGO that exhibits an electrical conductivity of 1425 S cm^{-1} using annealing in the presence of acetylene [37].

Thermal reduction methods may be classified into two types, viz., (i) hydrothermal, in which GO is dispersed in water or another solvent and placed in a reactor at medium temperature and high pressure, and (ii) using quartz tube type furnaces, which can be used in high pressure or vacuum, with an inert, reactive, or air atmosphere at temperatures that can exceed 1000 °C [38].

The hydrothermal process is a highly efficient reduction process. When subjected to high pressures, water exceeds its boiling point, in which case a supercritical water temperature becomes an efficient, environmentally friendly reduction method. As in chemical reduction methods, pH plays an essential role in the dispersion of GO when reduced to rGO. The thermal reduction mechanisms can be understood as the dehydration of the hydroxyl, ether, and epoxy groups, occurring either intermolecularly or intramolecularly [39]. Both at acidic and basic pH values, the epoxy groups can be broken down to form hydroxyls, but in different ways. In the case of an acidic medium, the groups containing oxygen are protonated, tending to agglomerate the GO sheets. Under these conditions, two hydroxyl groups present at the edges break up, releasing H_2O and forming an ether-containing structure without restoring the aromatic rings [39]. Under alkaline pH conditions, the leaves are under strong repulsion, and the supercritical water provides the H^+ necessary for the protonation of the hydroxyl groups, which are then eliminated, thus restoring the sp^2 domains of carbon [39].

Zheng et al. varied the reaction time and pH in the hydrothermal process and reported the possibility of partially restoring the sp^2 carbon bonds, in which NaOH was added, showing greater reduction efficiency at pH 11. The reduction effects were evaluated by surface-enhanced Raman scattering (SERS) [39]. Using an alkaline medium (ammonium hydroxide), Tetsuka et al. obtained graphene quantum dots (GQDs) by hydrothermal process [40]. In this reduction process, epoxy groups form primary amines and alcohols by nucleophilic substitution, opening the epoxide ring and functionalizing directly the edges of GO with NH_2 groups, and breaking the sheets down to sizes of c. 1–3 nm. The photoemission wavelength can be controlled by changing only the initial ammonia concentration and the reaction temperature. These parameters modify the average particle size, which is related to the optical properties of GQDs [40]. These are examples of the versatility of the hydrothermal technique, producing different materials with different properties even under similar conditions [40].

Unlike hydrothermal reduction, the thermal reduction of GO is composed of complex phenomena. Due to the energies involved, multiple processes of elimination of oxygen and water absorption occur, resulting in the release of gases such as CO and CO_2, in addition to H_2O. Different studies have been carried out to elucidate the thermodynamical mechanisms involved in these systems and the reduction kinetics [41]. A negative point of this technique is the elimination of carbon atoms, which causes damage to the crystal structure and consequently changing the electronic properties, decreasing the mobility of rGO. Still, these effects can be minimized in an inert and reactive atmosphere [41].

Ishida et al. have shown that the processing of GO in the presence of ethanol at high temperatures (near to 1800 °C), using a solar furnace, produced rGO multilayers by restoring the material structure with crystallinity comparable to CVD-grown graphene [42]. Raman spectroscopy analysis showed a significant decrease in the D band and an increase in the intensity of the 2D band, suggesting changes in the structure of graphene, with high crystallinity, comparable to that of CVD-grown graphene [42]. It is still possible to insert a heteroatom in the rGO crystal lattice. For example, Li and coworkers have shown that it is possible to obtain up to 5% N-doped rGO by thermal annealing of GO in ammonia [43].

As in other techniques, thermal reduction at high temperatures presents some inconveniences, including the self-energy cost in addition to the critical conditions in which the system is placed, occasionally using toxic gases such as ammonia and H_2, which are highly explosive [43]. When samples in the form of films are used, the heating rate must be slow enough to not expand, since gas is released, which can damage the film, making the reduction process time-consuming. Another concern is with regard to the substrate: when used at very high temperatures, substrates with low melting points cannot be used, including some metals, thus limiting the use of the technique [38].

3.2.3 Electrochemical Reduction

Another interesting approach to reducing GO via removal of oxygen groups is the use of electrochemistry techniques in which the reduction processes occur only through the exchange of electrons between GO and the electrode, making the method environmentally friendly. In general, there is no need to use special reagents or organic solvents. Due to the absence of toxic reagents, there is no incorporation of heteroatoms in the graphene structure, as it occurs in the chemical reduction method, and reduction can be made at room temperature or at a low temperature. Also, there is the possibility to reduce GO into rGO films directly on different substrates, such as indium-tin-oxide (ITO), glass, glassy carbon, gold, and polymeric substrates [44].

Different electrochemical techniques such as cyclic voltammetry (CV), linear scan voltammetry (LSV), among others can be used for the reduction of GO [44]. Generally, the reduction of GO is performed in a conventional electrochemical cell containing the electrolyte in a configuration of two or three electrodes. Two possibilities exist for electrochemically reducing GO. The first one is the dispersion of GO in a solution containing electrolytes, which is electrochemically reduced *in situ*, forming a film on the working electrode. The seçond is the deposition of a GO film on the working electrode before applying the potential. Then, the potential is applied relative to the reference electrode, according to the necessary conditions for reduction [44].

Lee et al. have presented a composite electrode of rGO and gold nanoparticle (AuNP) in glassy carbon electrode (GCE) to detect dopamine, ascorbic acid, and uric acid with one-step GO reduction. A GCE electrode was immersed in a buffer solution containing GO and gold(III) chloride solution ($HAuCl_4$). A potential range of −1.5 to 0.8 V (vs. Ag/AgCl) was applied with a scan rate of $10\,mV\,s^{-1}$ for three cycles. This reduction methodology allows a simple and efficient device fabrication method, both in the area of sensors/biosensors and in electronics (supercapacitors) [45].

Our research group has shown that GO can be electrochemically reduced and deposited onto interdigitated gold electrodes. GO was reduced via CV in a conventional three-electrode electrochemical cell containing NaCl (0.5 M) as electrolyte. Figure 3.4a shows the cyclic voltammograms performed in the range of −1.4 to −0.4 V. Only one cycle is needed to reduce the functional groups present in the GO film. With the expansion of the voltammogram (Figure 3.4b) of the first cycle, we can observe two peaks formed at −1.09 and −0.895 V, which indicate that the functional groups were reduced for the formation of rGO on the electrode surface. After reduction, a $I \times V$ curve was obtained by applying a voltage between the contacts of the interdigitated electrodes. The electrodes containing rGO displayed a resistance below

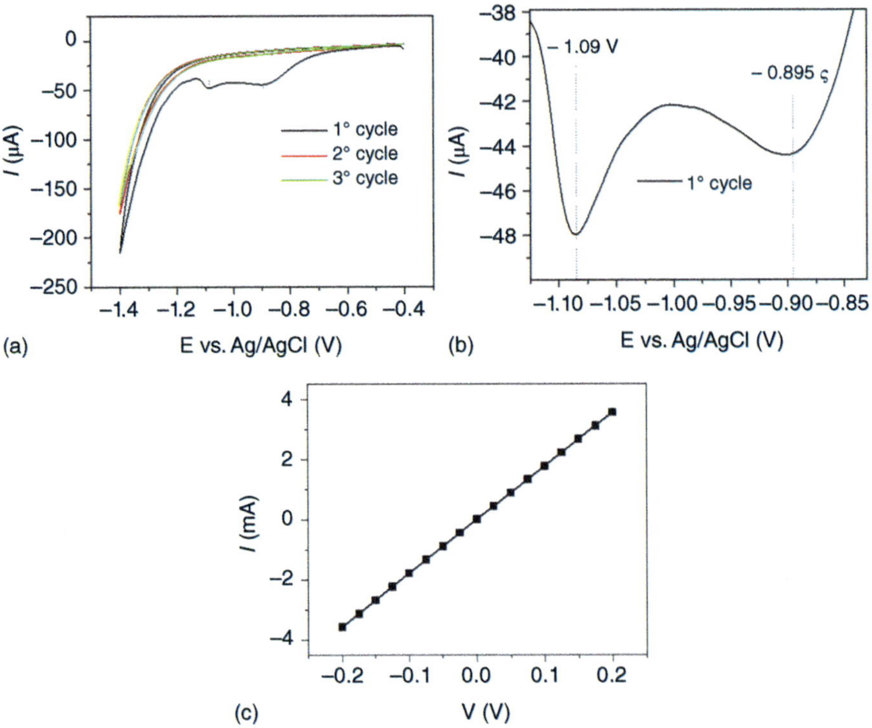

Figure 3.4 (a) Cyclic voltammograms for reducing the GO on interdigitated electrodes. (b) Zoom on the curve of the first voltammogram. Measurement conditions: 0.5 M NaCl, scan rate = 50 mV s^{-1}. (c) $I \times V$ curve performed between the contacts of the interdigitated electrodes after electrochemical reduction of the GO.

100 Ohms. Figure 3.4c shows a representative $I \times V$ curve, where the resistance of 56 Ohms was achieved, validating the results obtained from the reduction by CV.

3.3 rGO-Based Solution-Gated FETs

Because of its high sensitivity, low cost, and simple large-scale manufacturing process, rGO is a highly versatile material in the production of SG-FETs. The remaining functional groups derived from the GO reduction allow rGO to be easily dispersed in several different solvents, such as water. This specific characteristic facilitates the dispersion of rGO on many distinct substrates and surfaces. Furthermore, another advantage regarding the functional groups is their use to immobilize biomolecules (DNA, enzymes, antibodies, antigens, etc.). This allows SG-FETs to be applied in many different bioanalytical areas to detect a great range of analytes and with selectivity [35].

The first rGO SG-FET was reported by Sohn et al. for pH sensing [46]. Since then, remarkable progress has been made using different strategies for FET manufacturing according to the final application.

3.3.1 Manufacturing Strategies

When it comes to the manufacturing process of SG-FET devices for bioanalytical applications, different strategies can be adopted. From the chosen substrate for rGO deposition, the deposition of the electrical contacts (source and drain) above or underneath the rGO layer, and the reduction of GO in solution or directly on the substrate. Every choice will depend on the needs and objectives of each SG-FET. The substrates chosen for the FET fabrication can greatly impact the device's performance. The most common substrate employed for rGO FET-based biosensors is SiO_2/Si, which is one of the most common substrates used in electronic devices because of their semiconductor properties and low cost [47]. Another common type of substrate is glass that in addition to being low-cost is transparent, electrical insulator, and thermally stable substrate, similar that closely resembles Si substrates [48].

Flexible electronics represents one of the recent advances in analytical devices due to the possibility of developing wearable sensors and biosensors. For this reason, rGO FETs have been manufactured on substrates such as polyethylene terephthalate (PET), polyethylene naphthalate (PEN), and others. As shown in Furlan and colleagues' study, PET substrates are flexible, durable, and transparent, which makes it suitable for wearable electronics devices [48]. Similar to PET, but with an improved oxygen barrier property, PEN substrates are also chemically and physically stable, making them an important substrate for wearable electronic devices [49].

Another type of substrate used for FETs deposition are the PCB (printed circuit board) substrates, which represent a relatively new technology that can be applied in the SG-FETs manufacturing process. The PCB substrate is already well established in the electronics field, making it extremely compatible with different types of technology and allowing a vast array of applications [50].

The first manufacturing method implemented for rGO SG-FET fabrication was developed by Sohn and coworkers, using rGO films. The chosen substrate, a SiO_2 (100 nm)/Si wafer, was properly cleaned and immersed in an APTES solution, resulting in the formation of amino groups SiO_2. After rinsing the substrate, a GO dispersion was deposited on top of it and left for one hour. The substrate was then cleaned again for the removing the excess of GO. The GO film was reduced through hydrazine monohydrate vapor at 60 °C for a total of 12 hours and the resulting rGO film was annealed for 2 hours in an argon atmosphere. As the final step, the Au/Cr drain–source electrodes were deposited on top of the rGO film by thermal evaporator using a shadow mask to obtain a channel length of 400 μm with a width/length (W/L) ratio of 20 [46]. Kim and colleagues used a similar strategy, that is, the deposition of GO, via immersion in a 1 $mg\,ml^{-1}$ GO dispersion on a suitable substrate

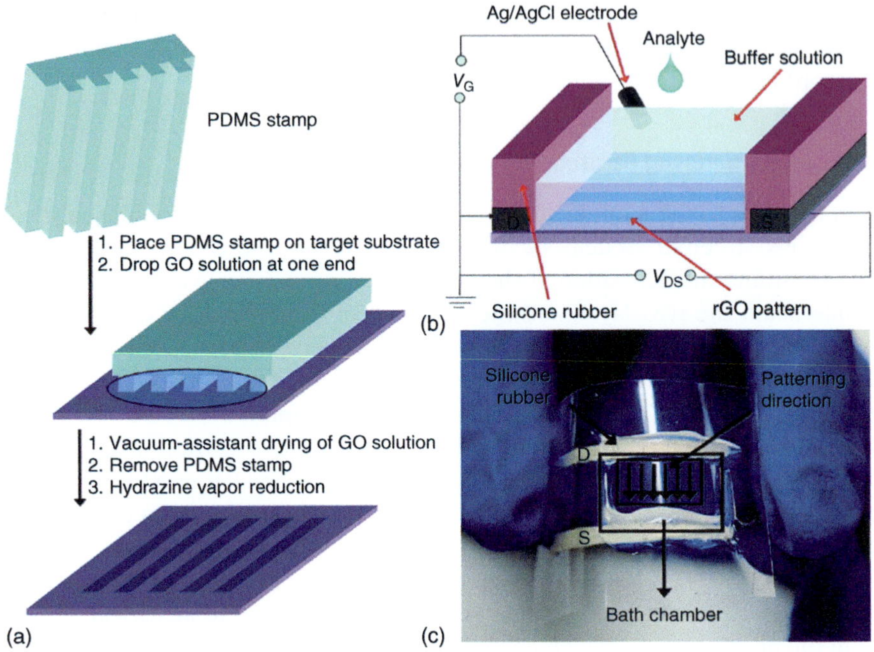

Figure 3.5 (a) Schematic illustration of patterning GO films by MIMIC and reduction of the patterned GO films. (b) Schematic illustration of the experimental setup of front-gate FET for sensing application. (c) Photograph of rGO pattern-based sensing device on PET film. Source: Reproduced with permission from He et al. [50]/American Chemical Society.

(glass) for four hours, followed by its reduction using hydrazine vapor, and subsequent deposition of the Ti/Au source and drain contacts. The resulting SG-FET system was applied as an electrical aptamer sensor for the detection of anthrax toxin at the attomolar (aM) level [51].

A viable alternative to the metallic deposition process, which can be quite expensive, is the use of conductive paint, usually made of silver (Ag). In He's paper, silver conductive paint was applied to prepare the source and drain electrodes using different substrates (SiO_2, quartz, and PET film) [50]. The SG-FET was fabricated as follows: The chosen substrate was modified with APTES. GO aqueous solution was then deposited through a MIMIC (micromolding in capillary) process, setting the GO patterns to 10 μm width using a PDMS (polydimethylsiloxane) stamp. The GO patterns were then reduced through a hydrazine vapor method. Following, the Ag source and drain electrodes were deposited using silver paint that covered the opposite sites of rGO patterns on the substrate. Figure 3.5 shows the scheme of the MIMIC used by He et al. for micropatterns of rGO films and SG-FETs [50].

Papamatthaiou et al. fabricated an SG-FET (referred to as electrolyte-gated FET) on a commercially PCB in a planar layout, as shown in Figure 3.6. This study is a good example of how metal deposition for source and drain contacts can be avoided. The authors first deposited the source and drain electrode alongside the reference electrode in the core board substrate and then cleaned it with acetone, isopropanol,

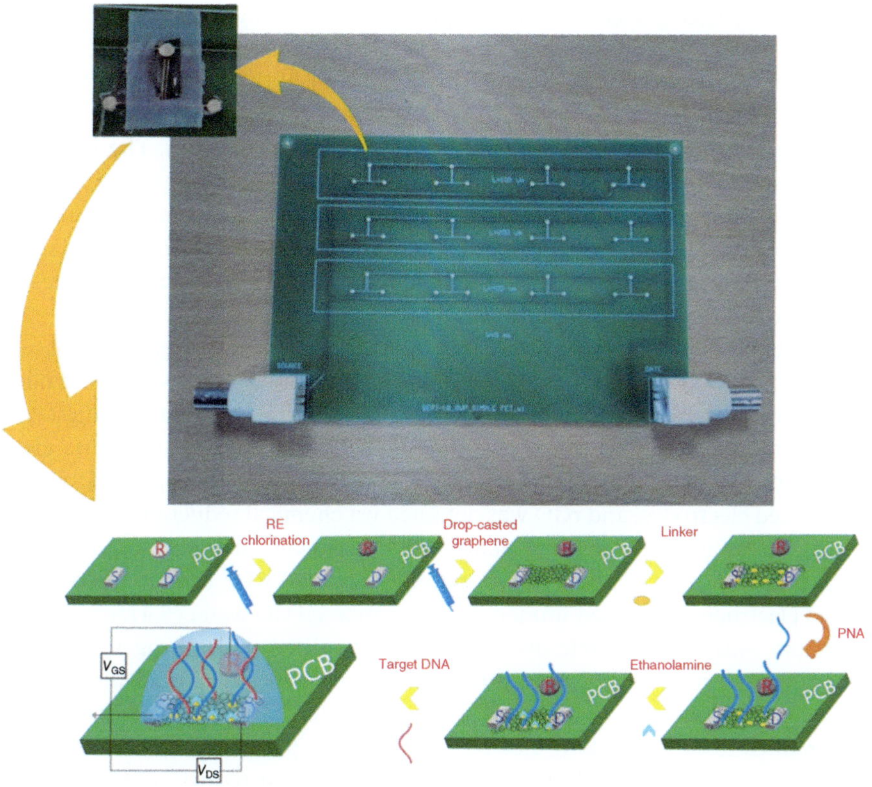

Figure 3.6 Top: Photograph of the printed circuit board (PCB) designed in Altium for evaluation of graphene as the channel material of electrolyte-gated FET. In magnification, one of the devices where the drain, source, gate electrodes are visible along with the drop-casted graphene channel. Self-adhesive feflon tape was attached to confine the electrolyte droplet. Bottom: Schematic of assay steps for PNA-DNA hybridization detection employing an electrolyte-gated graphene field-effect transistor. Source: Reproduced from Papamatthaiou et al. [52], Springer Nature/CC BY 4.0.

and distilled water. Following, the chosen silver PCB reference electrodes suffered the chlorination process, indicated by the darkening of the silver to a dark grey color, by drop casting 10 μl of sodium hypochlorite solution at 80 °C. The graphene ink solution (990 ± 130 nm flake length) was first sonicated, followed by the two drop-casting processes, which deposited 10 μl of graphene ink on the channel. Finally, the chip was dried at 80 °C [52].

The deposition of GO or rGO on the pre-patterned metallic contacts is more adequate than the deposition of these contacts on the rGO. In Chae and coworkers' research, pre-patterned gold electrodes were used to manufacture SG-FETs. In their study, 100 nm gold(Au) electrodes were patterned on the SiO_2 wafer using an e-beam evaporator. The Au-patterned SiO_2 substrates were cleaned using a Piranha solution ($H_2SO_4 : H_2O_2 = 3 : 1$) and immersed in 5% 3-(ethoxydimethylsilyl)propylamine diluted in ethanol. This procedure introduces amine groups in the substrate that aid in the further adhesion of GO sheets. Then, 5 ml of GO solution was deposited on the substrate by was spin coating. The deposited GO then was reduced using a HI acid vapor at 80 °C for three hours [53].

In another study, Cai et al. also demonstrated the use of the pre-patterned source and drain electrodes. The 50 nm Au electrodes were fabricated on a SiO_2 (285 nm)/Si substrate using photolithography and electron-beam evaporation. The rGO dispersion was prepared by adding 15 mg of GO to 10 ml of hydrazine solution. This solution was drop casted on top of the Au patterned electrodes and annealed at 150 °C. To obtain a few-layer rGO, the electrodes were immersed in a Piranha solution (7 : 3 v/v) and sonicated for 30 seconds, followed by a last cleanse using DI water and a drying process using nitrogen [54].

Interdigitated electrodes are a good choice for SG-FETs because their manufacturing process is already very well established, and a great variety of low-cost electrodes are commercialized. Aspermair et al. used commercially available interdigitated electrodes for the manufacture of rGO SG-FETs. After proper cleaning, the electrodes were modified with APTES. A GO dispersion was then deposited on the interdigitated electrodes, and rGO was obtained via chemical reduction with hydrazine [55]. Furlan and colleagues standardized interdigitated electrodes on different substrates (Si/SiO_2, glass, and PET) to manufacture SG-FETs. The modified interdigitated electrodes were then immersed in commercial GO water (0.4 mg ml^{-1}) for three hours, washed in double-distilled water, and dried with nitrogen. GO was transformed into rGO via electrochemical reduction by CV. In this case, the electrochemical configuration of two electrodes was used: interdigitated electrodes were short circuited to act as working electrodes, and an Au plate was used as a counter (CE)/reference (RE) single electrode. The electrochemical reduction experiments were performed in double-distilled water, and a few (1–4) voltammetric cycles were needed to GO reduction [48].

Vasilijevíc and coworkers have applied a different strategy for the GO deposition, in which a GO was inkjet-printed directly onto the Si substrate [56]. For this process, a GO dispersion (1 mg ml^{-1} GO in distilled water) was prepared to formulate de ink according to the Dimatix printer's rheological parameters (50% GO, 20% ethylene glycol, and 30% *n*-propanol). Then, the prepared ink was inkjet-printed on top of the interdigitated Au source and drain electrodes at 40 °C. After the ink deposition, the GO was electrochemically reduced at −1.3 V vs. Ag/AgCl [56].

According to Santos et al., another rGO SG-FET manufacturing method is the layer-by-layer (LbL) assembly of rGO sheets with opposite charges on interdigitated electrodes. The strategy consisted of using a modified Hummers method to obtain a GO aqueous dispersion followed by the modified Umrao method to acquire the negatively charged rGO (rGO-N) [35]. An APTES covalent bond was used to obtain the positively charged rGO (rGO-P), followed by a chemical reduction proceed using $NaBH_4$ [35]. To assemble the rGO-P/rGO-N sheets, the substrates were hydrophilized with a KOH solution (2%) for a total of five minutes and then cleaned with ethanol and ultrapure water. The Au interdigitated electrodes were manufactured using photolithography and contained 30 pairs of Cr/Au fingers (5 mm length, 50 μm width and 150 nm thickness, 20 nm of Cr, and 130 nm of Au). For the

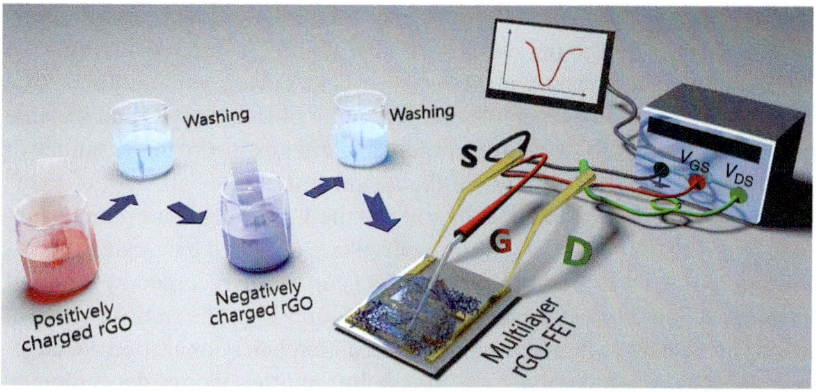

Figure 3.7 Schematic representation of the manufacturing method of rGO-SG-FETs by LbL. On the left, the interdigitated electrodes are alternately immersed in rGO dispersions, with intermediate washing steps. A schematic of an electrode during an electrical measurement using a computer-controlled source measure unit is shown on the right. Source: Reproduced with permission from Santos et al. [35]/Elsevier.

deposition processes, electrodes were first immersed in the positively charged rGO dispersion for one hour, forming the first layer, and then immersed in the negatively charged rGO solution for 15 minutes, forming the second layer. This process was repeated to achieve the number of layers required. Finally, rGO-P/rGO-N LbL films were dried at 150 °C for three hours. Figure 3.7 represents the manufacturing process utilized by Santos and colleagues, as well as the electrical measuring equipment used, i.e. a computer-controlled source measure unit [35].

As we have seen, there are several strategies for manufacturing rGO SG-FETs. Each one has its peculiarities, advantages, and disadvantages in terms of cost, ease of processing, involving simple steps, or a little more sophisticated. In general, all methods of making rGO SG-FETs are more straightforward than those involving pristine (mechanically exfoliated) or CVD-grown graphene. The fabrication of SG-FETs using graphene CVD involves several fabrication steps. The different manufacturing methods mentioned here show the versatility of using rGO in SG-FETs. This material is an excellent alternative for the manufacture of SG-FETs applicable as biosensors.

3.4 Applications of rGO SG-FETs as Biosensors

The possibility of monitoring a detection event using different parameters represents a great advantage of SG-FETs when applied as biosensors. In general, in a detection event, these biosensors are characterized by electrical measurements including transfer curves ($I_{DS} \times V_{GS}$ with constant V_{DS}), output curves ($I_{DS} \times V_{DS}$ with constant V_{GS}), and I_{DS} current variations over time (with V_{GS} voltages with constant V_{DS}) [57]. Figure 3.8 illustrates the most used analytical parameters in rGO

SG-FET biosensors. Analyte detection generally occurs by exposing the properly functionalized rGO SG-FET to the sample; usually, a buffer solution containing the analyte, biological sample (blood, plasma, saliva, and urine), among others. The detection of the interaction between the receptor immobilized on rGO and the analyte occurs by comparing the $I \times V$ curves before and after exposure to the sample or in real-time by analyzing the dynamics of the I_{DS} current.

$I_{DS} \times V_{GS}$ transfer curves are obtained by varying the V_{GS} voltage for different constant values of the V_{DS}. The $I_{DS} \times V_{DS}$ output curves are obtained by varying the V_{DS} voltage for different fixed values of V_{GS}. The $I_{DS} \times t$ curves are recorded by monitoring the current as a function of time at constant V_{GS} and V_{DS}. The transfer curve of graphene or rGO FET exhibits an ambipolar conduction behavior around a charge-neutrality point (CNP), also known as a Dirac point. In other words, depending on the voltage applied to the gate, there is conduction by positive or negative charges and a situation where the density of the two carriers equals [3]. The changes in the curves shown in Figure 3.8 depend on the effect generated in the rGO after the receptor–analyte interaction. In general, a chemical/biochemical reaction is responsible for changing the electric field close to the rGO and generating modulating the quantity/type of carriers, causing displacements in the CNP voltage value, variations in the slope of the output curve, and the dynamics of I_{DS} [8].

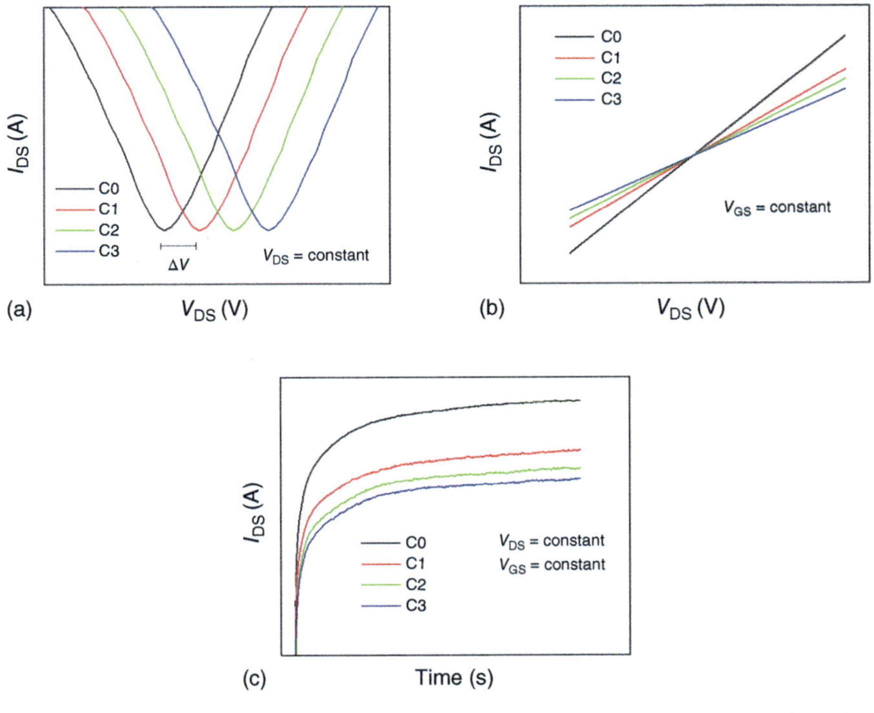

Figure 3.8 Idealized curves mot utilized as analytical parameters in biosensors based on rGO SG-FETs (a) $I_{DS} \times V_{GS}$ transfer curves. (b) $I_{DS} \times V_{DS}$ output curves. (c) $I_{DS} \times$ *time* curves at constant V_{GS} and V_{DS} voltages.

3.4.1 rGO Functionalization

The use of rGO has attracted significant interest in recent years for developing SG-FET biosensors as an alternative to the use of pristine graphene, especially due to its flat structure with high-surface area, biocompatibility, flexibility, high electrical conductivity, wide electrochemical potential window, and ease processing [58]. rGO can be functionalized with other materials, such as metallic nanoparticles, which promote the reduction of graphene network defects and increase the presence of functional groups [59, 60].

Since rGO contains some functional groups available for biomolecules immobilization, since GO reduction is not total, it is possible to improve its interaction with biological recognition materials (enzymes, antigens, antibodies, and DNA) from π–π and van der Waals interactions [61], allowing a more selective detection.

One of the most common strategies for the functionalization of rGO is the modification with 1-pyrenebutyric acid succinimidyl ester (PBSE). PBSE is a bifunctional agent that interacts with rGO via π–π interactions of the pyrene group [62]. The other end of PBSE contains NHS that binds to the amine groups of the antibody [62]. Therefore they are easy to direct into conjugation with other reactive groups using acylation or alkylation [63].

As already mentioned, rGO can be functionalized with metallic nanoparticles allowing the further immobilization of biomolecules capable of binding to the nanoparticles. For example, single-stranded DNA modified with thiol group can be ease immobilized on rGO/AuNPs. The latter was recently shown by Li et al. to detect a specific RNA biomarker of SARS-CoV-2, which tends to appear in the initial stage of the disease [59]. The authors used phosphorodiamidate morpholino oligos (PMO) probes, a DNA analog but uncharged, that exhibits high flexibility and long-term activity, which are exciting features for the rapid and unamplified identification of RNA sequences. PMO modified with thiol group was immobilized on AuNPs/rGO in one step [59]. Immobilization of antibodies can also be done using nanoparticles. Thakur et al. used specific *Escherichia coli* antibodies anchored on AuNPs. AuNPs were deposited in the rGO SEG-FET by sputtering. The next step involved the modification of the AuNPs with glutathione, activating the carboxyl groups with NHS/EDC, and finally, the immobilization of *E. coli* antibodies [64].

Finally, LbL for immobilization of biomolecules can be mentioned. A versatile and simple technique that is based on the alternating electrostatic adsorption of negatively/positively charged elements [65]. Piccinini et al. developed rGO SG-FETs for selective detection of urea in aqueous media [66]. Polyethylenimine (PEI) was used as polycation and urease enzyme as polyanion for LbL assembly. rGO SG-FETs were incubated first in PEI solution for 10 minutes and rinsed with deionized water. Next, PEI-modified rGO was incubated in urease solution for 30 minutes, followed by rinsing with deionized water, forming a PEI/urease bilayer [66]. The cycle can be repeated for the construction of a desired number of bilayers. LbL assembly for noncovalent immobilization of the enzymes provided accessibility to the active sites and greater biological activity, besides affecting less the sp^2 structure of rGO. At the same time, the use of PEI/urease bilayers favored a more significant increment of device surface coverage and better response for the detection of urea [66].

3.4.2 Enzymatic Biosensors

The use of nanomaterials is very stimulating in the development of enzymatic biosensors because they improve the electrocatalytic capability of the system. The use of rGO SG-FET may be an excellent alternative to conventional enzyme-modified FETs because it presents excellent biocompatibility and high electronic transport, which aid and facilitate the rapid detection of analytes. Despite the importance of these systems as analytical tools, there are only a few reports on rGO SG-FETs enzymatic biosensors. Piccinini et al. have developed an rGO SG-FET enzymatic biosensor modified with the PEI/urease LbL films for the detection of urea in aqueous media [66]. Figure 3.9a shows the transfer curves in the absence (blue line) and in the presence of urea (red line). The shift of the transfer curve to more positive values for urea indicates that urease catalyzed the hydrolysis of urea, producing NH_4^+, HCO_2^-, and OH^-, increasing the local pH value. Figure 3.9b,c show the I_{DS} vs. *time* response obtained upon increasing urea concentrations. According to the authors, a substantial improvement of the urea response was found for the PEI/urease films containing two and three bilayers [66]. The PEI/urease biosensor has shown excellent stability after six days of storage, 96% of the initial response, which is required for validation and application of biosensors in real samples of clinical interest, such as hemodialysis, for example [66].

In another study using the LbL technique, Berninger et al. developed an rGO-FET for ultrasensitive detection of arginine using cascade reactions of the enzymes arginase and urease [67]. In biosensors with enzyme cascades, the product of the first enzyme reaction is immediately consumed by another enzyme, thus favoring a greater sensitivity for the system [67]. In the first reaction phase, arginase hydrolyzes L-arginine to produce ornithine and urea. The urea then served as a substrate

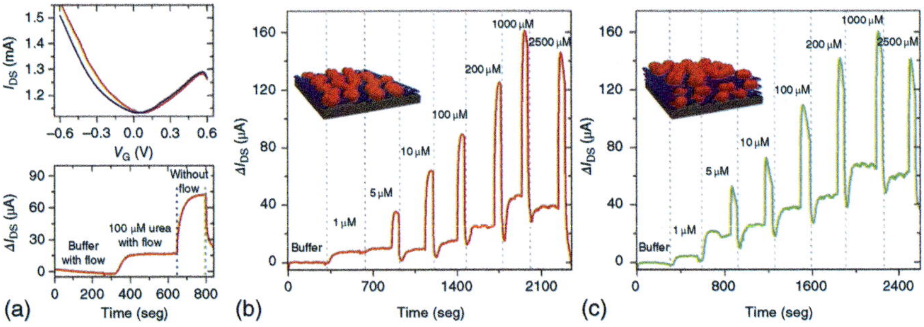

Figure 3.9 (a) Transfer characteristics of a (PEI/Urease)1 (one bilayer) assembly in the absence (blue line) and in the presence (red line) of 100 μM urea (top), and the change in the channel current with and without flow (bottom). Channel current response (ΔI_{DS}) monitored in real-time for (PEI/Urease)1 (b) and (PEI/Urease)3 (three bilayers) (c) LbL assemblies onto rGO FETs at different concentrations of urea. The marked line indicates the elution of the next concentration of urea. Experimental conditions: 10 mM KCl and 0.1 mM HEPES at pH 6, V_{GS} = −0.2 V and V_{DS} = 0.1 V. (For interpretation of the references to color in this figure legend, the reader is referred to the web version of the article.) Source: Reproduced with permission from Piccinini et al. [66]/Elsevier.

for the second enzymatic reaction, which was converted into NH_3 and CO^2, leading to a change in the pH of the rGO-FET system [67].

Chae et al. have developed an enzyme-modified rGO SG-FET for the evaluation of Alzheimer's disease (AD) progression and treatment [53]. The device consisted in the detection of pH changes, in which ions (H^+) are generated from the enzymatic reaction between acetylcholinesterase (AChE) and acetylcholine (ACh). Alterations in V_{Dirac} values were noted with increasing ACh concentration, thus showing a sensitivity of 13.9 mV dec^{-1}, which is interesting for therapeutic uses [53].

3.4.3 Affinity Biosensors

The use of affinity biosensors (restricted here as immunosensors and DNA sensors) has grown astonishingly in recent years because of their notorious applications in areas such as clinical monitoring, pharmaceuticals, and disease diagnostics. The use of antibodies and DNA in conjunction with rGO in SG-FET configuration has significantly advanced this field.

rGO also allows the development of biosensors with better performance for the diagnosis and treatment of diseases by using aptamers. This characteristic is due to the noncovalent adsorption process between the pyramidic bases of RNA or DNA aptamers and the rGO plane, caused by the stacking π–π interaction [68]. In this perspective, Aspermair et al. proposed an rGO-FET with a modified surface for detection of the E7 protein of HPV-16 (human papillomavirus), in a linear detection range of 30 and 1000 nM with a LOD of 1.75 nM [55]. It is worth mentioning that after RNA aptamer immobilization, the Dirac point was stabilized at 130 ± 10 mV, which is relatively low and interesting because it does not affect the electrical transport properties of rGO and the sensitivity of the biosensor [55]. Majd et al. demonstrated a flexible FET aptasensor modified with multiwalled carbon nanotubes (MWCNTs) for the detection of ovarian cancer biomarker (CA125) [69]. Anti-CA125 ssDNA aptamer forms enveloped in the nanotubes and stacked on the surface of the rGO-FET via π–π interactions, increasing the dI/dV from 1.04 to 2.82 μS [69]. The proposed aptasensor allowed detection of CA125, which was applied in real samples for medical diagnosis [69].

Yu et al. proposed an rGO-FET immunosensor for the label-free quantification of exosomes [63]. The Dirac point changes demonstrated that the biosensor modification steps occurred effectively, which was also observed with increasing exosome concentration. The LOD for the system was found to be 33 particles μl^{-1}, being possible to be applied in the analysis of clinical samples. In addition, the label-free quantification reduces the time and expenses with reagents, thus simplifying the whole process of assembling the biosensor [63].

rGO-FETs can also be used in point-of-care technologies, which favor the accurate and on-site rapid diagnosis of diseases. An exciting example is demonstrated by Chen et al. who proposed an rGO immunosensor for the rapid and precise detection of Ebola antigens (Ebola glycoprotein, EGP) in phosphate-buffered saline (PBS), human serum, and plasma [70]. The use of AuNPs deposited on the rGO channel favored the conjugation of antibodies increasing their bioactivity and potentially resulting in a more sensitive system. Figure 3.10 shows the dynamic response (I_{DS} vs. *time*) of this biosensor toward EGP in diluted PBS, human serum, and human plasma. Note that the biosensor had an almost instantaneous response to EGP (a few seconds). The device showed potential for EGP detection up to 1.0 ng ml^{-1} which makes it feasible for early stage disease identification [70].

The COVID-19 (Coronavirus Disease of 2019) pandemic propelled the development of a fantastic array of biosensors for the rapid detection of the SARS-CoV-2 virus [41]. On using rGO SG-FETs as a biosensor for COVID-19 identification, Li et al. developed an AuNP-modified rGO SG-FET and with PMO as capture probes, replacing single strands of DNA. The so-called PMO-functionalized G-FET nanosensor (Figure 3.12) allowed an accurate and unamplified analysis of the patient's samples within two minutes, making it very advantageous, compared to other methods for COVID-19 detection, such as genomic sequencing and CRISPR-based tests [59]. Tests with real samples have shown excellent agreement with RT-PCT tests (Kappa index = 0.92), and the biosensor could still distinguish between SARS-CoV-2 and SARS-CoV genes. Figure 3.11 shows the schematic diagram of the test for the identification of COVID-19 using the rGO SG-GET developed by Li et al. [59].

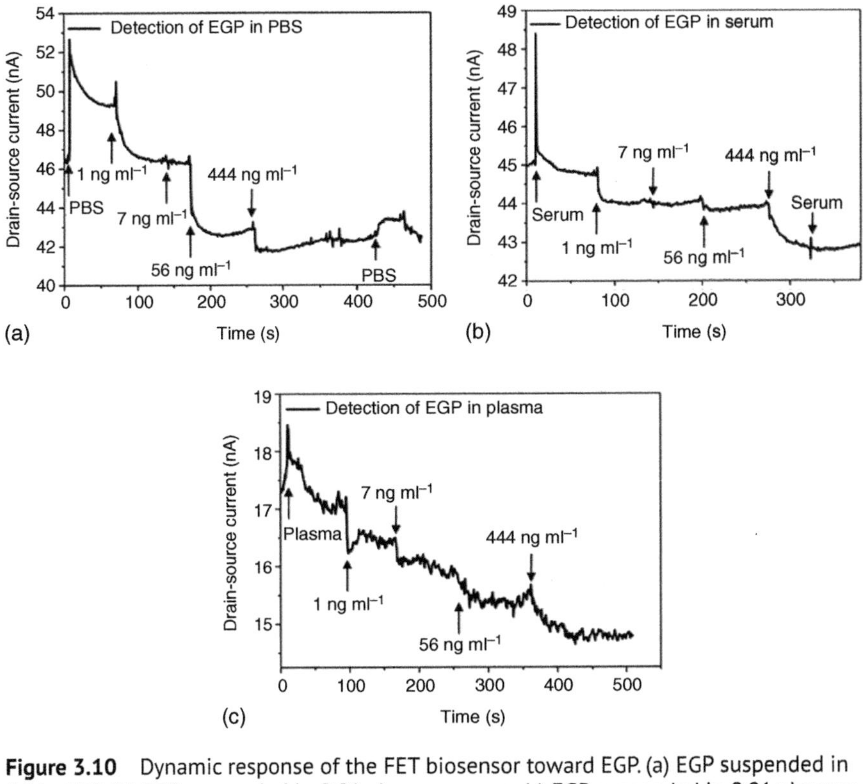

Figure 3.10 Dynamic response of the FET biosensor toward EGP. (a) EGP suspended in 0.01 × PBS. (b) EGP suspended in 0.01 × human serum. (c) EGP suspended in 0.01 × human plasma. The sensor's response to EGP is instantaneous. For (a–c), V_{DS} was fixed at 0.01 V. Source: Chen et al. [70]/Springer Nature/Public Domain.

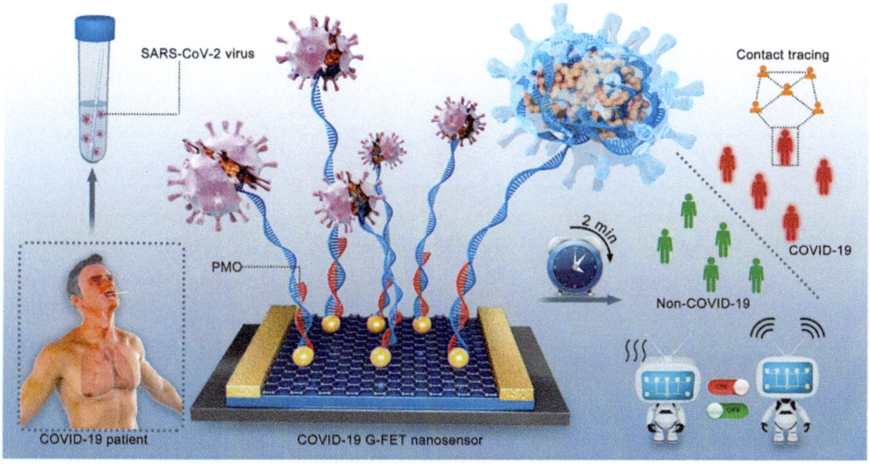

Figure 3.11 Schematic diagram of PCR-free rapid direct identification of COVID-19 using the PMO-functionalized G-FET nanosensors. Source: Reproduced with permission from Li et al. [59]/Elsevier.

3.4.4 Debye Length Screening and How to Overcome It

The development of SG-FET biosensors has shown to be of interest in medical and analytical diagnostics due to their excellent sensitivity, small size, and application as point-of-care devices. Despite these advantages, some points still deserve attention, such as the charge screening in biomolecular detection in SG-FETs applied as affinity biosensors. Because only the potential variations occurring within the EDL or Debye length (λ_D) from the surface of the FET can be detected by these devices, the counterions in the solution may create an electrostatic screening, which directly affects the development of useful platforms [72].

Most SG-FETs applied as affinity biosensors are engineered to perform the detection tests in solutions with low ionic strength, since λ_D increases considerably when the ionic strength decreases [73]. The latter represents a way to increase the Debye length and consequently create a detectable signal in SG-FETs. For example, Chen et al. performed rapid detection of Ebola antigen from PBS/serum/plasma dilution. For the dilution in 0.01 × PBS, the λ_D corresponded to about 7.4 nm. The AuNPs used were less than 5 nm in diameter, and the sizes of the Ebola antibodies and PGE (EGP) were 3–10 nm, as estimated by the supplier. The sum of all components lies within the Debye length. Thus, this SG-FETs biosensor exhibited high sensitivity and selectivity for PGE, with a detection limit of up to 1 ng ml^{-1}.

Other interesting strategies can be utilized to overcome the charge screening limitations, like the improvement of the structure and organization of the materials and reduction of the size of the recognition biomolecule. For example, the use of antibody fragments and aptamers has been proven to create a detectable signal in other nanosized FETs [74, 75]. Regarding rGO SG-FETs, Piccinini et al. developed an rGO-FET sensors using polyelectrolyte multilayers (PEMs), leading to significant improvement of the sensitivity limit of the device [71]. In the study,

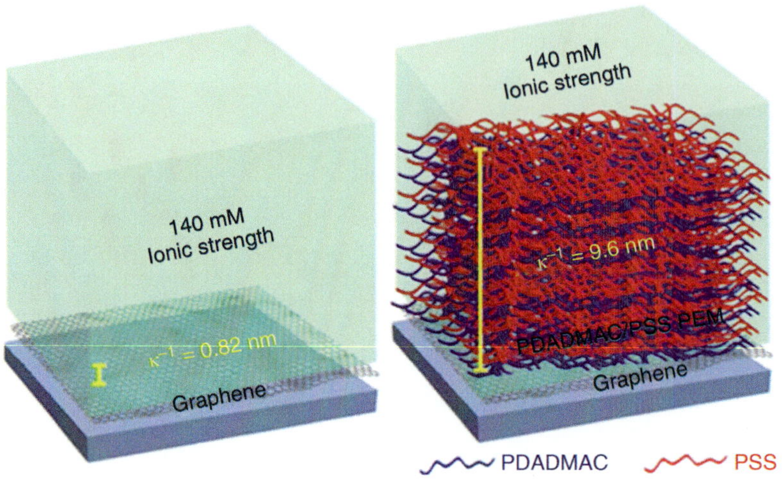

Figure 3.12 Representation of the Debye screening length (κ^{-1}) at a graphene interface for a 140 mM ionic strength solution (i.e. physiological ionic strength) without (Left) and with (Right) a PDADMAC/PSS polyelectrolyte multilayer (PEM) film. Source: Reproduced with permission from Piccinini et al. [71]/American Chemical Society.

the authors showed that poly(diallyldimethylammonium chloride)/poly(sodium 4-styrenesulfonate) (PDADMAC/PSS) PEMs are capable of overcoming the Debye screening in rGO SG-FETs. This was possible by estimating the correlation of the potential change at rGO with the thickness of the PDADMAC/PSS film, the electrostatic screening length, and the mobile ion concentration within the films, causing the Debye length to increase by more than one order of magnitude, as shown in Figure 3.12 [71]. The range of these points can be described by the entropy loss when molecules are trapped. This means that the electrostatic screening within the layers depends solely on the density of the polymer used and ionic strength of the solution. In this manner, it was possible to extend the Debye length from 0.8 to 10 nm, leading to the detection of small molecules and biomacromolecules such as DNA and proteins [71].

New strategies are emerging to overcome the Debye length limitations in SG-FETs [76]. For instance, Hwang et al. showed that the change from flat to crumpled CVD graphene allows the direct detection of nucleic acid sequences in buffer and human serum samples down to 600 zM and 20 aM [77]. Using computational simulations, the authors showed that the nanoscale deformations form "electrical hot spots" in the crumpled graphene, which will reduce the charge screening at the concave regions [77].

3.5 Final Remarks and Challenges

The use of GO and its derivatives in the biomedical fields has become attractive and very promising in the last few years since these materials can efficiently enhance the performance of analytical devices by improving sensitivity and selectivity, especially for biosensors in the SG-FET configuration. These devices are fast, low-cost, portable, and reliable bioanalytical sensors with the potential for large-scale manufacturing, to be used, for example, to screen the population in a pandemic situation. However, significant challenges persist regarding the large-scale production of these bioanalytical devices. Among the main limitations regarding the use of GO-based biosensors are the translational aspects and scale-up of the manufacturing methods, as well as the limitations imposed by the Debye length. We have shown that several studies have recently reported new strategies to overcome such limitations, including the use of reduced GO. Therefore, we believe that in the near future these biosensors will be available for real-applications on an industrial scale.

Acknowledgments

The authors are grateful for the financial support provided by CAPES (Finance Code 001 and Epidemias 88887.504861/2020-00), São Paulo Research Foundation (FAPESP: 18/07508-3 and 2017/21097-3), and CNPq (303338/2019-9).

References

1 Novoselov, K.S., Novoselov, K.S., Geim, A.K. et al. (2004). Electric field effect in atomically thin carbon films. *Science* 306 (5696): 666–669.

2 Zhan, B., Li, C., Yang, J. et al. (2014). Graphene field-effect transistor and its application for electronic sensing. *Small* 10 (20): 4042–4065.

3 Vieira, N.C.S., Borme, J., Machado, G. Jr. et al. (2016). Graphene field-effect transistor array with integrated electrolytic gates scaled to 200 mm. *J. Phys. Condens. Matter* 28 (8): https://doi.org/10.1088/0953-8984/28/8/085302.

4 Dreyer, D.R., Todd, A.D., Bielawski, C.W. et al. (2010). The chemistry of graphene oxide. *Chem. Soc. Rev.* 39 (1): 228–240.

5 Pei, S. and Cheng, H.-M. (2012). The reduction of graphene oxide. *Carbon* 50 (9): 3210–3228. https://doi.org/10.1016/j.carbon.2011.11.010.

6 Konios, D., Petridis, C., Kakavelakis, G. et al. (2015). Reduced graphene oxide micromesh electrodes for large area, flexible, organic photovoltaic devices. *Adv. Funct. Mater.* 25 (15): 2213–2221.

7 Wöbkenberg, P.H., Eda, G., Leem, D.-S. et al. (2011). Reduced graphene oxide electrodes for large area organic electronics. *Adv. Mater.* 23 (13): 1558–1562.

8 Yan, F., Zhang, M., and Li, J. (2014). Solution-gated graphene transistors for chemical and biological sensors. *Adv. Healthcare Mater.* 3 (3): 313–331.

9 Brodie, B. (1860). Sur le poids atomique du graphite. *Ann. Chim. Phys.* 59: 466–472.

10 Hummers, W.S. and Offeman, R.E. (1958). Preparation of graphitic oxide. *J. Am. Chem. Soc.* 80 (6): 1339. https://doi.org/10.1021/ja01539a017.

11 Staudenmaier, L. (1898). Verfahren zur Darstellung der Graphitsäure. *Ber. Dtsch. Chem. Ges.* 31 (2): 1481–1487. https://doi.org/10.1002/cber.18980310237.

12 Ye, S. and Feng, J. (2016). The effect of sonication treatment of graphene oxide on the mechanical properties of the assembled films. *RSC Adv.* 6 (46): 39681–39687. https://doi.org/10.1039/C6RA03996K.

13 Liang, H.F., Smith, C.T.G., Mills, C.A. et al. (2015). The band structure of graphene oxide examined using photoluminescence spectroscopy. *J. Mater. Chem. C* 3 (48): 12484–12491. https://doi.org/10.1039/C5TC00307E.

14 Taniselass, S., Arshad, M.K.M., and Gopinath, S.C.B. (2019). Current state of green reduction strategies: solution-processed reduced graphene oxide for healthcare biodetection. *Mater. Sci. Eng., C* 96: 904–914.

15 Tadyszak, K., Wychowaniec, J.K., and Litowczenko, J. (2018). Biomedical applications of graphene-based structures. *Nanomaterials* 8 (11): 944.

16 Bandi, S. and Srivastav, A.K. (2020). Graphene-based chemiresistive gas sensors. In: *Comprehensive Analytical Chemistry*, 1e. Elsevier B.V. https://doi.org/10.1016/bs.coac.2020.08.006.

17 Krsihna, B.V., Ahmadsaidulu, S., Teja, S.S.T. et al. (2021). Design and development of graphene FET biosensor for the detection of SARS-CoV-2. *Silicon* https://doi.org/10.1007/s12633-021-01372-1.

18 Mauro, N., Scialabba, C., Cavallaro, G. et al. (2015). Biotin-containing reduced graphene oxide-based nanosystem as a multieffect anticancer agent: combining hyperthermia with targeted chemotherapy. *Biomacromolecules* 16 (9): 2766–2775. https://doi.org/10.1021/acs.biomac.5b00705.

19 Wang, N., Wu, Y., Han, G. et al. (2020). Preparing electrodes with highly reduced graphene oxide load for supercapacitors by dropping-electrochemical reduction. *Diamond Relat. Mater.* 105: 107766. https://doi.org/10.1016/j.diamond.2020.107766.

20 Méndez-Romero, U.A., Pérez-García, S.A., Fan, Q. et al. (2020). Lateral size reduction of graphene oxide preserving its electronic properties and chemical functionality. *RSC Adv.* 10 (49): 29432–29440. https://doi.org/10.1039/D0RA04726K.

21 Texter, J. (2014). Graphene dispersions. *Curr. Opin. Colloid Interface Sci.* 19 (2): 163–174. https://doi.org/10.1016/j.cocis.2014.04.004.

22 Martinez Paino, I.M., Santos, F., and Zucolotto, V. (2017). Biocompatibility and toxicology effects of graphene oxide in cancer, normal, and primary immune cells. *J. Biomed. Mater. Res. A* 105 (3): 728–736. https://doi.org/10.1002/jbm.a.35946.

23 Li, D., Müller, M.B., Gilje, S. et al. (2008). Processable aqueous dispersions of graphene nanosheets. *Nat. Nanotechnol.* 3 (2): 101–105. http://www.nature.com/nnano/journal/v3/n2/suppinfo/nnano.2007.451_S1.html.

24 Janegitz, B.C., dos Santos, F.A., Faria, R.C. et al. (2014). Electrochemical determination of estradiol using a thin film containing reduced graphene oxide and dihexadecylphosphate. *Mater. Sci. Eng., C* 37: 14–19.

25 Agarwal, V. and Zetterlund, P.B. (2021). Strategies for reduction of graphene oxide – a comprehensive review. *Chem. Eng. J.* 405: 127018. https://doi.org/10.1016/j.cej.2020.127018.

26 Stankovich, S., Dikin, D.A., Piner, R.D. et al. (2007). Synthesis of graphene-based nanosheets via chemical reduction of exfoliated graphite oxide. *Carbon* 45 (7): 1558–1565. https://doi.org/10.1016/j.carbon.2007.02.034.

27 Chua, C.K. and Pumera, M. (2014). Chemical reduction of graphene oxide: a synthetic chemistry viewpoint. *Chem. Soc. Rev.* 43 (1): 291–312. https://doi.org/10.1039/c3cs60303b.

28 Eda, G. and Chhowalla, M. (2010). Chemically derived graphene oxide: towards large-area thin-film electronics and optoelectronics. *Adv. Mater.* 22 (22): 2392–2415. https://doi.org/10.1002/adma.200903689.

29 Ke, C.C., Yang, Y.C., and Tseng, W.L. (2016). Synthesis of blue-, green-, yellow-, and red-emitting graphene-quantum-dot-based nanomaterials with excitation-independent emission. *Part. Part. Syst. Char.* 33 (3): 132–139. https://doi.org/10.1002/ppsc.201500196.

30 Kim, M.C., Hwang, G.S., and Ruoff, R.S. (2009). Epoxide reduction with hydrazine on graphene: a first principles study. *J. Chem. Phys.* 131 (6): 64704. https://doi.org/10.1063/1.3197007.

31 Ren, P.-G., Yan, D.-X., Ji, X. et al. (2010). Temperature dependence of graphene oxide reduced by hydrazine hydrate. *Nanotechnology* 22 (5): 55705. https://doi.org/10.1088/0957-4484/22/5/055705.

32 Ambrosi, A., Chua, C.K., Bonanni, A. et al. (2012). Lithium aluminum hydride as reducing agent for chemically reduced graphene oxides. *Chem. Mater.* 24 (12): 2292–2298. https://doi.org/10.1021/cm300382b.

33 Shin, H.-J., Kim, K.K., Benayad, A. et al. (2009). Efficient reduction of graphite oxide by sodium borohydride and its effect on electrical conductance. *Adv. Funct. Mater.* 19 (12): 1987–1992. https://doi.org/10.1002/adfm.200900167.

34 Moon, I.K., Lee, J., Ruoff, R.S. et al. (2010). Reduced graphene oxide by chemical graphitization. *Nat. Commun.* 1 (1): 73. https://doi.org/10.1038/ncomms1067.

35 dos Santos, F.A., Vieira, N.C.S., Zambianco, N.A. et al. (2021). The layer-by-layer assembly of reduced graphene oxide films and their application as solution-gated field-effect transistors. *Appl. Surf. Sci.* 543 (December 2020): 148698. https://doi.org/10.1016/j.apsusc.2020.148698.

36 Cuong, T.V., Pham, V.H., Tran, Q.T. et al. (2010). Optoelectronic properties of graphene thin films prepared by thermal reduction of graphene oxide. *Mater. Lett.* 64 (6): 765–767. https://doi.org/10.1016/j.matlet.2010.01.009.

37 Liang, Y., Frisch, J., Zhi, L. et al. (2009). Transparent, highly conductive graphene electrodes from acetylene-assisted thermolysis of graphite oxide sheets and nanographene molecules. *Nanotechnology* 20 (43): 434007. https://doi.org/10.1088/0957-4484/20/43/434007.

38 Huh, S.H. (2011). Thermal reduction of graphene oxide. In: *Physics and Applications of Graphene-Experiments*, vol. 19 (ed. S. Mikhailov), 73–90. IntechOpen.

39 Zheng, X., Peng, Y., Yang, Y. et al. (2017). Hydrothermal reduction of graphene oxide; effect on surface-enhanced Raman scattering. *J. Raman Spectrosc.* 48 (1): 97–103. https://doi.org/10.1002/jrs.4998.

40 Tetsuka, H., Asahi, R., Nagoya, A. et al. (2012). Optically tunable amino-functionalized graphene quantum dots. *Adv. Mater.* 24 (39): 5333–5338. https://doi.org/10.1002/adma.201201930.

41 Yin, K., Li, H., Xia, Y. et al. (2011). Thermodynamic and kinetic analysis of lowtemperature thermal reduction of graphene oxide. *Nano-Micro Lett.* 3 (1): 51–55. https://doi.org/10.1007/BF03353652.

42 Ishida, T., Miyata, Y., Shinoda, Y. et al. (2016). Anomalous restoration of graphitic layers from graphene oxide in ethanol environment at ultrahigh temperature using solar furnace. *Appl. Phys Express* 9 (2): 25103. https://doi.org/10.7567/apex.9.025103.

43 Li, X., Wang, H., Robinson, J.T. et al. (2009). Simultaneous nitrogen doping and reduction of graphene oxide. *J. Am. Chem. Soc.* 131 (43): 15939–15944. https://doi.org/10.1021/ja907098f.

44 Toh, S.Y., Loh, K.S., Kamarudin, S.K. et al. (2014). Graphene production via electrochemical reduction of graphene oxide: synthesis and characterisation. *Chem. Eng. J.* 251: 422–434. http://dx.doi.org/10.1016/j.cej.2014.04.004.

45 Lee, C.-S., Yu, S.H., and Kim, T.H. (2018). One-step electrochemical fabrication of reduced graphene oxide/gold nanoparticles nanocomposite-modified electrode for simultaneous detection of dopamine, ascorbic acid, and uric acid. *Nanomaterials* https://doi.org/10.3390/nano8010017.

46 Sohn, I.Y., Kim, D.-J., Jung, J.-H. et al. (2013). pH sensing characteristics and biosensing application of solution-gated reduced graphene oxide field-effect transistors. *Biosens. Bioelectron.* 45 (1): 70–76. https://doi.org/10.1016/j.bios.2013.01.051.

47 Chen, P. (2008). *Integration of III-V Materials with Silicon for Optoelectronic Integrated Circuit Applications*. San Diego: University of California.

48 Furlan de Oliveira, R., Livio, P.A., Montes-García, V. et al. (2019). Liquid-gated transistors based on reduced graphene oxide for flexible and wearable electronics. *Adv. Funct. Mater.* 29 (46): https://doi.org/10.1002/adfm.201905375.

49 Tamai, T., Watanabe, M., Kobayashi, Y. et al. (2017). Surface modification of PEN and PET substrates by plasma treatment and layer-by-layer assembly of polyelectrolyte multilayer thin films and their application in electroless deposition. *RSC Adv.* 7 (53): 33155–33161. https://doi.org/10.1039/c7ra04880g.

50 He, Q., Sudibya, H.G., Yin, Z. et al. (2010). Centimeter-long and large-scale micropatterns of reduced graphene oxide films: fabrication and sensing applications. *ACS Nano* 4 (6): 3201–3208. https://doi.org/10.1021/nn100780v.

51 Kim, D.J., Park, H.-C., Sohn, I.Y. et al. (2013). Electrical graphene aptasensor for ultra-sensitive detection of anthrax toxin with amplified signal transduction. *Small* 9 (19): 3352–3360. https://doi.org/10.1002/smll.201203245.

52 Papamatthaiou, S., Estrela, P., and Moschou, D. (2021). Printable graphene BioFETs for DNA quantification in Lab-on-PCB microsystems. *Sci. Rep.* 11 (1): 1–9. https://doi.org/10.1038/s41598-021-89367-1.

53 Chae, M.S., Yoo, Y.K., Kim, J. et al. (2018). Graphene-based enzyme-modified field-effect transistor biosensor for monitoring drug effects in Alzheimer's disease treatment. *Sens. Actuators, B* 272 (March): 448–458. https://doi.org/10.1016/j.snb.2018.06.010.

54 Cai, B., Wang, S., Le Huang, Y.N. et al. (2014). Ultrasensitive label-free detection of PNA-DNA hybridization by reduced graphene oxide field-effect transistor biosensor. *ACS Nano* 8 (3): 2632–2638. https://doi.org/10.1021/nn4063424.

55 Aspermair, P., Mishyn, V., Bintinger, J. et al. (2020). Reduced graphene oxide–based field effect transistors for the detection of E7 protein of human papillomavirus in saliva. *Anal. Bioanal.Chem.* 779–787. https://doi.org/10.1007/s00216-020-02879-z.

56 Vasilijević, S., Mattana, G., Anquetin, G. et al. (2021). Electrochemical tuning of reduced graphene oxide in printed electrolyte-gated transistors. Impact on charge transport properties. *Electrochim. Acta* 371: 137819. https://doi.org/10.1016/j.electacta.2021.137819.

57 Béraud, A., Sauvage, M., Bazán, C.M. et al. (2021). Graphene field-effect transistors as bioanalytical sensors: design, operation and performance. *Analyst* 146: 403–428.

58 Balandin, A.A. (2011). Thermal properties of graphene and nanostructured carbon materials. *Nat. Mater.* 10 (8): 569–581. https://doi.org/10.1038/nmat3064.

59 Li, J., Wu, D., Yu, Y. et al. (2021). Rapid and unamplified identification of COVID-19 with morpholino-modified graphene field-effect transistor nanosensor. *Biosens. Bioelectron.* 183 (December 2020): https://doi.org/10.1016/j.bios.2021.113206.

60 Ma, M., Zhou, Y., Li, J. et al. (2020). Non-invasive detection of glucose via a solution-gated graphene transistor. *Analyst* 145 (3): 887–896.

61 Yavir, K., Marcinkowski, Ł., Marcinkowska, R. et al. (2019). Analytical applications and physicochemical properties of ionic liquid-based hybrid materials: a review. *Anal. Chim. Acta* 1054: 1–16. https://doi.org/10.1016/j.aca.2018.10.061.

62 Reiner-Rozman, C., Kotlowski, C., and Knoll, W. (2016). Electronic biosensing with functionalized rGO FETs. *Biosensors* 6 (2): 1–12. https://doi.org/10.3390/bios6020017.

63 Yu, Y., Jin, D., Yang, F. et al. (2019). Electrical and label-free quantification of exosomes with a reduced graphene oxide field effect transistor biosensor. *Anal. Chem.* 91 (16): 10679–10686. https://doi.org/10.1021/acs.analchem.9b01950.

64 Thakur, B., Zhou, G., Chang, J. et al. (2018). Rapid detection of single *E. coli* bacteria using a graphene-based field-effect transistor device. *Biosens. Bioelectron.* 110 (January): 16–22. https://doi.org/10.1016/j.bios.2018.03.014.

65 Zhao, H. and Ju, H. (2006). Multilayer membranes for glucose biosensing via layer-by-layer assembly of multiwall carbon nanotubes and glucose oxidase. *Anal. Biochem.* 350 (1): 138–144. https://doi.org/10.1016/j.ab.2005.11.034.

66 Piccinini, E., Bliem, C., Reiner-Rozman, C. et al. (2017). Enzyme-polyelectrolyte multilayer assemblies on reduced graphene oxide field-effect transistors for biosensing applications. *Biosens. Bioelectron.* 92: 661–667.

67 Berninger, T., Bliem, C., Piccinini, E. et al. (2018). Cascading reaction of arginase and urease on a graphene-based FET for ultrasensitive, real-time detection of arginine. *Biosens. Bioelectron.* 115 (May): 104–110. https://doi.org/10.1016/j.bios.2018.05.027.

68 Ping, J., Zho, Y., Wu, Y. et al. (2015). Recent advances in aptasensors based on graphene and graphene-like nanomaterials. *Biosens. Bioelectron.* 64: 373–385. https://doi.org/10.1016/j.bios.2014.08.090.

69 Mansouri Majd, S. and Salimi, A. (2018). Ultrasensitive flexible FET-type aptasensor for CA 125 cancer marker detection based on carboxylated multiwalled carbon nanotubes immobilized onto reduced graphene oxide film. *Anal. Chim. Acta* 1000: 273–282. https://doi.org/10.1016/j.aca.2017.11.008.

70 Chen, Y., Ren, R., Pu, H. et al. (2017). Field-effect transistor biosensor for rapid detection of Ebola antigen. *Sci. Rep.* 7 (1): 4–11. https://doi.org/10.1038/s41598-017-11387-7.

71 Piccinini, E., Alberti, S., Longo, G.S. et al. (2018). Pushing the boundaries of interfacial sensitivity in graphene FET sensors: polyelectrolyte multilayers strongly increase the Debye screening length. *J. Phys. Chem. C* 122 (18): 10181–10188.

72 Stern, E., Wagner, R., Sigworth, F.J. et al. (2007). Importance of the Debye screening length on nanowire field effect transistor sensors. *Nano Lett.* 7 (11): 3405–3409.

73 Debye, P. (1923). The theory of electrolytes. I. Lowering of freezing point and related phenomena. *Physikalische Zeitschrift* 24: 185–206.

74 Elnathan, R., Kwiat, M., Pevzner, A. et al. (2012). Biorecognition layer engineering: overcoming screening limitations of nanowire-based FET devices. *Nano Lett.* 12 (10): 5245–5254.

75 Nakatsuka, N., Yang, K.-A., Abendroth, J.M. et al. (2018). Aptamer–field-effect transistors overcome Debye length limitations for small-molecule sensing. *Science* 362 (6412): 319–324.

76 Zheng, Z., Zhang, H., Zhai, T. et al. (2021). Overcome Debye length limitations for biomolecule sensing based on field effective transistors. *Chin. J. Chem.* 39 (4): 999–1008.

77 Hwang, M.T., Heiranian, M., Kim, Y. et al. (2020). Ultrasensitive detection of nucleic acids using deformed graphene channel field effect biosensors. *Nat. Commun.* 11 (1): 1–11.

4

Graphene-Based Electronic Biosensors for Disease Diagnostics

Ahmar Hasnain and Alexey Tarasov

University of Applied Sciences Kaiserslautern, Faculty of Computer Sciences and Microsystems Technology, Amerikastr. 1, 66482, Zweibrücken, Germany

4.1 Introduction

4.1.1 A Promise for Diagnostics

A biosensor can be used in disease diagnosis by qualitative and quantitative analysis of relevant biomarkers associated with the presence or severity of a disease. Biomarkers are measurable indicators of pathological and physiological conditions and can be measured in blood, urine, or other bodily fluids or tissues. For example, blood glucose is a biomarker for diabetes and LDL cholesterol is a biomarker for heart disease. The search for new specific biomarkers for oncological, cardiovascular, infectious, and other diseases is an active research area. Detection of biomarkers can help in the prevention, diagnosis, monitoring, treatment, and prognosis of various diseases if appropriate analytical techniques are available. Currently, quantitative analysis is performed in highly automated, centralized laboratories. This requires that a sample be sent for analysis, which causes delays. Different analytical techniques are used, depending on the type of analyte. Examples include enzyme-linked immunosorbent assays (ELISAs) for antibody-antigen interactions and polymerase chain reactions (PCRs) for deoxyribonucleic acid (DNA) analysis. These techniques involve complex sample preparation, long experimental duration, and expensive reagents and instruments. By contrast, there are many qualitative tests that can be used at home or at the point of care (POC), such as lateral flow assays (LFAs). Examples include pregnancy tests or SARS-CoV-2 rapid antigen tests. There is a growing need for the combination of both: a technique that can be used outside the centralized laboratories while still providing lab-grade results.

Blood glucose biosensors are a prime example for this development. First enzyme-based biosensor for continuous monitoring in cardiovascular surgery was developed in 1962 by Leland C. Clark [1]. Later on, his idea was transformed into a glucose electrochemical sensor, which was then developed and commercialized.

Graphene Field-Effect Transistors: Advanced Bioelectronic Devices for Sensing Applications,
First Edition. Edited by Omar Azzaroni and Wolfgang Knoll.

Apart from conventional electrochemical biosensors, field-effect transistors (FETs) have been investigated as a possible technology platform for biosensing applications. First metal-oxide-semiconductor field-effect transistor (MOSFET) were demonstrated in 1960 by Bell Telephone Laboratories [2]. In 1970, Piet Bergveld replaced the metallic gate of a MOSFET with an electrolyte solution and a reference electrode, so the gate oxide can be in direct contact with nerves, muscles or biomolecules, resulting in an ion-sensitive field-effect transistor (ISFET) [3]. Early ISFET-based biosensors were frequently used as pH sensors [4]. Since the discovery of FET biosensors, various materials were used for their fabrication, including standard silicon [5, 6], semiconducting nanowires [7], carbon nanotubes [8], graphene [9], MoS_2 [10], and organic semiconductors.

With the discovery and isolation of graphene layers in 2004 [11], graphene has become the most used two-dimensional (2D) material in biosensors. It is well documented that mechanical, thermal, optical, and electrical properties of graphene make it a very good candidate for biosensors [12]. For instance, it is transparent and flexible, has very high charge carrier density, high surface-to-volume ratio, and sensitivity, and can be chemically functionalized to achieve targeted detection [13].

Graphene is used as a component in many types of modern biosensors, including fluorescence [14], surface plasmon resonance (SPR) [15], electrochemical [16], and surface enhanced Raman scattering (SERS) [17]. Among these biosensors, electrolyte-gated FETs are promising due to ease of miniaturization and integration with compact readout electronics, high sensitivity and selectivity, possibility of real time measurements with detection limits down to femtomolar range. Especially, the use of large area chemical vapor deposition (CVD)-grown monolayer graphene is quite promising due to its lower number of intrinsic defects [18].

In this chapter, we will focus on monolayer graphene-based FETs in biosensing applications. We will cover the operating principle, fabrication, and functionalization of these biosensors. We will primarily cover examples related to disease diagnostics, including targeted detection of bacteria, viruses, nucleic acids, proteins, and ions. The main idea is to provide a basic understanding of graphene-based FET biosensors.

4.1.2 Principle of Graphene FET Sensor

Electrolyte-gated graphene field-effect transistor (EG-GFET) is widely used in electrochemical biosensing applications. Structurally similar to a standard ISFET, it contains three main terminals: source, drain, gate, and a graphene channel. Figure 4.1a illustrates a simple (EG-GFET). Source and drain terminals are usually thin metal films. The gate electrode can be either a standard reference electrode such as Ag/AgCl or a pseudoreference electrode such as gold or platinum wire. Reference electrodes provide a stable and reliable potential over a long period of time, while pseudoreference electrodes can be more easily miniaturized and integrated on the chip.

There is also an important difference between ISFETs and conventional ISFETs: conventional ISFETs usually have an insulating oxide layer to protect the semiconductor channel and provide a spatially separated sensing interface. In EG-GFETs,

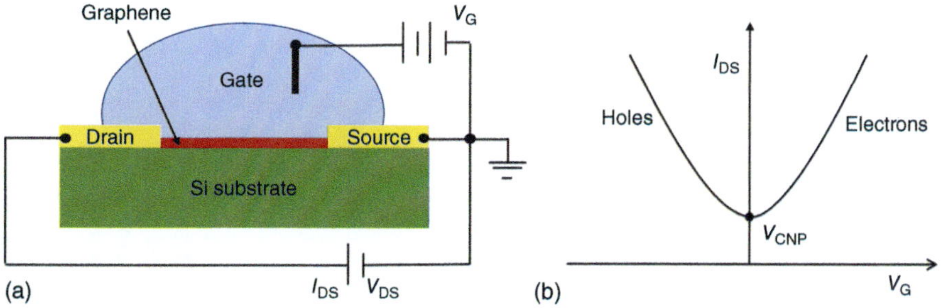

Figure 4.1 (a) Schematic and circuit model of electrolyte-gated graphene FET. (b) Typical transfer curve of a GFET.

graphene is in direct contact with the electrolyte, acting as both the sensing interface and the semiconductor channel.

The general working principle of EG-GFET sensors is that the charge carrier density in the graphene channel is altered by the target analyte in the proximity of graphene surface. This change is induced by electrostatic field generated by physical or chemical changes occurring in the vicinity of graphene channel. This change is transduced to an electrical signal, which can then be processed and displayed. As shown in a circuit model in Figure 4.1a, EG-GFETs are controlled by the voltages applied between source and drain and gate and source terminals. The source–drain voltage "V_{DS}" drives the current "I_{DS}" through the channel, while the gate(-source) voltage "V_G" modulates the charge carrier density inside the channel. By sweeping the gate voltage "V_G" at a fixed bias "V_{DS}", a transfer curve is obtained (I_{DS} vs. V_G), showing a typical parabolic shape (Figure 4.1b). The voltage at the minimum of the transfer curve is called the charge neutrality point "V_{CNP}".

The charged analyte species introduce an additional electric field that also influences the charge carrier density and causes a shift of the transfer curve.

In EG-GFET configuration, the gate voltage is applied through the reference electrode which is coupled through electrolyte solution to graphene channel. Hence, the interface between the graphene channel and the solution is very critical. This interface is described by the interfacial capacitance, which is a series combination of graphene quantum capacitance and capacitance of the electrical double layer (EDL) [19, 20]. The EDL also gives rise to the charge screening phenomenon, which is explained later in this chapter.

In a typical biosensor measurement, a fluidic chamber is used for delivery of biomolecules to the surface of graphene channel, and the graphene surface is usually functionalized with specific capture molecules (Figure 4.2a, upper panel). These capture molecules perform targeted detection of charged analyte species. Once captured, the analyte induces a change in the transfer curve.

"V_{CNP}" is a commonly used parameter for characterization of EG-GFET biosensors. Ideally, in the absence of any charged species in the proximity of graphene surfaces, the transfer characteristic curve has a "V_{CNP}" close to zero gate voltage, as

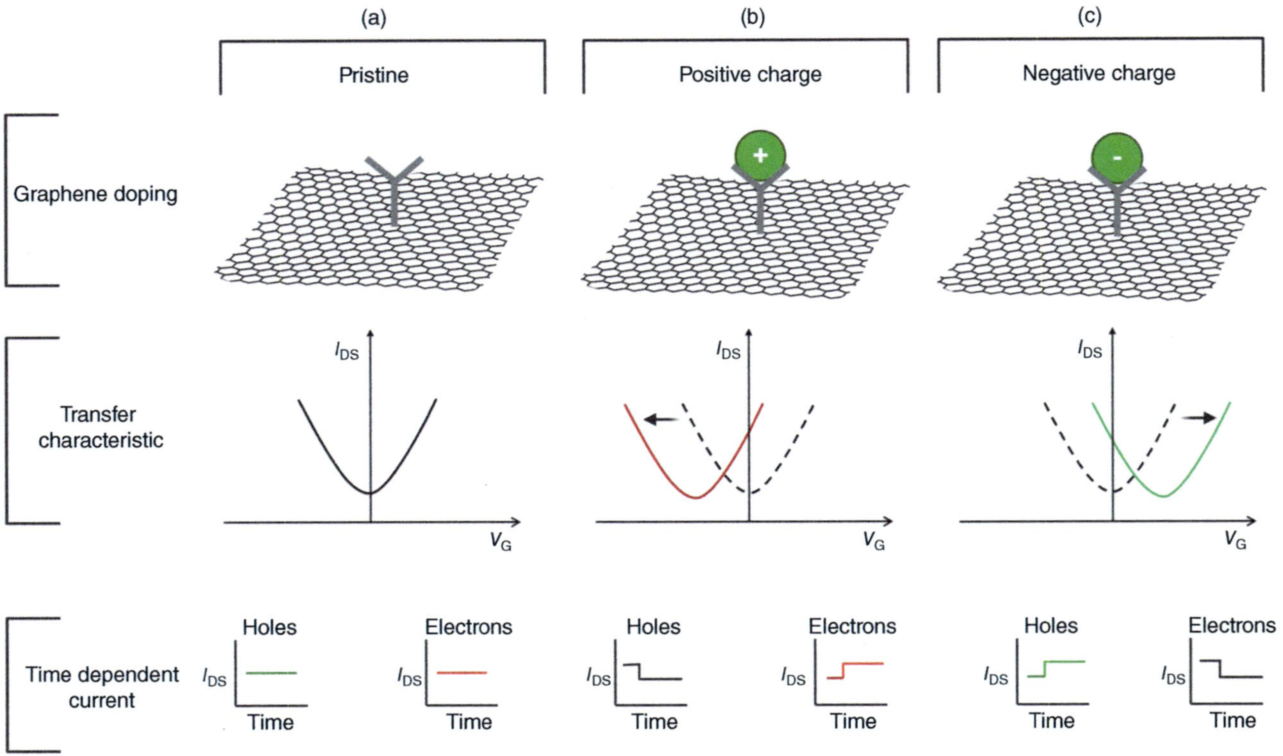

Figure 4.2 Principle of EG-GFET. (a) Upper panel: "pristine" refers to functionalized graphene without analyte; center: typical ambipolar behavior of pristine graphene with the transfer curve minimum (CNP) at $V_G = 0$ (idealized), lower panel: time-dependent signal without analyte provides a baseline. (b) Upper panel: binding of positively charged molecule; middle panel: negative shift of the transfer curve and CNP; lower panel: time-dependent current decreases or increases depending on the exact gate voltage, due to depletion or accumulation of charge carriers (holes or electrons). (c) Upper panel: binding of negatively charged molecule; middle panel: positive shift of the transfer curve and CNP; lower panel: time-dependent current increases or decreases depending on the exact gate voltage.

shown in the transfer curve of Figure 4.2a (middle panel). Upon binding of positively charged species (Figure 4.2b, upper panel), V_{CNP} shifts to more negative voltages, while negatively charged species shift the V_{CNP} to more positive voltages (Figure 4.2c, upper panel). These shifts are due to doping of the graphene channel by the captured charged species that act as electron donors or acceptors. If the capture event is reversible, the charge neutrality point (CNP) shifts back to its original position. The doping phenomenon is also known from conventional back-gated or top-gated GFETs, where unintentional dopants from fabrication process or intrinsic defects in CVD-grown graphene can introduce a shift in CNP [21].

In addition to CNP shifts, conductance or current I_{DS} can be monitored as a function of time at a fixed gate voltage V_G (Figure 4.2, lower panels). Depending on the choice of V_G, the device may be operated on the hole ($V_G < V_{CNP}$) or the electron regime ($V_G > V_{CNP}$). This choice determines whether the current increases or decreases upon binding of charged species, as indicated in Figure 4.2b,c (lower panels: time dependent current).

The shift of the transfer curve is not always the same over the voltage range considered (i.e. not entirely parallel). Sometimes, changes in the transfer curve slope (or transconductance) were also observed. This leads to a signal dependence on the exact gate voltage applied [22]. The transconductance depends (among other things) on the charge carrier mobility inside the graphene channel and the capacitance of the interface. The mobility can be influenced by charge scattering introduced by target analytes. Additional care must be taken when analyzing complex signals originating from a combination of different mechanisms. These and other electrical metrices are explained in detail in previous work [23, 24].

4.2 Device Fabrication Process

4.2.1 Graphene Synthesis

Synthesis of high-quality defects free graphene has been a major challenge in fabrication of GFET-based biosensors. Ideally, large area graphene consisting of single crystals of graphene with zero or negligible defects is best for graphene application in GFET biosensors. In general, there are top-down and bottom-up synthesis routes. Few-layer and single-layer graphene can be extracted by top-down mechanical exfoliation from graphite, renowned as the "Scotch tape method." [11]. This method yields high-quality graphene flakes with rather small lateral size. However, this method does not allow production of large-area graphene needed for biosensing applications. Reduction of chemically exfoliated graphene oxide (GO) is another top-down approach that attracted attention due to its large-scale and low-cost production. However, surface defects in reduced GO films and loss of electrical conductivity (and transparency) limit its applications in GFET-based biosensors.

Among bottom-up methods, epitaxial growth of graphene is worth mentioning, which is based on decomposition of silicon carbide (SiC) at very high temperatures. Due to high vapor pressure of Si compared to C, Si atoms desorb from the substrate and leave behind carbon atoms that rearrange to form (few-layer) graphene. This method yields multilayer graphene films containing 5–100 layers, depending on the process parameters and kinetics. The main advantage of this method is that graphene is grown on a nonmetallic substrate (SiC) and can be used as-is for device fabrication. However, this substrate is relatively expensive and can also have detrimental effects on the electronic properties of graphene [25].

CVD of graphene has proven to be an effective method for large-scale production of high-quality monolayer graphene [26]. Graphene obtained by this method gives high light transmittance, high charge carrier density, and fewer defects, which are highly desirable for the fabrication of GFET biosensors. In this bottom-up technique, gas species, mainly hydrocarbons, decompose and adsorb at high temperature (900–1000 °C) on a catalytic transition metal substrate (e.g. Cu and Ni). Copper is the most used substrate due to negligible miscibility of carbon in it. After formation of the initial graphene layer, the growth process stops because the copper is coated and can no longer act as a catalyst. Thus, under controlled conditions, large areas of single-layer graphene are obtained, which can be transferred from copper to other substrates [18].

4.2.2 Graphene Transfer Over Substrates

To fabricate GFETs, CVD-grown graphene must be transferred from metallic copper to an insulating substrate. Graphene transfer is a critical step during the fabrication of GFET-based biosensors. Various methods have been employed for reliable graphene transfer, e.g. wet transfer, electrochemical bubbling transfer, thermal release tape method, roll-to-roll (R2R) transfer, and support-free transfer [27]. Among all these methods, the wet transfer technique as shown in Figure 4.3, is a frequently used technique for biosensor applications.

In wet transfer technique, graphene on growth substrate (Figure 4.3a) is first coated with a polymer support layer (e.g. PMMA, Figure 4.3b). Ionic etchants such as a ferric chloride solution is then used to etch away growth substrates (Figure 4.3c). The remaining graphene layer supported by the polymer is transferred to target substrate while floating on transfer media (generally DI water, Figure 4.3d,e). Once transferred (Figure 4.3f), the support layer can be thermally removed at high temperature under inert atmosphere or washed away using different solvents, e.g. acetone and isopropanol (Figure 4.3g). Finally, depending on the type of substrate, the target substrate with monolayer graphene can be dried at different temperatures and duration.

Ionic etchants can unintentionally dope graphene; hence, use of aqueous ammonia solution as transfer medium was proposed to wash away undesired dopants [21]. Similarly, polymer support layers can also change intrinsic electrical properties of graphene. However, several support layers such as PMMA, Paraffin, Cytop, and EPDM were successfully used with minimal undesired doping and reproducible wrinkle-free graphene transfer [27].

To date, wet transfer is a widely used lab scale graphene transfer technique among researchers, as it is an easy and reproducible way to obtain high quality graphene

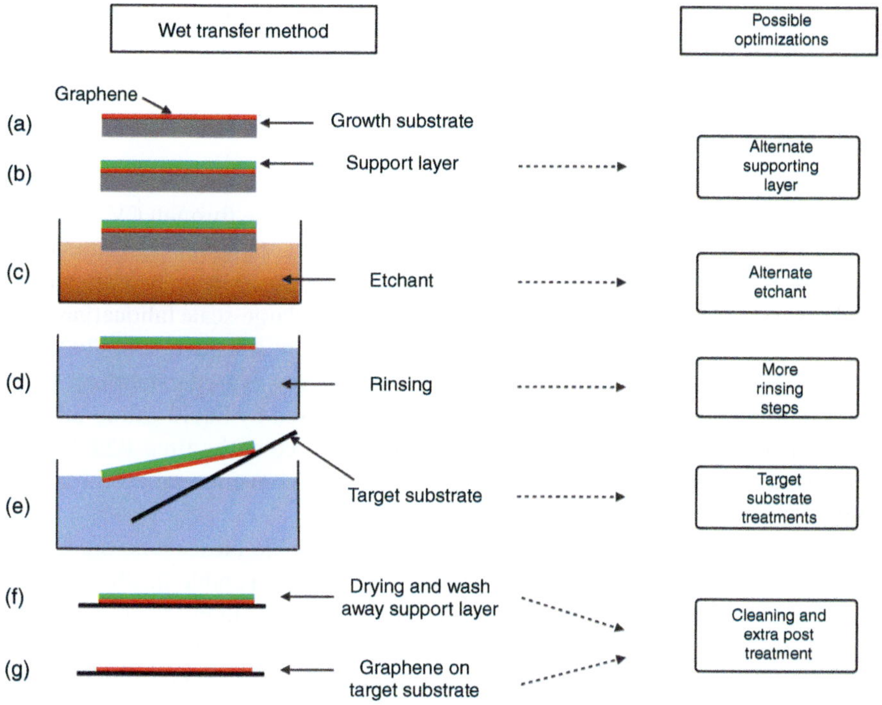

Figure 4.3 Basic graphene transfer method and possible optimization steps. (a) Starting material is graphene on growth substrate. (b) Graphene is coated with a support layer. (c) Growth substrate is removed by selective wet etching. (d) Graphene is rinsed to remove etchant residues. (e) Floating graphene with support layer are transferred onto target substrate. (f) Support layer is removed from transferred graphene. (g) Graphene is on target substrate. Source: Created with BioRender.com.

surface. However, optimization of etchants, supporting layers, cleaning steps, growth substrates, and solvents is still an ongoing quest to obtain clean wrinkle- and crack-free graphene monolayer with intrinsic electrical properties (Figure 4.3, right panel).

4.2.3 Fabrication of GFET

Photolithography is a commonly used technique to fabricate GFET devices. In this technique, CVD graphene is first transferred from growth substrate to target substrate. Then standard photolithography is used to pattern graphene channel and undesired graphene outside the channel is removed by oxygen plasma etching. Next, the metallic layer (e.g. gold) with an adhesion layer (e.g. titanium) is sputtered or evaporated on the substrate. Again, standard photolithography is used to shape source and drain electrodes, followed by lift-off process. Then, metallic electrodes are passivated with an insulation layer, such as oxides (e.g. SiO_2, Al_2O_3) or polymers, followed by lift-off process to partially remove insulation for actual electrical connections. Thus, the final structure of GFET is formed. The above-mentioned photolithography method was used frequently in GFET fabrication and biosensor applications [28, 29]. Graphene transfer onto pre-patterned substrates is also a common approach for easy lab-scale fabrication of GFET devices [21].

4.2.4 New Developments

Large-scale synthesis and graphene transfer are critical for upscale GFET-based biosensor fabrication. Large-scale graphene synthesis is possible through CVD growth process. While various options are available to transfer CVD graphene onto target substrates, most of these methods can only support fabrication up to wafer scale. R2R transfer of CVD grown graphene is an approach for large-scale fabrication. It is a dry transfer technique involving double-sided pressure-sensitive adhesive tape, allowing multiple graphene transfers [30]. This approach was used to fabricate PET-based interdigitated electrodes (IDEs), which can be used by liquid gating for biosensing applications. Due to dry graphene transfer process, the final R2R GFETs showed lower charge neutrality point compared to conventional wet transfer method [30].

In some applications, including wearable or implantable biosensors, the use of graphene on flexible substrate is desirable. For instance, wearable flexible sensors for real-time monitoring of cytokines [31] or for diabetes monitoring and therapy [32] were reported. Sensors printed on silk for bacteria detection [33], PET-based wearable corneal biosensor for dopamine detection from tears [34], or hydrophobic cotton-based sensor for lactate detection from human sweat [35] are some of the latest approaches towards graphene-based wearable or flexible biosensors. Implantable EG-GFET-based biosensors were reported for *in vivo* recording of brain activity [36, 37].

Reusability and shelf life of these biosensors are another critical consideration. For example, reusability was studied by performing regenerative cycles (up to 80 times) with a variation of 8.3% from first cycle to last cycle [38]. Shelf life of 42 days with 74.79% activity retention of initial response was recently reported for a graphene-based glucose sensor [39]. These results are quite promising, but much more work is needed to reach higher technology readiness levels.

Combining flexible substrates with R2R transfer of CVD graphene may be the way to go for large-scale GFET biosensor fabrication, if remaining issues can be solved. Alternatively, solution-processed reduced GO may become a viable option for large-scale biosensor fabrication if acceptable electrical performance can be achieved [40–42].

4.3 Functionalization and Passivation

Graphene channel is an integral part of GFET biosensor, and it plays a vital role in detection of target biomolecules. However, specificity of detection is also very critical, which can be obtained by immobilizing specific molecules on graphene surface. These molecules can detect and capture specific biomolecules in target analyte and

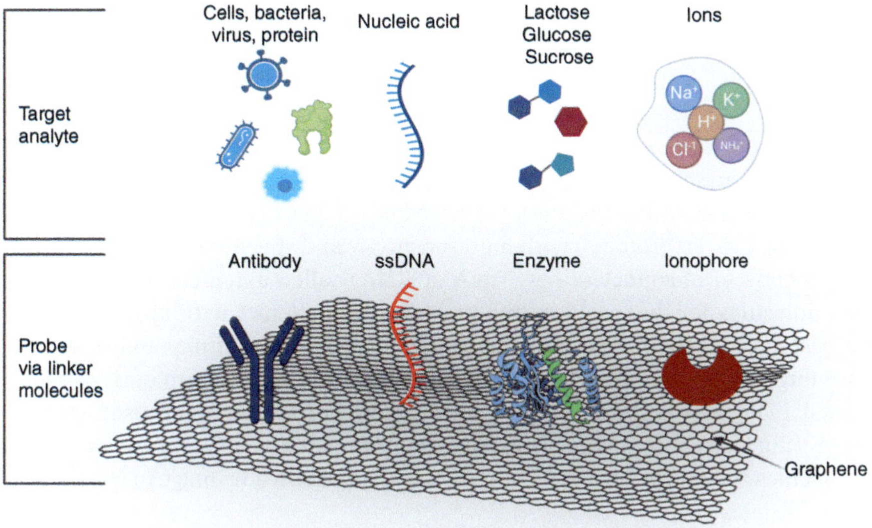

Figure 4.4 Main classes of target analytes and probe molecules used in GFET-based biosensors. Source: Created with BioRender.com.

are often referred to as probe molecules. The density of capture molecules also influences the signal and needs to be optimized. Depending on the target analyte, probe molecules can be, e.g. antibodies, nucleic acids, enzymes, or ionophores, as shown in Figure 4.4. Several examples of different probe molecules, as described in Section 4.3.4.

4.3.1 Probe Molecules

Nucleic acid sequences can serve as probe molecules to specifically detect complementary nucleic acid sequences. Most nucleic acid recognition studies [43–45] rely on synthetic (single-strand nucleotides [ssDNA)] probes of various lengths (22-44 mer) immobilized on graphene surfaces. The complementary sequence is present in the target solution, which is brought in contact with functionalized graphene during the sensing experiment [46]. For higher sensitivity and less background noise, single-strand PNA (peptide nucleic acid) probes were also reported [47]. These PNA probes have a neutral backbone resulting in less electrostatic repulsion between PNA and target nucleic acid [48]. Recently, a target-specific guide RNA (gRNA) complexed with CRISPR-associated (Cas) enzyme was used as a probe to differentiate between different mutations in target genomic DNA [49]. More complex probes, i.e. hairpin DNA and DNA tweezers, with better sensitivity, have also been reported [50, 51].

Due to very high specificity of antibody–antigen hybridization, antibodies are often used as probes in GFET-based immunosensors. Antibody probes on GFETs have also been used to detect bacteria [52, 53] and viruses [54–56]. For instance, Anti *Escherichia coli* antibodies were used to detect *E. coli* bacteria [52, 53]. Most recently, SARS-CoV-2 spike antibodies has been immobilized on graphene for a real-time accurate detection of COVID-19 virus through GFET-based immunosensor [54]. Monoclonal antibodies were immobilized as probe molecules to detect Zika virus, Japanese encephalitis, and Avian influenza virus [55, 56]. Short antibody fragment $F(ab')_2$ have also been used as probe molecules to detect thyroid stimulating hormone (TSH) (proteins) with high sensitivity and specificity [57].

A new class of oligonucleotides (DNA or RNA) called aptamers is also used as probe molecules for diagnostic purposes. Aptamers interact with target biomolecules based on 3D folding patterns, which gives them high affinity and selectivity toward target molecules. Aptamers can capture bacteria, viruses, proteins, peptides, and small molecules [58]. For instance, 8–17 DNAzyme is a commonly used aptamer for capturing lead (Pb^{+2}) [59]. Due to small size of aptamers, the capturing of target biomolecules occurs within Debye-length, which is a main advantage to use them as probe molecules [20].

For highly qualitative and quantitative detection of small molecules (i.e. glucose, sucrose, and lactose) present in human body the most common probe molecules are enzymes. Various approaches have been reported where glucose oxidase (GOD) was functionalized on graphene surface for glucose detection in blood and sweat of diabetic patients [60–63].

For target ion detection, the commonly used strategy is the use of self-assembled membranes or ionophore membranes as probe elements. Several studies have been reported where ion-selective probe membranes, e.g. (1-octadecanethiol, valinomycin, crown ether, and Al_2O_3) were used for detection of pH and other electrolyte ions [64–70].

4.3.2 Immobilization of Probe Molecules

Probe molecules can be immobilized on graphene surface through covalent and noncovalent functionalization strategies. Graphene functionalization through covalent moieties introduces structural changes to carbon lattice. Intrinsically, graphene has out-of-plane defects, which are prone to attack from reactive species. Carbon atoms at these defect points face a change of hybridization from sp^2 to sp^3. These structural changes allow opening of a bandgap and thus impart significant changes to electronic properties of graphene [71]. Contrary to covalent functionalization, noncovalent functionalization introduces less severe changes to electronic properties of graphene.

Several approaches for covalent immobilization of graphene have been reported. For instance, NHS/EDC (1-ethyl-3-(3-dimethylaminopropyl) carbodiimide hydrochloride) activation is widely used to activate carboxylic groups for binding of primary amines present on various biomolecules. This EDC/NHS approach was also used to attach antibodies to graphene surface to enable detection of viral antigens [55, 56]. This implies that graphene has some carboxylic groups available for activation (defect sites). It was also reported that covalent functionalization of graphene with aryl diazonium salts is possible [72]. For instance, –COOH groups were covalently anchored on graphene surface through electrochemical reduction of 4-carboxyphenyl diazonium salt. These carboxyl groups were then activated through EDC/NHS chemistry to immobilize ovalbumin antibodies for detection of egg white allergen ovalbumin [73].

Most popular approach to noncovalently functionalize graphene with probe molecules is through intermediary pyrene-based linker molecules, i.e. pyrenebutanoic acid succinimidyl ester (PASE), 1-pyrenebutanoic acid (PBA), 1-pyrenebutyric acid succinimidyl ester (PBSE), 1-pyrenebutyric acid *N*-hydroxysuccinimide ester (PBASE) [43–45, 47, 50–55, 60, 74–79]. In this approach, aromatic pyrene end binds to plane of graphene through noncovalent π–π stacking interaction. Succinimidyl ester ends are capable of coupling with amine end groups on side chains of probe molecules via nucleophilic substitution reactions to form amide bond [80–82].

At times, nonspecific interactions of graphene with probe molecules were also used in biosensors without any linker molecules. Such strategies were reported for direct physical adsorption of probe DNA [83, 84]. Also, thin membranes containing probe molecules, e.g. ionophores were deposited on graphene surface [63–65, 69, 70, 85].

4.3.3 Debye Length

In GFET biosensors, the sensor signal relies heavily on the surface charge due to the layer of captured target molecules. However, the target analyte is typically present in a buffer solution containing other ions as well. The ions of the opposite sign are attracted by the surface charge via Coulomb interaction, forming a so-called EDL. Different EDL models exist that describe the distribution of ions within this layer [19, 86]. As a result, the analyte charge is at least partially neutralized (or "screened"), reducing the measurable sensor signal. The relevant length scale used to characterize the screening effect is called the Debye screening length (λ_D), which can be calculated as follows [19, 87].

$$\lambda_D = \sqrt{\frac{\varepsilon_0 \varepsilon_r k_B T}{2 N_a e^2 I}} \tag{4.1}$$

where ε_r is the relative permittivity of water and ε_0 is the vacuum permittivity, T is the temperature, k_B is the Boltzmann's constant, e is the elementary charge, N_a is the Avogadro constant, and I is the ionic strength of the solution. For a monovalent electrolyte, λ_D is approximately 0.304 nm [19]. For commonly used phosphate-buffered saline (PBS) (1–0.1 × PBS), λ_D is approximately (0.7–2.2 nm) [88], which is less than the average size of an antibody (5–10 nm) [89], a protein (6.6 nm) [90], and a cell (1 μm). Most of the analyte charge located further away than the Debye length is screened by the counter ions and cannot be detected by the sensor. That is why Debye screening length is a major challenge in realization of highly sensitive GFET biosensors. Several strategies were proposed to circumvent this issue.

The most straightforward strategy is to extend the Debye length by reducing the ionic strength of the solution (see Eq. (4.1)). However, this approach has only limited practical value because many relevant physiological solutions have high ionic strengths. While the relevant solutions can be diluted to some extent, this requires additional liquid handling steps and dilutes the analyte concentration.

Alternatively, the probe size can be reduced so that the bioconjugation can take place within the Debye length limit. For instance, immunoglobulin aptamers with 3 nm height instead of antibodies, which have around 10 nm height, was used for IgE protein detection with different ionic concentrations within EDL [91]. The use of antigen-binding fragments which are shorter than a regular antibody can also help to reduce the Debye length problem in GFET-based immunosensors [57, 92]. However, even small capture molecules are still larger than the Debye length of undiluted physiological solutions (<1 nm). Therefore, additional strategies are required to overcome this issue.

Polyethylene glycol (PEG) is a blocking agent against nonspecific binding on graphene channel, as described in the following section 4.3.4 [55, 93]. Interestingly, it was also shown to increase the effectiveness sensor signals in high-ionic-strength solutions [94]. For instance, DNA aptamers were immobilized along with PEG spacers to detect prostate specific antigen (PSA). The authors claimed an increase in Debye length and sensitivity by using the mixed layer of PEG and probe molecules [95]. Instead of a mixed layer, a pyrene PEG linker was also used as an intermediate step to couple a probe aptamer. This modification was used to capture biomarkers with high sensitivities in serum [96]. The combination of short capture molecules (i.e. antibody fragments) and long PEG spacers in a mixed layer yielded highly sensitive protein sensors with femtomolar detection limits in undiluted serum [57]. The exact mechanism of PEG action and its relation to the Debye length are described in detail elsewhere [97].

Finally, a deformed graphene channel, with different morphology (i.e. concave and convex surface) compared to flat graphene film, was shown to have higher sensitivities toward small number of charges. Due to its large surface area, crumpled graphene exhibited high hydrophobicity, which was directly linked to EDL and increase in Debye length [98, 99]. This approach was used for ultrasensitive detection of SARS-CoV-2 proteins and dopamine [100]. The same approach was also applied to nucleic acid detection [101].

4.3.4 Passivation

After immobilization of probe molecules, graphene surface is still partially exposed and prone to unspecific binding of unwanted molecules, in addition to targeted analytes. This issue becomes more severe if complex solutions such as serum or sweat are used instead of idealized buffer solutions. Hence, surface passivation is required to prevent the unintentional binding of non-target species. A popular strategy is the use of blocking agents, i.e. Tween 80 [102], Tween 20 [74], PEG [103], or BSA [104] to cover free spaces between probe linker molecules on graphene surface. Apart from spaces, some linker molecules are also unreacted and unused on graphene surface, hence quenching of these molecules is also an effective passivation approach. For instance, amino-PEG5-alcohol was used to block unreacted linker molecules [55, 77]. Quenching of unreacted succinimidyl ester ends on pyrene-based linker through glycine [105] or ethanolamine was also reported [31, 52, 104, 106].

4.4 CVD GFETs for Diagnostics

4.4.1 Graphene-Based FET Biosensors for Nucleic Acids

DNA is an important macromolecule than encodes entire information all known organisms and many viruses need to function. DNA consists of two single strands arranged in a double helix structure. DNA is a biopolymer built up from monomeric nucleotides, each consisting of the sugar deoxyribose, phosphate backbone, and one of the four different bases (adenine = A, guanine = G, thymine = T, and cytosine = C). DNA detection can be applied to environmental monitoring of different pathogens [46] or diagnostics of diseases, including viral or bacterial infections and cancer. The detection principle of such DNA biosensors is based on the highly specific interaction of two complementary DNA strings (hybridization).

Typically, a single-stranded probe DNA is immobilized on the sensor surface, which can identify and detect its complementary target nucleic acid. The hybridization can result in duplexes or triplexes [46, 107]. Alternatively, artificial PNA can be used as a probe, with the added benefits of higher stability toward pH changes or degradation by enzymes. Also, PNA does not have a charged phosphate backbone, which reduces unintentional charge on graphene prior to hybridization [108]. Among various methods available to functionalize graphene surface with nucleic acid probes, physisorption, and chemical linking are prominent. Pyrene-based linker molecules i.e. PASE and PBASE are used to attach probe molecules on graphene surface as shown in Figure 4.5 [45, 47].

The intrinsic charge of the DNA phosphate backbone makes it an attractive target for label-free charge-based FET biosensors. The use of graphene promises very low detection limits required for DNA sensing. For example, PNA- functionalized graphene biosensor was used to detect DNA sequences with a detection limit of 10 fM [47]. In another study, direct physisorption of nucleobases by van der Waals forces on graphene channel was studied. To differentiate among different nucleobases the charge neutrality point shift caused by graphene-nucleobase interaction was observed [83]. Also, various lengths of DNA with different mismatches were successfully detected in back-gated graphene biosensors with detection limit of 1 fM [45]. In a recent study, a combined GFET-CRISPR (clustered regularly interspaced short palindromic repeats) technology was used to study genetic mutations in Duchenne muscular dystrophy (DMD) patients. Dystrophin gene from DMD patients was studied, and specific mutations (deletion of two sequences) were detected in whole target genome with a lower detection limit of 1.7 fM. The authors proposed this biosensor as a rapid tool for POC diagnostics in genetic mutation

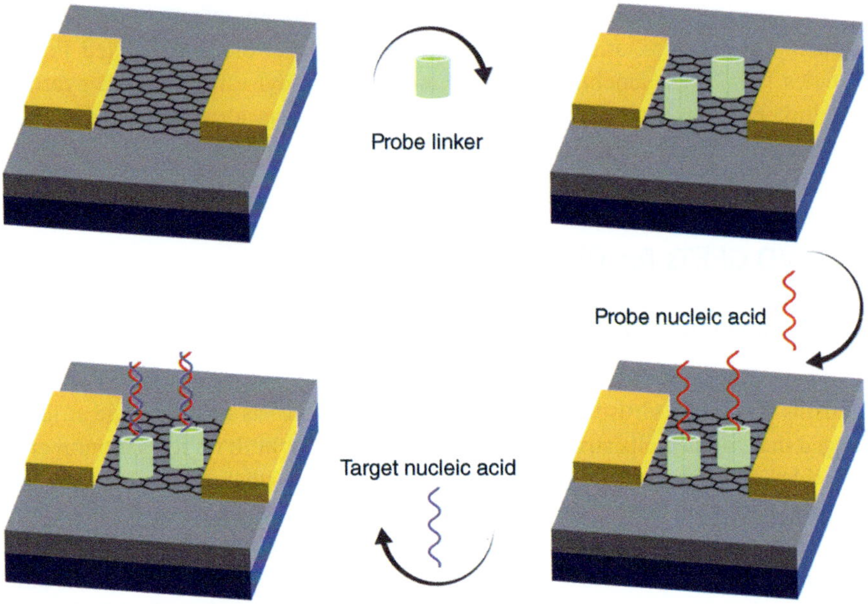

Figure 4.5 Schematic illustration of the preparation of G-FET based biosensors for nucleic acid detection. Source: Created with BioRender.com.

related diseases at early stages [77]. Liquid-gated GFET was also reported for targeted detection of different types of cancer related DNA sequences with higher selectivity [84].

Additionally, graphene FETs can be integrated with other analytical techniques for improved performance. For instance, a combined opto-electric biosensor was developed and attached to the face-end of an optical fiber for targeted detection of DNA. The combination of FET and fluorescence in this dual-mode biosensor significantly improved the accuracy and reliability of DNA detection [43].

DNA methylation is an important epigenetic mechanism that controls gene expression. Knowledge of DNA methylation status can provide critical information related to human health and progression of diseases, e.g. brain tumors. Recently, a GFET biosensor with a specific DNA tweezer probe was established to differentiate between methylated and non-methylated DNA targets with picomolar detection limits [51].

As shown in Table 4.1, graphene-based nucleic acid biosensors are quite promising in disease diagnostics with very low detection limits. They have very high selectivity due to unique nature of nucleotides. However, they suffer from operational limitations because of temperature and pH sensitivities of hybridization reaction [107, 109]. A recent study showed how concentration, salts, and pH of analyte medium (buffer) influence the sensitivity of DNA detection. The authors claimed that additional salts in PBS give prominent DNA detection and higher sensitivity compared to phosphate buffer (PB), irrespective of the longer Debye length in case of PB [44]. Furthermore, because these biosensors largely depend on graphene

Table 4.1 Overview of graphene FET-based nucleic acid biosensors.

References	Target analyte	Analyte condition	Probe immobilization mechanism	Probe	Response unit	LOD
[47]	Complementary Non-Complementary Single-base mismatch DNA	PBS buffer	PASE linker	Peptide nucleic acid PNA	Current (I_{DS})	10 fM
[83]	DNA nucleobases	Direct physisorption	N/A	N/A	Charge carrier density (Δn)	10^9 molecules
[45]	DNA sequences	Aqueous solution	PBASE linker	DNA	(ΔV_{CNP})	1 fM
[77]	Dystrophin gene from DMD patients	$MgCl_2$ Solution	PBA linker	Single guide RNA	Current ($I\%$)	1.7 fM
[84]	TP53 cancer DNA	PBS buffer	Physisorption	DNA	Current (I_{DS})	1 nM
[43]	DNA	Solution	PBASE linker	DNA	Current (I_{DS})	10 nM
[51]	Methyltransferase (MGMT), Brain tumor DNA	—	PBASE linker	DNA tweezers (DTP)	Current (I_{DS})	10 pM
[44]	DNA	PB, PBS buffer	PBSE linker	DNA	Current (I_{DS}) (ΔV_{CNP})	~10 aM

channel, so an even immobilization of probe molecules on graphene surface is highly desired [110]. With some extra quenching of probe molecules [51] and passivation of unfunctionalized graphene surface, these biosensors can become a leading analytical tool in disease diagnostics.

4.4.2 Graphene-Based FET Biosensors for Antibody–Antigen Interactions

Antibodies have the ability to bind to a broad set of antigens (immunogens) with remarkable specificity. This antibody–antigen (Ab–Ag) conjugation has been widely used as a basis for molecular recognition studies [111, 112] and in analytical platforms, called immunosensors. These platforms have a wide range of potential applications, including medical diagnosis of viral and bacterial infections and the identification of protein biomarkers of various diseases. Antibody (Ab) molecules are Y-shaped molecules composed of four domains; two light chains also called "arms" and two heavy chains which are also called "body," connected through disulfide bonds.

There are two regions of antibodies: fragment antigen-binding (Fab fragment) and fragment crystallizable region (Fc region), which contains carboxy and amino terminals. Each part contributes to specific functionality of antibody. The variable domain (Fab fragment) is responsible for (Ab–Ag) conjugation, which has two Ag binding sites. Distinguished by heavy chains there are further five classes of antibodies, i.e. IgM, IgD, IgG, IgA, and IgE [113, 114].

GFET-based immunosensors are developed by immobilizing antibodies on graphene surfaces through linking chemistry, as shown in Figure 4.6. Commonly used immobilization strategies of probe (Ab) is through linker molecules, i.e. PASE and PBASE. EDC/NHS (*N*-Hydroxysuccinimide) chemical reaction on graphene surface is also a frequently used functionalization route, in which amide bond is formed between amine terminal of antibody and immobilized carboxyl terminal on graphene surface.

Once graphene surface is functionalized with probe antibodies, the next step is the detection. Target antigens are brought into contact with binding sites on antibodies, where conformational changes due to conjugation occur with high affinity and specificity. As the antigens are typically charged at physiological pH, their binding causes a change in the surface charge close to graphene surface. This change is transduced as an electrical signal.

An important application of GFET immunosensors is the detection of pathogen antigens. For example, *E. coli* bacteria is the most common cause of food poisoning. Rapid sensing of *E. coli* and its metabolic activity was done using GFET-based

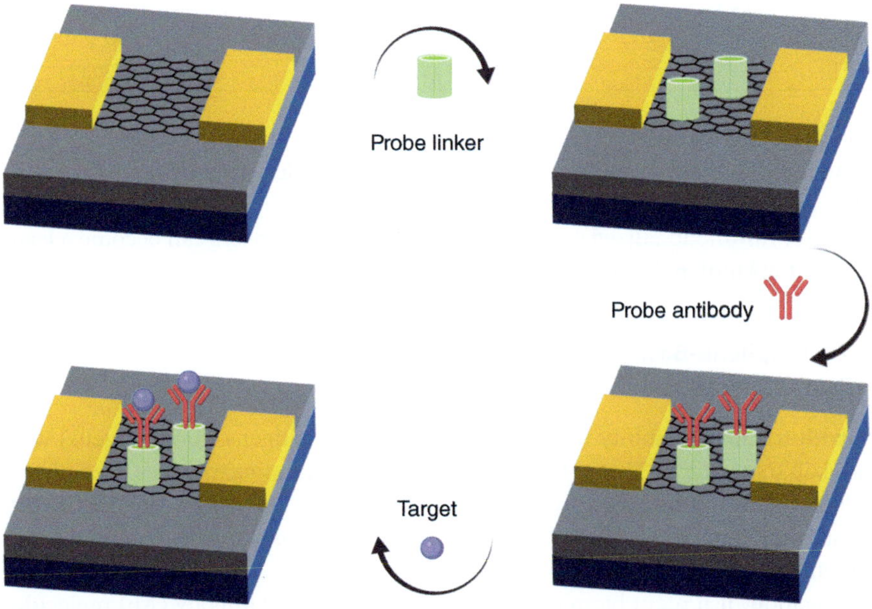

Figure 4.6 Schematic illustration of the preparation of G-FET based immunosensor. Source: Created with BioRender.com.

immunosensors [52, 53]. Similar sensor was proposed for the detection of the Zika virus in clinical samples, and its selectivity was also cross-checked with another viral antigen [55]. GFET immunosensors can also be used to combat HIV, and this was done by developing sensors for qualitative and quantitative detection of antiviral protein (microbicide extracted from soya beans) for HIV [74]. An early POC diagnostic tool has also been proposed for detection of Japanese encephalitis and avian influenza virus by detecting their viral antigens [56]. Recently, a graphene immunosensor was also proposed for early detection of COVID-19. SARS-CoV-2 spike antibodies were used to rapidly detect virus in nasopharyngeal swab specimens from COVID-19 patients with high sensitivity and without any sample pretreatments [54].

In addition to infectious diseases, detection of protein biomarkers is a relevant application field. For instance, early diagnosis of cancer is vital to the development of effective clinical treatments. This hugely depends on accurate detection of cancer biomarkers, which was done in real time using a graphene-based immunosensor [76]. Many protein biomarkers are available only in low concentrations and need to be detected with high precision from a complex sample (e.g. blood), placing stringent requirements on novel sensing platforms. As a representative example of a challenging yet well-characterized analyte, TSH was measured from complex whole serum with sub-picomolar detection limits and a wide measurement range [57].

Table 4.2 summarizes the discussed antibody-based graphene biosensors. Most of proposed graphene biosensors have reported a picomolar level of lower detection limit, which is quite remarkable compared to other label-free diagnostic tools. Especially, the immunosensor for COVID-19 has a lower detection limit of 16 pfu ml^{-1} in clinical samples with remarkable accuracy. Overall, there are several promising reports highlighting the potential of graphene for rapid and early diagnosis of various diseases that can also help in guiding treatment decisions.

Although many immunosensors have been proposed in the past decade, there are still some limitations. Debye screening and complexity of analyte medium are among the major challenges. Recent strategies to tackle these challenges are by using only the antigen binding part (Fab fragment) of the antibody on graphene surface in combination with PEG molecules [57]. Fragment is shorter compared to whole (Ab) thus antibody–antigen reaction occurs at a smaller distance from the surface of immunosensor [115]. The PEG molecules help to reduce both the non-specific binding and the Debye screening effect [95, 97].

4.4.3 Graphene-Based FET Biosensors for Enzymatic Biosensors

Enzymes are complex proteins that can catalyze biochemical reactions with high specificity. The specificity comes from chemical and electrical properties of certain active sites present on enzyme surfaces, which allow specific target substrate binding. During this binding reaction, enzymes convert the target substrate to an intermediate complex, which has lower activation energy and further diverges into smaller molecules [116]. This specificity has been used as a principle for enzyme-based electrochemical biosensors. Several studies have been performed to employ GFET as a sensing platform for enzymatic reactions, exploiting catalytic properties of enzymes [110]. In GFET

Table 4.2 Overview of graphene FET-based immunosensors.

References	Target analyte	Analyte condition	Probe immobilization mechanism	Probe	Response unit	LOD
[52]	*E. coli* bacteria	PBS buffer	PBASE linker	Anti *E. coli* antibody	Current (I_{DS})	10 cfu-ml
[53]	*E. coli* O157:H7	—	PASE linker	Anti *E. coli* antibody	Current (I_{DS})	10 cfu-ml
[76]	Carcino-embryonic antigen protein (Cancer)	PBS buffer	PASE linker	Antibodies	Current (I_{DS})	0.5 pm
[55]	Zika Virus antigens	PBS buffer	PBASE linker	Anti-Zika antibody	Capacitance (C%)	450 pM
[74]	Recombinant cyanovirin-N (anti-HIV resource)	PBS buffer	PASE linker	Antibody	Resistance (R)	0.45 pg ml^{-1}
[56]	Japanese encephalitis and Avian influenza virus	PB buffer	EDC-sulfo NHS	Antibody	Resistance (R)	1 fM for JEV 10 fM for AIV
[54]	SARS-CoV-2	PBS buffer	PBASE linker	SARS-CoV-2 spike antibody	Current (I_{DS})	16 pfu ml^{-1}
[57]	Thyroid stimulating hormone	Serum and buffer	PBA-PyMal EDC-NHS	Antibody fragment	Conductance ($\Delta G/G_o$)	0.8 fM

biosensors, enzymes are first immobilized on graphene surface through different methods. Then analyte solution is added, leading to the catalytic reaction close to the surface and transforming the analyte into a detectable reaction product (Figure 4.7). The transducer reads the change in electrical signals due to diffusion of the (charged) reaction product to the graphene surface. As shown in Table 4.3, the immobilization can be done through chemical bonding, i.e. EDC/NHS chemistry, or nonspecific binding through entrapment as a self-assembled monolayer (SAM) of enzymes.

Glucose monitoring gives direct information about diabetes, and its detection through electrochemical biosensors has been reported already in 1960s [1]. The GOD oxidizes glucose into detectable H_2O_2 which can alter electrical properties of graphene [118]. This approach has been applied in various biosensors for POC diagnostic of diabetes. Using above-mentioned principle, glucose and glutamate have been detected in millimolar concentrations in PBS buffer medium, by using PBSE linker molecules to functionalize graphene surface with GOD and glutamic dehydrogenase [60, 78]. Another approach to functionalizing graphene with enzyme molecules is through thin film deposition. GOD is entrapped in silk film coated on graphene; silk film is also used as a substrate, which gives biocompatibility aspects to proposed biosensors [63].

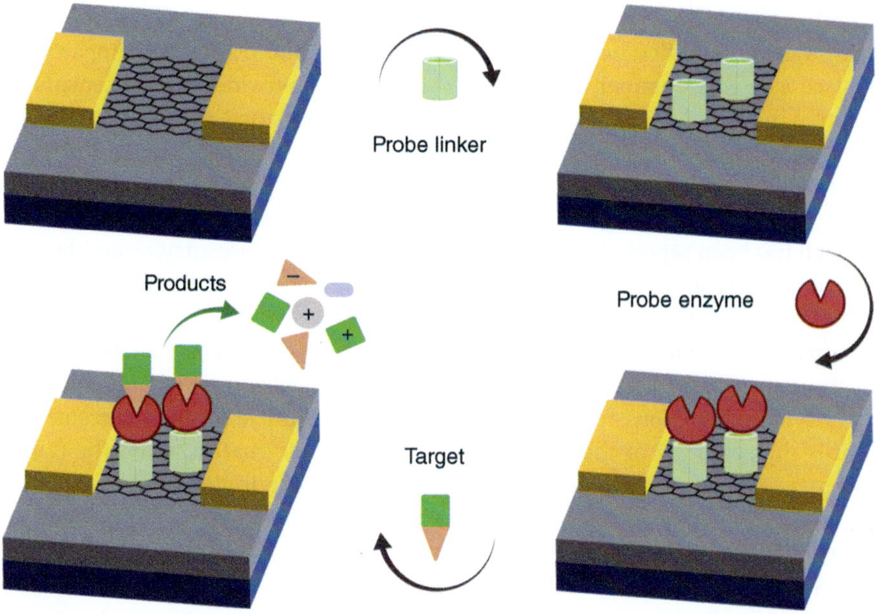

Figure 4.7 Schematic illustration of the preparation of G-FET based enzymatic biosensor. Source: Created with BioRender.com.

Table 4.3 Overview of graphene FET based enzymatic biosensors.

References	Target analyte	Reaction products	Probe immobilization mechanism	Probe	Response unit	LOD
[78]	Glucose, Glutamate	H_2O_2, NH_4OH	PBSE linker	Glucose oxidase, glutamic dehydrogenase	Current (I_{DS})	0.1–5 mM
[60]	Glucose	H_2O_2	PBSE linker	Glucose oxidase	Current (I_{DS})	3.3–10.9 mM
[63]	Sweat glucose	H_2O_2	Enzyme entrapped silk film	Glucose oxidase	Current (I_{DS})	0.1–10 mM
[61]	Glucose	H_2O_2	Enzyme entrapped chitosan film	Glucose oxidase	Current (I_{DS}) Gate voltage (ΔV_{geff})	0.5–300 μM
[117]	Penicillin	Penicilloic acid (H^+)	Encapsulation in hydrogel	Penicillinase	Current (I_{DS})	0.2 mM

Another sensor was proposed for glucose detection in human saliva or sweat. Here graphene modified with GOD, chitosan, Nafion, and platinum nanoparticles was used as a FET channel. This sensor configuration provided higher sensitivity compared to traditional Si-based biosensors and was suggested as a convenient tool for clinical diagnostic of glucose [61].

Hydrogel layers are also used to encapsulate enzymes on graphene surfaces to restrict unspecific binding and increase timespan for enzymatic activity. This approach has been reported to detect penicillin by penicillinase enzyme and hydrogel increased sensor's shelf life to seven days compared to few hours of catalytic activity of enzyme without hydrogel [117].

Compared to previously explained DNA and antibody-based electrochemical biosensors, enzyme-based sensors are not contained by disadvantage of Debye screening. Enzymatic reactions result in molecules, which diffuse to graphene surface, and it is the detection of these molecules that gives information about the target molecules [79]. Enzyme-based biosensors also have some limitations, i.e. pH and temperature, which can affect the stability of enzymes. But graphene is proven to be an excellent material which is highly compatible with enzymatic reactions and stability [110].

4.4.4 Graphene-Based FET Biosensors for Sensing of Small Ions

In medical diagnostics, analysis of ion content in patient samples is very important. When ions come directly into contact with graphene surface, they induce doping, resulting in a shift in the material's electrical properties [67]. In GFET-based pH sensors, graphene channel measures the concentration of (H^+) ions. Therefore, graphene quality and its stability against surface dopants are key to reliable pH measurements.

In a recent study, dual-gated graphene-based pH sensor was proposed. Here, oxygenated graphene was used as channel, while fluorinated graphene was used as gating electrode. This approach resulted in an on-chip gate electrode with higher and more stable pH sensitivity. Solution deposition of materials with high dielectric constants is also reported in graphene pH sensors. The Al_2O_3 membrane has shown better chemical stability and pH sensitivity [64]. Hydroxyl (OH)-containing aromatic molecules, i.e. phenol can also be functionalized on graphene surface, and showed promising pH sensitivity [119].

Table 4.4 summarizes ion-sensitive GFET sensors. To achieve selectivity for ions of interest, ionophores are frequently used. These molecules can be attached to graphene via linkers (Figure 4.8) or as a component within a membrane. Ionophore membranes are composed of molecules, which allow binding and diffusion of specific ions; these membranes can be used on platforms such as ISFET. Such membranes were used for real-time measurement of ionic strength of selective ions, i.e. (K^+, Na^+, $NH_4{}^+$, $NO_3{}^-$, $SO_4{}^{-2}$, $HPO_4{}^{-2}$, and Cl^-). Crown ether and valinomycin are highly selective ionophores for sodium and potassium cations, and when used in

Table 4.4 Graphene-based FETs for pH and other ions.

References	Target analyte	Analyte condition	Probe immobilization mechanism	Probe	Response unit	LOD
[70]	Hg^{2+}	—	Adsorbed SAM	1-octa decanethiol	(ΔV_{CNP})	10 ppm
[69]	Na^{+}	Tris–HCl buffer	—	Gr	Current (I_{DS})	1 nM–1 mM
[68]	K^{+}, Na^{+}	Tris–HCl buffer	Spin coated	Valinomycin (ionophore membrane)	Dirac point (V_{DP}) Current (I_{DS})	10 nM–1.0 mM
[119]	H^{+} (pH)	Buffer solution	π-bonding	Phenol	Conductance G_S Charge mobility	4 ppm
	K^{+}	Aqueous solution	π–π stacking	Crown ether	Conductance G_S Charge mobility	100 μM–1 M
[59]	Pb^{2+}	Blood sample	π–π stacking	8–17 DNAzyme	Dirac point (V_{DP})	0.5 nM
[67]	K^{+}, Na^{+}, NH_4^{+}, NO_3^{-}, SO_4^{-2}, HPO_4^{-2}, Cl^{-}	Aqueous solution	Drop casting	Ionophore membrane	Current (I_{DS})	100 mM
[64]	pH	Buffer solution	Spin coated	Al_2O_3 membrane	Current (I_{DS})	36.8 mV/pH
[65]	pH	Buffer solution	Plasma treatment	Oxygenated graphene	Current (I_{DS}) Dirac point (V_{DP})	36.8 mV/pH

combination with ISFET, they have proven to be an excellent tool for targeted detection of (K^{+} and Na^{+}) at various concentrations [68, 119]. Mercury (Hg^{2+}) is a toxicant which contributes to lethal diseases, i.e. Alzheimer, Parkinson, neurological impairment and autism [120]. Its detection in environment was done using GFET.

A well-defined SAM of 1-octadecanethiol on monolayer graphene gives target detection of mercury, as mercury attaches to thiol end of SAM. Lead (Pb) is a dangerous toxicant for child's nervous system, and its detection in children's blood was realized by using GFET-based aptasensor. Lead binding ability and high selectivity in real blood samples are complex features, but highly selective layer of 8–17 DNAzyme was stacked through π–π bonding on graphene surface [59].

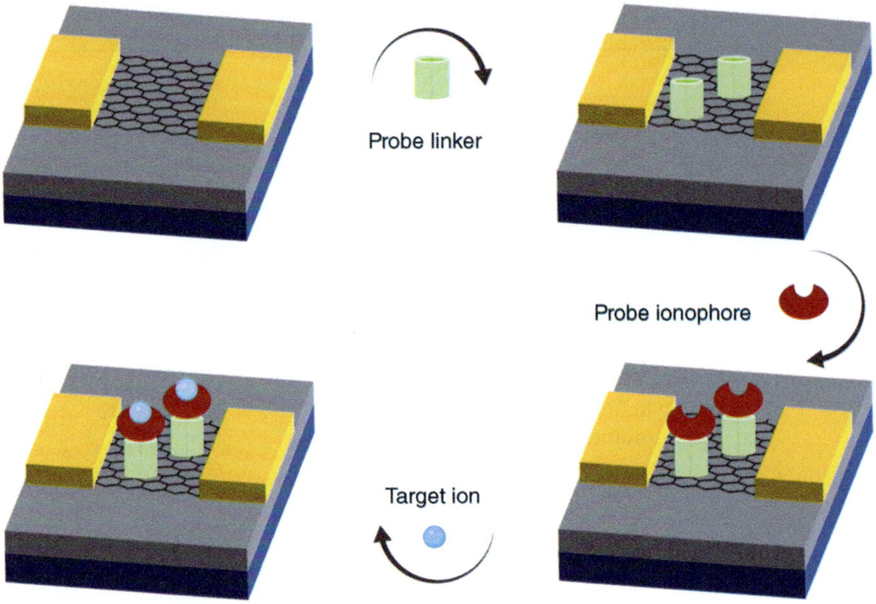

Figure 4.8 Schematic illustration of the preparation of G-FET-based chemical sensor for ions. Source: Created with BioRender.com.

4.5 Discussion

4.5.1 Summary

In this chapter, we introduced biosensors based on GFETs with a particular emphasis on applications in diagnostics. We briefly described the working principle and the measurement scheme, the fabrication and chemical functionalization of these devices. While several types of graphene and graphene-related materials are currently used in biosensors, we mostly focused here on CVD-grown graphene transferred via wet transfer method. GFETs are investigated as a possible new tool for the analysis of a wide range of clinically relevant analytes. Based on representative examples, we provided an overview of the research related to the detection of DNA hybridization, antibody–antigen interactions, enzymatic reactions, small molecules, and ions. These and other studies may pave the way for future development of graphene-based biosensors toward commercial applications.

4.5.2 Challenges

Although many proof-of-principle studies exist showing the promise of GFET-based biosensors for specific detection of clinically relevant molecules, various challenges are still to be overcome. One of the major challenges is the performance of these sensors in real biological solutions, as non-specific interactions with species other than the target analyte can mask the specific signals. Another major challenge is bringing these laboratory-scale sensors to the mass market. As of now, the fabrication of these biosensors involves various manual steps at laboratory scale. Many of these fabrication steps, such as graphene transfer and the chemistry related to immobilization of probe molecules, are difficult to upscale. Factors such as stability, orientation, and specificity of probe molecules on graphene channels are key contributors toward successful surface chemistry. Hence, standardization and automation of surface chemistry are significant for upscaling these devices. Graphene transfer can cause unintentional doping and degrade the performance of transistors. As of now these biosensors are used in combination with a fluidic chamber, which is responsible for delivering target analytes to graphene channel. The fluid dynamics occurring in small channels and overall miniaturized design of this chamber deserve more attention to achieve scalable biosensors with accurate sensing abilities. The quality of the transferred graphene and the linking chemistry contribute toward the detection limit, sensitivity, precision, and other key factors necessary to evaluate quantitative and qualitative ability of these novel biosensors. All these challenges need attention to demonstrate an efficient, commercially viable, and industrially feasible way of producing these graphene-based biosensors. Additionally, theoretical modeling and simulation are required to understand the exact nature of interfacial processes between biological solutions and graphene channel. This can help in conceiving robust fabrication processes, predicting shelf life, reproducibility, and overall performance. Also, artificial intelligence can help to design more robust sensors and enable multiparametric measurements.

4.5.3 Future Perspectives

There is still plenty of room for graphene-based transistors in biomedical research and clinical applications beyond detection of analytes discussed here. Graphene can also be used in monitoring of heart rates, respiration, blood pressure, and muscular motions. For instance, graphene biosensors can be used to monitor muscle and joint movements related to incurred stress in athlete's body and help to minimize or prevent severe injuries. Certain, more invasive, applications related to the nervous system and cardiovascular system can be game changers in medicine. For example, electrical stimulation of neurons or muscles could be performed using implanted GFETs on biocompatible substrates, enabling new treatment options. Further down the road, electrical stimulation could be combined with *in vivo* biosensing of relevant molecules, e.g. neurotransmitters, or even with controlled drug release to realize a complete closed loop system of diagnosis and therapy.

References

1 Clark, L.C. Jr. and Lyons, C. (1962). Electrode systems for continuous monitoring in cardiovascular surgery. *Ann. N.Y. Acad. Sci.* 102 (1): 29–45. https://doi.org/10.1111/j.1749-6632.1962.tb13623.x.

2 Kahng, D. (1963). Electric field controlled semiconductor device. US Patent 3,102,230, filed 31 May 1960 and granted 27 August 1963, https://patents.google.com/patent/US3102230A/en.
3 B.P. (1972). Development, operation, and application of the ion-sensitive field-effect transistor as a tool for electrophysiology. *IEEE Trans. Biomed. Eng.* 19 (5): 342–351.
4 Shitashima, K., Kyo, M., Koike, Y., and Henmi, H. (2002). Development of in situ pH sensor using ISFET, *Proceedings of the 2002 International Symposium on Underwater Technology (Cat. No.02EX556)*, Tokyo, Japan, 2002, pp. 106–108, https://doi.org/10.1109/UT.2002.1002403.
5 Jane, A., Dronov, R., Hodges, A., and Voelcker, N.H. (2009). Porous silicon biosensors on the advance. *Trends Biotechnol.* 27: 230–239.
6 Parce, J.W., Owicki, J.C., Kercso, K.M. et al. (1989). Detection of cell-affecting agents with a silicon biosensor. *Science* 246 (4927): 243–247.
7 Nair, P.R. and Alam, M.A. (2007). Design considerations of silicon nanowire biosensors. *IEEE Trans. Electron Devices* 54 (12): 3400–3408. http://ieeexplore.ieee.org.
8 Wang, J. (2005). Carbon-nanotube based electrochemical biosensors: a review, *Electroanalysis*, 17: 7–14. https://doi.org/10.1002/elan.200403113.
9 Shao, Y., Wang, J., Wu, H. et al. Graphene based electrochemical sensors and biosensors: a review. (2010), *Electroanalysis*, 22: 1027–1036. https://doi.org/10.1002/elan.200900571.
10 Sarkar, D., Liu, W., Xie, X. et al. (2014). MoS_2 field-effect transistor for next-generation label-free biosensors. *ACS Nano* 8 (4): 3992–4003.
11 Geim, A.K. and Novoselov, K.S. (2007). The rise of graphene. *Nat. Mater.* 6 (3): 183–191. https://doi.org/10.1038/nmat1849.
12 Kuila, T., Bose, S., Khanra, P. et al. (2011). Recent advances in graphene-based biosensors. *Biosens. Bioelectron.* 26: 4637–4648.
13 Pumera, M. (2011). Graphene in biosensing. *Mater. Today* 14 (7): 308–315. https://www.sciencedirect.com/science/article/pii/S1369702111701602.
14 Chang, H., Tang, L., Wang, Y. et al. (2010). Graphene fluorescence resonance energy transfer aptasensor for the thrombin detection. *Anal. Chem.* 82 (6): 2341–2346. https://doi.org/10.1021/ac9025384.
15 Daher, M.G., Taya, S.A., Colak, I. et al. (2022). Surface plasmon resonance biosensor based on graphene layer for the detection of waterborne bacteria. *J. Biophotonics* 15 (5): e202200001. https://doi.org/10.1002/jbio.202200001.
16 Bakker, E. and Telting-Diaz, M. (2002). Electrochemical sensors. *Anal. Chem.* 74 (12): 2781–2800. https://doi.org/10.1021/ac0202278.
17 Qu, H., Liu, H., Wang, B. et al. (2022). A three dimensional porous diamond-multilayer graphene nanohybrid film for surface-enhanced Raman spectroscopy. *Diamond Relat. Mater.* 121: 108737. https://www.sciencedirect.com/science/article/pii/S0925963521005008.
18 Plutnar, J., Pumera, M., and Sofer, Z. (2018). The chemistry of CVD graphene. *J. Mater. Chem. C* 6 (23): 6082–6101. http://dx.doi.org/10.1039/C8TC00463C.

19 Israelachvili, J. (2011). *Intermolecular and Surface Forces*. Academic Press, https://doi.org/10.1016/C2009-0-21560-1.

20 Chen, F., Qing, Q., Xia, J. et al. (2009). Electrochemical gate-controlled charge transport in graphene in ionic liquid and aqueous solution. *J. Am. Chem. Soc.* 131 (29): 9908–9909. https://doi.org/10.1021/ja9041862.

21 Hassanpour Amiri, M., Heidler, J., Hasnain, A. et al. (2019). Doping free transfer of graphene using aqueous ammonia flow. *RSC Adv.* 10 (2): 1127–1131.

22 Andoy NM, Filipiak MS, Vetter D, Gutiérrez-Sanz Ó, Tarasov A, Andoy NM, et al. Graphene-based electronic immunosensor with femtomolar detection limit in whole serum. 2018; https://doi.org/10.1002/admt.201800186.

23 Fu, W., Jiang, L., van Geest, E.P. et al. (2017). Sensing at the surface of graphene field-effect transistors. *Adv. Mater.* 29: 1603610.

24 Béraud, A., Sauvage, M., Bazán, C.M. et al. (2021). Graphene field-effect transistors as bioanalytical sensors: design, operation and performance. *Analyst* 146: 403–428.

25 Yazdi, G.R.; Iakimov, T.; Yakimova, R. *Crystals* Epitaxial graphene on SiC: a review of growth and characterization. 6, 53. https://doi.org/10.3390/cryst6050053 2016.

26 Xuesong, L., Weiwei, C., Jinho, A. et al. (2009). Large-area synthesis of high-quality and uniform graphene films on copper foils. *Science* 324 (5932): 1312–1314. https://doi.org/10.1126/science.1171245.

27 Ullah S, Yang X, Ta HQ, Hasan M, Bachmatiuk A, Tokarska K, et al. Graphene transfer methods: a review. 2014; https://doi.org/10.1007/s12274-021-3345-8.

28 Lerner, M.B., Matsunaga, F., Hee Han, G. et al. (2014). Scalable production of highly sensitive nanosensors based on graphene functionalized with a designed G protein-coupled receptor. *Nano Lett.* 14: 19. https://pubs.acs.org/sharingguide lines.

29 Lvova, L., Lagarde, F., Lyon, C. et al. (2018). Graphene FET array biosensor based on ssDNA aptamer for ultrasensitive Hg Graphene FET array biosensor based on ssDNA aptamer for ultrasensitive Hg^{2+} detection in environmental pollutants. *Front. Chem.* 1: 333. www.frontiersin.org.

30 Hong N, Kireev D, Zhao Q, Chen D, Akinwande D, Li W, et al. Roll-to-roll dry transfer of large-scale graphene. 2021; https://doi.org/10.1002/adma.202106615.

31 Wang Z, Hao Z, Yu S, Huang C, Pan Y, Zhao X. A wearable and deformable graphene-based affinity nanosensor for monitoring of cytokines in biofluids. *Nanomaterials* 10, 1503. https://doi.org/10.3390/nano10081503 2020.

32 Lee, H., Choi, T.K., Lee, Y.B. et al. (2016). A graphene-based electrochemical device with thermoresponsive microneedles for diabetes monitoring and therapy. *Nat. Nanotechnol.* 11 (6): 566–572. https://doi.org/10.1038/nnano.2016.38.

33 Mannoor, M.S., Tao, H., Clayton, J.D. et al. (2012). Graphene-based wireless bacteria detection on tooth enamel. *Nat. Commun.* **3**: 763 https://doi.org/10.1038/ncomms1767.

34 Zhang W, Dong G, Feng H, Shan S, Huang L, Yuan F, et al. 2000682 (1 of 7) wearable corneal biosensors fabricated from PEDOT functionalized sulfur-doped graphene for use in the early detection of myopia. 2020. https://doi.org/10.1002/admt.202000682.

35 Khan, A., Winder, M., and Hossain, G. (2022). Modified graphene-based nanocomposite material for smart textile biosensor to detect lactate from human sweat. *Biosens. Bioelectron.* 10: 100103. https://www.sciencedirect.com/science/article/pii/S259013702100039X.

36 Masvidal-Codina, E., Illa, X., Dasilva, M. et al. (2019). High-resolution mapping of infraslow cortical brain activity enabled by graphene microtransistors. *Nat. Mater.* 18 (3): 280–288. https://doi.org/10.1038/s41563-018-0249-4.

37 Blaschke, B.M., Tort-Colet, N., Guimerà-Brunet, A. et al. (2017). Mapping brain activity with flexible graphene micro-transistors. *2D Mater.* 4 (2): 025040. http://dx.doi.org/10.1088/2053-1583/aa5eff.

38 Wang Z, Hao Z, Wang X, Huang C, Lin Q, Zhao X, et al. A flexible and regenerative aptameric graphene-nafion biosensor for cytokine storm biomarker monitoring in undiluted biofluids toward wearable applications. 2005; https://doi.org/10.1002/adfm.202005958.

39 Akin Anakok, D. and Cete, S. (2021). A new surface based on graphene modified with nanoparticles and nafion for the detection of glucose. *Russ. J. Electrochem.* 57 (12): 1186–1195.

40 Eda G, Fanchini G, Chhowalla M. Large-area ultrathin films of reduced graphene oxide as a transparent and flexible electronic material. *Nat. Nanotechnol.* **3**, 270–274, https://doi.org/10.1038/nnano.2008.83 2008.

41 Li, D., Müller, M., Gilje, S. et al. (2008). Processable aqueous dispersions of graphene nanosheets. *Nat. Nanotechnol.* **3**, 27: 101–105 https://doi.org/10.1038/nnano.2007.451.

42 Wang, J., Liang, M., Fang, Y. et al. (2012). Rod-coating: towards large-area fabrication of uniform reduced graphene oxide films for flexible touch screens. *Adv. Mater.* 24: 2874–2878. www.advmat.de.

43 Zhang, Y., Ding, Y., Li, C. et al. (2021). An optic-fiber graphene field effect transistor biosensor for the detection of single-stranded DNA. *Anal. Methods* 13 (15): 1839–1846. http://dx.doi.org/10.1039/D1AY00101A.

44 Purwidyantri, A., Domingues, T., Borme, J. et al. (2021). Influence of the electrolyte salt concentration on DNA Detection with graphene transistors. *Biosensors* 11 (1): 24.

45 Ping, J., Vishnubhotla, R., Vrudhula, A., and Johnson, A.T.C. (2016). scalable production of high-sensitivity, label-free DNA biosensors based on back-gated graphene field effect transistors. *ACS Nano* 10 (9): 8700–8704. https://doi.org/10.1021/acsnano.6b04110.

46 Wang, J., Rivas, G., Cai, X. et al. (1997). DNA electrochemical biosensors for environmental monitoring. A review. *Anal. Chim. Acta* 347 (1): 1–8. https://www.sciencedirect.com/science/article/pii/S0003267096005983.

47 Zheng, C., Huang, L., Zhang, H. et al. (2015). Fabrication of ultrasensitive field-effect transistor DNA biosensors by a directional transfer technique based on CVD-grown graphene. *ACS Appl. Mater. Interfaces* 7 (31): 16953–16959. https://doi.org/10.1021/acsami.5b03941.

48 Nielsen, P.E., Egholm, M., and Buchardu, O. (1994). Reviews peptide nucleic acid (PNA). A DNA mimic with a peptide backbone. *Bioconjugate Chem.* 5: https://pubs.acs.org/sharingguidelines.

49 Balderston S, Taulbee JJ, Celaya E, Fung K, Jiao A, Smith K, et al. Discrimination of single-point mutations in unamplified genomic DNA via Cas9 immobilized on a graphene field-effect transistor. 2021; https://doi.org/10.1038/s41551-021-00706-z.

50 Gao, Z., Xia, H., Zauberman, J. et al. (2018). Detection of Sub-fM DNA with target recycling and self-assembly amplification on graphene field-effect biosensors. *Nano Lett.* 18 (6): 3509–3515. https://doi.org/10.1021/acs.nanolett.8b00572.

51 Ban, D.K., Liu, Y., Wang, Z. et al. (2020). Direct DNA methylation profiling with an electric biosensor. *ACS Nano* 14 (6): 6743–6751. https://doi.org/10.1021/acsnano.9b10085.

52 Huang, Y., Dong, X., Liu, Y. et al. (2011). Graphene-based biosensors for detection of bacteria and their metabolic activities. *J. Mater. Chem.* 21 (33): 12358–12362. http://dx.doi.org/10.1039/C1JM11436K.

53 Pourasl, A.H., Ahmadi, M.T., Rahmani, M., and Ismail, R. (2017). Graphene based biosensor model for *Escherichia coli* bacteria detection. *J. Nanosci. Nanotechnol.* 17 (1): 601–605.

54 Seo, G., Lee, G., Kim, M.J. et al. (2020). Rapid detection of COVID-19 causative virus (SARS-CoV-2) in human nasopharyngeal swab specimens using field-effect transistor-based biosensor. *ACS Nano* 14 (4): 5135–5142. https://doi.org/10.1021/acsnano.0c02823.

55 Afsahi, S., Lerner, M.B., Goldstein, J.M. et al. (2018). Novel graphene-based biosensor for early detection of Zika virus infection. *Biosens. Bioelectron.* 100: 85–88. https://www.sciencedirect.com/science/article/pii/S0956566317305869.

56 Roberts, A., Chauhan, N., Islam, S. et al. (2020). Graphene functionalized field-effect transistors for ultrasensitive detection of Japanese encephalitis and Avian influenza virus. *Sci. Rep.* 10 (1): 14546. https://doi.org/10.1038/s41598-020-71591-w.

57 Andoy, N.M., Filipiak, M.S., Vetter, D. et al. (2018). Graphene-based electronic immunosensor with femtomolar detection limit in whole serum. *Adv. Mater. Technol.* 3 (12): 1800186. https://onlinelibrary.wiley.com/doi/abs/10.1002/admt.201800186.

58 Wang, L., Wu, A., and Wei, G. (2018). Graphene-based aptasensors: from molecule–interface interactions to sensor design and biomedical diagnostics. *Analyst* 143 (7): 1526–1543. http://dx.doi.org/10.1039/C8AN00081F.

59 Wang C, Cui X, Li Y, Li H, Huang L, Bi J, et al. A label-free and portable graphene FET aptasensor for children blood lead detection. *Sci. Rep.* **6**, 21711. https://doi.org/10.1038/srep21711 2016.

60 Kwak, Y.H., Choi, D.S., Kim, Y.N. et al. (2012). Flexible glucose sensor using CVD-grown graphene-based field effect transistor. *Biosens. Bioelectron.* 37 (1): 82–87.

61 Zhang, M., Liao, C., Hin Mak, C. et al. (2015). Highly sensitive glucose sensors based on enzyme-modified whole-graphene solution-gated transistors. *Sci. Rep.* **5**: 8311, https://doi.org/10.1038/srep08311.

62 Viswanathan, S., Narayanan, T.N., Aran, K. et al. (2015). Graphene–protein field effect biosensors: glucose sensing. *Mater. Today* 18 (9): 513–522. https://www.sciencedirect.com/science/article/pii/S1369702115001194.

63 You, X. and Pak, J.J. (2014). Graphene-based field effect transistor enzymatic glucose biosensor using silk protein for enzyme immobilization and device substrate. *Sens. Actuators, B* 202: 1357–1365. https://www.sciencedirect.com/science/article/pii/S0925400514004900.

64 Bae, T.-E., Kim, H., and Jung, J. (2014). Fabrication of high-performance graphene field-effect transistor with solution-processed Al_2O_3 sensing membrane. *Appl. Phys. Lett.* 104: 153506. https://doi.org/10.1063/1.4871865.

65 Kim, D.H.; Park, W.H.; Oh, H.G.; Jeon, D.C.; Lim, J.M.; Song, K.S. Two-channel graphene pH sensor using semi-ionic fluorinated graphene reference electrode. *Sensors 20*, 4184. https://doi.org/10.3390/s20154184 2020.

66 Zeng, Z., Wei, W., Li, B. et al. (2021). CVD polycrystalline graphene as sensing film of extended-gate ISFET for low-drift pH sensor. *J. Electrochem. Soc.* 168 (6): 067520. https://iopscience.iop.org/article/10.1149/1945-7111/ac0ab1.

67 Fakih I, Durnan O, Mahvash F, Napal I, Centeno A, Zurutuza A, et al. Selective ion sensing with high resolution large area graphene field effect transistor arrays. 2020; https://doi.org/10.1038/s41467-020-16979-y.

68 Maehashi, K., Sofue, Y., Okamoto, S. et al. (2013). Selective ion sensors based on ionophore-modified graphene field-effect transistors. *Sens. Actuators, B* 187: 45–49. https://www.sciencedirect.com/science/article/pii/S0925400512009458.

69 Sofue, Y., Ohno, Y., Maehashi, K. et al. (2011). Highly sensitive electrical detection of sodium ions based on graphene field-effect transistors. *Jpn. J. Appl. Phys.* 50 (6): 06GE07. http://dx.doi.org/10.1143/JJAP.50.06GE07.

70 Zhang, T., Cheng, Z., Wang, Y. et al. (2010). Self-assembled 1-octadecanethiol monolayers on graphene for mercury detection. *Nano Lett.* 10 (11): 4738–4741. https://doi.org/10.1021/nl1032556.

71 Criado A, Melchionna M, Marchesan S, Prato M. Supported graphene the covalent functionalization of graphene on substrates. *Angew. Chem. Int. Ed.*, 54: 10734–10750. https://doi.org/10.1002/anie.201501473 2015.

72 Paulus, G.L.C., Wang, Q.H., and Strano, M.S. (2013). Covalent electron transfer chemistry of graphene with diazonium salts. *Acc. Chem. Res.* 46 (1): 160–170. https://doi.org/10.1021/ar300119z.

73 Eissa, S., L, hocine, L., Siaj, M., and Zourob, M. (2013). A graphene-based label-free voltammetric immunosensor for sensitive detection of the egg allergen ovalbumin. *Analyst* **138**: 4378–4384.

74 de Almeida, P.R., Murad, A.M., Silva, L.P. et al. (2020). Development of a graphene-based biosensor for detecting recombinant cyanovirin-N. *Biosensors* 10 (12): https://www.mdpi.com/2079-6374/10/12/206.

75 Zhao, W., Xing, Y., Lin, Y. et al. (2020). Monolayer graphene chemiresistive biosensor for rapid bacteria detection in a microchannel. *Sens. Act. Rep.* 2 (1): 100004. https://www.sciencedirect.com/science/article/pii/S2666053920300011.

76 Zhou, L., Mao, H., Wu, C. et al. (2017). Label-free graphene biosensor targeting cancer molecules based on non-covalent modification. *Biosens. Bioelectron.* 87: 701–707. https://www.sciencedirect.com/science/article/pii/S0956566316308971.

77 Hajian, R., Balderston, S., Tran, T. et al. (2019). Detection of unamplified target genes via CRISPR–Cas9 immobilized on a graphene field-effect transistor. *Nat. Biomed. Eng.* 3 (6): 427–437. https://doi.org/10.1038/s41551-019-0371-x.

78 Huang, Y., Dong, X., Shi, Y. et al. (2010). Nanoelectronic biosensors based on CVD grown graphene. *Nanoscale* 2 (8): 1485–1488. http://dx.doi.org/10.1039/C0NR00142B.

79 Ono, T., Kanai, Y., Inoue, K. et al. (2019). Electrical biosensing at physiological ionic strength using graphene field-effect transistor in femtoliter microdroplet. *Nano Lett.* 19 (6): 4004–4009. https://doi.org/10.1021/acs.nanolett.9b01335.

80 Hermanson, G.T. (2013). Chapter 3 – The reactions of bioconjugation. In: *Bioconjugate Techniques*, 3e (ed. G.T. Hermanson), 229–258. Boston: Academic Press https://www.sciencedirect.com/science/article/pii/B9780123822390000030.

81 Chen, R.J., Zhang, Y., Wang, D., and Dai, H. (2001). Noncovalent sidewall functionalization of single-walled carbon nanotubes for protein immobilization. *J. Am. Chem. Soc.* 123 (16): 3838–3839. https://doi.org/10.1021/ja010172b.

82 Fu, W., Nef, C., Tarasov, A. et al. (2013). High mobility graphene ion-sensitive field-effect transistors by noncovalent functionalization. *Nanoscale* 5 (24): 12104–12110. http://dx.doi.org/10.1039/C3NR03940D.

83 Dontschuk, N., Stacey, A., Tadich, A. et al. (2015). A graphene field-effect transistor as a molecule-specific probe of DNA nucleobases. *Nat. Commun.* 6 (1): 6563. https://doi.org/10.1038/ncomms7563.

84 Kim, H.E., Schuck, A., Lee, J.H., and Kim, Y.-S. (2019). Solution-gated graphene field effect transistor for TP53 DNA sensor with coplanar electrode array. *Sens. Actuators, B* 291: 96–101. https://www.sciencedirect.com/science/article/pii/S092540051930437X.

85 Her, J.L., Pan, T.M., Lin, W.Y. et al. (2013). Label-free detection of alanine aminotransferase using a graphene field-effect biosensor. *Sens. Actuators, B* 182: 396–400.

86 Butt, H.-J., Graf, K., and Kappl, M. (2003). *Physics and Chemistry of Interfaces*. Wiley-VCH Verlag GmbH & Co. KGaA https://doi.org/10.1002/3527602313.

87 Zheng, Z., Zhang, H., Zhai, T., and Xia, F. (2021). Overcome Debye length limitations for biomolecule sensing based on field effective transistors. Critical review. *Chin. J. Chem.* 39: 999–1008. https://mc.manuscriptcentral.com/cjocFor publishedarticles:https://onlinelibrary.wiley.com/journal/16147065*.

88 Vacic, A., Criscione, J.M., Rajan, N.K. et al. (2011). Determination of molecular configuration by Debye length modulation. *J. Am. Chem. Soc.* 133 (35): 13886–13889. https://doi.org/10.1021/ja205684a.

89 Stern, E., Vacic, A., Rajan, N.K. et al. (2010). Label-free biomarker detection from whole blood. *Nat. Nanotechnol.* 5 (2): 138–142.

90 Erickson, H.P Size and shape of protein molecules at the nanometer level determined by sedimentation, gel filtration, and electron microscopy, *Biol. Proced. Online* **11**, 32 (2009). https://doi.org/10.1007/s12575-009-9008-x.

91 Ohno, Y., Maehashi, K., and Matsumoto, K. (2010). Label-free biosensors based on aptamer-modified graphene field-effect transistors. *J. Am. Chem. Soc.* 132 (51): 18012–18013. https://doi.org/10.1021/ja108127r.

92 Okamoto, S., Ohno, Y., Maehashi, K. et al. (2012). Immunosensors based on graphene field-effect transistors fabricated using antigen-binding fragment. *Jpn. J. Appl. Phys.* 51: 06FD08. http://dx.doi.org/10.1143/JJAP.51.06FD08.

93 Liu, B., Huang, P.-J.J., Zhang, X. et al. (2013). Parts-per-million of polyethylene glycol as a non-interfering blocking agent for homogeneous biosensor development. *Anal. Chem.* 85 (21): 10045–10050. https://doi.org/10.1021/ac4024654.

94 Gao, N., Zhou, W., Jiang, X. et al. (2015). General strategy for biodetection in high ionic strength solutions using transistor-based nanoelectronic sensors. *Nano Lett.* 15 (3): 2143–2148. https://doi.org/10.1021/acs.nanolett.5b00133.

95 Gao, N., Gao, T., Yang, X. et al. (2016). Specific detection of biomolecules in physiological solutions using graphene transistor biosensors. *Proc. Natl. Acad. Sci.* 113 (51): 14633. http://www.pnas.org/content/113/51/14633.abstract.

96 Wang, X., Zhu, Y., Olsen, T.R. et al. (2018). A graphene aptasensor for biomarker detection in human serum. *Electrochim. Acta* 290: 356–363. https://www.sciencedirect.com/science/article/pii/S0013468618319431.

97 Haustein, N., Gutiérrez-Sanz, Ó., and Tarasov, A. (2019). Analytical model to describe the effect of polyethylene glycol on ionic screening of analyte charges in transistor-based immunosensing. *ACS Sens.* 4 (4): 874–882. https://doi.org/10.1021/acssensors.8b01515.

98 Kwon, S.S., Choi, J., Heiranian, M. et al. (2019 Jul 10). Electrical double layer of supported atomically thin materials. *Nano Lett.* 19 (7): 4588–4593. https://doi.org/10.1021/acs.nanolett.9b01563.

99 Choi, J., Mun, J., Wang, M.C. et al. (2017). Hierarchical, dual-scale structures of atomically thin MoS_2 for tunable wetting. *Nano Lett.* 17 (3): 1756–1761. https://doi.org/10.1021/acs.nanolett.6b05066.

100 Hwang MT, Park I, Heiranian M, Taqieddin A, You S, Faramarzi V, et al. Ultrasensitive detection of dopamine, IL-6 and SARS-CoV-2 proteins on crumpled graphene FET biosensor. 2021; https://doi.org/10.1002/admt.202100712.

101 Taeyoung Hwang M, Heiranian M, Kim Y, You S, Leem J, Taqieddin A, et al. Ultrasensitive detection of nucleic acids using deformed graphene channel field effect biosensors. 2020; https://doi.org/10.1038/s41467-020-15330-9.

102 Hao Z, Luo Y, Huang C, Wang Z, Song G, Pan Y, et al. An intelligent graphene-based biosensing device for cytokine storm syndrome biomarkers detection in human biofluids. *Small.* 17(29):e2101508. https://doi.org/10.1002/smll.202101508 2021.

103 Figueroa-Miranda, G., Wu, C., Zhang, Y. et al. (2020). Polyethylene glycol-mediated blocking and monolayer morphology of an electrochemical aptasensor for malaria biomarker detection in human serum. *Bioelectrochemistry* 136: 107589. https://www.sciencedirect.com/science/article/pii/S1567539420301080.

104 Danielson, E., Sontakke, V.A., Porkovich, A.J. et al. (2020). Graphene based field-effect transistor biosensors functionalized using gas-phase synthesized gold nanoparticles. *Sens. Actuators, B* 320: 128432. https://www.sciencedirect.com/science/article/pii/S0925400520307772.

105 Kwong D, Tsang H, Lieberthal J, Watts C, Dunlop IE, Ramadan S, et al. Chemically functionalised Graphene FET biosensor for the label-free sensing of exosomes. 2019; https://doi.org/10.1038/s41598-019-50412-9.

106 Oshin O, Kireev D, Hlukhova H, Idachaba F, Akinwande D, Atayero A, et al. Graphene-based biosensor for early detection of iron deficiency. *Sensors, 20*, 3688. https://doi.org/10.3390/s20133688 2020.

107 Wetmur, J.G. (1991). DNA probes: applications of the principles of nucleic acid hybridization. *Crit. Rev. Biochem. Mol. Biol.* 26 (3–4): 227–259. https://doi.org/10.3109/10409239109114069.

108 Forsyth, R., Devadoss, A., and Guy, O.J. (2017). Graphene field effect transistors for biomedical applications: current status and future prospects. *Diagnostics* 7 (3): https://www.mdpi.com/2075-4418/7/3/45.

109 Bhushan, B., Luo, D., Schricker, S.R. et al. (2014). *Handbook of Nanomaterials Properties*. Heidelberg: Springer Berlin https://doi.org/10.1007/978-3-642-31107-9.

110 Peña-Bahamonde, J., Nguyen, H.N., Fanourakis, S.K., and Rodrigues, D.F. (2018). Recent advances in graphene-based biosensor technology with applications in life sciences. *J. Nanobiotechnol.* 16: 75.

111 Webster, D.M., Henry, A.H., and Rees, A.R. (1994 Jan 1). Antibody-antigen interactions. *Curr. Opin. Struct. Biol.* 4 (1): 123–129.

112 Wilson, I.A. and Stanfield, R.L. (1993). Antibody-antigen interactions. *Curr. Opin. Struct. Biol.* 3 (1): 113–118.

113 Chiu ML, Goulet DR, Teplyakov A, Gilliland GL. Antibody structure and function: the basis for engineering therapeutics. *Antibodies (Basel)*;8(4):55. https://doi.org/10.3390/antib8040055 2019.

114 Janeway, C. (2012). *Immunobiology*, 5e. Elsevier.

115 Matsumoto, K., Maehashi, K., Ohno, Y., and Inoue, K. (2014). Recent advances in functional graphene biosensors. *J. Phys. D: Appl. Phys.* 47 (9): 094005. http://dx.doi.org/10.1088/0022-3727/47/9/094005.

116 Nguyen, H.H., Lee, S.H., Lee, U.J. et al. (2019). Immobilized enzymes in biosensor applications. *Materials* 12 (1): https://www.mdpi.com/1996-1944/12/1/121.

117 Bay, H.H., Vo, R., Dai, X. et al. (2019). Hydrogel gate graphene field-effect transistors as multiplexed biosensors. *Nano Lett.* 19 (4): 2620–2626. https://doi.org/10.1021/acs.nanolett.9b00431.

118 Rajesh, G.Z., Charlie Johnson, A.T., Puri, N. et al. (2021). Scalable chemical vapor deposited graphene field-effect transistors for bio/chemical assay. *Appl. Phys. Rev.* 8 (1): 011311. https://doi.org/10.1063/5.0024508.

119 Fu, W., Nef, C., Tarasov, A. et al. (2013). High mobility graphene ion-sensitive field-effect transistors by noncovalent functionalization. *Nanoscale Commun.* 26 (5): www.rsc.org/nanoscale.

120 Hyman, M. (2004). The impact of mercury on human health and the environment. *Altern. Therap. Health Medicine* 10: 70–75, https://drhyman.com/downloads/Impact-of-Hg.pdf.

5

Graphene Field-Effect Transistors: Advanced Bioelectronic Devices for Sensing Applications

Kyung Ho Kim[1], Hyun Seok Song[2], Oh Seok Kwon[3], and Tai Hyun Park[4]

[1]Infectious Disease Research Center, Korea Research Institute of Bioscience and Biotechnology (KRIBB), 125, Gwahak-ro, Yuseong-gu, Daejeon, 34141, Republic of Korea
[2]Sensor System Research Center, Korea Institute of Science and Technology (KIST), 5, Hwarang-ro 14-gil, Seongbuk-gu, Seoul 02792, Republic of Korea
[3]Sungkyunkwan University, SKKU Advanced Institute of Nanotechnology (SAINT), Department of Nano Engineering, 2066 Seobu-ro, Jangan-gu, Suwon 16419, Republic of Korea
[4]Ewha Womans University, Department of Nutritional Science & Food Management, 52, Ewhayeodae-gil, Seodaemun-gu, Seoul 03760, Republic of Korea

5.1 Introduction

Carbon-based nanomaterials, such as graphite, carbon nanotubes, fullerene, carbon dots, graphene, and graphene oxide (GO), have been discovered and fabricated for decades (Figure 5.1) [1]. The nanomaterials of carbon allotropes showed various properties and have been applied in diverse fields. Among nanomaterials, graphene has been regarded as theoretical, but it was discovered and fabricated by Andre Geim and Konstantin Novoselov in 2005 [2]. In addition, graphene was shown to consist of a two-dimensional (2D) sheet of sp^2-hybridized carbon atoms and has been investigated extensively with other 2D nanomaterials, including MoS_2, TiS_2, WS_2, $MoSe_2$, and WSe_2 [3]. These 2D nanomaterials have various properties, such as ultrathin thickness and unique electrical, mechanical, chemical, and optical properties, and have been utilized in various fields due to the previously mentioned properties. In particular, graphene of 2D nanomaterials is most commonly used for electrical fields due to its properties; high capacity, high carrier mobility, tunable conductance, ambipolar field effect, adjustable bandgap, and quantum Hall effect [4, 5]. Moreover, these characteristics of graphene are applied in the fields of physics, chemistry, materials, biology, biomedicine, electronics, and energy resources.

To apply to diverse fields, graphene has been fabricated by various methods for decades and can be divided into top-down and bottom-up methods. Various methods of graphene fabrication exist, as shown in Figure 5.2.

The top-down procedure for graphene fabrication generally involves exfoliation and reduction methods from graphite powder [6]. There are detailed methods such

Graphene Field-Effect Transistors: Advanced Bioelectronic Devices for Sensing Applications,
First Edition. Edited by Omar Azzaroni and Wolfgang Knoll.

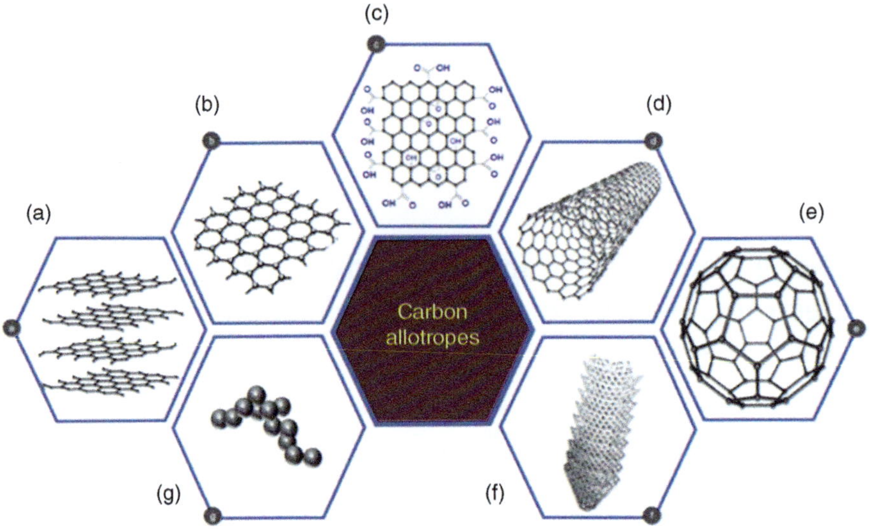

Figure 5.1 Different allotropic forms of carbon: (a) graphite, (b) graphene, (c) graphene oxide, (d) carbon nanotubes, (e) fullerene, (f) carbon nanofibers, (g) carbon dot. Source: Celso E.D. Cardoso (2019)/MDPI/CC BY 4.0.

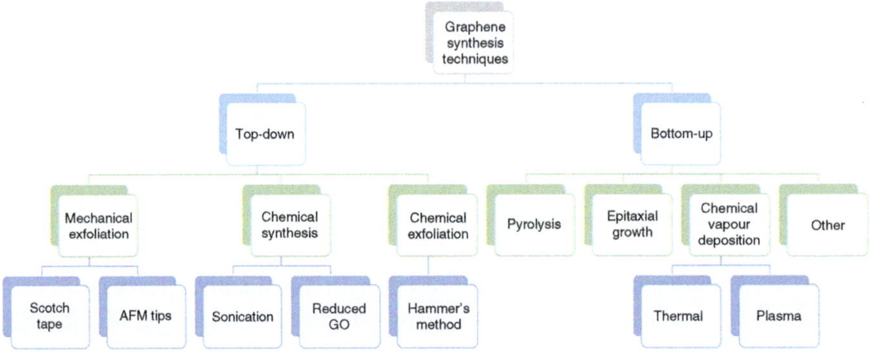

Figure 5.2 A flow sheet of methods of graphene synthesis.

as mechanical exfoliation, liquid-phase exfoliation (LPE), electrochemical exfoliation, and chemical oxidation–reduction methods [7]. Graphene fabrication with the Nobel Prize-winning method can be categorized as mechanical exfoliation, which operates by mechanically peeling graphene using general Scotch tape from highly oriented pyrolytic graphite (HOPG) [8]. Based on the Langmuir–Blodgett method, GO can be fabricated by using an atomic force microscopy (AFM) tip. The LPE method is used to separate graphite interlayers using a reduction reaction, and then the energy required to heat the material (sonication, high shear forces) is applied to obtain graphene sheets [9–11]. The electrochemical method is an additional synthesis method using chemical oxidation–reduction, such as Brodie's method, Staudenmaier's method, Hofmann's method, and Hummers' method [12, 13].

In particular, oxidized graphene can be fabricated by various methods, and there are diverse strategies, such as electrochemical, chemical, plant extract, microwave, laser irradiation, photoreduction, and thermal techniques [13, 14]. However, these approaches have limited yields and are unsuitable for production applications on a large scale [11].

Bottom-up syntheses have progressed using precursors of hydrocarbon compounds, and the methods include chemical vapor deposition (CVD), epitaxial growth, laser-assisted synthesis, thermal pyrolysis, and organic synthesis [6, 10, 15]. In particular, CVD methods are commonly utilized for graphene synthesis due to their simple and reasonable control in the bottom-up procedure. Among the CVD methods, the thermal CVD technique can obtain high-quality graphene and can easily control the graphene layer [16]. In addition, approaches based on the CVD method were developed by using various technologies, including microwave plasma chemical vapor deposition (MP-CVD), low-pressure chemical vapor deposition (LP-CVD), atmospheric-pressure chemical vapor deposition (AP-CVD), radiofrequency PE-chemical vapor deposition (RFP-CVD), plasma-enhanced chemical vapor deposition (PECVD), surface-wave plasma chemical vapor deposition (SWP-CVD), inductively coupled PE-chemical vapor deposition (ICP-CVD), and ultrahigh vacuum chemical vapor deposition (UHV-CVD) [17]. These methods use carbon sources in diverse phases, such as gaseous precursors, liquid precursors, and solid precursors. For the fabrication of CVD graphene, metal films (Ni, Fe, Pt, Si, Ga, and Cu) are mostly used as substrates for graphene growth [18]. In addition, the technology development of CVD growth graphene on metal films has solved scale-up problems, simplifying the transfer process and the energy-efficient procedure [19].

Monolayer graphene was obtained by using the CVD method. To fabricate graphene with micrometer thickness, Cu foil was loaded in a hot furnace as high as 1000 °C in the presence of CH_4, H_2, and Ar gases. When the furnace was heated to 1000 °C, the H_2 pressure was maintained at 120 mTorr, followed by introduction of 80 sccm CH_4 for more than 30 minutes at a pressure of approximately 1 Torr. Next, the furnace was allowed to cool, and when the furnace reached room temperature, the graphene was coated using poly(methyl methacrylate) (PMMA). The wet-transfer method was used to transfer the graphene onto a SiO_2 wafer before the microelectromechanical system (MEMS) procedure began. The photoresist-patterned monolayer graphene was patterned uniformly using photolithography (PL) and reactive-ion etching (RIE) systems. An electron beam (E-beam) evaporator was used for Au deposition to fabricate the electrode composition (source and drain) and the passivation layer (Figure 5.3). The electrodes were in contact with the graphene micropatterend (GM) channel. Finally, electrode passivation was conducted to prevent nonspecific binding and lessen the noise current level [20].

As monolayer graphene fabrication technology is widely known, its applications have been continuously developed by using various electrical properties of graphene, especially in the biosensor field. Among electrical biosensors, field-effect transistor (FET) platforms have shown excellent performance. Active research on FETs and graphene field-effect transistors (GFETs) in particular is on-going due to their interesting characteristics. It is well known that graphene is a semiconductor material

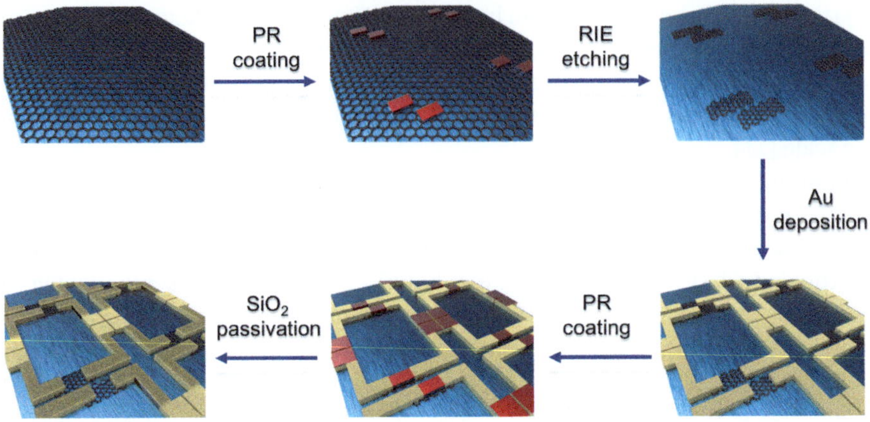

Figure 5.3 A schematic illustration for GFET fabrication through MEMS process. Source: K. H. Kim (2020)/with permission from Elsevier.

with a zero bandgap. With this, the electrode channel can be fabricated over a large area, and it is also possible to modify the channel size [21]. The MEMS are being introduced to fabricate GFETs to demonstrate their semiconductor properties.

The GFET platform allows for high sensitivity for the sensing devices. For the development of bioelectronic sensors, the utilization of bioprobes should enhance their selectivity to detect biological analytes. In this chapter, advances of the GFET-based bioelectronic devices for sensing applications are discussed.

5.1.1 Bioelectronic Nose Using Olfactory Receptor-Conjugated Graphene

The olfactory system of living organisms offers the ability to identify and discriminate smells [22]. Our nose can recognize and discriminate thousands of odor molecules at very low concentrations [23]. Many efforts have been made to develop sensing devices that mimic the olfactory system, allowing for high sensitivity and selectivity for the detection of specific odors [24]. Bioelectronic noses have been developed by the integration of olfactory receptors (ORs) originating from the natural olfactory system and various sensing devices [25–27]. These biolectronic noses hold promise for application in a wide range of fields, including environmental monitoring, food safety, anti-bioterrorism, and disease diagnosis [26, 27]. Recently, nanomaterial-based FET sensor platforms that allow for high sensitivity have been applied as sensing elements for bioelectronics noses [28–30]. In this section, the utilization of graphene-based FETs for the development of bioelectronic noses is discussed.

The bioelectronic noses that utilize nanomaterial-based sensor platforms as sensing elements are called nanobioelectronic noses (NBNs). NBSs are normally fabricated by the combination of ORs produced using heterologous cell systems and nanomaterial-based secondary transducers [27]. FET platforms based on nanomaterials have been efficiently applied for the fabrication of NBNs.

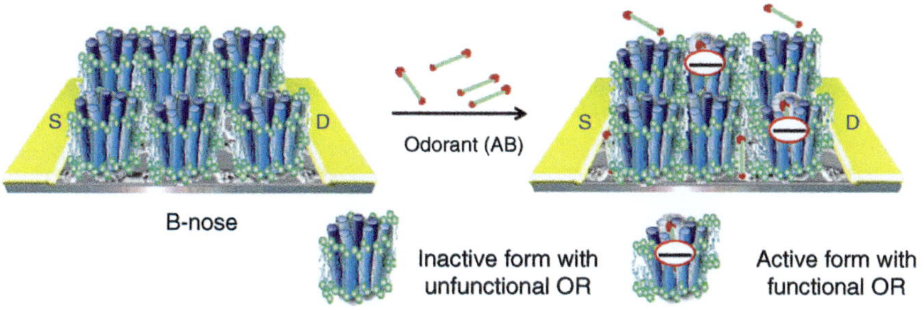

Figure 5.4 The plausible detection mechanism of the OR-conjugated FET platform. Source: S. J. Park (2012)/with permission from American Chemical Society.

The FET-type platforms hold advantages in NBN devices due to their sensing mechanism [29, 31]. Figure 5.4 illustrates the plausible detection mechanism of the OR-conjugated FET platform [31]. ORs contain ionizable amino acid residues and undergo a reversible conformational change between the active and inactive states [32]. The electrical charge of these materials can be changed by their state. The activation of ORs by the stimulation of odor molecules results in the charge shift of ORs toward the negative state, which affects the gate potential of the nanomaterial-based channel of the FET. Finally, the specific binding of odor molecules can be measured by monitoring the source–drain current.

The production of ORs should be a critical process for the development of bioelectronic noses. ORs, belonging to the G-protein-coupled receptor family, are expressed in olfactory sensory neurons and recognize a number of odor molecules with high selectivity [22, 23, 33, 34]. The binding event of odorants and ORs triggers olfactory signal transduction in neurons [33]. ORs are membrane proteins and are composed of seven transmembrane- spanning domains [35]. Due to their complicated protein structures and high hydrophobicity, ORs are poorly expressed in heterologous cell systems; thus, it is difficult to produce ORs, especially using bacterial cell systems [36]. Even with these difficulties, the production of ORs using the *Escherichia coli* (*E. coli*) system for the fabrication of bioelectronic noses has been demonstrated [28, 29, 37, 38].

The GFET-based NBN was first demonstrated by Park et al. as an OR-conjugated flexible GFET system [31]. Figure 5.5 shows the fabrication process of the OR-conjugated flexible GFET system. The bilayer graphene (BLG) was grown by the CVD process and transferred to a flexible polyethylene terephthalate (PET) film. Microelectrodes were then thermally deposited by the MEMS process. The pristine BLG was doped by foreign atoms through O_2 and NH_3 plasma treatment to allow for p- and n-type behavior, respectively. The human olfactory receptor 2AG1 (hOR2AG1), which specifically interacts with amyl butyrate (AB), was produced using an *E. coli* system and conjugated to the surface of plasma-treated BLG. 1,5-diamoninaphalene (DAN) was stacked on the side plane of BLG by π–π stacking interactions to minimize graphene defects and modify the NH_2 functional group. hOR2AG1 was comically conjugated to BLG by a Schiff-base reaction by glutaraldehyde (GA) treatment.

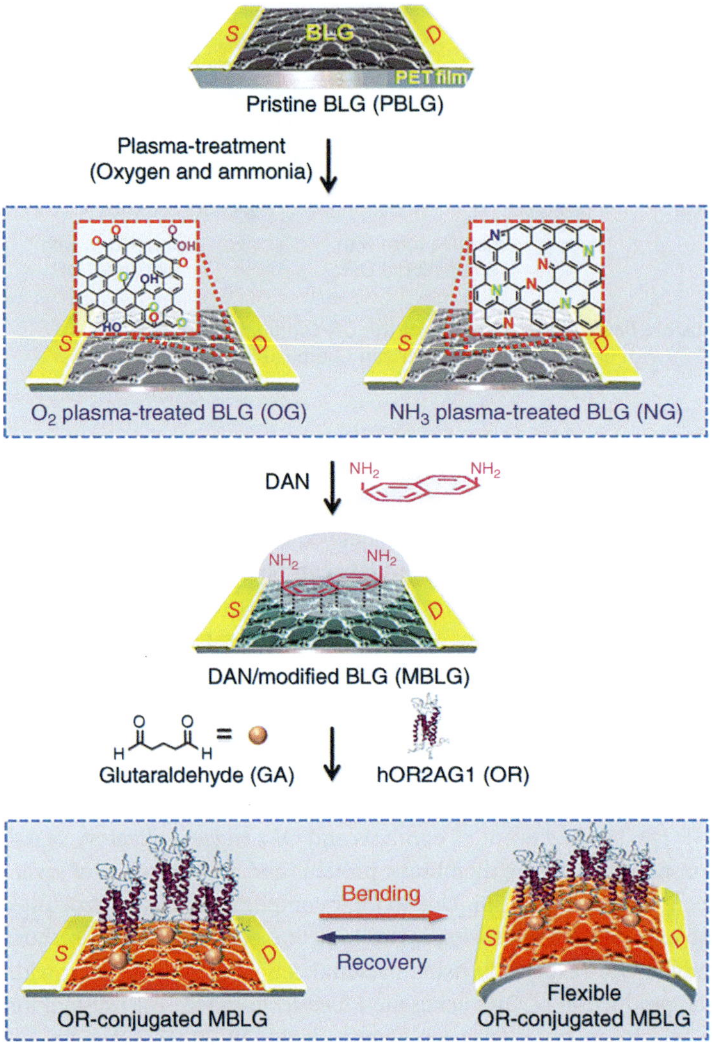

Figure 5.5 The fabrication process of the OR-conjugated flexible GFET system. Source: S. J. Park (2012)/with permission from American Chemical Society.

After the conjugation of OR, GFET-based NBN was assembled under a liquid-gating configuration (Figure 5.6a). The real-time repose of OR-GFET was measured upon AB stimulation. The limit of detection (LOD) was determined to be 0.04 fM for the oxygen plasma-treated OR-GFET system and 0.4 fM for the ammonia-treated OR-GFET system (Figure 5.6b). In addition, GFET-based NBN showed high selectivity with atomic-level resolutions at the level of femtomolar concentrations. No significant signal change was observed after the addition of nontarget odorants (hexyl butyrate, HB; propyl butyrate, PB; and butyl butyrate, BB), which only differed by the number of carbon atoms, in contrast to the target molecule AB (Figure 5.6c). The lifetimes of OR-GFETs were investigated and showed efficient

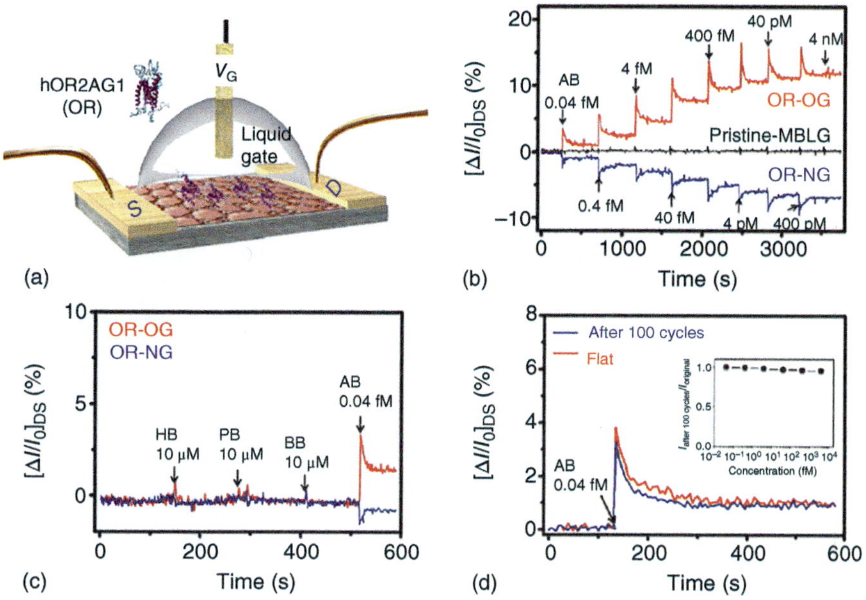

Figure 5.6 (a) Illustration of a liquid-ion gated FET using OR-conjugated graphene (the capital "S" "D," and "V_G" indicate source/drain electrodes and gating voltage). (b) Real-time responses and a calibration curve of GFET based on OR-conjugated OG and NG upon the stimulation of different concentration of AB. (c) Selective responses of OR OR-conjugated OG and NG upon addition of nontargets (hexyl butyrate, HB; propyl butylate, PB; and butyl butyrate, BB) and target (amyl butyrate, AB) odorant. (d) Sensing response of OR-conjugated OG in the flat (blue). The fatigue test (red) was carried out after 100 cycles of bending/relaxing and measured it on a flat state. Source: S. J. Park (2012)/with permission from American Chemical Society.

long-term stability. A decrease of only 5–6% in sensitivity was observed after storage of the sensor devices for 10 days. The GFET-based NBN allowed for strong mechanical stability in terms of bending ability and durability. The response intensity of the flexible NBN devices decreased by only approximately 9% after 100 bending/relaxing cycles (Figure 5.6d).

Humans can identify smell based on a combination of olfactory signals generated by different olfactory sensory neurons. To mimic this combinatorial perception in the human olfactory system closely, Kwon et al. demonstrated a multiplexed NBN using an FET system based on GMs [37]. Figure 5.7a illustrates the functional anatomy of the human olfactory system and components of the multiplexed NBNs corresponding to each functional stage of the human olfactory system. In the human nose, the olfactory neuroepithelium contains OSNs, and the axons of OSNs are separately connected to glomeruli in the olfactory bulb. Each OSN expresses only one of approximately 390 functional OR genes (Figure 5.7b). The olfactory signals triggered by the binding of odor compounds (Figure 5.7c,d) and ORs are transferred to the olfactory bulb, and finally, the forebrain recognizes smell by the combination of multidimensional signals. The multiplexed NBN using GM was systematically constructed and mimicked each stage of the human olfactory system: (i) hOR

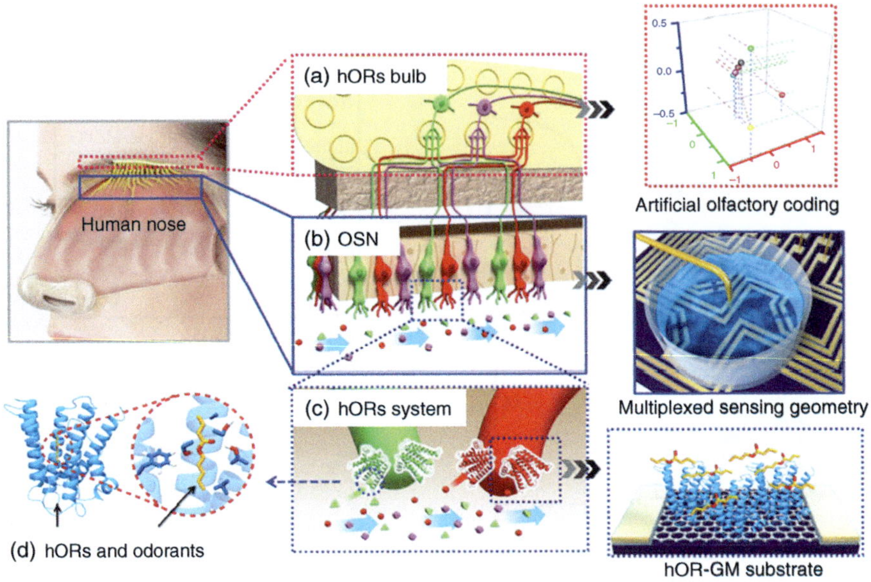

Figure 5.7 Schematic diagram of functional anatomy of human olfactory system and components of multiplexed NBN based on GMs simulating each functional stages of human olfactory system. (a) Olfactory bulbs generate combinational olfactory codes by combining of the olfactory signals generated by OSNs and matching them with artificial olfactory codes generated by multiplexed NBN. (b) OSNs generating olfactory signals triggered by the specific binding of hORs and odorants, matching with GMs functionalized with hORs. (c) hORs for the specific recognition of odorants. (d) Illumination of specific interaction between hOR and odorant. Source: O. S. Kwon (2015)/with permission from American Chemical Society.

immobilized GM substrate recognize odor compounds; (ii) multiplexed sensing geometry generate multiple sensing signals; and (iii) the signals combined as an artificial olfactory code, mimicking the combinatorial olfactory perception in the human olfactory bulb.

Figure 5.8a illustrates the fabrication process of a multiplexed NBN based on GMs. The GMs were constructed by conventional PL and RIE processes after the preparation of a single layer of graphene on the silicon wafer. Cr/Au microelectrodes, which were used as the source (S) and drain (D) electrodes, were then gradually deposited on the GM channels. After the fabrication of multiplexed GMs, two types of hORs (hOR2AG1 and hOR3A1), which were produced using the *E. coli* system (Figure 5.8b), were immobilized by the Schiff-base reaction using DAN and GA (Figure 5.8c).

hOR-conjugated GMs were constructed as the liquid-ion gating system. Figure 5.9a shows the performance of hOR-conjugated GMs. hOR2AG1 and hOR3A1 are ORs that interact with the target odorants AB and helional (HE), respectively. Both hOR2AG1- and hOR3A1-conjugated NBNs showed an LOD level of approximately 0.1 fM. The hOR-conjugated GMs exhibited single-atomic resolution of selectivity. The hOR2AG1-conjugated GM only responds to AB among the similar odorant

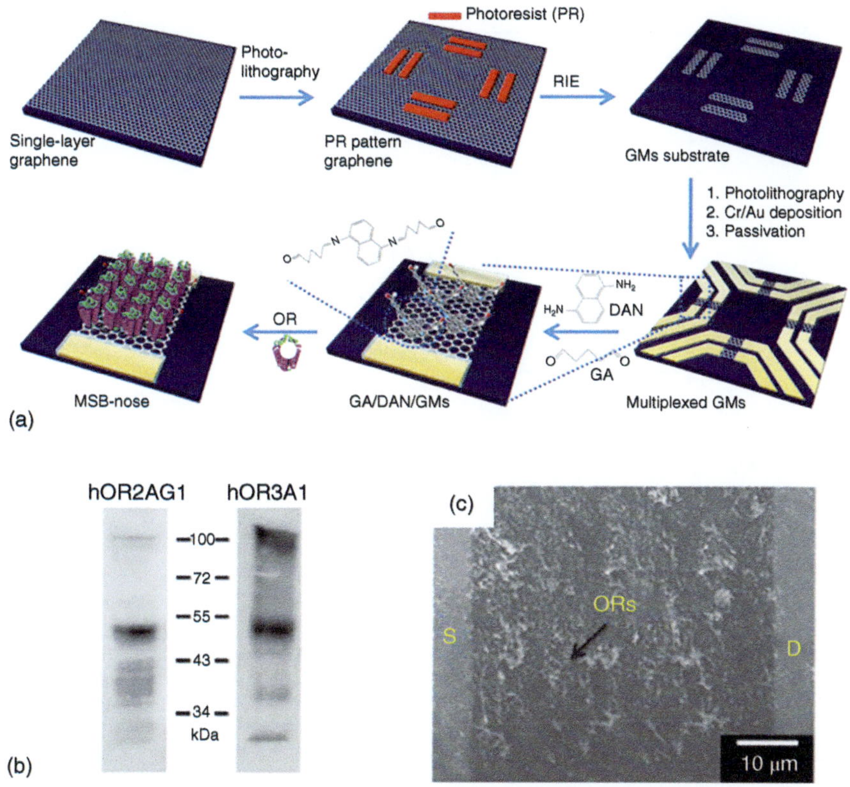

Figure 5.8 (a) illustrates the fabrication process of multiplexed NBN based on GMs. (b) Western blot analysis of hOR2AG1 and hOR3A1 produced using *E. coli*. (c) Electron microscope image of OR conjugated GM electrode. Source: O. S. Kwon (2015)/with permission from American Chemical Society.

structures with different numbers of carbon atoms. In addition, hOR3A1-conjugated GM only responds to HE among similar molecular structures with different terminal groups. Furthermore, conducting polymer nanotubes (CPNTs) were conjugated with hOR2AG1 and compared against that of the hOR-conjugated GMs. Multiple channels of NBN were constructed using four different combinations of hORs and nanomaterials, including GM, carboxylated poly(3,4-ethylenedioxythiophene) nanotubes (CPEDOT NTs) and carboxylated polypyrrole nanotubes (CPPy NTs). The signals were measured from each channel of multiplexed NBN exposed to odorants and volatile organic compounds (VOCs) (Figure 5.9b,c). The multiple sensing results were subjected to principal component analysis (PCA), and the first three principal component scores are plotted in Figure 5.9d. The analytes exhibited unique signatures in the PCA plot, and components of AB and HE were clearly distinguished from VOCs, indicating that the odorants can be individually identified in a mixture. Therefore, multiplexed NNNs based on GMs offer a useful tool for odor identification with an artificial olfactory code mimicking combinatorial human olfactory perception occurring in the human brain.

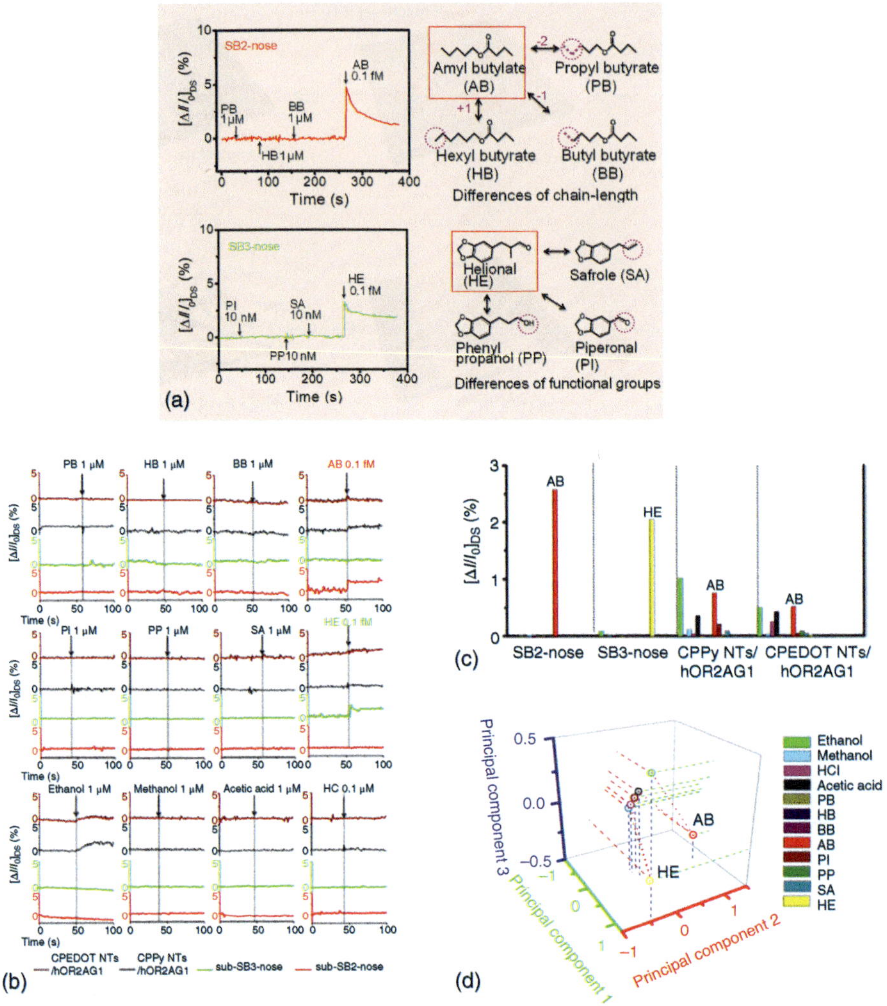

Figure 5.9 (a) Real-time response of hOR2AG1 (SB2-nose) and hOR3A1 (SB3-nose)-conjugated GMs to nontarget odorants and target odorants (AB and HE). (b) The real-time signals measured from each channels of multiplexed NBN were exposed to odorants and volatile organic compounds (VOCs). (c) Comparison histogram of response values characterized from (b). (d) Principle component analysis plots of the data set of response intensities inputted from signals generated by the multiplexed NBN. Source: O. S. Kwon (2015)/with permission from American Chemical Society.

5.1.2 Bioelectronics for Diagnosis Using Bioprobe-Modified Graphene

It is well known that antibiotic medications kill or restrain bacterial infection, which is crucial to preventing and treating diseases caused by bacterial infections. Misuse or overuse of antibiotics in humans and animals can cause serious problems, such as untreatable infections due to the spread of antibiotic-resistant bacteria [39]. In addition, delayed or estimated diagnosis of pathogenic bacteria can cause illnesses or even lead to death [40]. To prevent the life-threatening situation that is a consequence of the abuse of antibiotics, it is highly important to clearly differentiate between gram-positive bacteria (GPB) and gram-negative bacteria (GNB) [20, 41]. Polymerase chain reactions (PCRs), gene microarrays, enzyme-linked immunosorbent assays (ELISAs), gram strains, and cultures are the methods used currently for pathogen diagnosis as early as possible and for gram typing. However, these methods are labor intensive and time consuming when early detection is crucial. Although many biosensors, such as cyclic voltammetry (CV), fluorescence-based fluidic sensors, and colorimetric methods, have been developed, these methods also face limitations, such as low accuracy with aged bacterial samples, expensive fluorescent readers or analysis materials, and time-consuming discrimination processes [42]. A novel dual antibiotic-conjugated graphene micropattern field-effect transistor (ABX-GMFET) device was introduced to overcome the abovementioned problems with a rapid response time of approximately five seconds and high sensitivity with a detection limit of 10^0 CFU ml^{-1} [20].

As shown in Figure 5.10a,b, vancomycin and magainin I were identified as clinical GNB and GPB. Bis(2-aminoethylene)perylene-3,4,9,10-tetracarboxyldiimide

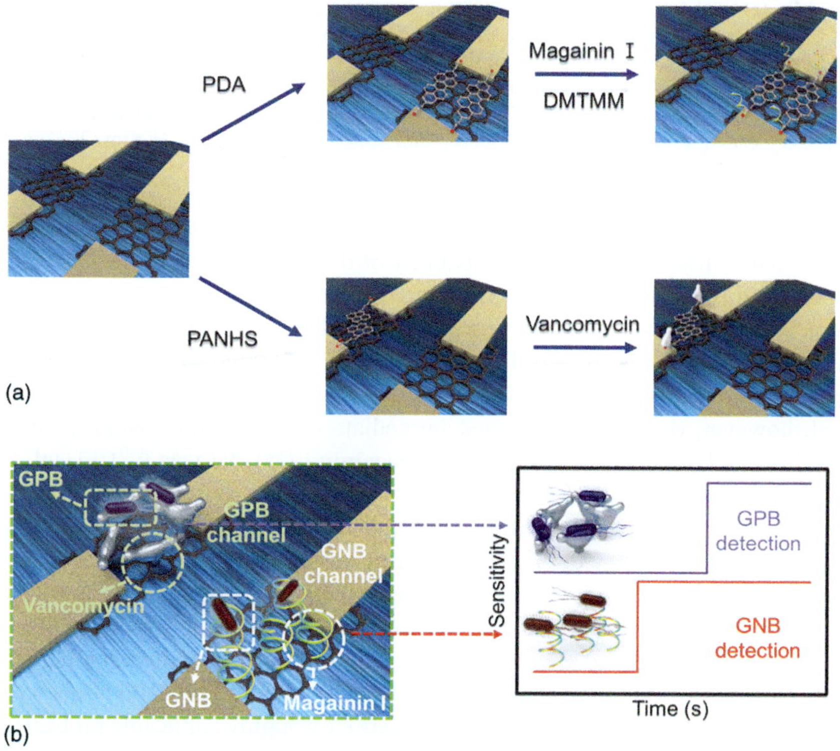

Figure 5.10 Schematic illustrations of the (a) surface treatment processes for dual channel bacteria sensor (b) GFET responses showing the electrical signals of the specific detection between bioprobes and GPB/GNB. Source: K. H. Kim (2020)/with permission from Elsevier.

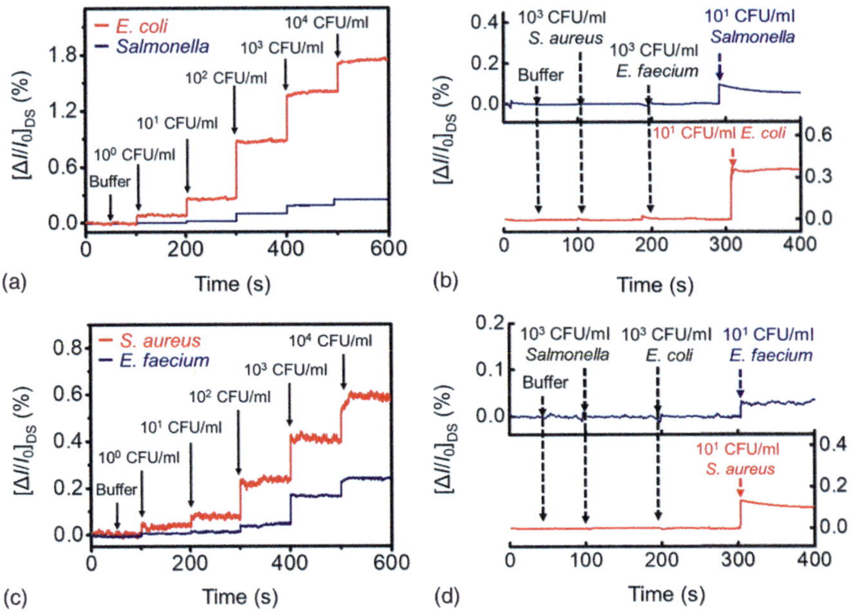

Figure 5.11 (a) Real-time response of the liquid-ion-gated magainin I-GMFET toward various GNB concentrations (V_{DS} = −1 mV and V_G = −0.1 V). (b) Specificity of the magainin I-GMFET toward target GNB and nontarget GPB. (c) Real-time responses of the liquid-ion-gated vancomycin-GMFET toward various GPB concentrations (V_{DS} = −1 mV and V_G = −0.1 V). (d) Specificity of the vancomycin-GMFET toward target GPB (*Staphylococcus aureus* and *Enterococcus faecium*) and nontarget GNB (*E. coli* and *Salmonella*). Source: K. H. Kim (2020)/ with permission from Elsevier.

(PDA) and 1-pyrenebutyric acid *N*-hydroxysuccinimide ester (PANHS) were synthesized to enhance the immobilization capability of the bioreceptors on the GM and were used as the chemical linkers for GPB and GNB, respectively.

Figure 5.11a presents the real-time responses of the magainin I-GMFET using different GNB concentrations from 10^0 to 10^4 CFU ml^{-1} of living cultured *E. coli* (red line) and *Salmonella* (blue line). There were no changes when the buffer was dropped; however, the current changed immediately (within approximately five seconds) bacterial injection. The LOD of the magainin I-GMFET was 10^0 CFU ml^{-1}. Figure 5.11b shows the selectivity of GNB when a higher concentration of the nontarget GPB (10^3 CFU ml^{-1}) was introduced, and a significant response was not shown until GNB was injected. Figure 5.11c displays the real-time responses of the vancomycin-GMFET with various GPB concentrations. The signal was observed within approximately five seconds, and the LOD was observed to be 10^0 CFU ml^{-1} as well. Figure 5.11d shows the selectivity of the GPB GFET by introducing a higher concentration of the nontarget, which showed no response until the GPB target was removed. These results showed that the ABX-GMFET is highly applicable for bacterial diagnosis due to its great sensitivity and selectivity.

As the world is suffering from the COVID-19, the importance and demand for high-performance sensors for respiratory viruses have increased greatly

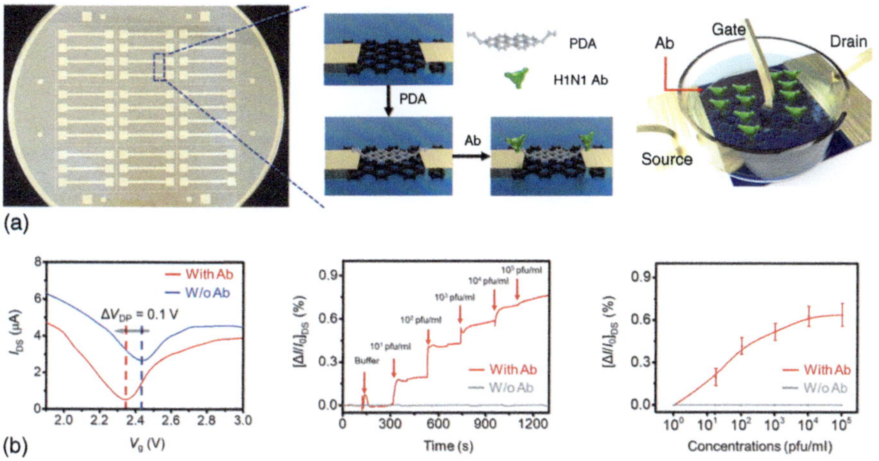

Figure 5.12 (a) Schematic images of H1N1 GFET transistor surface treatment using PDA as a linker, H1N1 antibody as a probe, and the entire sensing schematic illustration of H1N1 GFET sensor via liquid-gated system. (b) H1N1 GFET sensor characterized by the transfer curve, real-time monitoring, and the selectivity test. Source: K. H. Kim (2021)/The Korean Vacuum Society.

worldwide [43]. Respiratory viruses cause different symptoms and various diseases, such as cough, bronchitis, pneumonia, and fever. As coronaviruses and influenza viruses show similar symptoms, an enhanced biosensor to detect and differentiate each virus at an early stage is significant. Different methods or devices for virus diagnosis have been studied, but they are still considered a challenge since the sensitivity and accuracy of PCR are not yet competitive [44]. Recently, the lateral flow assay (LFA) has been used well because of its comfortability; however, it is not suitable for practical use [45]. Here, a high-performance and convenient GFET sensor for H1N1 is discussed. Figure 5.12a presents schematic images of the sensor preparation using the electrode fabricated through the MEMS process. The channel surface was treated with PDA and GA, and H1N1 antibodies were attached to the surface for immobilization and detection of the target viruses when they were introduced. The transfer curve, real-time response, and selectivity test were monitored as shown in Figure 5.12b in order. The transfer curve was monitored with and without the antibody. The transition voltage decreased from 2.43 to 2.33 V. This observation can be explained by change in the electrical properties of the graphene surface due to the existence of antibodies. The real-time responses of the H1N1 GFET with (red) and without (gray) antibody treatment using various ranges of the H1N1 virus from 10^1 to 10^5 pfu ml^{-1}. Here, it can be clearly observed that no significant signal was shown without antibody treatment. In contrast, the sensor with H1N1 antibody treatment showed a significant current change from 10^1 pfu ml^{-1}. Last, the selectivity test conducted with nontargets such as the SARS-COV-2 spike protein and influenza B viruses showed current change only after the injection of the target (influenza A) virus [46].

Recently, whole-gene sequencing has been implemented to analyze and identify biomarkers of various diagnoses [47]. With this, studies on clustered regularly

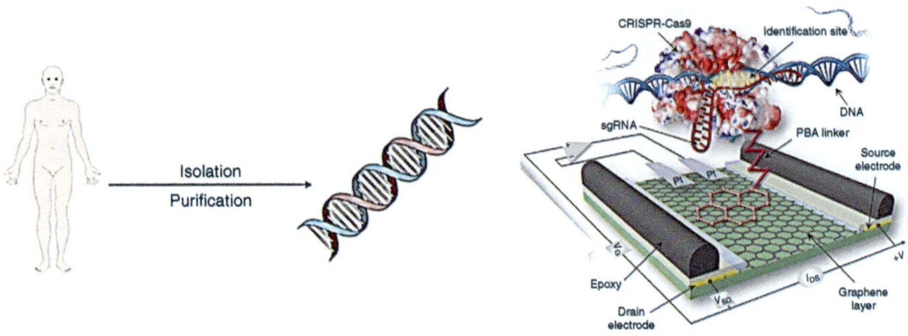

Figure 5.13 Schematic illustration of CRISPR GFET system for rapid detection of a gene target from the entire genomic sample without amplification. Source: R. Hajian (2019)/with permission from Springer Nature.

interspaced short palindromic repeats (CRISPRs)-associated nucleases (Cas) are actively ongoing [48]. CRISPR-Cas proteins are guided by a single-stranded RNA, which is a very specific and great tool for sequence specification, targeting, and even detection. Previous studies reported that GFET is known to have excellent selectivity and sensitivity with unamplified genomic samples [49]. Figure 5.13 shows the CRISPR GFET study conducted and applied to CRISPR-Chip. This CRISPR GFET system was designed with the gene-targeting capacity of CRISPR–Cas9 with highly discriminative detection power. To enable the fast and selective detection of the target sequence, undamaged genomic DNA was included. This CRISPR GFET chip was used to analyze DNA samples from HEK293T cells, which are also called HEK-BFP cells, a derivative of human cells that express a mutant. The real-time response of the CRISPR GFET chip showed no significant signal when no DNA was applied to the electrode without any treatment. On the other hand, when the CRISPR GFET chip was treated by genomic sample incubation, it took five minutes for signal saturation. This CRISPR GFET system took approximately 15 minutes for the whole procedure, including the incubation and washing (rinse) processes and an LOD as low as 2.3 fM [50].

5.1.3 Biosensors for Environmental Component Monitoring Using Graphene

The Earth has suffered serious environmental problems, such as the desertification of many areas, continuous droughts, and the increase in seawater levels due to global warming. Issues related to water, diatoms, green algae and blue–green algae (cyanobacteria) are environmental problems that influence drinking water [51]. In particular, geosmin, with a musty aroma, can contaminate the water supply system [52]. Therefore, technologies are required for rapid detection at low concentrations and for real-time monitoring. However, the present devices for geosmin detection have limitations, such as time consumption and impossible real-time monitoring [53]. To overcome these limitations, a GFET was designed by aptamer conjugation, and the system showed a high level of performance with an LOD of approximately 10 pM for real-time monitoring [54].

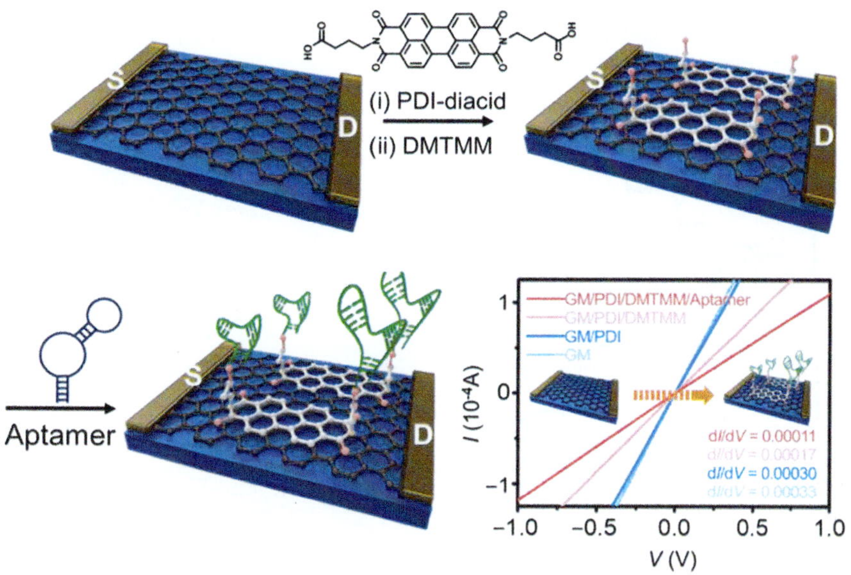

Figure 5.14 Schematic illustration of aptamer immobilization process onto graphene electrodes and I–V curve depending on the surface modification. Source: S. J. Park (2021)/ with permission from Elsevier.

The electrodes were in contact with the GM channel, and the passivation layer was placed on the electrodes to prevent nonspecific binding. To fabricate the aptasensor, the bis(2-aminoethylene) perylene-3,4,9,10-tetracarboxyldiimide (PDI) linker was introduced for immobilization of the aptamer, which was conjugated using π–π stacking between the perylene backbone and graphene sp^2 plane (Figure 5.14) [4]. The carboxyl group of PDI was activated by the DMTMM coupling agent, and then the aptamer was immobilized by forming an amide bond between the carboxyl group and amine group. The linear current–voltage (I–V) curve for the aptasensor demonstrated excellent electrical properties and ohmic behavior in the voltage range of −1 to 1 V.

The aptasensor was composed of a liquid-ion gated system. Phosphate-buffered saline (pH 7.4) was used to surround the GM channel, which was utilized as an electrolyte. The response of the aptasensor was compared with the response of pristine GFET to demonstrate the target binding of the aptamer in real-time monitoring, as shown in Figure 5.15a, where I_{DS} showed an instant increase after exposure to geosmin due to the change in charge density on the GM channel surface by reorientation of the aptamer, following the schematic illustrations (Figure 5.15b). In addition, the specific binding was confirmed by exposing various odorous compounds, such as 2-methylisoborneol (MIB), 2-isopropyl-3-methoxypyrazine (IPMP), and 2-isobutyl-3-methoxypyrazine (IBMP) (Figure 5.15c). The response showed no signal after exposure to nontarget odorous compounds; however, the current was immediately changed after geosmin exposure. Therefore, this biosensor platform can be utilized for detecting and screening harmful materials in water.

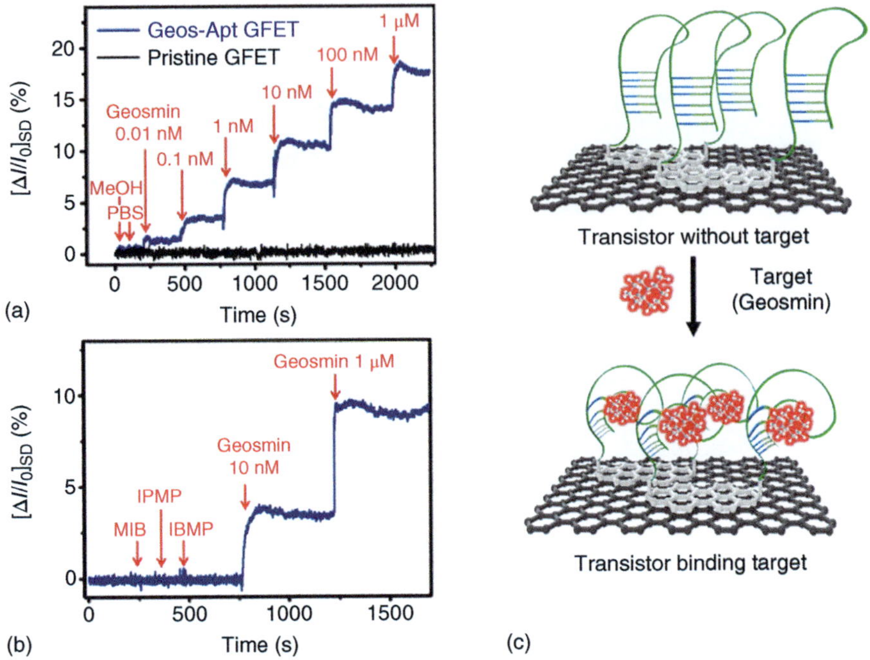

Figure 5.15 Real-time response of (a) various geosmin concentrations (10 pM–1 μM) and (b) specificity of the geosmin aptasensor toward other odorous compounds. (c) Schematic illustration toward the mechanism of geosmin aptasensor. Source: S. J. Park (2021)/with permission from Elsevier.

Food safety is a main issue in the healthcare field due to increasing food poisoning, which occurs due to biogenic amines such as cadaverine (CV) and putrescine (PT). Biogenic amines are difficult to remove from meats using various methods [55, 56]. Therefore, technologies for the detection of CV and PT are required for early monitoring, and various sensor platforms have been developed using surface-enhanced Raman scattering (SERS), LFAs, fluidic chips, and chemiresistive sensors [56, 57]. However, various sensor platforms suffer from limitations, including being time-consuming, having low selectivity, and having large instruments. To overcome the various limitations, the electronic nose (E-nose) platform was developed using CV and PT receptors based on GM. These E-noses showed LODs for CV and PT of 27.04 and 7.29 ppb, respectively, and exhibited excellent selectivity toward various spoilage biomarkers.

Graphene was used to fabricate the E-nose, and the electrodes consisted of single-layer graphene, as condensed in the previous part. For specific detection, the receptors 2,7-bis(3-morpholinopropyl)benzo[lmn] [3, 8] phenanthroline-1,3,6,8(2H,7H)-tetraone (NaPhdiMor [NPM]) for CV detection and 2,7-bis(2-((3-morpholinopropyl)amino)ethyl) benzo[lmn][3,8]phenanthroline-1,3,6,8(2H,7H)-tetraone (NaPhdiEtAmMor [NPEAM]) for PT detection were immobilized onto GM using π–π interactions. The graphene modified using receptors was investigated using high-resolution transmission electron microscopy (HR-TEM), and the morphologies were confirmed by the changed shape of a lattice, as shown in Figure 5.16a. To prove the conjugation, *I–V* curves were analyzed

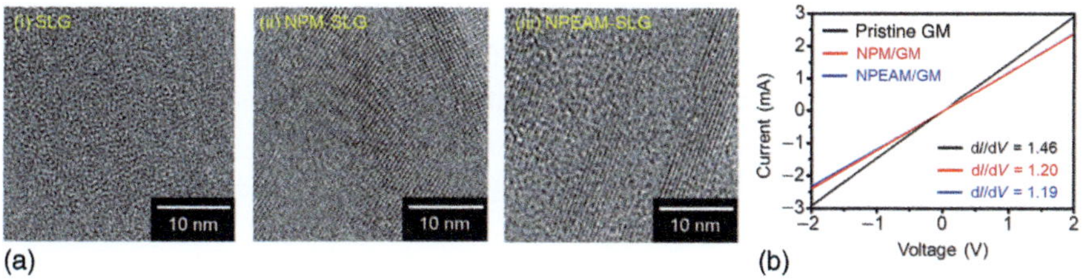

Figure 5.16 (a) TEM images of pristine graphene and receptor-conjugated graphene for comparison of surface morphologies and (b) $I-V$ curves. Source: K. H. Kim (2022)/with permission from Elsevier.

by measuring the electrical properties. The linear plots showed similar conductivity toward each receptor and exhibited excellent ohmic behavior (Figure 5.16b).

Real-time monitoring of CV and PT was performed with receptor-introduced graphene, and the responses were observed to instantly increase R_{DS} after exposure to each target gas at diverse concentrations, ranging from 27.04 to 38.09 ppb for CV and 7.29 to 11.88 ppb for PT (Figure 5.17a,b). Each E-nose showed different detection performance depending on each concentration and exhibited high performance with an LOD of 27.04 ppb for CV and 7.29 ppb for PT. To confirm the reusability, the nitrogen gas was blown into each E-nose to remove the residual gases, and the R_{DS} was recovered to its initial state. In addition, the selectivity performances toward each E-nose were determined by exposure to the various spoilage markers, H_2S and NH_3 (Figure 5.17c,d). Two types of E-noses were investigated for specific detection

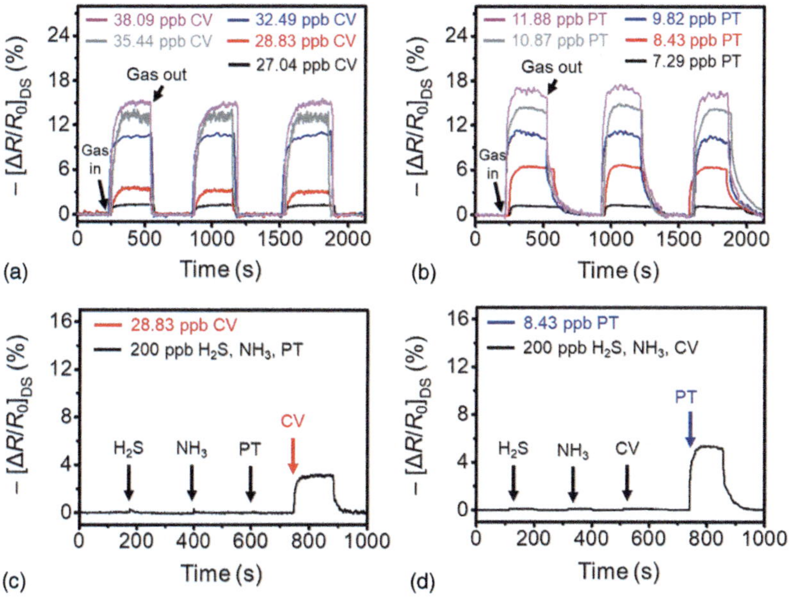

Figure 5.17 Real-time monitoring upon exposure to (a) CV and (b) PT at various gas concentrations. (c, d) Specificity test of E-noses to target (CV or PT) and nontarget (hydrogen sulfide (H2S), ammonia (NH_3)). Source: K. H. Kim (2022)/with permission from Elsevier.

of the target gas and displayed excellent detection performances at concentrations under approximately 7 times and 23 times for CV and PT, respectively. This E-nose platform can be applied for the on-site monitoring of spoilage degree from food samples.

5.2 Conclusion

This chapter discusses interesting nanomaterials, graphene, and their application in various sensor applications, such as in bioelectronics noses, bioprobe-conjugated biosensors for medical diagnosis, and even in environmental monitoring systems. In addition, significant information about the excellent technologies from the fabrication of the GFET sensors to the characterization and results are also presented. By sharing this article showing the great selectivity and high sensitivity of the graphene-based biosensor, it is hoped that this chapter will inspire and encourage the further development of abundant novel biosensors.

Acknowledgments

This study was supported by the Technology Innovation Program (20020801, Construction of international standardization basis for the biodigital biosense) funded by the Ministry of Trade, Industry & Energy (MOTIE), Republic of Korea; Korea Institute of Planning and Evaluation for Technology in Food, Agriculture and Forestry (IPET) and Korea Smart Farm R&D Foundation (KosFarm) through Smart Farm Innovation Technology Development Program, funded by Ministry of Agriculture, Food and Rural Affairs (MAFRA) and Ministry of Science and ICT (MSIT), Rural Development Administration (RDA) (421020-03).

References

1 Faruque, M.A.A., Syduzzaman, M., Sarkar, J. et al. (2021). *Nanomaterials* 11 (9): 2414. https://doi.org/10.3390/nano11092414.
2 Novoselov, K.S., Geim, A.K., Morozov, S.V. et al. (2005). *Nature* 438: 197.
3 Park, C.S., Yoon, H., and Kwon, O.S. (2016). *J. Ind. Eng. Chem.* 38: 13.
4 Kwon, O.S., Lee, S.H., Park, S.J. et al. (2013). *Adv. Mater.* 25: 4177.
5 Gomez De Arco, L., Zhang, Y., Schlenker, C.W. et al. (2010). *ACS Nano* 4: 2865.
6 Lim, J.Y., Mubarak, N.M., Abdullah, E.C. et al. (2018). *J. Ind. Eng. Chem.* 66: 29.
7 Bhuyan, M.S.A., Uddin, M.N., Islam, M.M. et al. (2016). *Int. Nano Lett.* 6: 65.
8 Kumar, N., Salehiyan, R., Chauke, V. et al. (2021). *FlatChem* 27: 100224.
9 (a) Cote, L.J., Kim, F., and Huang, J. (2009). *J. Am. Chem. Soc.* 131: 1043.
(b) Giesbers, A.J.M., Zeitler, U., Neubeck, S. et al. (2008). *Solid State Commun.* 147: 366.

10 Schniepp, H.C., Li, J.L., McAllister, M.J. et al. (2006). *J. Phys. Chem. B* 110: 8535.
11 Liu, W.W., Chai, S.P., Mohamed, A.R., and Hashim, U. (2014). *J. Ind. Eng. Chem.* 20: 1171.
12 (a) Achee, T.C., Sun, W., Hope, J.T. et al. (2018). *Sci. Rep.* 8: 14525. (b) Verma, S., Mungse, H.P., Kumar, N. et al. (2011). *Chem. Commun.* 47: 12673. (c) Dreyer, D.R., Park, S., Bielawski, C.W., and Ruoff, R.S. (2010). *Chem. Soc. Rev.* 39: 228.
13 (a) Marcano, D.C., Kosynkin, D.V., Berlin, J.M. et al. (2010). *ACS Nano* 4: 4806. (b) Staudenmaier, L. (1899). *Ber. Dtsch. Chem. Ges.* 32: 1394.
14 (a) Thakur, S. and Karak, N. (2012). *Carbon* 50: 5331. (b) Guex, L.G., Sacchi, B., Peuvot, K.F. et al. (2017). *Nanoscale* 9: 9562. (c) Chen, W. and Yan, L. (2010). *Nanoscale* 2: 559. (d) Chen, W., Yan, L., and Bangal, P.R. (2010). *Carbon* 48: 1146.
15 (a) Kim, Y., Song, W., Lee, S.Y. et al. (2011). *Appl. Phys. Lett.* 98: 263106. (b) Antonatos, N., Ghodrati, H., and Sofer, Z. (2020). *Appl. Mater. Today* 18: 100502.
16 (a) Hintsho, N., Shaikjee, A., Masenda, H. et al. (2014). *Nanoscale Res. Lett.* 9: 387. (b) Hlekelele, L., Franklyn, P.J., Dziike, F., and Durbach, S.H. (2018). *New J. Chem.* 42: 4531.
17 (a) Ge, X., Zhang, Y., Chen, L. et al. (2018). *Carbon* 139: 989. (b) Lee, J.K., Lee, S., Kim, Y.I. et al. (2014). *Sci. Rep.* 4: 5682. (c) Qi, J., Zhang, L., Cao, J. et al. (2012). *Chin. Sci. Bull.* 57: 3040. (d) Woo, Y.S., Seo, D.H., Yeon, D.H. et al. (2013). *Carbon* 64: 315.
18 (a) Cushing, G.W., Johánek, V., Navin, J.K., and Harrison, I. (2015). *J. Phys. Chem. C* 119: 4759. (b) Weatherup, R.S., Dlubak, B., and Hofmann, S. (2012). *ACS Nano* 6: 9996. (c) Mukanova, A., Tussupbayev, R., Sabitov, A. et al. (2017). CVD graphene growth on a surface of liquid gallium. *Materials Today Proc.* 4 (3): 4548–4554.
19 Lee, J.H., Lee, E.K., Joo, W.J. et al. (2014). *Science* 344: 286.
20 Kim, K.H., Park, S.J., Park, C.S. et al. (2020). *Biosens. Bioelectron.* 167: 112514.
21 Schwierz, F. (2010). *Nat. Nanotechnol.* 5: 487.
22 Malnic, B., Hirono, J., Sato, T., and Buck, L.B. (1999). *Cell* 96: 713.
23 (a) Krautwurst, D., Yau, K.-W., and Reed, R.R. (1998). *Cell* 95: 917. (b) Mori, K., Nagao, H., and Yoshihara, Y. (1999). *Science* 286: 711.
24 Persaud, K.C., Khaffaf, S.M., Payne, J.S. et al. (1996). *Sens. Actuators, B* 36: 267.
25 Son, M., Lee, J.Y., Ko, H.J., and Park, T.H. (2017). *Trends Biotechnol.* 35: 301.
26 Lee, S.H. and Park, T.H. (2010). *Biotechnol. Bioprocess Eng.* 15: 22.
27 Kwon, O.S., Song, H.S., Park, T.H., and Jang, J. (2018). *Chem. Rev.* 119: 36.
28 (a) Oh, J., Yang, H., Jeong, G.E. et al. (2019). *Anal. Chem.* 91: 12181. (b) Yang, H., Kim, D., Kim, J. et al. (2017). *ACS Nano* 11: 11847. (c) Yoon, H., Lee, S.H., Kwon, O.S. et al. (2009). *Angew. Chem. Int. Ed.* 121: 2793.
29 Kim, T.H., Lee, S.H., Lee, J. et al. (2009). *Adv. Mater.* 21: 91.
30 Lee, S.H., Kwon, O.S., Song, H.S. et al. (2012). *Biomaterials* 33: 1722.
31 Park, S.J., Kwon, O.S., Lee, S.H. et al. (2012). *Nano Lett.* 12: 5082.
32 (a) Rubenstein, L.A. and Lanzara, R.G. (1998). *Theochem* 430: 57. (b) Rubenstein, L.A., Zauhar, R.J., and Lanzara, R.G. (2006). *J. Mol. Graph. Modell.* 25: 396.
33 Firestein, S. (2001). *Nature* 413: 211.
34 Malnic, B., Godfrey, P.A., and Buck, L.B. (2004). *Proc. Nat. Acad. Sci.* 101: 2584.

35 Buck, L. and Axel, R. (1991). *Cell* 65: 175.

36 (a) Kiefer, H., Krieger, J., Olszewski, J.D. et al. (1996). *Biochemistry* 35: 16077. (b) Sarramegn, V., Muller, I., Milon, A., and Talmont, F. (2006). *Cell. Mol. Life Sci.* 63: 1149.

37 Kwon, O.S., Song, H.S., Park, S.J. et al. (2015). *Nano Lett.* 15: 6559.

38 Song, H.S., Lee, S.H., Oh, E.H., and Park, T.H. (2009). *Curr. Microbiol.* 59: 309.

39 (a) Kennedy, A.D., Otto, M., Braughton, K.R. et al. (2008). *Proc. Natl. Acad. Sci. U.S.A.* 105: 1327. (b) Kiraly, D.D., Walker, D.M., Calipari, E.S. et al. (2016). *Sci. Rep.* 6: 1. (c) Li, W., Dong, K., Ren, J., and Qu, X. (2016). *Angew. Chem. Int. Ed.* 55: 8049. (d) Currie, J., Lin, W., and Meng, J. (2014). *J. Dev. Econ.* 110: 39.

40 (a) García-Aljaro, C., Bangar, M.A., Baldrich, E. et al. (2010). *Biosens. Bioelectron.* 25: 2309. (b) Singh, S., Moudgil, A., Mishra, N. et al. (2019). *Biosens. Bioelectron.* 136: 23. (c) Ugolini, M., Gerhard, J., Burkert, S. et al. (2018). *Nat. Immunol.* 19: 386. (d) Vaisocherová-Lísalová, H., Víšová, I., Ermini, M.L. et al. (2016). *Biosens. Bioelectron.* 80: 84. (e) Wang, R., Kim, K., Choi, N. et al. (2018). *Sens. Actuators, B* 270: 72.

41 Smith, P.A., Koehler, M.F.T., Girgis, H.S. et al. (2018). *Nature* 561: 189.

42 (a) Ouyang, Z., Zhai, Z., Qin, A. et al. (2015). *Exp. Therap. Med.* 9: 1857. (b) Moscicki, B., Shafer, M.A., Millstein, S.G. et al. (1987). *Am. J. Obstet. Gynecol.* 157: 65. (c) Romero, R., Emamian, M., Quintero, R. et al. (1988). *Am. J. Obstet. Gynecol.* 159: 114.

43 Sachs, J.D., Abdool Karim, S., Aknin, L. et al. (2020). *Lancet* 396: 1102.

44 (a) Corman, V.M., Landt, O., Kaiser, M. et al. (2020). *Eurosurveillance* 25 (3): 2000045. https://doi.org/10.2807/1560-7917.ES.2020.25.3.2000045. (b) Cui, F. and Zhou, H.S. (2020). *Biosens. Bioelectron.* 165: 112349. https://doi.org/10.1016/j.bios.2020.112349.

45 (a) Carter, L.J., Garner, L.V., Smoot, J.W. et al. (2020). *ACS Central Sci.* 6: 591. (b) Zhu, X., Wang, X., Han, L. et al. (2020). *Biosens. Bioelectron.* 166: 112437. https://doi.org/10.1016/j.bios.2020.112437.

46 Kim, K.H., An, J.E., Kim, J.S. et al. (2021). *Appl. Sci. Convergence Technol.* 30: 111.

47 (a) Iyer, G., Hanrahan, A.J., Milowsky, M.I. et al. (2012). *Science* 338: 221. (b) Waddell, N., Pajic, M., Patch, A.M. et al. (2015). *Nature* 518: 495. (c) Yuen, R.K.C., Merico, D., Bookman, M. et al. (2017). *Nat. Neurosci.* 20: 602.

48 (a) Gootenberg, J.S., Abudayyeh, O.O., Lee, J.W. et al. (2017). *Science* 356: 438. (b) Pardee, K., Green, A.A., Takahashi, M.K. et al. (2016). *Cell* 165: 1255. (c) Li, S.Y., Cheng, Q.X., Li, X.Y. et al. (2018). *Cell Disc.* 4: 18.

49 Balderston, S., Taulbee, J.J., Celaya, E. et al. (2021). *Nat. Biomed. Eng.* 5: 713.

50 Zheng, C., Huang, L., Zhang, H. et al. (2015). *ACS Appl. Mater. Interfaces* 7: 16953.

51 (a) Urase, T. and Sasaki, Y. (2013). *Water Sci. Technol.* 68: 1969. (b) Suurnäkki, S., Gomez-Saez, G.V., Rantala-Ylinen, A. et al. (2015). *Water Res.* 68: 56.

52 (a) Parinet, J., Rodriguez, M.J., and Sérodes, J. (2010). *Water Res.* 44: 5847. (b) Watson, S.B., Monis, P., Baker, P., and Giglio, S. (2016). *Harmful Algae* 54: 112.

53 Lvova, L., Jahatspanian, I., Mattoso, L.H.C. et al. (2020). *Sensors* 20.Bristow, R.L., Young, I.S., Pemberton, A. et al. (2019). *TrAC, Trends Anal. Chem.* 110: 233.

54 Park, S.J., Seo, S.E., Kim, K.H. et al. (2021). *Biosens. Bioelectron.* 174: 112804.

55 (a) Makhamrueang, N., Sirilun, S., Sirithunyalug, J. et al. (2021). *Foods* 10: 996. (b) Naila, A., Flint, S., Fletcher, G. et al. (2010). *J. Food Sci.* 75: R139.

56 Ruiz-Capillas, C. and Herrero, A.M. (2019). *Foods (Basel, Switzerland)* 8: 62.

57 (a) Ma, Z., Chen, P., Cheng, W. et al. (2018). *Nano Lett.* 18: 4570. (b) Xu, X.-Y., Lian, X., Hao, J.-N. et al. (2017). *Adv. Mater.* 29: 1702298.

6

Thin-Film Transistors Based on Reduced Graphene Oxide for Biosensing

Kai Bao[1], Ye Chen[2], Qiyuan He[1], and Hua Zhang[3,4,5]

[1] *City University of Hong Kong, Department of Materials Science and Engineering, Tat Chee Avenue, Kowloon, 999077, Hong Kong, China*
[2] *The Chinese University of Hong Kong, Department of Chemistry, Central Ave, Shatin, 999077, Hong Kong, China*
[3] *City University of Hong Kong, Department of Chemistry, Tat Chee Avenue, Kowloon, 999077, Hong Kong, China*
[4] *City University of Hong Kong, Hong Kong Branch of National Precious Metals Material Engineering Research Center (NPMM), Tat Chee Avenue, Kowloon, 999077, Hong Kong, China*
[5] *City University of Hong Kong, Shenzhen Research Institute, Yuexing 1st Road, Shenzhen, 518057, China*

6.1 Introduction

Graphene and its derivatives are ideal materials for various sensing platforms owing to their many unique and superior properties [1–3]. Electronic sensors based on graphene field-effect transistors (GFETs) are particularly attractive and widely explored in chemical and biosensing mainly due to the atomic thickness, extremely high surface area, high conductivity and mobility, and high chemical and thermal stabilities of graphene. Graphene-based electronic sensors can be divided into two main categories based on the graphene used as the active channel in the FETs, i.e. mechanically exfoliated nanosheets and large-area thin film. Mechanically exfoliation produces pristine graphene from graphite flake with high crystallinity. It is favored for many proof-of-concept applications, such as the early demonstration of graphene-based gas detection [4] and pH sensors in aqueous environment [5]. On the other hand, graphene thin film is chemically synthesized graphene with high structural uniformity across large areas (from cm to wafer scale). The success in synthesizing various graphene thin films promises their practical applications in biosensing.

There are two types of graphene thin films used in biosensing: the chemical vapor deposition (CVD)-grown graphene and solution-processable graphene. The CVD synthesis of graphene has shown great success in the last decade and is gradually approaching the industrial scale. However, CVD synthesis is energy-intensive and requires specific thin-film processing technology. Alternatively, solution-processable

Graphene Field-Effect Transistors: Advanced Bioelectronic Devices for Sensing Applications,
First Edition. Edited by Omar Azzaroni and Wolfgang Knoll.

graphene refers to exfoliated graphene nanosheets from graphite that can remain relatively stable while suspended in solvents. Solution-processable graphene attracts considerable interest due to its facile processing, wide availability, and cost-effectiveness.

Reduced graphene oxide (rGO) is an extensively explored type of solution-processable graphene. rGO can be obtained from the reduction of graphene oxide (GO) via various reduction methods, which remove a large number of functional groups and restore the electrical conductivity. The GO source is generally produced by the Hummers method and its many variations. The property of rGO varies significantly depending on the synthetic method of GO and the subsequent reduction method. However, it is impossible to completely remove the oxygen-containing functional groups by any reduction method [6], which inevitably compromises its electrical performance to a certain degree. On the other hand, the presence of remaining functional groups on the rGO surface can contribute to the facile functionalization and promote strong surface interaction with many chemical and biological species, making them more suitable candidates in chemical and biological sensing compared to the single-crystalline graphene prepared by mechanical exfoliation or CVD [7]. Many types of rGO-based biosensors have been developed, including electrochemical, fluorescent, and FET-based electronic biosensors. In particular, thin-film transistors (TFTs) based on rGO are desirable in electronic biosensing for facial fabrication, easy surface functionalization, and high biocompatibility.

In this chapter, we first introduce the preparation and working principles of TFT-based biosensors. Then, we summarize the research works of rGO-based TFTs for the detection of various biospecies, including proteins, metal ions, nucleic acids, small biomolecules, and living cells. We elaborate on many of our previous related research works. Finally, we envision the challenges and opportunities of rGO-based FET for biosensing.

6.2 Working Principle of TFT-Based Biosensing

Different from traditional metal-oxide-semiconductor field-effect transistors (MOSFETs), TFTs based on solution-processable rGO can be easily prepared on insulated substrates, even flexible substrates, in large scale at very low cost, which makes them very attractive in practical biosensing [8]. Figure 6.1 shows the schematic illustration of a typical rGO-based TFT operating as a biosensor. The rGO thin films prepared by various methods serve as the active channels in the TFTs.

The drain and source electrodes are connected with the rGO channel through lithography techniques. It is worth mentioning that the drain and source electrodes are usually identical and interchangeable in graphene-based thin-film devices. A polymer reservoir holds the electrolyte within the rGO channel area. The electrodes and their contact areas with the rGO channel are also passivated with polymer to confine the biological interactions at the rGO channel. The functionalization of

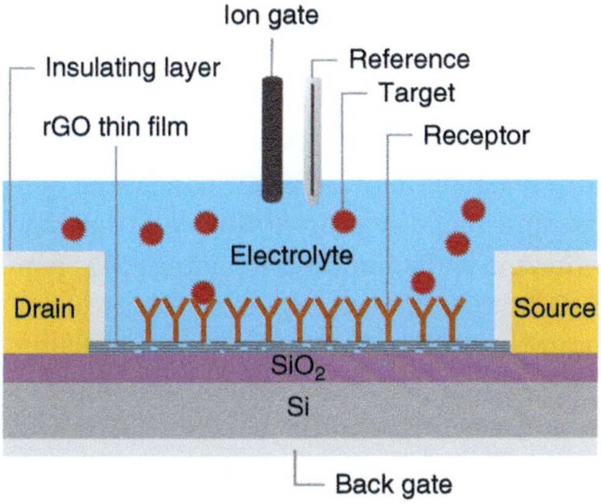

Figure 6.1 Schematic illustration of a typical rGO-based thin-film transistor used as a biosensor in an electrolyte solution.

bioreceptors, a necessary step for highly selective biodetection, is realized via physical adsorption (π–π interaction) or covalent bonding through various linker molecules or materials. For example, pyrene-based molecules, such as 1-pyrenebutanoic acid succinimidyl ester (PASE), are widely used linker molecules in graphene and rGO-based sensors [9, 10], which spontaneously adsorb on the sp^2-basal plane of rGO through π–π interaction. Back-gate modulation is sometimes not presented in rGO-based thin-film devices due to the diverse substrates used, and more importantly, the possible gate leakage resulting from the large channel area. Instead, ion gate, also referred to as solution gate or electrolyte gate, can be used to modulate the TFT in electrolyte solution held by the polymer reservoir. Thus, the carrier concentration (conductance) of the rGO channel is modulated by the interfacial potential of the electrical double layer at the solid-electrolyte interface. In some works, a reference electrode (e.g. Ag/AgCl) is used to further regulate the gate potential to avoid influences from interfacial faradaic processes (electrochemical reactions). Owing to the high biocompatibility and chemical stability of rGO, a typical rGO-based TFT can operate well in physiological environments. There are mainly two types of detection modes, i.e. real-time and nonreal-time detections. In the nonreal-time detection, the electrical properties of the TFT are measured before and after the biological interactions in a back-gate configuration, usually both in a dry condition by completely removing the electrolytes. This mode is usually employed in proof-of-concept applications. In the real-time detection, the drain–source current (I_{DS}) is continuously monitored while the analytes are added into the reservoir. The specific interaction between target biomolecules and the bioreceptor-modified rGO surface causes detectable changes in the channel conductance (drain–source current, I_{DS}). Real-time detection represents the major advantage of FET-based electronic biosensing compared to many conventional biosensing techniques, and is therefore generally favored.

6.3 TFTs Based on rGO for Biosensing

rGO has been a superior nanomaterial for various thin-film applications. In 2007, Gilje et al. [11] reported the early success in fabricating TFTs based on large-scale rGO thin film with good electrical performance. In their work, GO nanosheets synthesized by the Hummers method were spray-deposited on Si/SiO_2 substrate and consequently chemically reduced by hydrazine to obtain the conducting rGO channel. Later, Eda et al. [12] reported an rGO-based TFT with improved electrical performance based on the nanoscale control over the film thickness via vacuum filtration-based methods. Zhou et al. [13] also reported patterned rGO thin films on various substrates using spray coating and electrochemical reduction. However, rGO generally suffers from small lateral sizes and low structural integrity due to the strong chemical oxidation process used for the preparation of GO based on the Hummers method. As a result, the electrical property of rGO thin films is significantly diminished from that of pristine or CVD-grown graphene. Further modifications of the Hummers methods could yield higher-quality rGO nanosheets with larger size and lower defects, making them more favored in thin-film devices [14]. In addition, our group also reported high-quality thin-film devices based on patterned rGO channels and demonstrated their promising performances for various biodetections, which will be elaborated later [15].

6.3.1 Protein Detection

Fast and accurate protein detection is essential in many fundamental research areas and practical applications. For example, detecting SARS-CoV-2 (the coronavirus that causes COVID-19) antibody and antigen proteins has become one of the critical protocols in fighting the COVID-19 pandemic [16]. For example, Mao et al. [17] demonstrated an early rGO-based protein sensor based on individual thermally reduced graphene oxide (TRGO) nanosheets (Figure 6.2). A GO nanosheet was physically adsorbed between two Au electrodes and reduced by thermal annealing to form the rGO-based FET. Successful protein detection relies on the functionalization of specific receptors on the rGO channel. In this early study, anti-immunoglobulin G (anti-IgG), serving as the receptor to bind the target immunoglobulin G (IgG) protein, was conjugated with Au nanoparticles (NPs) and then functionalized on the rGO surface by electrospray and electrostatic force directed assembly (Figure 6.2). In addition, a solution mixture of 0.1% Tween 20, 0.1% fish gelatin, and 1% bovine serum albumin (BSA) was coated on the anti-IgG-functionalized rGO surface to block the nonspecific adsorption of protein. Note that this is not a real-time detection. The protein detection is realized by measuring the transfer curves and output curves of the rGO-based FET before and after capturing target IgG protein in a back-gate configuration.

In 2011, our group reported an all-rGO based TFT as a flexible and transparent biosensor for real-time protein detection [18]. rGO thin film inherits many important merits of graphene, such as high flexibility, high conductivity, and

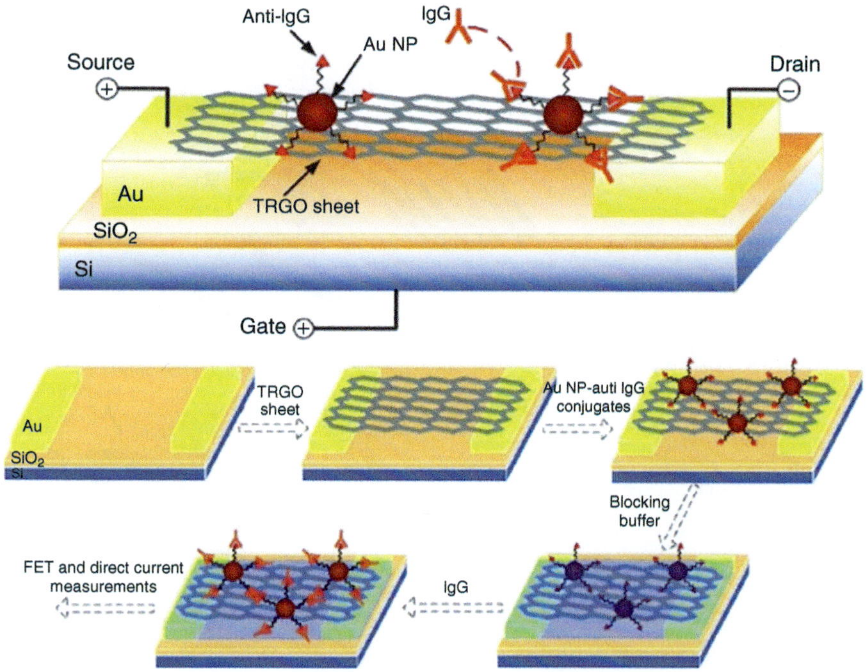

Figure 6.2 Schematic illustration for the fabrication of rGO-based-FET and its application in detecting immunoglobulin G (IgG) protein. Source: Reproduced with permission from Mao et al. [17]/John Wiley & Sons.

thickness-dependent electronic properties, making it suitable as both the active channel and the electrode contacts in a flexible biosensor.

As shown in Figure 6.3a, GO suspension was first spin-coated onto a polyethylene terephthalate (PET) film, which is modified with a self-assembled monolayer of 3-aminopropyltriethoxysilane (APTES) for better adhesion between the GO nanosheets and the polymer substrate. Two relatively thick GO strips (~8–10 nm) were then created by a previously developed scratch method [19]. Next, a micromolding-based technique (Figure 6.3a) was developed based on patterned polydimethylsiloxane (PDMS) stamp to create long and ultrathin GO micropatterns (2–4 nm) as the active channel between the two thick GO strips. The GO strips and patterns were chemically reduced in hydrazine vapor to obtain the rGO electrodes and channels, respectively. As the final step, silicone rubber was used to insulate the rGO electrodes and the electrical-contact area while forming a solution reservoir. It was found that both the conductivity and light transmittance of rGO thin film could be tuned by the film thickness. Highly transparent and functional all-rGO TFTs were fabricated by choosing the optimized thickness for both the rGO channel and the electrode. The all-rGO-based TFT exhibited high transmittance, stable ambipolar properties in ion-gate configuration, and more importantly, exceptional flexibility, which could endure thousands of bending at a small radius without losing performance. The all-rGO-based TFT was first used to detect fibronectin's nonspecific adsorption in real-time,

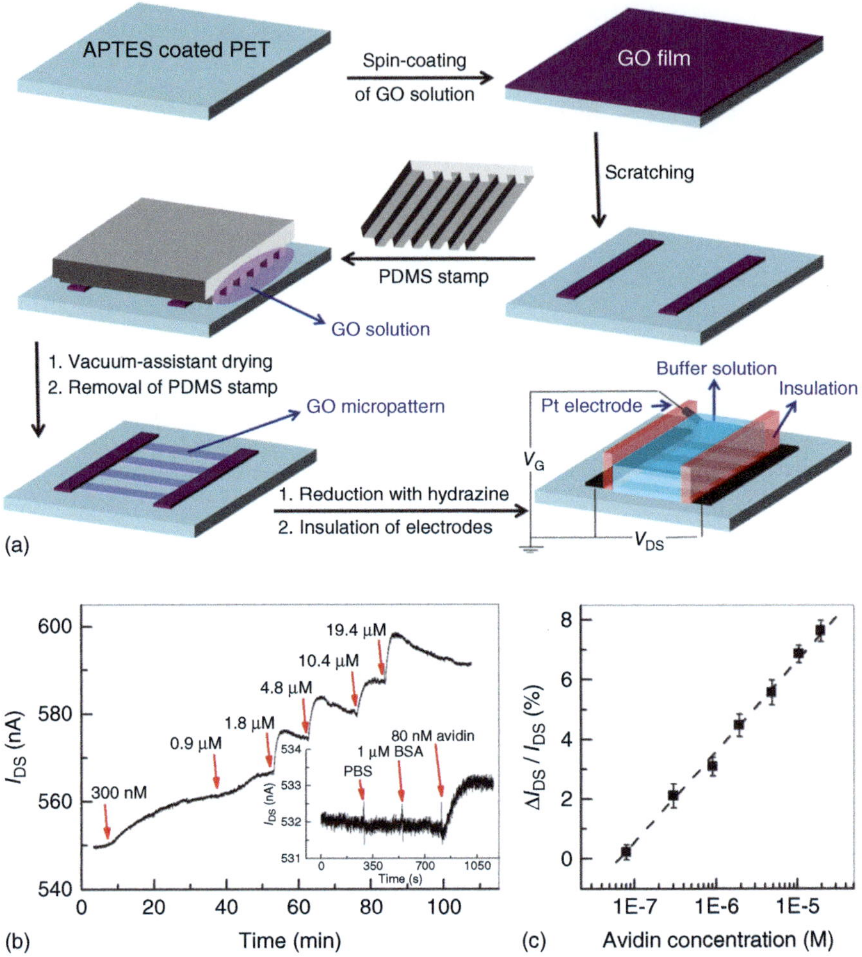

Figure 6.3 All-rGO TFT for the real-time detection of proteins. (a) Schematic illustration of the fabrication of an all-rGO TFT. APTES: 3-aminopropyltriethoxysilane; PDMS: polydimethylsiloxane. (b) Real-time detection of avidin with increasing concentration in PBS buffer by using a biotinylated all-rGO TFT. Inset: Current response of a biotinylated all-rGO TFT to the addition of PBS buffer, 1 μM bovine serum albumin (BSA), and 80 nM avidin, respectively. (c) The plot of current change vs. avidin concentration in (b), showing good linearity. The standard error in each concentration was obtained by measuring five samples. Source: Reproduced with permission from He et al. [18]/American Chemical Society.

showing high sensitivity down to 0.5 nm. Selective protein detection was then demonstrated based on the specific biotin–avidin interaction. Specifically, rGO channel was noncovalently functionalized with polyethyleneimine (PEI) as a linker for the biotin molecules and polyethylene glycol (PEG) as passivation to prevent the nonspecific adsorption of other proteins. As shown in Figure 6.3b,c, real-time detection of avidin was demonstrated with good linearity. No drain–source current (I_{DS}) change was observed with the addition of phosphate-buffered saline (PBS) or even at a high concentration of BSA protein (1 μM), confirming the high selectivity of the protein sensor. Further investigation of the transfer curves suggested that the conductance change was likely caused by a direct doping effect from the protein to the rGO channel. Lastly, the all-rGO-based TFTs maintained highly transparent and fully operational under bent conditions, demonstrating their potential in *in vivo* biodetection.

The detections of practical protein targets such as disease biomarkers were also demonstrated on rGO-based-TFTs. For instance, Myung et al. [20] reported a TFT device based on rGO nanosheets supported on SiO_2 NPs. The TFT was used to detect epidermal growth factor receptor (EGFR), which is considered a biomarker for breast cancer. In their work, SiO_2 NPs were used to support the assembly of GO nanosheets into a continuous thin film between two Au electrodes on a SiO_2/Si substrate. After reduction of GO, a continuous rGO thin film supported on SiO_2 NPs was obtained and used as the active channel in the TFT. The rGO channel was modified with EGFR antibody to detect the binding of EGFR in physiological environments. The incorporation of SiO_2 NPs significantly increased the surface-to-volume ratio of the rGO thin film, leading to an impressive 100 pM detection limit of EGFR. In another example, Lei et al. [21] reported an rGO-based-TFT modified with Pt NPs for brain natriuretic peptide (BNP) detection. The rGO thin film was obtained by chemical reduction of GO film formed on SiO_2/Si substrate. Then, Pt NPs were deposited on the rGO surface via UV light irradiation-assisted reduction of $PtCl_4^{-2}$. Thioglycolic acid, which could provide carboxylic groups to immobilize BNP antibodies, was modified on Pt NPs through Pt—S bond. BNP antibodies were immobilized on thioglycolic acid-modified Pt NPs via *N*-hydroxysuccinimide (NHS)/1-ethyl-3-(3-dimethylaminopropyl)carbodiimide (EDC) coupling. The functionalization of Pt NPs provides high surface area and stable covalent functionalization for the BNP antibodies [22]. A real-time detection limit of 100 fM was achieved. Furthermore, the device was also applied to detect BNP in a whole blood sample via the measurement of Dirac-point shift in an ion-gate configuration (nonreal-time).

In short, rGO-based-TFTs are ideal for the real-time detection of proteins. The high selectivity is achieved by the modification of rGO with both specific receptors for the protein target and passivation layer to prevent nonspecific adsorption. In addition to the proof-of-concept protein detection such as BSA and avidin detections, rGO-based-TFTs have also been employed in practical diagnosis of diseases, including viral infections [23], cancer [24], and heart disease [25]. However, economic ways toward large-scale fabrication remain a challenge. In addition, adaption and long-term operation in real sensing environments (such as blood and living cells) are challenging and require further research efforts.

6.3.2 Metal-Ion Detection

Metal-ion detection plays essential roles in disease diagnosis, health monitoring, and environmental protection. Metal ions are involved in various bioactivities in living organisms, such as electron transfer mediates, oxygen carriers, and active centers of enzymes. For example, the balance of metal ions in human body is related to many diseases such as Alzheimer's disease [26], neurodegenerative spongiform encephalopathies [27], and familial amyotrophic lateral sclerosis (FALS) [28].

Additionally, heavy metals such as copper, zinc, cadmium, and mercury, which living organisms cannot chemically degrade, can accumulate in food chains, posing a significant risk to ecosystems [29]. Conventional detection methods for metal ions, like inductively coupled plasma mass spectrometry (ICP-MS), are often complex and time-consuming [30]. Therefore, facile and real-time detection of metal ions is highly desirable.

Zhang et al. [31] demonstrated the early graphene-based detection of heavy-metal ions. In this work, individual graphene nanosheet obtained from mechanical exfoliation was used as the active channel in a FET. The graphene channel was functionalized with 1-octadecanethiol based on van der Waals interaction between the hydrophobic graphene and the long alkyl chains. Consequently, the selective detection of mercury ions is achieved via the capture of Hg^{2+} by forming Hg—S bond with the thiol group in the 1-octadecanethiol receptor. The detection signal is based on the measurement of Dirac-point shift in the transfer curves (back-gate configuration) upon mercury capture, which is not a real-time detection. In 2011, our group [10] reported an rGO-based-TFT for the real-time detection of metal ions, including Ca^{2+} and Hg^{2+}. Patterned rGO thin film fabricated based on previously reported method [15] was used as the active channel. Selective detection of Ca^{2+} was realized via the surface functionalization of calmodulin, a specific type of protein that selectively binds to Ca^{2+}. Calmodulin was functionalized on rGO surface using PASE as the linker molecule, which spontaneously adsorbs on rGO via π–π interaction. Upon capturing Ca^{2+}, the protein deformed, which led to a detectable conductance change (I_{DS}) of the rGO channel. Real-time detections of Ca^{2+} down to 1 μM were achieved with a high signal-to-noise ratio (Figure 6.4a). Furthermore, no response was observed with the addition of Na^+ and K^+ ions, proving the high selectivity. In addition, by using surface functionalization of a different protein (Metallothionein type-II protein), which specifically binds to heavy metal ions, selective detection of Hg^{2+} was achieved. As shown in Figure 6.4b, real-time detection of Hg^{2+} down to 1 nM was achieved, surpassing most fluorescence-based methods. Owing to the bipolar nature of graphene, the sensor response was tunable by operating the TFT at different ion-gate potentials, and therefore could be amplified, as shown in Figure 6.4c,d. The detection mechanism is based on the protein conformational change upon the capture of heavy metal ions, which changes the distance between the rGO surface and the charged protein surface, equivalent to a change of the interfacial gate potential at the interface. More importantly, highly sensitive and selective detections were also realized in a practical lake-water sample, showing minimal interference from the soil, dirt, and living organisms.

Sun et al. [32] reported another all-rGO based-TFT by using a layer-by-layer (LbL) assembly method. In their work, GO nanosheets were covalent LbL assembled into a thin film and reduced via thermal annealing to obtain the rGO channel. The all-rGO-based TFT was obtained by selectively functionalizing the rGO thin film with different chemical agents, (3-aminopropyl) trimethoxymethylsilane (APTMS) and 5,10,15,20-tetrakis (4-aminophenyl)-21H,23H-porphine (TAPP). rGO thin film functionalized with electrically inert APTMS served as the semiconductor channel. Meanwhile, rGO thin film functionalized with both APTMS and conductive TAPP

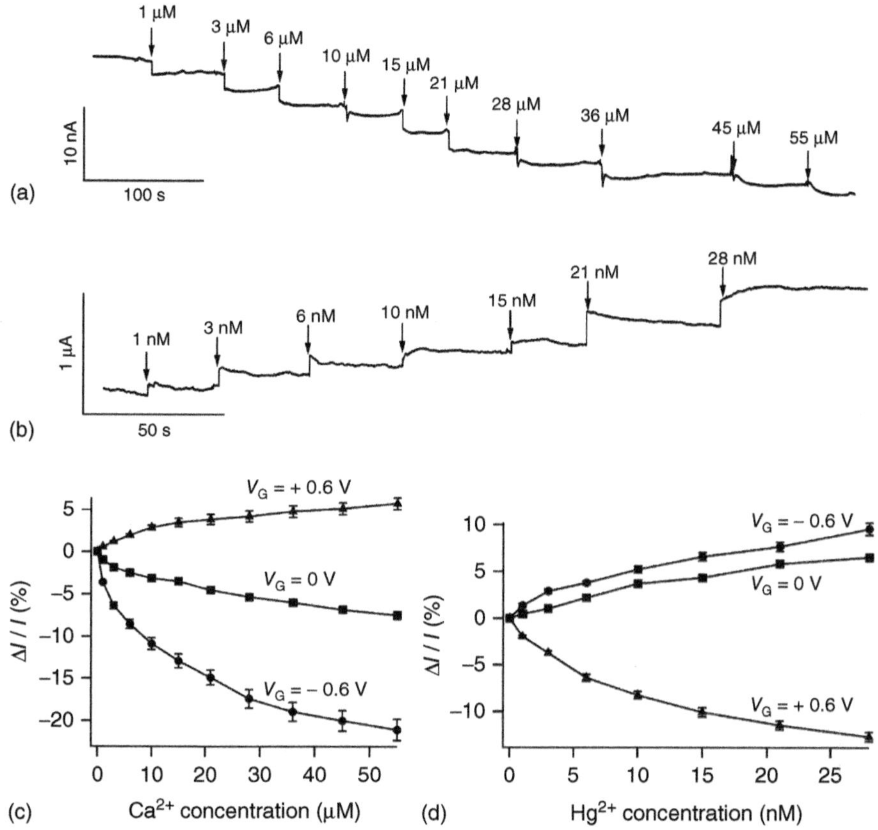

Figure 6.4 rGO-based-TFT for the real-time detection of metal ions. Typical real-time recording of I_{DS} with the addition of (a) Ca^{2+} and (b) Hg^{2+} ions. The change of I_{DS} measured in rGO-FETs (tested sample number $n = 6$) with the addition of (c) Ca^{2+} and (d) Hg^{2+} ions at $V_{DS} = 0.4$ V and $V_G = -0.6$ (circle), 0 (square), and +0.6 V (triangle). Source: Reproduced with permission from Sudibya et al. [10]/American Chemical Society.

served as electrodes due to the enhanced electrical conductivity. The reported all-graphene TFT achieved an ultra-low detection limit of 16 pM for Hg^{2+}, much more sensitive than the control TFTs with Au electrodes. Such all-rGO configuration benefits from enhanced electrical contact and, more importantly, enhanced electrolyte stability compared to metal electrodes.

The direct anchoring of large-biomolecular receptors on the rGO surface could be unstable and time-consuming. To simplify the functionalization process, Chen et al. [33] demonstrated the facial functionalization of rGO TFT with the addition of Au NPs. In their work, the rGO thin film was obtained via the thermal reduction of GO film in Ar atmosphere in a tube furnace. TRGO generally bears fewer defects and oxygen-containing functional groups, thus exhibiting better electrical properties. Au NPs were then directly deposited on the rGO surface via electrospray. The rGO channel was further covalently functionalized with thioglycolic acid via the Au—S bond. The selective detection of Hg^{2+} ions was demonstrated based on the formation of chemical bonds between target metal ions and the abundant –COOH functional groups on thioglycolic acid. As a result of the enhanced active surface area from the Au NPs, the rGO-based-TFT demonstrated the real-time detection of Hg^{2+} ions with a detection limit of 2.5 nM.

In short, rGO-based-TFTs endure high-performance real-time detection of metal ions. Unlike proteins that spontaneously adsorb on rGO surface, metal ions generally show weak interactions with rGO surface. Therefore, specific chemical and physical functionalization of the rGO surface is essential for the selective detection of metal ions, while nonspecific adsorption is generally not considered.

6.3.3 Nucleic Acid Detection

The detection of nucleic acid, deoxyribonucleic acid (DNA), and ribonucleic acid (RNA), holds great importance in clinical [34], environmental [35], diagnostics, [36] and food applications [37]. Many types of nucleic acid biosensors have been developed based on the highly specific hybridization between two complementary single-stranded deoxyribonucleic acid (ssDNA) or single-stranded ribonucleic acid (ssRNA) chains [38], such as Northern blot analyses [39], polymerase chain reaction (PCR), [40] and microarray [41]. However, most of them require complex instruments, long operation time, and highly trained operators. Therefore, they are not suitable for rapid and point-of-care diagnosis [42]. On the other hand, FET-based nucleic-acid detection is favored for the rapid and real-time response, convenient setup and data analysis, and miniaturization [43]. The early FET-based DNA detection was reported in 1997 using silicon-based FETs [44]. A thin oxide layer (~100 Å) was coated on the semiconducting silicon channel to prevent channel degradation in electrolyte and meanwhile facilitate the functionalization of the ssDNA receptor. Significant change of V_G could be observed in a real-time constant current mode (V_G-time), corresponding to the hybridization of target ssDNA. However, biosensors based on traditional FETs generally suffer from low sensitivity, instable semiconductor, bulky device, and high cost. On the other hand, graphene, a highly stable 2D material, strongly interacts with ssDNA and ssRNA via the π–π interaction, making it the ideal channel material in nucleic acid detection. The early rGO-based-TFT for DNA detection was reported by Stine et al. [45] in 2010. In this work, GO nanosheets were deposited on SiO_2/Si substrate functionalized with an amino-silane film for better adhesion and subsequently reduced to rGO thin film in hydrazine. It is worth mentioning that during the reduction process, some oxygen-containing functional groups on the GO nanosheets are preserved during the mild chemical reaction with ethylenediamine (EDA), which help to stabilize the rGO on amino-silane film via covalent bonds [46]. Real-time detection of target ssDNA sequence down to 10 nM was achieved, which is about the detection limit of surface plasmon resonance-based methods.

Our group reported another rGO-based-TFT for real-time DNA detection [47]. In this work, Langmuir–Blodgett (LB) method [48] was used to fabricate the rGO thin film with high uniformity and ultralow thickness. Then, Pt NPs, as the linker to ssDNA probe, were attached to rGO surface by reducing chloro-platinum complexes via photochemical reduction. Thiolated probe ssDNA was immobilization on the Pt

NPs as the receptor to selectively detect the complementary ssDNA. The left shift of Dirac point in I_{DS}–V_G curve in an ion-gate configuration confirmed that the detection is based on the n-doping effect induced by the hybridization of target ssDNA. Due to the lack of passivation in this work, a small electrical signal corresponding to the nonspecific adsorption of random ssDNA was also observed owing to π–π interaction between ssDNA and rGO surface.

In addition to ssDNA probes, peptide nucleic acids (PNAs) can also hybridize with ssDNA and serve as the receptors. The neutral character of PNA may help to reduce electrostatic repulsion during the hybridization process and achieve a better signal-to-noise ratio [49]. For example, Cai et al. [50] developed an rGO-based-TFT by employing PNA–DNA hybridization. PASE was used as the linker molecule to immobilize the PNA probes through the covalent bond between amino group on PNA and amide group on PASE. The rGO channel was further passivated with ethanolamine to prevent nonspecific binding of ssDNA. The PNA-based TFT demonstrated a superior detection limit of 100 fM, showing the great potential of rGO-based TFT in DNA sensing.

In short, rGO-based TFTs could realize real-time DNA detection with high sensitivity and selectivity. The detection is generally based on the hybridization between target ssDNA and complementary ssDNA probes. The sensing performance could be significantly affected by the fabrication methods of rGO thin films and the choice of link molecules for the ssDNA probes. Due to the strong interaction between the ssDNA chain and the rGO surface, a passivation layer is generally required for the highly specific detection.

6.3.4 Small Biomolecular Biosensor

The rapid and accurate detection of small biomolecules plays vital roles in health monitoring and environmental protection. For example, harm caused by small toxic biomolecules like aflatoxins, fumonisins, and phycotoxins [51, 52] could be amplified through the food chain and pose serious health risks. Meanwhile, small biomolecules like glucose, amino acids, and dopamine have essential roles in human bioactivities, such as signaling, expression of gene signals, and metabolism [53–55]. rGO-based-TFTs have demonstrated excellent performance in the real-time detections of many small biomolecules such as amino acid [56], dopamine, [57] and H_2O_2 [58].

In an early example, TFT based on rGO composite was used in H_2O_2 detection [58]. H_2O_2 is considered a biosignal for cell metabolism and immune system [59]. The development of H_2O_2 in physiological environments could facilitate clinical diagnostics such as Alzheimer's disease [60]. In the aforementioned work, GO was composited with polypyrrole nanotubes through the π–π interaction and was reduced in high temperature to form the active channel. In the real-time detection in PBS buffer, the introduction of H_2O_2 resulted in p-dope of the rGO composite, leading to an increase of I_{DS}. The rGO-based-TFT exhibited an excellent detection limit of 100 pM, which could be explained by the enhanced p-type behavior and increased surface area of the rGO channel, as a result of the incorporation of polypyrrole nanotubes. Vasu et al. [61] reported a non-enzymatic rGO-based TFT for glucose detection. In their work, aminophenylboronic acid (APBA) was functionalized on rGO surface through the π–π interaction by simply immersing the chip in 4-APBA solution (10 mM in ethanol) at 100 °C. As a comparison, covalent functionalization of APBA was also performed through bonding between the amino groups (–NH_2) on APBA and carboxy groups (–COOH) on rGO via the NHS/EDC coupling. Compared to the covalent functionalization, the noncovalent functionalized rGO preserved more –COOH functional groups, which led to more APBA immobilization and hence more binding sites for glucose. As a result, the rGO-based-TFT with noncovalent functionalized APBA exhibited much better sensitivity (~19.1% conductance change/glucose decade) and detection limit (2 nM) than that of covalent functionalization (~5.5%/decade and 500 nM, respectively).

In addition, rGO could be a suitable material for pH sensors due to the strong interaction between protons and the oxygen-containing functional groups on rGO surface such as hydroxyl (–OH), carboxyl (–COOH), and epoxy groups (–COC) [62]. Since proton is often a by-product of enzymatic reactions, rGO-based pH sensor could also be used to locally monitor enzymatic reactions. Chae et al. [63] reported rGO-based-TFTs for monitoring enzymatic kinetics between acetylcholinesterase (AChE) and acetylcholine (ACh). In their work, AChE was immobilized on the rGO surface by simple physical absorption. The concentration of AchE could be directly related to the Dirac-point shift of transfer curves in an ion-gate configuration, which could be attributed to the pH change resulting from the proton release due to enzymatic catalysis of AChE. Based on the mechanism, two AChE inhibitors, donepezil and rivastigmine, were modified on ACh biosensor to study the function of inhibitors in enzyme kinetics. As a result, IC_{50} values, the inhibitor concentration required to reduce the enzymatic activity to 50% of the original value, for donepezil and rivastigmine were identified to be 47 and 130 nM, respectively. LbL assembly provides another efficient way to functionalize rGO channel with large biomolecule receptors. For example, Piccinini et al. [64] reported an LbL assembly of polyethylenimine and urease on rGO channel for selective urea detection. Before the LbL process, sodium 1-pyrenesulfonate (SPS) was attached to bare rGO surface via the π–π interactions to provide the charged surface needed in the LbL assembly. In the LbL process, through electrostatic interactions, PEI layer was formed on the rGO surface, followed by the assembly of urease. These two steps were repeated several times to form a densely packed functionalization layer on rGO surface. The LbL assembly of urease facilitated the urea hydrolysis that led to the change of pH value. As a result, a wide detection range of 0.001–1 mM urea was achieved. In another example, Berninger et al. [56] reported the real-time detection of L-arginine on rGO-based-TFT using the LbL assembly. Similar to the previous work, two enzymes, urease and arginase, were immobilized on rGO surface via electrostatic LbL assembly with the assist of PEI. During the detection process, L-arginine was firstly hydrolyzed by arginase to release urea and ornithine. The released urea was further decomposed by urease into NH_3 and CO_2, which could lead to a pH response detectable by the rGO-based-TFT. As a result, $33\,\mu A\,\mu M^{-1}$ sensitivity was achieved. In a recent work, Fenoy et al. [65] presented another rGO-based-TFT for ACh detection based on the noncovalent functionalization of AChE receptors on rGO surface. In this work, poly(3-amino-benzylamine-*co*-aniline) (PABA) was directly deposited on rGO surface through electrodeposition, followed by the immobilization of AChE through electrostatic interactions. AChE greatly facilitated the hydrolysis of ACh, leading to variation of pH value. As a result, real-time detection of ACh with high sensitivity ($-27.5 \pm 0.8\,\mu A$/ACh decade), rapid response, and great selectivity was achieved.

In short, with proper surface functionalization, rGO-based-TFTs perform well in real-time detection of many small biomolecules. However, there are still many challenges, such as low detection range and interference from competitive interactions among the biomolecules, functionalized receptors, and the rGO surface.

6.3.5 Living-Cell Biosensor

Detection of biosignals from living cells is vital for understanding its fundamental biological functions and medicine development [66, 67]. Our group reported the first rGO-based-TFT for the real-time monitor of living cell activities [15]. The fabrication of the patterned rGO thin film was similar to the previously mentioned method [18]. In our work, the width of rGO pattern was precisely controlled to $10\,\mu m$ to match the size of the living PC12 neuron cells for the *in vivo* detection. As shown in Figure 6.5a, PC12 cells could be directly cultured on the rGO thin film

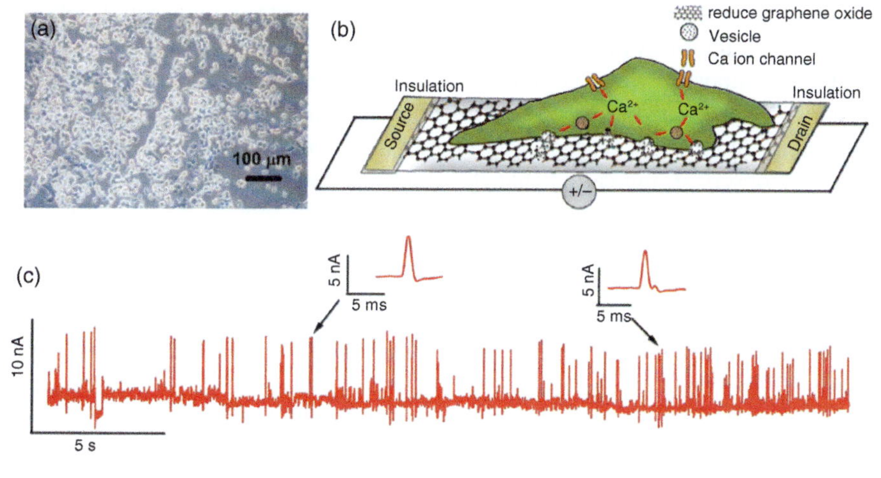

Figure 6.5 rGO-based-TFT for monitoring living cell activities. (a) Optical image of PC12 cells grown confluently on poly-L-lysine coated rGO-based-TFT on polyethylene terephthalate substrate. (b) Schematic illustration of the interface between a PC12 cell and rGO FET, showing the vesicular secretion process. (c) Real-time response of rGO-based-TFT to the dynamic vesicular secretion of catecholamines from PC12 cells stimulated by high K^+ solution. Source: Reproduced with permission from He et al. [15]/American Chemical Society.

owing to its high biocompatibility. The rGO-based-TFT could detect the nonspecific adsorption of dopamine molecules in real-time on both quartz and flexible PET substrates under bending. Consequently, we demonstrated the real-time monitoring of the vesicular secretion of catecholamines from PC12 cells induced by the addition of high K^+ solution, as illustrated in Figure 6.5b. Figure 6.5c shows the I_{DS}-time curve of the device during the high-K^+ stimulation.

We believed that each current spike corresponds to dopamine release from an individual living cell. Furthermore, the device retained high sensitivity under bending, showing potential in *in vivo* biodetection. Later, Xie et al. [68] reported the real-time monitor of NO released by living cells using an rGO-based-TFT. Fe(III) meso-tetra(4-carboxyphenyl)porphyrin (FeTCP) was deposited on rGO surface as the receptor to interact with NO molecules. Similar to our previously mentioned work, the human umbilical vein endothelial cells (HUVECs) were successfully cultured on a highly biocompatible rGO surface. L-Arginine was used to stimulate the generation of NO in HUVECs. The released NO molecule could donate electrons to the Fe(III) center in porphyrin, which could weaken the interaction between graphene and the Fe(III) center, leading to electron transfer from the Fe(III) center to the valence band of graphene. As a result, a decrease of I_{DS} was observed after introducing L-Arginine. In a recent work, Zheng et al. [69] reported the real-time detection of H_2O_2 released from living cells by using an rGO composite-based TFT. In their work, MoS_2 suspension was drop-casted on the rGO channel to form a layer of MoS_2 thin film, which served as the receptor for H_2O_2. MoS_2 was chosen because it can function as a catalyst to decompose H_2O_2 according to previous research [70]. HeLa cells were directly cultured on the rGO composite, which could release H_2O_2 upon stimulation by phorbol 12-myristate 13-acetate (PMA). The released H_2O_2 could be decomposed on the composite channel, resulting in *n*-dope of the rGO channel and a decrease in I_{DS} in the real-time detection. In a control experiment, when catalase, an H_2O_2 scavenger, was added in the test solution, no noticeable I_{DS} change was observed, indicating that the signal indeed came from the release of H_2O_2 by HeLa cells.

In short, rGO-based-TFTs are promising platforms to monitor living cell activities in real time, mainly attributed to the high biocompatibility and the high chemical stability of rGO nanosheets in physiological environments. Compared with the detection of biomaterials, it is much more challenging to isolate the signals from complex living cells.

6.3.6 Gas Detection

In addition to biosensing, we also reported TFT-based on rGO and other 2D materials for real-time gas detection. Although gas detection cannot be strictly classified into biosensing, it is sometimes directly related to many important bioactivities, such as toxic detection and health monitoring. For example, it is reported that analysis of human breath may help to diagnose disease [71].

In the last century, metal oxides such as SnO_2, ZnO, and In_2O were extensively studied in FET-based gas detection [72]. However, FETs based on these bulky semiconductors suffered from small surface area [73] and high operation temperature (over 100°C), which limit their practical applications [74]. In 2007, Schedin et al. [75] reported the room-temperature gas detection based on individual, mechanically exfoliated graphene nanosheets. Very impressively, the adsorption of individual gas molecule on graphene surface could be detected. However, the detection is based on the measurement of the carrier concentration of graphene by using Hall measurement in highly specified equipment. In recent years, more practical TFT-based gas sensors utilizing various 2D materials were demonstrated [76]. For example, Lu et al. [77] reported the rGO-based-TFT as a NO_2 sensor. rGO thin film with a high C/O atomic ratio was synthesized by thermal reduction. The high C/O ratio indicates more sp^2 carbons, which could adsorb small molecules better than the oxygen functional groups [78, 79]. Besides graphene, 2D semiconductors such as MoS_2 are also promising for gas sensing but suffer from complex preparation methods and low yield. Our group first reported the proof-of-concept gas sensor based on single- to few-layer MoS_2 nanosheets for NO detection [80]. Later, we also reported a room-temperature NO_2 detector based on rGO and MoS_2 thin films as the electrode and active channel, respectively [81]. Contact resistance, which usually reduces sensing performance, was negligible in this TFT due to the long MoS_2 channel that dominated the overall charge transport of the device. To further enhance the sensing performance, Pt NPs were mounted on surface of MoS_2, resulting in significantly enhanced sensitivity by up to ~3 times toward NO_2. More importantly, the 2D-material-based TFT shows high mechanical flexibility which is promising for future wearable sensing technologies.

TFTs based on rGO and other 2D materials are ideal in detecting toxic gases due to their low cost, high sensitivity, and mild operation temperature. However, they generally suffer from long current recovery times due to the slow gas desorption process on rGO surface, which limits their practical applications. In one extreme case of TaS_2-based-TFT for NO detection, no current recovery was observed even after a long-time purge by N_2 at elevated temperature. Our group reported a room-temperature dosimeter for toxic NO gas based on TaS_2 thin film [76]. As shown in Figure 6.6a, high-quality TaS_2 suspension, obtained from electrochemical intercalation [82], was used to fabricate functional TaS_2 thin film via a filtration-based method. As shown in Figure 6.6b, the device conductance decreased upon exposure to NO gas and was stable under N_2 purge, showing an irreversible electrical response. As a comparison, devices based on MoS_2 and TiS_2 thin films were also fabricated and tested, exhibiting pronounced current recovery during N_2 purge (Figure 6.6c). Such irreversible current response is unique in 2D materials and is likely a result of the irreversible adsorption of the NO gas molecules on TaS_2. As a result, our thin-film device could monitor the total dosage of chronicle exposure to toxic gas, which could be very useful for field works and risk assessments [83].

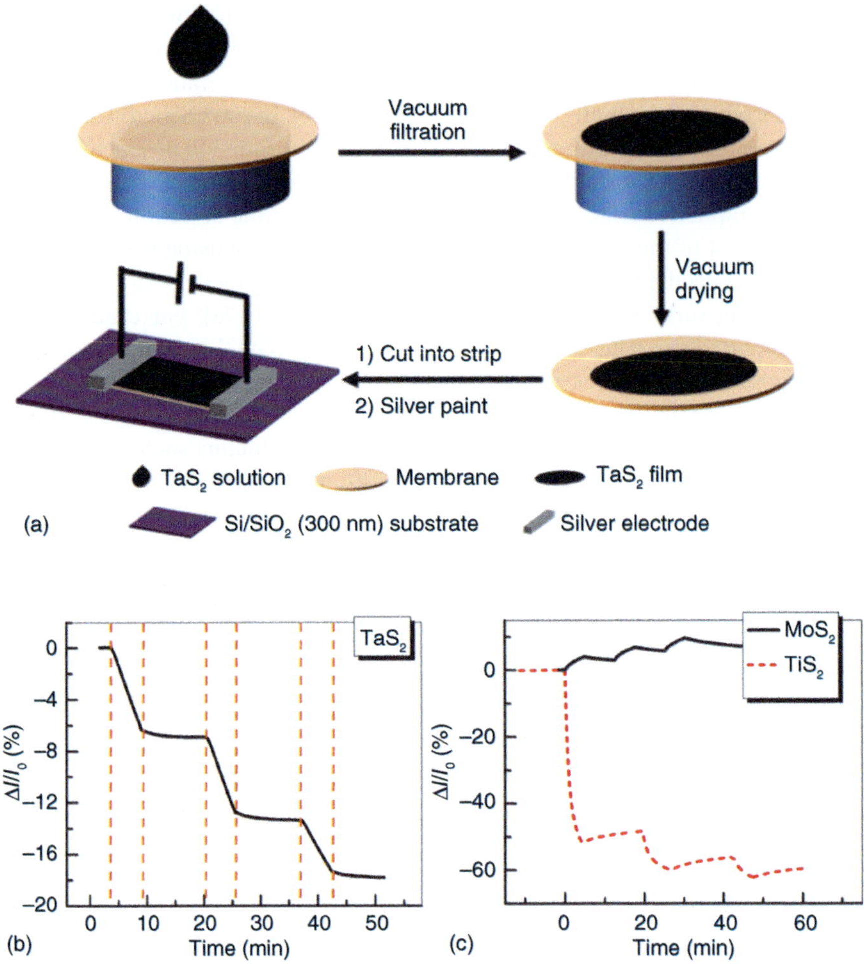

Figure 6.6 TaS_2-based-TFT as a dosimeter for toxic gas. (a) Schematic illustration of the fabrication of TaS_2 device by vacuum filtration. (b) The typical current response of TaS_2 film upon NO exposure at 1.96 ppm. (c) The typical current responses of MoS_2 (black) and TiS_2 (red) films upon exposure of 1.96 ppm NO at room temperature with N_2 as purge gas. Source: He et al. [76]/© AIP Publishing/CC BY 3.0.

6.4 Conclusion

TFTs based on rGO thin films have shown great promise as biosensing platforms in a variety of applications, mainly attributing to their cost-effectiveness, scalable fabrication, miniaturization, and real-time detection. We have summarized some representative research works on high-performance biosensing by using rGO-based-TFTs for proteins, metal ions, nucleic acids, and small biomolecules. The highly specified covalent and noncovalent functionalization of rGO surface is crucial for the high selectivity towards different biospecies. Though tremendous progress has been made in the last decade, several key challenges remain.

1) Most rGO thin films are derived from the reduction of GO nanosheets produced by the Hummers method. Although many modifications have been made over the Hummers method, rGO still suffers from a large amount of defects and low structural integrity, which compromise the electrical performance and stability. Furthermore, it is not easy to prepare large-scale GO thin films with uniform thickness and morphology, and free of agglomeration and restacking. Further development on the chemical processes as well as the thin-film processing is still desired.
2) Although high selectivity and low detection limit have been achieved in many proof-of-concept rGO-based TFTs, their response time, recovery profile and reliability are often not satisfactory. Prolonged response time can cause inaccuracy in real-time detection, while the poor recovery, resulting from irreversible adsorption of biomolecules, results in the high-cost one-time usage. More importantly, the benchmark repeatability of sensing performance and the reliability under long-term operation in physiological environments demand further studies.
3) As a vital step for selective biosensing, the immobilization methods of bioreceptors on rGO surface are still far from the desired efficiency and reliability. Although noncovalent functionalization is very convenient, it often suffers from weak binding to maintain the attachment of receptors during prolonged sensing operation. On the contrary, covalent functionalization may break the conjugated sp^2-carbon network, further compromising the electrical performance of rGO. Furthermore, the intrinsic interactions between rGO and different biospecies also need further investigation to avoid false signals due to the nonspecific interaction.
4) The devices introduced in most research works reported to date operate in artificial electrolyte environments such as physiological buffer solutions. However, the biodetection is carried out in practical sensing environments such as human blood, cell culture media, and groundwater remain scarce. The complex solution conditions in these environments, involving ions, organic molecules, biomolecules, and contaminations, may significantly alter the sensing performance due to nonspecific interactions. Continuous effort on the practical deployment of rGO-based-TFT in practical samples and applications is encouraged.
5) Eco-friendliness is a critical indicator of next-generation biosensors. The chemical processes to produce rGO thin film and perform surface functionalization may involve highly toxic chemicals, thus requiring further optimization. Meanwhile, the intrinsic biocompatibility of rGO nanosheets with various biological systems still needs to be clarified. For instance, based on previous report, the layered morphology of rGO nanosheets may result in the damage of cell membranes and absorption of lipid molecules [84]. Besides, the high dispersibility and abundance of functional groups on the rGO surface may lead to more substantial toxicity to human cells [85]. More research efforts are required in both the optimization of device fabrication and the toxicity assessment of rGO for the further development in this field.
6) Since the initial development of graphene and rGO-based biosensors, researchers have continuously pushed the boundaries toward other 2D material systems

including transition metal dichalcogenides (TMDs), metal organic frameworks (MOFs), covalent organic frameworks (COFs) and many more. The large family of 2D materials possesses diverse physiochemical properties, which offer unique advantages in various chemical and biodetective applications. Like the initial development of graphene-based FET sensors, most reported works have focused on FETs based on individual 2D nanosheets, while the continuous development thin-film technologies is critical toward practical applications.

7) The global efforts against the COVID-19 pandemic highlight the critical role of rapid and accurate virus detection technologies [2]. FET sensors based on 2D materials have demonstrated high performance in the real-time detection of SARS-COV-2, the virus that causes the coronavirus disease. For example, graphene-based FETs have demonstrated exceptional performance in detecting the SARS-COV-2 spike protein in PBS buffer, and more importantly, the virus particles in clinical samples [86]. More recently, FET sensors based other 2D semiconductors such as WSe_2 and $Ti_3C_2T_x$ Mxene have also shown good performance in detection the nucleic acid, antigens, and antibodies of SARS-COV-2 [87, 88]. Therefore, we can foresee more research works and commercial utilization of 2D material-based biosensors in the near future.

References

1 He, Q., Wu, S., Yin, Z., and Zhang, H. (2012). Graphene-based electronic sensors. *Chem. Sci.* 3 (6): 1764–1772.

2 Wang, W., Zhai, W., Chen, Y. et al. (2022). Two-dimensional material-based virus detection. *Sci. China Chem.* 65 (3): 497–513.

3 Chang, C., Chen, W., Chen, Y. et al. (2021). Recent progress on two-dimensional materials. *Acta Phys. Chim. Sin.* 37 (12): 2108017.

4 Dan, Y., Lu, Y., Kybert, N.J. et al. (2009). Intrinsic response of graphene vapor sensors. *Nano Lett.* 9 (4): 1472–1475.

5 Cheng, Z., Li, Q., Li, Z. et al. (2010). Suspended graphene sensors with improved signal and reduced noise. *Nano Lett.* 10 (5): 1864–1868.

6 Shin, H.J., Kim, K.K., Benayad, A. et al. (2009). Efficient reduction of graphite oxide by sodium borohydride and its effect on electrical conductance. *Adv. Funct. Mater.* 19 (12): 1987–1992.

7 Zhang, L., Zhang, A., Du, D., and Lin, Y. (2012). Biosensor based on Prussian blue nanocubes/reduced graphene oxide nanocomposite for detection of organophosphorus pesticides. *Nanoscale* 4 (15): 4674–4679.

8 Myny, K. (2018). The development of flexible integrated circuits based on thin-film transistors. *Nat. Electron.* 1 (1): 30–39.

9 Reiner-Rozman, C., Larisika, M., Nowak, C., and Knoll, W. (2015). Graphene-based liquid-gated field effect transistor for biosensing: theory and experiments. *Biosens. Bioelectron.* 70: 21–27.

10 Sudibya, H.G., He, Q., Zhang, H., and Chen, P. (2011). Electrical detection of metal ions using field-effect transistors based on micropatterned reduced graphene oxide films. *ACS Nano* 5 (3): 1990–1994.

11 Gilje, S., Han, S., Wang, M. et al. (2007). A chemical route to graphene for device applications. *Nano Lett.* 7 (11): 3394–3398.

12 Eda, G., Fanchini, G., and Chhowalla, M. (2008). Large-area ultrathin films of reduced graphene oxide as a transparent and flexible electronic material. *Nat. Nanotechnol.* 3 (5): 270–274.

13 Zhou, M., Wang, Y., Zhai, Y. et al. (2009). Controlled synthesis of large-area and patterned electrochemically reduced graphene oxide films. *Chem. Eur. J.* 15 (25): 6116–6120.

14 Marcano, D.C., Kosynkin, D.V., Berlin, J.M. et al. (2010). Improved synthesis of graphene oxide. *ACS Nano* 4 (8): 4806–4814.

15 He, Q., Sudibya, H.G., Yin, Z. et al. (2010). Centimeter-long and large-scale micropatterns of reduced graphene oxide films: fabrication and sensing applications. *ACS Nano* 4 (6): 3201–3208.

16 Liu, W., Liu, L., Kou, G. et al. (2020). Evaluation of nucleocapsid and spike protein-based enzyme-linked immunosorbent assays for detecting antibodies against SARS-CoV-2. *J. Clin. Microbiol.* 58 (6): e00461–e00420.

17 Mao, S., Lu, G., Yu, K. et al. (2010). Specific protein detection using thermally reduced graphene oxide sheet decorated with gold nanoparticle-antibody conjugates. *Adv. Mater.* 22 (32): 3521–3526.

18 He, Q., Wu, S., Gao, S. et al. (2011). Transparent, flexible, all-reduced graphene oxide thin film transistors. *ACS Nano* 5 (6): 5038–5044.

19 Li, B., Cao, X., Ong, H.G. et al. (2010). All-carbon electronic devices fabricated by directly grown single-walled carbon nanotubes on reduced graphene oxide electrodes. *Adv. Mater.* 22 (28): 3058–3061.

20 Myung, S., Solanki, A., Kim, C. et al. (2011). Graphene-encapsulated nanoparticle-based biosensor for the selective detection of cancer biomarkers. *Adv. Mater.* 23 (19): 2221–2225.

21 Lei, Y.-M., Xiao, M.-M., Li, Y.-T. et al. (2017). Detection of heart failure-related biomarker in whole blood with graphene field effect transistor biosensor. *Biosens. Bioelectron.* 91: 1–7.

22 Roushani, M. and Valipour, A. (2016). Using electrochemical oxidation of rutin in modeling a novel and sensitive immunosensor based on Pt nanoparticle and graphene-ionic liquid-chitosan nanocomposite to detect human chorionic gonadotropin. *Sens. Actuators, B* 222: 1103–1111.

23 Jin, X., Zhang, H., Li, Y.-T. et al. (2019). A field effect transistor modified with reduced graphene oxide for immunodetection of ebola virus. *Microchim. Acta* 186 (4): 223.

24 Mansouri Majd, S. and Salimi, A. (2018). Ultrasensitive flexible FET-type Aptasensor for CA 125 cancer marker detection based on carboxylated multiwalled carbon nanotubes immobilized onto reduced graphene oxide film. *Anal. Chim. Acta* 1000: 273–282.

25 Park, D., Kim, J.H., Kim, H.J. et al. (2020). Multiplexed femtomolar detection of Alzheimer's disease biomarkers in biofluids using a reduced graphene oxide field-effect transistor. *Biosens. Bioelectron.* 167: 112505.

26 Huang, X., Atwood, C.S., Hartshorn, M.A. et al. (1999). The Aβ peptide of Alzheimer's disease directly produces hydrogen peroxide through metal ion reduction. *Biochemistry* 38 (24): 7609–7616.

27 Perera, W.S.S. and Hooper, N.M. (2001). Ablation of the metal ion-induced endocytosis of the prion protein by disease-associated mutation of the octarepeat region. *Curr. Biol.* 11 (7): 519–523.

28 Goto, J.J., Zhu, H., Sanchez, R.J. et al. (2000). Loss of in vitro metal ion binding specificity in mutant copper-zinc superoxide dismutases associated with familial amyotrophic lateral sclerosis. *J. Biol. Chem.* 275 (2): 1007–1014.

29 Leyval, C., Turnau, K., and Haselwandter, K. (1997). Effect of heavy metal pollution on mycorrhizal colonization and function: physiological, ecological and applied aspects. *Mycorrhiza* 7 (3): 139–153.

30 Ekrami, E., Pouresmaieli, M., Shariati, P., and Mahmoudifard, M. (2021). A review on designing biosensors for the detection of trace metals. *Appl. Geochem.* 127: 104902.

31 Zhang, T., Cheng, Z., Wang, Y. et al. (2010). Self-assembled 1-octadecanethiol monolayers on graphene for mercury detection. *Nano Lett.* 10 (11): 4738–4741.

32 Sun, M., Zhang, C., Chen, D. et al. (2021). Ultrasensitive and stable all graphene field-effect transistor-based Hg^{2+} sensor constructed by using different covalently bonded RGO films assembled by different conjugate linking molecules. *SmartMat* 2 (2): 213–225.

33 Chen, K., Lu, G., Chang, J. et al. (2012). Hg(II) ion detection using thermally reduced graphene oxide decorated with functionalized gold nanoparticles. *Anal. Chem.* 84 (9): 4057–4062.

34 Justino, C.I.L., Rocha-Santos, T.A., Duarte, A.C., and Rocha-Santos, T.A. (2010). Review of analytical figures of merit of sensors and biosensors in clinical applications. *TrAC, Trends Anal. Chem.* 29 (10): 1172–1183.

35 Palchetti, I. and Mascini, M. (2008). Nucleic acid biosensors for environmental pollution monitoring. *Analyst* 133 (7): 846–854.

36 Choi, J.R., Hu, J., Tang, R. et al. (2016). An integrated paper-based sample-to-answer biosensor for nucleic acid testing at the point of care. *Lab Chip* 16 (3): 611–621.

37 Huo, B., Hu, Y., Gao, Z., and Li, G. (2021). Recent advances on functional nucleic acid-based biosensors for detection of food contaminants. *Talanta* 222: 121565.

38 Rashid, J.I.A. and Yusof, N.A. (2017). The strategies of DNA immobilization and hybridization detection mechanism in the construction of electrochemical DNA sensor: a review. *Sens. Bio-Sens. Res.* 16: 19–31.

39 Streit, S., Michalski, C.W., Erkan, M. et al. (2009). Northern blot analysis for detection and quantification of RNA in pancreatic cancer cells and tissues. *Nat. Protoc.* 4 (1): 37–43.

40 Gilliland, G., Perrin, S., Blanchard, K., and Bunn, H.F. (1990). Analysis of cytokine mRNA and DNA: detection and quantitation by competitive polymerase chain reaction. *Proc. Natl. Acad. Sci.* 87 (7): 2725–2729.

41 Johnson, B.N. and Mutharasan, R. (2014). Biosensor-based microRNA detection: techniques, design, performance, and challenges. *Analyst* 139 (7): 1576–1588.

42 Teles, F.R.R. and Fonseca, L.P. (2008). Trends in DNA biosensors. *Talanta* 77 (2): 606–623.

43 Sadighbayan, D., Hasanzadeh, M., and Ghafar-Zadeh, E. (2020). Biosensing based on field-effect transistors (FET): recent progress and challenges. *TrAC, Trends Anal. Chem.* 133: 116067.

44 Souteyrand, E., Cloarec, J.P., Martin, J.R. et al. (1997). Direct detection of the hybridization of synthetic homo-oligomer DNA sequences by field effect. *J. Phys. Chem. B* 101 (15): 2980–2985.

45 Stine, R., Robinson, J.T., Sheehan, P.E., and Tamanaha, C.R. (2010). Real-time DNA detection using reduced graphene oxide field effect transistors. *Adv. Mater.* 22 (46): 5297–5300.

46 Kim, M.C., Hwang, G.S., and Ruoff, R.S. (2009). Epoxide reduction with hydrazine on graphene: a first principles study. *J. Chem. Phys.* 131 (6): 064704.

47 Yin, Z., He, Q., Huang, X. et al. (2012). Real-time DNA detection using Pt nanoparticle-decorated reduced graphene oxide field-effect transistors. *Nanoscale* 4 (1): 293–297.

48 Cote, L.J., Kim, F., and Huang, J. (2009). Langmuir-Blodgett assembly of graphite oxide single layers. *J. Am. Chem. Soc.* 131 (3): 1043–1049.

49 Zhang, G.-J., Chua, J.H., Chee, R.-E. et al. (2008). Highly sensitive measurements of PNA-DNA hybridization using oxide-etched silicon nanowire biosensors. *Biosens. Bioelectron.* 23 (11): 1701–1707.

50 Cai, B., Wang, S., Huang, L. et al. (2014). Ultrasensitive label-free detection of PNA-DNA hybridization by reduced graphene oxide field-effect transistor biosensor. *ACS Nano* 8 (3): 2632–2638.

51 Zaccaroni, A. and Scaravelli, D. (2008). Toxicity of sea algal toxins to humans and animals. In: *Algal Toxins: Nature, Occurrence, Effect and Detection* (ed. V. Evangelista, L. Barsanti, A.M. Frassanito, et al.), 91–158. Dordrecht: Springer.

52 Torres, O., Matute, J., Gelineau-van Waes, J. et al. (2015). Human health implications from co-exposure to aflatoxins and fumonisins in maize-based foods in Latin America: Guatemala as a case study. *World Mycotoxin J.* 8 (2): 143–159.

53 Holeček, M. (2018). Branched-chain amino acids in health and disease: metabolism, alterations in blood plasma, and as supplements. *Nutr. Metabol.* 15 (1): 33.

54 Prasad, R. (1998). Fertilizer urea, food security, health and the environment. *Curr. Sci.* 75 (7): 677–683.

55 Hartmann, P.C., Bürgi, D., and Giger, W. (2004). Organophosphate flame retardants and plasticizers in indoor air. *Chemosphere* 57 (8): 781–787.

56 Berninger, T., Bliem, C., Piccinini, E. et al. (2018). Cascading reaction of arginase and urease on a graphene-based FET for ultrasensitive, real-time detection of arginine. *Biosens. Bioelectron.* 115: 104–110.

57 Oh, J., Lee, J.S., Jun, J. et al. (2017). Ultrasensitive and selective organic FET-type nonenzymatic dopamine sensor based on platinum nanoparticles-decorated reduced graphene oxide. *ACS Appl. Mater. Interfaces* 9 (45): 39526–39533.

58 Park, J.W., Park, S.J., Kwon, O.S. et al. (2014). Polypyrrole nanotube embedded reduced graphene oxide transducer for field-effect transistor-type H_2O_2 biosensor. *Anal. Chem.* 86 (3): 1822–1828.

59 Guyton, K.Z., Liu, Y., Gorospe, M. et al. (1996). Activation of mitogen-activated protein kinase by H_2O_2. Role in cell survival following oxidant injury. *J. Biol. Chem.* 271 (8): 4138–4142.

60 Opazo, C., Huang, X., Cherny, R.A. et al. (2002). Metalloenzyme-like activity of Alzheimer's disease beta-amyloid. Cu-dependent catalytic conversion of dopamine, cholesterol, and biological reducing agents to neurotoxic H_2O_2. *J. Biol. Chem.* 277 (43): 40302–40308.

61 Vasu, K.S., Sridevi, S., Sampath, S., and Sood, A.K. (2015). Non-enzymatic electronic detection of glucose using aminophenylboronic acid functionalized reduced graphene oxide. *Sens. Actuators, B* 221: 1209–1214.

62 Salvo, P., Melai, B., Calisi, N. et al. (2018). Graphene-based devices for measuring pH. *Sens. Actuators, B* 256: 976–991.

63 Chae, M.-S., Yoo, Y.K., Kim, J. et al. (2018). Graphene-based enzyme-modified field-effect transistor biosensor for monitoring drug effects in Alzheimer's disease treatment. *Sens. Actuators, B* 272: 448–458.

64 Piccinini, E., Bliem, C., Reiner-Rozman, C. et al. (2017). Enzyme-polyelectrolyte multilayer assemblies on reduced graphene oxide field-effect transistors for biosensing applications. *Biosens. Bioelectron.* 92: 661–667.

65 Fenoy, G.E., Marmisollé, W.A., Azzaroni, O., and Knoll, W. (2020). Acetylcholine biosensor based on the electrochemical functionalization of graphene field-effect transistors. *Biosens. Bioelectron.* 148: 111796.

66 Fang, Y., Ferrie, A.M., Fontaine, N.H. et al. (2006). Resonant waveguide grating biosensor for living cell sensing. *Biophys. J.* 91 (5): 1925–1940.

67 Cohen-Karni, T., Qing, Q., Li, Q. et al. (2010). Graphene and nanowire transistors for cellular interfaces and electrical recording. *Nano Lett.* 10 (3): 1098–1102.

68 Xie, H., Li, Y.-T., Lei, Y.-M. et al. (2016). Real-time monitoring of nitric oxide at single-cell level with porphyrin-functionalized graphene field-effect transistor biosensor. *Anal. Chem.* 88 (22): 11115–11122.

69 Zheng, C., Jin, X., Li, Y. et al. (2019). Sensitive molybdenum disulfide based field effect transistor sensor for real-time monitoring of hydrogen peroxide. *Sci. Rep.* 9 (1): 759.

70 Lei, J., Lu, X., Nie, G. et al. (2015). One-pot synthesis of algae-like MoS_2/PPy nanocomposite: a synergistic catalyst with superior peroxidase-like catalytic activity for H_2O_2 detection. *Part. Part. Syst. Char.* 32 (9): 886–892.

71 Di Francesco, F., Fuoco, R., Trivella, M.G., and Ceccarini, A. (2005). Breath analysis: trends in techniques and clinical applications. *Microchem. J.* 79 (1): 405–410.

72 Batzill, M. (2006). Surface science studies of gas sensing materials: SnO_2. *Sensors* 6 (10): 1345–1366.

73 Bag, A. and Lee, N.-E. (2019). Gas sensing with heterostructures based on two-dimensional nanostructured materials: a review. *J. Mater. Chem. C* 7 (43): 13367–13383.

74 Zhang, L., Khan, K., Zou, J. et al. (2019). Recent advances in emerging 2D material-based gas sensors: potential in disease diagnosis. *Adv. Mater. Interfaces* 6 (22): 1901329.

75 Schedin, F., Geim, A.K., Morozov, S.V. et al. (2007). Detection of individual gas molecules adsorbed on graphene. *Nat. Mater.* 6 (9): 652–655.

76 He, Q., Ma, Q., Chen, B. et al. (2014). TaS_2 nanosheet-based room-temperature dosage meter for nitric oxide. *APL Mater.* 2 (9): 092506.

77 Lu, G., Park, S., Yu, K. et al. (2011). Toward practical gas sensing with highly reduced graphene oxide: a new signal processing method to circumvent run-to-run and device-to-device variations. *ACS Nano* 5 (2): 1154–1164.

78 Yang, D., Velamakanni, A., Bozoklu, G. et al. (2009). Chemical analysis of graphene oxide films after heat and chemical treatments by X-ray photoelectron and micro-Raman spectroscopy. *Carbon* 47 (1): 145–152.

79 Acharyya, D., Acharyya, S., Huang, K. et al. (2017). Highly sensitive ppb level methanol sensor by tuning C:O ratio of rGO-TiO_2 nanotube hybrid structure. *IEEE Trans. Nanotechnol.* 16 (6): 1122–1128.

80 Li, H., Yin, Z., He, Q. et al. (2012). Fabrication of single- and multilayer MoS_2 film-based field-effect transistors for sensing NO at room temperature. *Small* 8 (1): 63–67.

81 He, Q., Zeng, Z., Yin, Z. et al. (2012). Fabrication of flexible MoS_2 thin-film transistor arrays for practical gas-sensing applications. *Small* 8 (19): 2994–2999.

82 Zeng, Z., Yin, Z., Huang, X. et al. (2011). Single-layer semiconducting nanosheets: high-yield preparation and device fabrication. *Angew. Chem.* 123 (47): 11289–11293.

83 Groß, A., Beulertz, G., Marr, I. et al. (2012). Dual mode NO_x sensor: measuring both the accumulated amount and instantaneous level at low concentrations. *Sensors* 12 (3): 2831–2850.

84 Akhavan, O. and Ghaderi, E. (2010). Toxicity of graphene and graphene oxide nanowalls against bacteria. *ACS Nano* 4 (10): 5731–5736.

85 Zhao, J., Wang, Z., White, J.C., and Xing, B. (2014). Graphene in the aquatic environment: adsorption, dispersion, toxicity and transformation. *Environ. Sci. Technol.* 48 (17): 9995–10009.

86 Seo, G., Lee, G., Kim, M.J. et al. (2020). Rapid detection of COVID-19 causative virus (SARS-CoV-2) in human nasopharyngeal swab specimens using field-effect transistor-based biosensor. *ACS Nano* 14 (4): 5135–5142.

87 Fathi-Hafshejani, P., Azam, N., Wang, L. et al. (2021). Two-dimensional-material-based field-effect transistor biosensor for detecting COVID-19 virus (SARS-CoV-2). *ACS Nano* 15 (7): 11461–11469.

88 Chen, W.Y., Lin, H., Barui, A.K. et al. (2022). DNA-functionalized $Ti_3C_2T_x$ MXenes for selective and rapid detection of SARS-CoV-2 nucleocapsid gene. *ACS Appl. Nano Mater.* 5 (2): 1902–1910.

7

Towards Graphene-FET Health Sensors: Hardware and Implementation Considerations

Nicholas V. Apollo[1] *and Hualin Zhan*[2]

[1]*AMPED Consultants, Riverchase Drive, Valencia, CA 91355, USA*
[2]*The Australian National University, College of Engineering, Computing and Cybernetics, School of Engineering, North Road, Canberra, ACT 2601, Australia*

7.1 Introduction to Health Sensing

Wearable sensors have advanced significantly in recent years, transcending the realm of consumer electronics (e.g. fitness trackers) to become critical, life-saving technologies. For example, continuous glucose sensors such as those produced by Dexcom and Abbott have achieved regulatory approval and significant commercial success, not to mention substantial health and quality-of-life benefits for diabetic users. In addition to glucose, there are many other biomolecules that are collected during routine clinical laboratory testing, indicating a great opportunity to replace tedious, single-time point laboratory tests with continuous, convenient measurements using wearable health sensors.

The ability to continuously collect physiological data has already had a powerful impact on healthcare, and this trend is expected to continue given the large investment and interest in this sector [1, 2]. For example, the study of cardiac arrhythmia and epilepsy requires the frequent and long-term (i.e. >24 hours) collection of electrophysiological data from the heart and brain, respectively. Holter monitors and other similar wearable systems have been developed over the past decade to meet this need, but there have been a number of advancements in terms of both hardware and software to improve patient outcomes and comfort. For example, Medtronic's LINQ II is a miniaturized fully implantable cardiac recording system, surgically implanted near the clavicle, which can acquire electrophysiological data on a continuous basis to help monitor and diagnose abnormal heart function [3, 4]. Similar technology, ranging from invasive to noninvasive, has been developed for the continuous monitoring and forecasting of epileptic seizures and was recently shown to decrease seizure-related injuries [5, 6]. In addition to electrophysiological data, there is significant interest in the continuous monitoring of biomolecules such as glucose,

Graphene Field-Effect Transistors: Advanced Bioelectronic Devices for Sensing Applications,
First Edition. Edited by Omar Azzaroni and Wolfgang Knoll.

lactate, and many others which are of clinical significance for a broad patient population [1]. Standard clinical laboratory measurements typically requested by a physician include the complete blood count, metabolic panels (e.g. glucose, calcium, sodium, potassium, carbon dioxide, albumin, bilirubin, and more), lipid panels, thyroid panels, cardiac biomarkers, and other biomolecules more specific to a disease or organ of interest. However, these tests are slow to yield results, recorded at a single time point (i.e. when a patient visits the clinic), and require on-site visits to the clinic, the latter being especially challenging during public health crises such as the COVID-19 pandemic.

Continuous analyte monitors provide a solution to many of the aforementioned limitations of standard clinical laboratory tests. One of the most commercially successful wearable analyte sensors is the continuous glucose monitor (CGM). For diabetic patients, the ability to monitor blood glucose levels is imperative for both short- and long-term health [7, 8]. Furthermore, the use of an implantable device for this application relieves patients of burdensome, often painful, blood draws from their fingertips. Fortunately, there are several regulatory-approved CGM devices, including Dexcom's G5/G6/G7, Abbott's FreeStyle Libre, and Medtronic's Guardian Connect. All of these devices are examples of a subclass of sensors ("biosensors,") defined as analytical devices which contain biological sensing, or biorecognition, an element that is part of a signal transduction system. Examples of biorecognition elements include enzymes, antibodies, and strands of nucleic acids ("aptamers") among others (Figure 7.1).

The primary goal of this biorecognition element is to improve the selectivity of the biosensor, thus reducing the likelihood of false positives. Given the large number of biomolecules present in body fluids, many of which have similar chemical structures or electrochemical properties, achieving molecular specificity is one of the greatest challenges faced by biosensor engineers and scientists [9, 10].

In practice, there is more to a biosensor than just the biorecognition element. As mentioned above, a transduction mechanism is required to generate a measurable signal. Generally, this is a chemical-to-electrical (i.e. electrochemical), chemical-to-optical, or chemical-to-mechanical energy transduction. A classic enzyme-based biosensor, known as the "biocatalytic" type, achieves chemical-to-electrical transduction

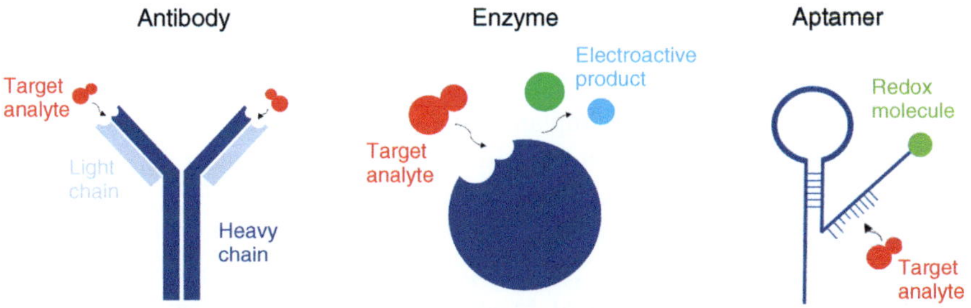

Figure 7.1 Examples of biorecognition elements commonly utilized to achieve selectivity in biosensors. In all three elements, a target analyte binds with high specificity and affinity and leads to a measurable change in the sensor system.

when the analyte of interest binds to the enzyme and either produces or consumes an electroactive molecule [11]. This electroactive molecule is then directly oxidized or reduced at an electrode, generating a measurable electrical current that correlates to the analyte concentration changes. Active research in this field has led to several innovative transduction pathways, including the use of mediators and even direct electron transfer with an enzyme's redox center [12, 13]. In the case of "affinity" type biosensors such as the aptamer- and antibody-based sensors shown in Figure 7.1, the analyte of interest is generally not changed chemically, and, thus, no product is produced for electrochemical detection. However, the binding event must induce a measurable change, typically at the electrode–electrolyte interface. In the case of an antibody-based biosensor to measure human serum albumin (HSA), the analyte binding event induced a drop in the double-layer capacitance [14]. For further reading on the transduction mechanisms of affinity-type biosensors, readers are directed to a review by Holford et al. [15].

7.2 Graphene-FET in Liquid for Sensing

In general, the sensitivity of a biosensor indicates its capability to detect a certain amount of biomolecules in a given environment, which is usually a physiological solution for medical biosensors. An ideally sensitive biosensor can detect a single molecule, providing the ultimate accuracy for subsequent medical procedures such as pathogen analysis or early disease diagnosis. Popular highly sensitive biosensing techniques include electrochemistry, field-effect transistors (FETs), fluorescence, and plasmonics, among others [16–18]. Each of these techniques has its own advantages and disadvantages during implementation. For example, the electrochemical biosensors can change the molecular species after detection, and the application of fluorescence- or plasmonics-based biosensors could be limited by the availability of optical instruments. The use of FETs has gained wide interest in the biosensor community [16, 19, 20]. Graphene transistor-based sensors can operate at a relatively low voltage, which would have advantages in reducing the power requirement and minimizing the oxidation of interfering molecules such as uric acid, ascorbic acid, and commonly used drugs such as acetaminophen [21]. The operation voltage of graphene can be further modulated by engineering its doping level. In addition, graphene can be transferred to or grown directly on dielectric substrates [22], making it advantageous compared with other materials like diamond by facilitating the process of transistor fabrication. In this section, we will focus on the sensitivity discussion of graphene-transistor-based biosensors.

From a technical perspective, the discussion of sensitivity involves two aspects. First, we are interested in how much signal can be yielded by the biosensor for a given concentration of analyte. Here, the analyte is the biomolecule that we intend to detect. The solution containing the analyte can be called liquid biopsy or simply the sample. For example, the sensitivity of an electrochemical dopamine sensor might be quoted in units of nA per M ($mol\,l^{-1}$). In the case of direct electrochemical

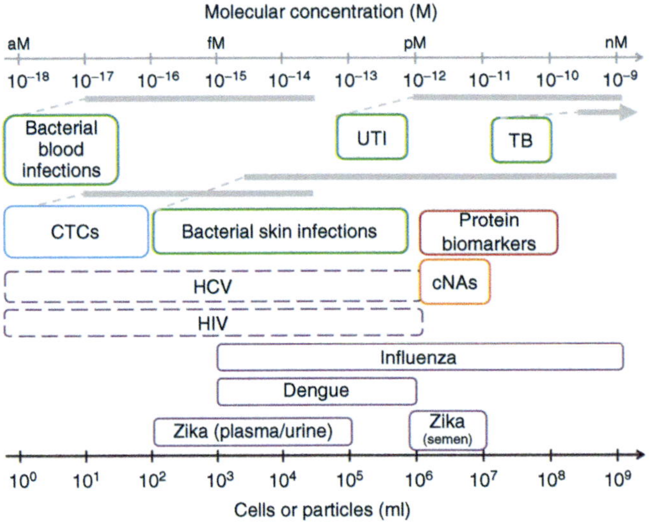

Figure 7.2 Concentration range of typical biomolecular analytes in clinical samples. The gray bars are the biomarkers of diseases. The green and purple boxes mark the bacterial and viral targets, respectively. The dashed lines denote that quantitative analysis is necessary. The cancer biomarkers are highlighted in blue (circulating tumor cells), red (protein biomarkers), and orange (circulating nucleic acids). UTI: urinary tract infections, TB: tuberculosis, CTC: circulating tumor cells, HCV: hepatitis C virus, HIV: human immunodeficiency virus, cNA: circulating nucleic acids. Source: Reproduced with permission from Kelley [24]/American Chemical Society.

oxidation of dopamine by a suitable carbon electrode (e.g. glassy carbon), the sensitivity is a function of several system parameters, including those which dictate the diffusion of dopamine to the sensor surface, sp^2/sp^3 ratio of the carbon material [23], the electronic density of states of materials, surface area, and the surface functional groups which determine electron transfer rate. Second, we want to reduce the minimum detectable quantity of the analyte by the biosensor, i.e. the limit of detection (LOD), to a value as small as possible, such as pM or even a single molecule. Figure 7.2 shows the concentration range of typical medical biomolecules. It is crucial for the illustrated species that the LOD reach a level below nM [24].

Table 7.1 demonstrates the typical LOD and the sensitivity of graphene sensors for electrophysiological and biomolecular detection. Note that both the transistor-based and the Hall-structure-based sensors are included due to their similar structures and fabrication processes, except that a magnetic field is also used for Hall-structure-based sensors [33, 35, 36]. The graphene sensors have demonstrated a LOD below nM for the detection of H_2O_2, DNA, proteins, and amino acids. As shown in the table, the LOD and the sensitivity can be affected by the quality and functionalization of graphene, the sensor structure, and the sensing method; therefore, a systematic understanding of the working principle of these sensors is required.

Table 7.1 Graphene transistor-based and Hall-structure-based health sensors.

Sensor	Structure	Graphene type	LOD	Sensitivity	References
pH sensor	Transistor	Mechanically exfoliated graphene	0.025 pH	~0.9 μA/pH (when $V_G = -0.1$ V, $V_{DS} = 0.1$ V)[a]	[25]
		CVD graphene covered by Al_2O_3	—	17 ± 2 mV/pH	[26]
Glucose sensor	Transistor	CVD graphene	0.1 mM	7.6%/mM of the conductance	[27]
		CVD graphene	3 mM	0.6 μA mM^{-1} when $V_G = 0.1$ V	[28]
H_2O_2 sensor	Transistor	Reduced graphene oxide -- polypyrrole nanocomposites	10 pM	~0.17 μA pM^{-1} between 10 and 100 pM	[29]
DNA sensor	Transistor	CVD graphene functionalized by DNA	1 pM	~2 mV/pM between 1 and 10 pM	[30]
Protein sensor	Transistor	Reduced graphene oxide functionalized by 1-Pyrenebutyric acid	0.5 nM	~2.5 nA nM^{-1} between 0.5 and 2.5 nM	[31]
Cortisol sensor	Transistor	CVD graphene functionalized by cortisol monoclonal antibody	10 pg ml^{-1}	1.84 ng ml^{-1} per 1% of the change in its resistance	[32]
Amino acid sensor	Hall structure	Un-functionalized CVD graphene	1 pM	~39 $(\Omega \cdot T^{-1})$/pM between 1 and 10 pM	[33]
Cellular activities detection	Transistor	Mechanically exfoliated graphene	S/N[b] ≥ 4	2290 μS V^{-1} when $V_G = 0.15$ V	[34]

[a] In this table, the notation of "~" indicates that the values followed are estimated based on the results reported in the reference.
[b] S/N: signal to noise ratio.

7.2.1 Graphene Transistors

Recently, graphene transistors have been extensively employed to develop high-performance biosensors [18, 20, 37]. In order to further improve its sensitivity, we need to first understand the working principle of transistors and transistor-based biosensors. Transistors, in particular logic transistors, can function as miniaturized switches which switch on and off a small electrical current passing through itself [38]. The on/off state of the switch can be controlled by an external electric field applied to the transistor itself, and hence the name FET.

In computer chips, this external electric field in transistors is applied directly using a capacitor-like configuration made by solid-state materials. Figure 7.3a shows the schematics of typical top-gated graphene transistors commonly used in the research of solid-state electronics, in which the electrons flow through the graphene from the source to the drain electrodes. The electronic current in graphene can be switched on/off by a straightforward application of the external electric field between the graphene and the gate electrode, where an oxide layer is sandwiched in between to form the capacitor-like configuration.

When used as a biosensor, the biomolecules exert an electric field that can switch on and off the electrical current passing through the transistor. We will then be able to determine if such biomolecules are present, i.e. whether the transistor-based biosensor is functional. However, these biomolecules are often present in a liquid environment. A practical graphene-transistor-based biosensor must be functional in such an environment for medical use. In 1970, Bergveld showed that the oxide layer in the transistor can be replaced by a liquid solution and form an ion-sensitive field-effect-transistor (ISFET) for biosensing [39]. The graphene-based ISFET is illustrated in Figure 7.3b, where the electric field can be between the graphene and the gate electrode through the electrical double layer formed at the graphene/solution interface [22]. This configuration is also referred to as a liquid-gated graphene transistor in literature [33, 40].

In research, the reference electrodes (e.g. Ag/AgCl electrode) used in electrochemical experiments are often adopted as the gate electrode as their known electrode potential is convenient for analysis. In implantable miniaturized biosensors, we can use a chemically inert electrode (e.g. Pt wire) as the gate for practicality [41]. This could be another advantage of transistor-based biosensors over other sensing techniques. For simplicity and cost reasons, it is common to produce biosensors with only two electrodes: the working and counter-ref electrodes. The counter-ref electrode, typically made of AgCl, can get consumed over time if using other

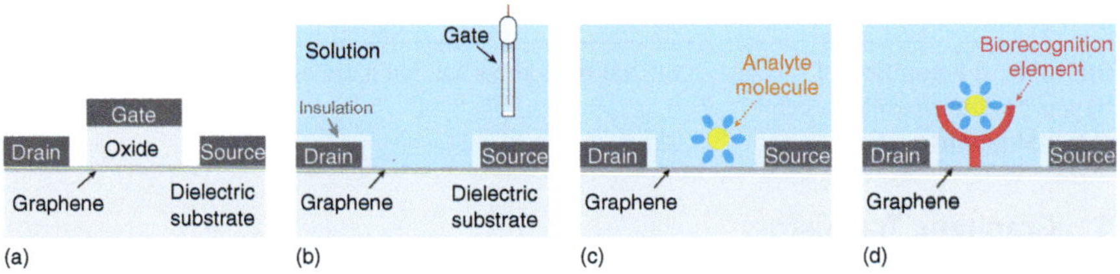

Figure 7.3 Graphene transistors of different structures. The electrical current flows from the drain to the source electrodes. The gate electrode is used to apply the electrical field on graphene to switch on/off the current. (a) The top-gated graphene transistor, which adopts the classical metal-oxide-semiconductor structure, is commonly used in the field of electronics. (b) The solution-gated graphene transistor, in which the gate electrode is immersed in solution, is used in the research of neural stimulation/recording and biosensing. (c) When the analyte molecule functions as the gate electrode, the graphene transistor can be used to detect its presence. (d) Biorecognition element can be used to selectively capture the analyte molecules through specific binding. This element is commonly included by graphene functionalization. The gate electrode can be included, but not necessary in some cases.

biosensing techniques such as amperometric detection. However, as high current will not pass through the gate electrode in a FET configuration – no electrodes will be consumed.

In practical biosensors, it is also important to improve the signal-to-noise ratio. The signal-to-noise ratio is highly relevant to the size of graphene for liquid-gated graphene transistors. Experimental studies have discovered that the charge noise increases with the area of graphene, and a bilayer graphene device exhibits lower noise than that made of monolayer graphene [42]. Researchers attribute the noise to the fluctuating charges and the scattering in the transistor channel when the carrier density is low and high, respectively. It can then be inferred that, for those liquid-gated graphene transistors whose sensitivity is improved by defect introduction, the signal-to-noise ratio may need to be evaluated. Therefore, when developing graphene-based health sensors, it is important to take into account the size, the quality, and the number of layers of graphene [42].

It is worth noting that, even in the absence of an externally applied electric field from the gate electrode, the graphene transistor can detect the presence of biomolecules such as glucose (see Zhang et al. [43] where the gate voltage V_G is 0). This is partly because the analyte molecule can act as a gate and apply an electric field to the graphene channel, where the current between drain and source is then changed (Figure 7.3c). Graphene, as a zero-bandgap semiconductor (or semi metal), should be sufficiently sensitive to capture such a change. The nature of zero bandgap allows smooth, continuous transition between the p-doped and n-doped graphene, enabling higher electrical sensitivity than the semiconductors with bandgaps. In addition, the inclusion of the gate electrode in transistor-based biosensors provides an additional means to tune their sensitivity [34, 43]. This is because the differential transconductance is usually higher when $V_G \neq 0$ [22]. These features, in principle, can make graphene-based transistors very sensitive biosensors.

However, the sensitivity of the transistor-based biosensors made from pristine or unfunctionalized graphene has been a debatable topic [26, 44, 45]. While some researchers claim that pristine graphene can be used for fast biosensing applications such as DNA sequencing [45], other groups found that pristine graphene is insensitive in some cases [26, 44]. The sensitivity of graphene-based electrical biosensors can be improved using two general approaches. (i) Modification of graphene. This includes doping, defect introduction, and chemical functionalization [46, 47]. (ii) Combine the electrical measurements with extra techniques, such as introducing a magnetic field and performing a Hall structure measurement [33, 35, 36]. These approaches either change the sensing material or improve the detection technique, which might be integrated in the future to further enhance the sensitivity. While biorecognition elements are commonly included in the transistors through modification of graphene, the second approach avoids the use of biorecognition elements which could be an advantage for the cases where graphene functionalization is not possible or the corresponding biorecognition elements for the analyte molecules have not been discovered yet. Inclusion of biorecognition elements is commonly used to increase the sensor selectivity through their specific binding with the analyte elements (Figure 7.3d). In Sections 7.2.2 and 7.2.3, we will explore cutting-edge sensing concepts using transistor-like structures which do not require the use of biorecognition elements.

7.2.2 Graphene Hall Structures in Liquid

Combining the electrical measurements with techniques including magnetic perturbation is particularly interesting as it allows us to use pristine graphene for high-sensitivity applications where graphene functionalization is not possible, such as electrophysiological sensors, and could be useful for retaining high signal-to-noise ratio. The magnetic field is applied perpendicularly to the conduction channel, i.e. graphene for graphene transistors (Figure 7.4a,b). In comparison with the transistor structure, additional electrodes are fabricated on the side of the transistor channel to detect the accumulation of the charge carriers, i.e. electrons or holes, due to the

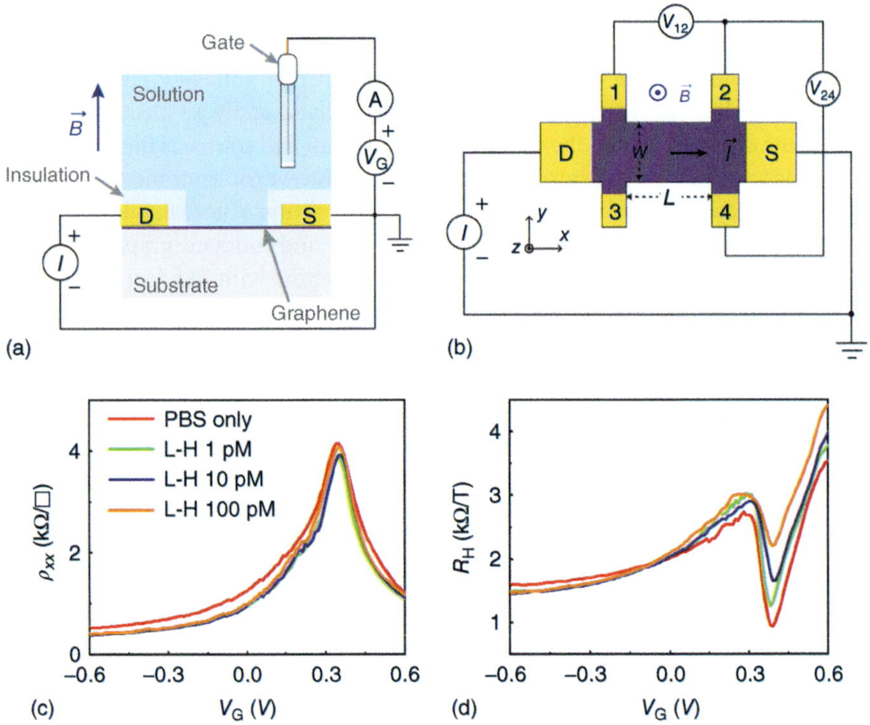

Figure 7.4 Hall-structure-based sensors of graphene in liquid. (a) The side view of a Hall-structure-based sensor in liquid. D and S are the drain and the source electrodes, respectively. This configuration is similar to that of liquid-gated transistors, except that a magnetic field ($\vec{B}$) is applied perpendicularly to the graphene. Note that the side electrodes are not illustrated here. (b) The top view of the graphene Hall structure, including all four side electrodes for the measurement of the Hall and the sheet resistances. (c) In an example study of L-histidine detection in a phosphate-buffered saline solution, the sheet resistance shows negligible changes. (d) In contrast, the Hall resistance changes evidently with the addition of L-histidine of pM concentration.

presence of the magnetic field. Since this accumulation is highly sensitive to both the density and mobility of the charge carriers, which can all be affected by the chemicals in the environment, graphene Hall structures provide extra information for sensing applications. This extra information can improve the sensitivity of graphene-transistor-based sensors. Note that the Hall effect can be measured based on both the Hall structure shown in Figure 7.4b and the van der Pauw structure [22]. In this section, we focus on the discussion using the Hall structure for simplicity.

In graphene transistor-based sensors, the chemicals or electrophysiological signals are detected by measuring the conductivity (σ) or resistivity change of the graphene. This conductivity-dependent mechanism can be insensitive in the cases where the biomolecules-induced change in the charge carrier density (n) cancels the mobility change (μ) of the graphene because $\sigma = qn\mu$ where q is the electric charge; for example, if n increases by a factor of α ($n_2 = \alpha n_1$) while μ changes inversely by the same factor α ($\mu_2 = \mu_1/\alpha$), the electrical conductivity remains the same ($\sigma_2 = \sigma_1$). Previous reports have demonstrated that the conductance change of transistors made of pristine graphene may not be sensitive to some low-concentration solutions [26, 48], although high-quality pristine graphene has been used and should provide high sensitivity in principle.

Hall-structure-based sensors can detect the changes in n and μ separately. However, n and μ are not directly measurable in Hall effect measurements; they are derived from the measured Hall resistance (R_H) and sheet resistance (ρ_{xx}) using semiconductor models. Note that the sheet resistance ρ_{xx} is similar to the resistance measured in transistor-based sensors, except that ρ_{xx} does not include the contact resistance between the electrode and graphene but the resistance measured in transistors does. Since graphene can have both electrons and holes as charge carriers when the Fermi level is close to the Dirac point, a two-carrier model, rather than the single carrier model used for conventional heavily doped semiconductors should be used for the derivation of n and μ in graphene [22]. In this two-carrier model, n and μ for electrons (e) and holes (h) are considered separately to calculate the total resistance matrix,

$$\rho_t = \begin{pmatrix} \rho_{xx} & \rho_{xy} \\ \rho_{yx} & \rho_{yy} \end{pmatrix}$$

$$= R_0 \begin{pmatrix} n_h\mu_h\left[1+(\mu_e B)^2\right] - n_e\mu_e\left[1+(\mu_h B)^2\right] & B\left\{n_e\mu_e^2\left[1+(\mu_h B)^2\right] - n_h\mu_h^2\left[1+(\mu_e B)^2\right]\right\} \\ B\left\{n_h\mu_h^2\left[1+(\mu_e B)^2\right] - n_e\mu_e^2\left[1+(\mu_h B)^2\right]\right\} & n_h\mu_h\left[1+(\mu_e B)^2\right] - n_e\mu_e\left[1+(\mu_h B)^2\right] \end{pmatrix}$$

where $R_0 = \dfrac{1}{e\left[(\mu_e\mu_h B)^2 (n_e - n_h)^2 + (n_e\mu_e - n_h\mu_h)^2\right]}$. R_H is simply $\dfrac{\rho_{xy}}{B}$. In Hall structures (Figure 7.4b), the ρ_{xx} and R_H are obtained by measuring the voltage and current across different terminals/electrodes of the graphene sheet,

$$\begin{cases} \rho_{xx} = \frac{V_{12}}{I} \cdot \frac{W}{L} \\ R_{\mathrm{H}} = -\frac{V_{24}}{IB} \end{cases},$$

where W and L are the width and the length of the graphene conduction channel (Figure 7.4c). V_{12} and V_{24} are the voltage between terminals 1 and 2 and terminals 2 and 4, respectively. I is the measured current. n_e, n_h, μ_e, and μ_h can then be derived conveniently from the voltage and current measurement of the Hall structure.

In the past, Hall effect measurements can be used to determine the carrier density of graphene immersed in liquid [49] and to investigate the charge scattering mechanism [50]. In this case of insensitive transistor-based sensors due to the inverse change of ρ and μ, Hall-structure-based sensors can detect the chemicals in the environment and demonstrate an improved sensitivity.

Figure 7.4c,d demonstrate the detection of L-histidine in a phosphate-buffered saline (PBS) solution, a solution made to mimic biological fluid, using a Hall-structure-based sensor. As shown in Figure 7.4c, the sheet resistance ρ_{xx} of graphene shows negligible changes when adding L-histidine of various concentrations into the PBS solution, suggesting a transistor-based sensor is insensitive in this case. This is because the resistance measured by transistors is mainly contributed by ρ_{xx}, the contacts between the graphene and the electrode are insulated from the solution, so their resistance should also remain constant. R_{H}, on the other hand, demonstrates evident changes (Figure 7.4d). In this example, while the transistor-based sensing mechanism is insensitive to pM L-histidine in the PBS solution, the Hall-structure-based sensors can detect L-histidine at a concentration as low as 1 pM.

The reasons why Hall-structure-based sensors can demonstrate improved sensitivity may be associated with the broken symmetry between the electrons and holes near the Dirac point in graphene, induced by the interaction of ions in liquid. As inferred from the peak of ρ_{xx} in Figure 7.4c, the Dirac point of the graphene should be closed to the point where $V_{\mathrm{G}} \approx 0.35\,\mathrm{V}$. In the Dirac point, the electrons and holes coexist and their behavior is symmetric in the absence of external perturbation. The evident changes in R_{H} occur at a position slightly above the Dirac point where $V_{\mathrm{G}} \approx 0.39\,\mathrm{V}$, in which R_{H} is clearly asymmetric. This indicates that this symmetry between electrons and holes is broken due to ion–electron interactions [33]. These interactions can happen in several ways. (i) Direct electrical interactions, which include changing the mobility of electrons/holes by affecting the scattering mechanism, producing electron/hole puddles, and inducing charge transfer. (ii) Molecules with different dipole moments, dielectric screening, and capacitive coupling may have different effects on electrons and holes. (iii) Graphene may become inhomogeneous or even get damaged in the presence of ions in liquid and/or gate voltage. This inhomogeneity could affect the measured R_{H}.

In order to make practical sensors based on transistors or Hall structures, we also need to enhance the selectivity while retaining the improved sensitivity. In addition to functionalizing graphene with biorecognition elements, a novel method to achieve this goal is to integrate transistor-based structure with nanoconfined ion dynamics, or what is known as nanofluidics. By selecting different ions by their sizes and providing a nanoconfined space for intimate interaction between ions in an electrolyte and electrons in graphene, this new method provides a promising avenue to improve both selectivity and sensitivity.

7.2.3 Graphene Membrane Transistors

Laminar graphene membranes, which are constructed by layering graphene sheets, form interconnected nanoslits, allowing the selection of ions by their sizes and providing nanosized spaces for intimate ion–electron interactions. This material could serve as the ideal platform for the development of novel sensors with enhanced sensitivity and selectivity. However, the quality of graphene sheets in laminar graphene membranes is very low among all graphene growth methods: the charge carrier mobility is too low, each graphene sheet is not necessarily monolayer, and the graphene sheets are corrugated and have many defects and functional groups [51]. The low quality mainly originates from the synthesis process of the laminar graphene membranes. In this process, graphite is first oxidized in liquid to obtain graphite oxide, in which the interlayer distance increases. These layers are then separated by sonication and reduced to form a reduced graphene oxide colloidal solution. This solution can be processed to produce laminar graphene membranes, or laminar reduced graphene oxide membranes. Many defects can be created in this process, each monolayer is not necessarily well-separated, and the reduction process cannot completely remove all oxide groups.

Fabrication of a transistor commonly requires graphene sheets of good quality, or at least with semiconductive properties. In laminar graphene membranes, the low quality and the uncertain number of layers of graphene sheets raise questions about the semiconductive nature of the membranes and present challenges in making transistors.

However, a recent study shows that the trans-membrane conductance of the laminar graphene membrane readily responds to the varying gate voltage (V_G) in liquid, clearly demonstrating a liquid-gated transistor behavior [52]. In addition, the conductance-V_G relation changes significantly when different electrolytes are used, showing a potential that the graphene membrane-based transistors can be used to differentiate the types of electrolyte. We must note that the concentration of the electrolyte is quite high in the demonstrated study as it was initially designed to analyze the ions in supercapacitors, which commonly have highly concentrated electrolytes for high energy storage capacity.

The graphene membrane transistors provide a new route to achieve selectivity rather than relying on the specific binding between the analyte molecules and the functional groups on graphene. This could be important for the case where functionalization of graphene is not possible. In future, the sensitivity for detecting ions/molecules at low concentrations must be tested on the graphene membrane transistors used to make health sensors.

7.3 Device Implementation Considerations

In Section 7.2, we introduced the basic operating principles of biosensors and graphene-FET biosensors, specifically. In Section 7.3, we will discuss the requirements and considerations for transitioning graphene-FET biosensors from the laboratory to practical daily use as a wearable health sensor. As part of this discussion, we will cover the hardware and instrumentation required to build a functional, wearable continuous health sensing device. Furthermore, we will discuss the biocompatibility and biostability of the invasive portions of graphene-FET devices, with a specific focus on graphene materials. Finally, we will discuss other considerations related to medical device implementation, including packaging, sterilization, and medical imaging compatibility. While noninvasive sensing approaches are a topic of recent interest, we will not discuss those approaches to continuous health sensing in this chapter. However, we direct interested readers to several recent reviews on the topic of non-invasive biosensors [53–55].

7.3.1 Hardware and Instrumentation

A majority of the biosensors described in the scientific literature were built to achieve proof-of-concept demonstrations of a novel sensing modality or application. As a result, the experimental setup utilized to study these biosensor systems was generally not designed with an end user or standalone product in mind. Once a biosensing modality has attracted commercial interest, a more practical implementation of the technology must be pursued. There are two primary commercial uses for biosensor systems: (i) clinic-based testing and (ii) continuous wearable sensors. In this chapter, we are primarily interested in (ii) and, thus, from here on, we will describe the practical considerations for transitioning biosensor systems from the laboratory to a wearable, continuous health sensor. Furthermore, this section will specifically consider the adaptation of graphene-FET systems to a wearable medical device platform.

To begin, it is helpful to identity the basic components of a continuous health sensor, which are listed below and illustrated in Figure 7.5:

1. Fluid-sensor interface.
2. Signal acquisition instrumentation.
3. Power source.
4. Algorithm; online processing.
5. Data transmitter and receiver.

The fluid-sensor interface is where the analyte of interest binds to the biorecognition element. This "fluid" may be sweat, tears, blood, interstitial fluid (ISF), or even cerebrospinal fluid (CSF). There are several ways to access the aforementioned biofluids, including electrical or chemical treatments of skin to extract sweat [56], microdialysis in which fluids are extracted from an implanted tube and analyzed offline [57], or direct implantation of the sensor into the tissue, the latter of which

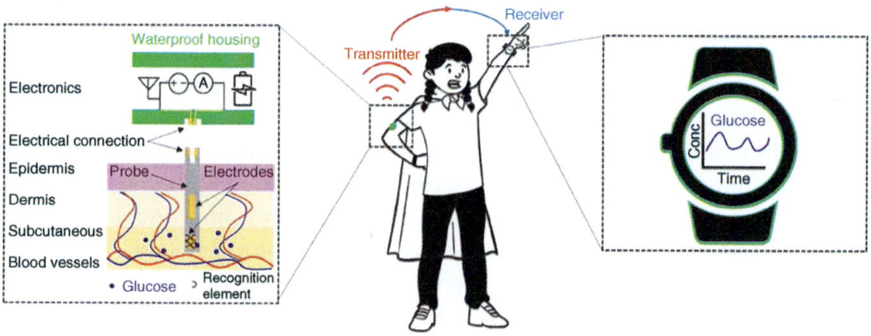

Figure 7.5 Schematic of a wearable health sensor. The sensor probe with electrodes and biorecognition molecules is implanted into the tissue to provide rapid and continuous access to the analyte of interest, which is glucose in this example. Once glucose binds to the biorecognition element, a current is produced, measured by the ammeter, and ultimately transmitted from the wearable via telemetry to a receiver device, which is shown as a watch in this example. Much of this form factor would be the same for a FET-based biosensor, given that changes in electrical current are the measurable output.

is done for several commercially available continuous glucose monitoring devices and, thus, is our focus here. These invasive interfaces can be further subdivided into short (10–14 days) and long (up to one year) implantation times, each requiring unique engineering and material approaches. Maintenance of this fluid-sensor or tissue-sensor interface is critical if any reasonable estimation of the physiological parameter of interest is to be obtained. However, it is well known that phenomena such as biofouling and the foreign body response can diminish sensor performance [58]. These aspects will be covered in Section 7.3.2, particularly as they pertain to the use of graphene-FET devices.

Once the analyte of interest has bound to the biorecognition element on the implanted sensor, a chemical change occurs, which is monitored using an electrical circuit. In the case of sensors utilizing an electrochemical transduction mechanism, this instrument may be a potentiostat in which a voltage is held or swept to drive a redox reaction which results in a measurable electrical current due to oxidation or reduction of a molecular species. For FET-based devices, the instrumentation is similar, though the electrode arrangement may be different, and more similar to those shown in Figure 7.3c above. For graphene-FET devices, a change in measured current may be driven by the electric field or dopant effects in the graphene gate produced by either the analyte or enzyme by-product rather than the direct oxidation of those molecules [28].

Transmitting data to a receiving device is now commonly done for numerous wearable health technologies. Fitness trackers, such as FitBit and Garmin watches, transmit data to and from smartphones via Bluetooth. In the case of a continuous sensor, the primary considerations for data transmission are electrical component size and power efficiency. For devices in which the electronics and biosensor are both fully implanted into the tissue [59], these considerations are even more critical given that larger implants (0.3 vs. 0.8 mm width) may lead to more traumas and, consequently, more aggressive foreign body reactions [32]. Recently, a graphene-FET cortisol biosensor used a miniaturized coil system to enable nearfield communication (NFC) with a mobile device [32]. In addition to data transfer, NFC was also used to power the graphene-FET device, relieving the requirement for a battery. However, this means that a measurement is only possible when the mobile device is in close proximity to the sensor, making this approach less continuous than the glucose sensors and other devices we have described previously. Achieving low power usage and efficient data telemetry is an active area of research in continuous health sensor space. As mentioned in Section 7.2, the possibility of applying lower voltages with graphene-FET devices may reduce some of the power requirements typically seen for direct electrochemical biosensors.

In this section, we have described the basic hardware components of a continuous health sensor. Our intention was to relay the message that only minor, if any, adjustments to standard health sensor hardware systems are required to accommodate graphene-FET sensors. In Sections 7.3.2 and 7.3.3, we will describe some of the practical safety, packaging, and clinical concerns around graphene-FET health sensors.

7.3.2 Biostability and Biocompatibility

As described in Section 7.2, at least the graphene layer (gate) of graphene-FET health sensors will be in direct contact with the body fluids and tissues. As a result, we will briefly review the biocompatibility and biostability of graphene materials in this section. While we must consider the safety of any implanted material lacking precedent use, it is important to note that an implantable graphene-FET device would most likely have the graphene affixed to a substrate and possibly surrounded by a biocompatible polymeric membrane. Furthermore, it is important to consider how the graphene-based materials were derived or processed since solvents or metal catalysts might show cytotoxicity where native graphene would not. The safety of graphene and other nanomaterials has been studied extensively using both *in vitro* and *in vivo* preparations [60, 61]. It is important to note that there are many varieties of graphene-based materials that have been explored for biomedical applications, including graphene oxide, reduced graphene oxide, and single- or multisheet graphene derived from CVD growth or mechanical exfoliation. For this review, we will focus on the single sheet of CVD graphene given its frequent use in FET-based biosensors [27].

For readers who may not be familiar with regulatory processes in the medical device industry, we will first provide some background on the types of evaluations that are performed to assess safety and biocompatibility. The International Organization for Standardization (ISO) issues guidance for the aforementioned testing in the ISO 10993 document series [62]. As of 2018, there has been a shift toward rigorous *in vitro* testing in combination with preexisting material information to reduce *in vivo* studies and, thus, animal usage (ISO 10993-1). As a result, a device may be considered safe without *in vivo* studies if it passes toxicity and leachable testing according to ISO 10993-18. For this testing, devices are soaked in physiological fluids under relevant use conditions (e.g. CSF for a neurochemical sensor), and possibly accelerated aging. At various time points, the soaking fluid is sampled to measure the presence of toxic compounds known as "leachables." If the concentrations of these leachables are acceptable or similar in concentration to other devices that have already proven to be safe, it is likely that no further *in vivo* testing is required though review by a regulatory body is still required. In the case of carbon-based materials like graphene, the leaching of toxic chemicals into physiological solutions is unlikely given the stability of the carbon bonds at body temperature, intrinsic hydrophobicity, and resistance to cyclic mechanical loading [63]. Furthermore, graphene films have even been used to prevent corrosion and, thus, improve the safety of other materials and medical devices [64, 65], in addition to demonstrating biocompatibility in the very delicate retinal tissues during evaluation of a novel graphene-based retinal prosthesis device [66]. However, there are a few safety considerations that arise from the CVD growth and subsequent delamination steps. Generally, graphene is grown on copper substrates, which are then etched chemically to liberate the graphene film. As a result, it would be important to test these films for the presence of leachable copper ions or copper-containing compounds. It is of interest to consider an alternative graphene film liberation method that uses electrochemical current rather than chemical etchants, which should reduce the probability of toxic leachable compounds present in the graphene film [67].

7.3.3 Medical Imaging Compatibility

Given that patients using continuous health sensors will be wearing devices for several days, possibly weeks at a time, it is possible that he or she will undergo a medical evaluation which requires imaging. We will consider two primary medical imaging modalities in this assessment: magnetic resonance imaging (MRI) and computed tomography (CT). MRI uses a strong magnetic field, usually 1.5–3 T, to acquire high-resolution images of biological tissues. The safety and compatibility of various medical devices is an active area of research, particularly in deep brain stimulation [68]. Safety risks are predominantly driven by the interaction of metallic components with the magnetic field, leading to eddy currents which can lead to device heating, unintended electrical stimulation of tissues, and possibly damage to integrated circuits and other electronic components. The aforementioned risks are generally a concern for metals such as platinum or gold, with much more severe risks for ferrous metals such as stainless steel. In the case of carbon-based materials, the risk of interaction with magnetic fields is extremely low given the low positive (weakly diamagnetic) magnetic susceptibility, specifically at physiologically relevant temperatures. Several examples of carbon-based electrodes were shown to be safe and produced very little or no artifact during MRI imaging [69, 70]. In the case of CT, an X-ray based imaging modality, the low density of carbon-based electrodes means that their interference with imaging will be minor compared with higher density metals such as platinum. However, both X-rays [71] and strong magnetic fields [72] have been shown to interact with graphene and graphene-FET devices specifically. As a result, it will be important to consider these interactions during the design of any graphene-FET health sensor devices intended for continuous or long-term wear.

References

1 Clifford, A., Das, J., Yousefi, H. et al. (2021). Strategies for biomolecular analysis and continuous physiological monitoring. *J. Am. Chem. Soc.* 143 (14): 5281–5294.

2 BCCresearch (2022). Sensors: Technologies and Global Markets, IAS006N.

3 LINQ II Insertable Cardiac Monitor (2023). Medtronic. https://www.medtronic.com/us-en/healthcare-professionals/products/cardiac-rhythm/cardiac-monitors/linq-ii.html.

4 Bezzerides Vassilios, J., Walsh, A., Martuscello, M. et al. (2019). The real-world utility of the LINQ implantable loop recorder in pediatric and adult congenital heart patients. *JACC Clin. Electrophysiol.* 5 (2): 245–251.

5 Swarup, O., Waxmann, A., Chu, J. et al. (2021). Long-term mood, quality of life, and seizure freedom in intracranial EEG epilepsy surgery. *Epilepsy Behav.* 123: 108241.

6 Hadady, L., Klivényi, P., Fabó, D., and Beniczky, S. (2022). Real-world user experience with seizure detection wearable devices in the home environment. *Epilepsia* 00: 1–6.

7 Tamborlane, W.V., Beck, R.W., Bode, B.W. et al. (2008). Continuous glucose monitoring and intensive treatment of type 1 diabetes. *N. Engl. J. Med.* 359 (14): 1464–1476.

8 Lin, R., Brown, F., James, S. et al. (2021). Continuous glucose monitoring: a review of the evidence in type 1 and 2 diabetes mellitus. *Diabet. Med.* 38 (5): e14528.

9 Jia, W.-Z., Wang, K., and Xia, X.-H. (2010). Elimination of electrochemical interferences in glucose biosensors. *TrAC, Trends Anal. Chem.* 29 (4): 306–318.

10 Basu, A., Veettil, S., Dyer, R. et al. (2016). Direct evidence of acetaminophen interference with subcutaneous glucose sensing in humans: a pilot study. *Diabetes Technol. Ther.* 18 (S2): S243–S247.

11 Clark, L.C. Jr. and Lyons, C. (1962). Electrode systems for continuous monitoring in cardiovascular surgery. *Ann. N.Y. Acad. Sci.* 102: 29–45.

12 Heller, A. (1999). Implanted electrochemical glucose sensors for the management of diabetes. *Ann. Rev. Biomed. Eng.* 1 (1): 153–175.

13 Wang, J. (2001). Glucose biosensors: 40 years of advances and challenges. *Electroanalysis* 13 (12): 983–988.

14 Mirsky, V.M., Riepl, M., and Wolfbeis, O.S. (1997). Capacitive monitoring of protein immobilization and antigen–antibody reactions on monomolecular alkylthiol films on gold electrodes. *Biosens. Bioelectron.* 12 (9): 977–989.

15 Holford, T.R.J., Davis, F., and Higson, S.P.J. (2012). Recent trends in antibody based sensors. *Biosens. Bioelectron.* 34 (1): 12–24.

16 Wadhera, T., Kakkar, D., Wadhwa, G., and Raj, B. (2019). Recent advances and progress in development of the field effect transistor biosensor: a review. *J. Electron. Mater.* 48 (12): 7635–7646.

17 Akkilic, N., Geschwindner, S., and Höök, F. (2020). Single-molecule biosensors: recent advances and applications. *Biosens. Bioelectron.* 151: 111944.

18 Fan, Q., Wang, L., Xu, D. et al. (2020). Solution-gated transistors of two-dimensional materials for chemical and biological sensors: status and challenges. *Nanoscale* 12 (21): 11364–11394.

19 Wang, C.-F., Sun, X.-Y., Su, M. et al. (2020). Electrochemical biosensors based on antibody, nucleic acid and enzyme functionalized graphene for the detection of disease-related biomolecules. *Analyst* 145 (5): 1550–1562.

20 Nehra, A. and Pal, S.K. (2015). Current trends in nanomaterial embedded field effect transistor-based biosensor. *Biosens. Bioelectron.* 74: 731–743.

21 Ohara, T.J., Rajagopalan, R., and Heller, A. (1994). "Wired" enzyme electrodes for amperometric determination of glucose or lactate in the presence of interfering substances. *Anal. Chem.* 66 (15): 2451–2457.

22 Zhan, H. (2020). *Graphene–Electrolyte Interfaces: Electronic Properties and Applications*. Jenny Stanford Publishing.

23 Purcell, E.K., Becker, M.F., Guo, Y. et al. (2021). Next-generation diamond electrodes for neurochemical sensing: challenges and opportunities. *Micromachines* 12 (2): 128.

24 Kelley, S.O. (2017). What are clinically relevant levels of cellular and biomolecular analytes? *ACS Sens.* 2 (2): 193–197.

25 Ohno, Y., Maehashi, K., and Matsumoto, K. (2010). Chemical and biological sensing applications based on graphene field-effect transistors. *Biosens. Bioelectron.* 26 (4): 1727–1730.

26 Fu, W., Nef, C., Knopfmacher, O. et al. (2011). Graphene transistors are insensitive to pH changes in solution. *Nano Lett.* 11 (9): 3597–3600.

27 Huang, Y., Dong, X., Shi, Y. et al. (2010). Nanoelectronic biosensors based on CVD grown graphene. *Nanoscale* 2 (8): 1485–1488.

28 Kwak, Y.H., Choi, D.S., Kim, Y.N. et al. (2012). Flexible glucose sensor using CVD-grown graphene-based field effect transistor. *Biosens. Bioelectron.* 37 (1): 82–87.

29 Naorem, M., Singh, R., and Paily, R. (2021). Detection of hydrogen peroxide using rGO/PPy nanocomposites in silicon dioxide trench embedded field effect transistor. *IEEE Sens. J.* 21 (20): 22426–22433.

30 Chen, T.-Y., Loan, P.T.K., Hsu, C.-L. et al. (2013). Label-free detection of DNA hybridization using transistors based on CVD grown graphene. *Biosens. Bioelectron.* 41: 103–109.

31 He, Q., Wu, S., Gao, S. et al. (2011). Transparent, flexible, all-reduced graphene oxide thin film transistors. *ACS Nano* 5 (6): 5038–5044.

32 Ku, M., Kim, J., Won, J.-E. et al. (2020). Smart, soft contact lens for wireless immunosensing of cortisol. *Sci. Adv.* 6 (28): eabb2891.

33 Zhan, H., Cervenka, J., Prawer, S., and Garrett, D.J. (2018). Molecular detection by liquid gated Hall effect measurements of graphene. *Nanoscale* 10 (3): 930–935.

34 Cohen-Karni, T., Qing, Q., Li, Q. et al. (2010). Graphene and nanowire transistors for cellular interfaces and electrical recording. *Nano Lett.* 10 (3): 1098–1102.

35 Cui, N., Wang, F., and Ding, H. (2020). Graphene-based hall effect biosensor for improved specificity and sensitivity of label-free DNA detection. *Nano* 15 (07): 2050088.

36 Collomb, D., Li, P., and Bending, S. (2021). Frontiers of graphene-based Hall-effect sensors. *J. Phys. Condens. Matter.* 33 (24): 243002.

37 Reina, G., González-Domínguez, J.M., Criado, A. et al. (2017). Promises, facts and challenges for graphene in biomedical applications. *Chem. Soc. Rev.* 46 (15): 4400–4416.

38 Schwierz, F. (2010). Graphene transistors. *Nat. Nanotechnol.* 5 (7): 487–496.

39 Bergveld, P. (1970). Development of an ion-sensitive solid-state device for neurophysiological measurements. *IEEE Trans. Biomed. Eng.* 17 (1): 70–71.

40 Reiner-Rozman, C., Larisika, M., Nowak, C., and Knoll, W. (2015). Graphene-based liquid-gated field effect transistor for biosensing: theory and experiments. *Biosens. Bioelectron.* 70: 21–27.

41 Schöning, M.J. and Poghossian, A. (2002). Recent advances in biologically sensitive field-effect transistors (BioFETs). *Analyst* 127 (9): 1137–1151.

42 Heller, I., Chatoor, S., Männik, J. et al. (2010). Charge noise in graphene transistors. *Nano Lett.* 10 (5): 1563–1567.

43 Zhang, M., Liao, C., Mak, C.H. et al. (2015). Highly sensitive glucose sensors based on enzyme-modified whole-graphene solution-gated transistors. *Sci. Rep.* 5 (1): 8311. ISSN = 2045-322, https://doi.org/10.1038/srep08311.

44 Kumar, B., Min, K., Bashirzadeh, M. et al. (2013). The role of external defects in chemical sensing of graphene field-effect transistors. *Nano Lett.* 13 (5): 1962–1968.

45 Min, S.K., Kim, W.Y., Cho, Y., and Kim, K.S. (2011). Fast DNA sequencing with a graphene-based nanochannel device. *Nat. Nanotechnol.* 6 (3): 162–165.

46 Guy, O.J. and Walker, K.-A.D. (2016). Chapter 4 – Graphene functionalization for biosensor applications. In: *Silicon Carbide Biotechnology* (Second Edition) (ed. S.E. Saddow), 85–141. Elsevier.

47 Tite, T., Chiticaru, E.A., Burns, J.S., and Ioniţă, M. (2019). Impact of nano-morphology, lattice defects and conductivity on the performance of graphene based electrochemical biosensors. *J. Nanobiotechnol.* 17 (1): 101.

48 Moss, S.D., Janata, J., and Johnson, C.C. (1975). Potassium ion-sensitive field effect transistor. *Anal. Chem.* 47 (13): 2238–2243.

49 Ye, J., Craciun, M.F., Koshino, M. et al. (2011). Accessing the transport properties of graphene and its multilayers at high carrier density. *Proc. Natl. Acad. Sci.* 108 (32): 13002–13006.

50 Newaz, A.K.M., Puzyrev, Y.S., Wang, B. et al. (2012). Probing charge scattering mechanisms in suspended graphene by varying its dielectric environment. *Nat. Commun.* 3 (1): 734.

51 Zhan, H., Xiong, Z., Cheng, C. et al. (2020). Solvation-involved nanoionics: new opportunities from 2D nanomaterial laminar membranes. *Adv. Mater.* 32 (18): 1904562.

52 Xiao, J., Zhan, H., Wang, X. et al. (2020). Electrolyte gating in graphene-based supercapacitors and its use for probing nanoconfined charging dynamics. *Nat. Nanotechnol.* 15 (8): 683–689.

53 Tang, L., Chang, S.J., Chen, C.-J., and Liu, J.-T. (2020). Non-invasive blood glucose monitoring technology: a review. *Sensors* 20 (23): 6925.

54 Xu, J., Fang, Y., and Chen, J. (2021). Wearable biosensors for non-invasive sweat diagnostics. *Biosensors* 11 (8): 245.

55 dos Santos, C.C., Lucena, G.N., Pinto, G.C. et al. (2021). Advances and current challenges in non-invasive wearable sensors and wearable biosensors – a mini-review. *Med. Devices Sens.* 4 (1): e10130.

56 Bariya, M., Nyein, H.Y.Y., and Javey, A. (2018). Wearable sweat sensors. *Nat. Electron.* 1 (3): 160–171.

57 Samper, I.C., Gowers, S.A.N., Booth, M.A. et al. (2019). Portable microfluidic biosensing system for real-time analysis of microdialysate in transplant kidneys. *Anal. Chem.* 91 (22): 14631–14638.

58 Klueh, U., Kaur, M., Montrose, D.C., and Kreutzer, D.L. (2007). Inflammation and glucose sensors: use of dexamethasone to extend glucose sensor function and life span in vivo. *J. Diabetes Sci. Technol.* 1 (4): 496–504.

59 Gough, D.A., Kumosa, L.S., Routh, T.L. et al. (2010). Function of an implanted tissue glucose sensor for more than 1 year in animals. *Sci. Transl. Med.* 2 (42): 42ra53.

60 Pinto, A.M., Gonçalves, I.C., and Magalhães, F.D. (2013). Graphene-based materials biocompatibility: a review. *Colloids Surf., B* 111: 188–202.

61 Bullock, C.J. and Bussy, C. (2019). Biocompatibility considerations in the design of graphene biomedical materials. *Adv. Mater. Interfaces* 6 (11): 1900229.

62 Biological Evaluation of Medical Devices. https://www.iso.org/ics/11.100.20/x/.

63 Seethamraju, S., Kumar, S., Bharadwaj, B.K. et al. (2016). Million-fold decrease in polymer moisture permeability by a graphene monolayer. *ACS Nano* 10 (7): 6501–6509.

64 Lu, C.-H., Shang, K.-M., Lee, S.-R. et al. (2022). Graphene on nanoscale-thick Au films: implications for anticorrosion in smart wearable electronics. *ACS Appl. Nano Mater.* 5 (3): 4343–4349.

65 Araujo, A.F., Ferreira, M.V.F., Felisberto, M.D.V. et al. (2022). Corrosion resistance of a superelastic NiTi alloy coated with graphene–based coatings. *Prog. Org. Coat.* 165: 106727.

66 Nguyen, D., Valet, M., Dégardin, J. et al. (2021). Novel graphene electrode for retinal implants: an in vivo biocompatibility study. *Front. Neurosci.* 15: 615256.

67 Wang, Y., Zheng, Y., Xu, X. et al. (2011). Electrochemical delamination of CVD-grown graphene film: toward the recyclable use of copper catalyst. *ACS Nano* 5 (12): 9927–9933.

68 Boutet, A., Chow, C.T., Narang, K. et al. (2020). Improving safety of MRI in patients with deep brain stimulation devices. *Radiology* 296 (2): 250–262.

69 Zhao, S., Liu, X., Xu, Z. et al. (2016). Graphene encapsulated copper microwires as highly MRI compatible neural electrodes. *Nano Lett.* 16 (12): 7731–7738.

70 Zhao, S., Li, G., Tong, C. et al. (2020). Full activation pattern mapping by simultaneous deep brain stimulation and fMRI with graphene fiber electrodes. *Nat. Commun.* 11 (1): 1788.

71 Cazalas, E., Sarker, B.K., Moore, M.E. et al. (2015). Position sensitivity of graphene field effect transistors to X-rays. *Appl. Phys. Lett.* 106 (22): 223503.

72 Goerbig, M.O. (2011). Electronic properties of graphene in a strong magnetic field. *Rev. Mod. Phys.* 83 (4): 1193–1243.

8

Quadratic Fit Analysis of the Nonlinear Transconductance of Disordered Bilayer Graphene Field-Effect Biosensors Functionalized with Pyrene Derivatives

Sung Oh Woo[1,2], *Sakurako Tani*[1], *and Yongki Choi*[1,3,4]

[1]*North Dakota State University, Department of Physics, 1211 Albrecht Blvd., Fargo, ND 58108, USA*
[2]*Texas A&M University, Department of Electrical and Computer Engineering, AggieFab Nanofabrication Facility, 1617 Research Parkway, College Station, TX 77843, USA*
[3]*North Dakota State University, Materials and Nanotechnology Program, 1735 NDSU Research Park Dr., Fargo, ND 58108, USA*
[4]*North Dakota State University, Cellular and Molecular Biology Program, 1735 NDSU Research Park Dr., Fargo, ND 58108, USA*

8.1 Introduction

Graphene is an atomic layer, two-dimensional (2D) material made up of carbon atoms in a honeycomb lattice. Since the first realization of the atomically thin graphene field effect transistor (FET) device in a laboratory [1, 2], much interest has been shown in understanding its fundamental electrical, mechanical, chemical, and relativistic quantum physical properties and developing potential applications such as sensors [2–4]. In addition, novel properties of graphene, including ultrafast charge carrier mobility [5–7], high transparency [8, 9], and ultra-sensitivity to atomic and molecular adsorption [10–12], as well as its compatibility with semiconductor fabrication processes and infrastructure, make it the most promising 2D material that could potentially replace conventional silicon-based electronic devices. A broad range of graphene applications, such as ultrafast transistors [13], transparent electrodes for display devices [14], sea-water desalination [15], DNA sequencing devices [16], and chemical and biological sensors, have already demonstrated such promises [12, 17].

The unique and novel characteristics of graphene originate from its honeycomb crystal structure, where two carbon atoms are in a unit cell. In the crystal, a single carbon atom covalently binds to three carbon neighbors in a plane, forming a mechanically strong planar lattice. One π electron at each carbon site remains unoccupied and electrons are hopping through the unfilled orbitals, which enables graphene to be conductive [18]. The one-atom layer structure, along with excellent

Graphene Field-Effect Transistors: Advanced Bioelectronic Devices for Sensing Applications, First Edition. Edited by Omar Azzaroni and Wolfgang Knoll.

electrical conductivity, makes graphene an extremely high surface-to-volume ratio material, which becomes a distinct advantage for sensing atoms and molecules.

In a sensing platform, graphene is exposed to mixed gases or biofluids containing target and nontarget atoms and molecules, both of which could be adsorbed on the graphene surface nonspecifically. In principle, graphene can detect various atoms and molecules with no selectivity. To achieve selective adsorption of target molecules from mixtures and increase the specificity of graphene-based biosensors, the chemical functionalization of graphene with bifunctional linker molecules has been introduced. One end of the linker molecules sticks to the graphene surface, and the other end selectively interacts with the target molecules through specific chemical and biochemical reactions [12, 19–21]. Such surface modification of graphene allows selective detection of many different molecules, such as proteins, biopolymers, and viruses, demonstrating the tremendous potential for disease diagnostic device applications [22].

However, the functionalization of graphene also leads to structural changes in monolayer graphene and the charge carriers' behavior in graphene. As a result, conventional sensing parameters associated with the monolayer graphene structure were ineligible to analyze or distinguish from sensing signals of the functionalized graphene sensors. Using a disordered bilayer graphene (BLG) model, a quadratic fit analysis of the nonlinear electrical signals of the sensors has been introduced. This model has provided a reliable and accurate analysis of the sensing data obtained from the functionalized graphene sensors.

This chapter reviews the technique of fabricating graphene-based field effect biosensors and discusses the sensing parameters of the biosensors. Section 8.2 briefly describes the fabrication of graphene and graphene transistors. Section 8.3 discusses the electronic properties of graphene transistors. Section 8.4 describes the nonlinear transconductance of the graphene transistors associated with the formation of disordered BLG when functionalizing with pyrene-based bifunctional linker molecules. Section 8.5 introduces a quadratic fit model to analyze the nonlinear transconductance and demonstrates it as a promising parameter for biosensing of pyrene-functionalized graphene biosensors.

8.2 Fabrication of Graphene-Based Field-Effect Biosensors

The revolutionary advances in graphene research have been enabled by the development of the chemical vapor deposition (CVD) method. Initially, graphene was obtained by the mechanical exfoliation technique, which is also known as the Scotch tape method [1, 2]. Using a sticky tape, a few tens of micron-scale graphene flakes were transferred and randomly distributed onto substrates such as silicon. This method is simple, but it is difficult to fabricate identical monolayer graphene FETs, ultimately limiting the device's dimension and scalable production. As an alternative, graphene was grown on a catalytic metal foil in a CVD reactor filled with the gas of a carbon precursor. In the CVD method, the size of metal substrates determines the size of graphene, allowing synthesizing of large-scale, monolayer graphene and the mass production of various graphene devices at a silicon wafer scale [14, 23–28].

The CVD system consists of precisely controlled gas supplies, a heating furnace, and a vacuum pumping system, as shown in Figure 8.1. While supplying the atomic carbon precursor in a gas form, such as methane, graphene grows on a catalytic metal substrate at high temperatures in the CVD reactor. It has been reported that graphene can be grown on various metals, such as Ni, Ir, Cu, and SiC [23, 29, 30]. Of these, a thin Cu foil is widely used because of its low cost and low carbon solubility as the graphene temperature rises, which is a critical aspect of the Cu foil to grow monolayer graphene [25, 31].

There are three stages for graphene synthesis. During the initial stage, the temperature of the CVD reactor increases up to the target synthesis temperature of 1000 °C with hydrogen flow under a base pressure of $P = 5\times10^{-2}$ Torr. Hydrogen gas acts as a reduction agent, cleaning organic materials and removing the native oxide on the Cu surface [31]. Next, the CVD reactor temperature is kept around 1000 °C for 30 minutes to further remove the native oxide, and then methane replaces hydrogen to begin graphene synthesis. Carbon atoms decompose from methane (CH_4) near the catalytic Cu foil, diffuse on the Cu surface and into the bulk of the Cu substrate, and start to form the honeycomb network. Therefore, the solubility of the metal substrate is one of the critical factors for growing monolayer graphene. Upon completion of

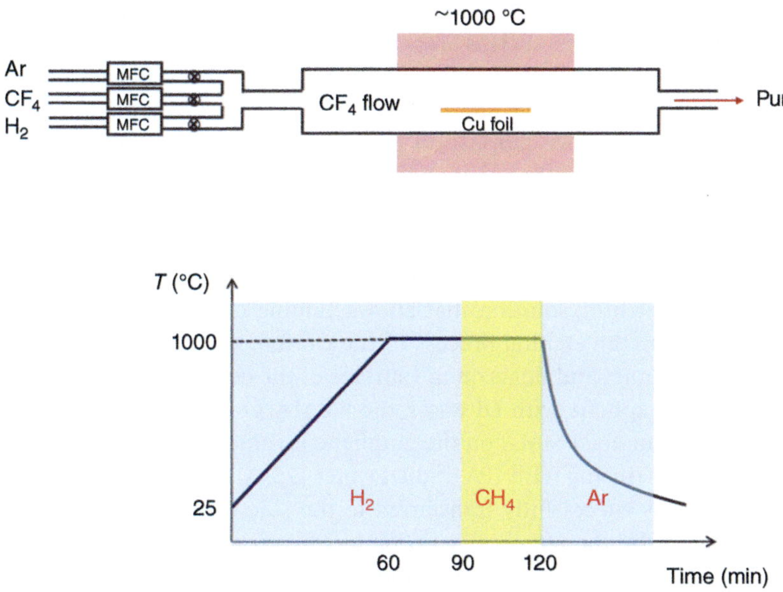

Figure 8.1 Schematic diagram of the CVD reactor and gas flow system controlled by mass flow controllers (MFCs), and temperature, time, and gas profiles during the graphene synthesis.

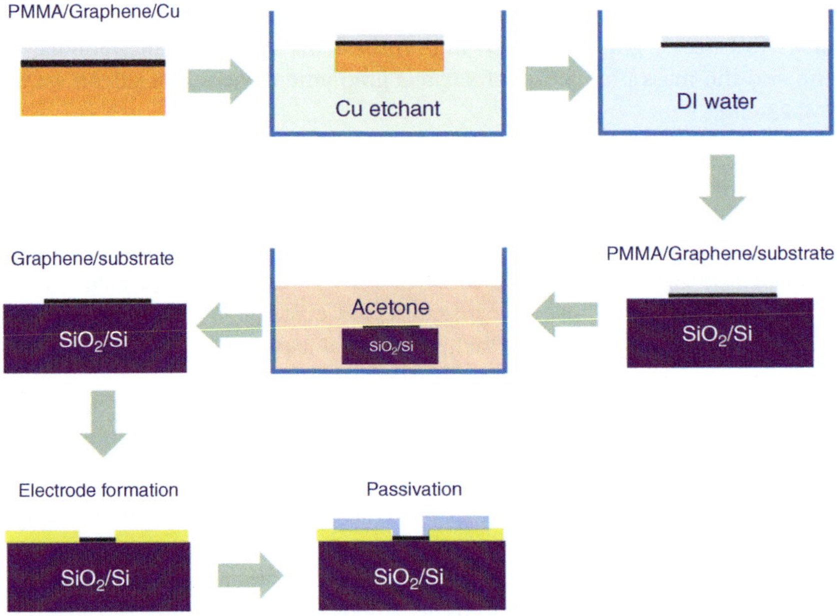

Figure 8.2 Graphene FET fabrication process includes the wet transfer of graphene onto the silicon substrate using PMMA and ionic etchant and electrode patterning using standard lithography techniques.

the graphene growth process, the CVD system is finally cooled down by switching the processing gas to an inert gas, such as argon, to keep the graphene free from contamination. Since graphene is synthesized on both sides of the metal foil, each side of the graphene can be easily removed by an oxygen plasma etching method.

The next step is a FET device fabrication process, which requires a graphene transfer from the Cu foil to the silicon substrate. The typical transfer method using a polymer such as polymethylmethacrylate (PMMA) is depicted in Figure 8.2. The polymer coating is employed to protect pristine graphene from breaking, overlapping, and folding during the metal etching process. A thin PMMA/graphene/Cu film floats on the Cu etching solution that allows etching on the Cu foil underneath graphene. During the etching process of the Cu foil, the PMMA/graphene bilayer itself remains intact and floats on the surface of the etching solution. After cleaning the PMMA/graphene with DI water, the PMMA/graphene are placed on the silicon substrate. Finally, PMMA on the graphene is removed by immersing it in acetone, followed by rinsing with IPA and DI water [25, 31].

Once the graphene is successfully transferred to the silicon substrate, the FET fabrication is followed by the standard semiconductor processes including lithography, thin film depositions, and reactive ion etching (RIE) [12]. Initially, the graphene channel can be shaped using photolithography, followed by the RIE. After defining the channel, the FET structure is fabricated by patterning metal electrodes using additional lithography, metal deposition, and lift-off processes. An additional passivation layer of either Si_3N_4, photoresists, or PMMA is frequently needed to cover all electrodes except the graphene channel for biosensing applications because the measurements are typically performed in an aqueous and physiological solution.

8.3 Fundamental Sensing Parameters of Graphene-Based Field-Effect Biosensors

The operational principle of graphene-based biosensors relies on the electronic properties of graphene. For example, the gate dependence of the source–drain current $I–V_G$, the charge carrier concentration dependence of the conductivity $\sigma–n$, the Dirac point V_D, and the transconductance g_m are the common sensing parameters to detect target biomolecules attached to the graphene.

The Dirac point is where the conduction band and valence band are touching in the momentum-energy space, and therefore the net charge carrier concentration is zero. However, graphene reveals nonzero conductivity at the Dirac point because of the electrons-hole puddle formations [32, 33]. Thus, the Dirac point is specified by the gate voltage V_G at which the conductivity is the minimum (Figure 8.3). As holes or electrons are doped externally, the Dirac point shifts to compensate for the additional charges. The shift is given by $\Delta V_D = \Delta Q/C = e\Delta n/C$, which allows measuring the charge doping induced by the adsorbates and estimating the charge concentration associated with the molecules adsorbed [12, 31, 34–37]. Here, ΔQ is the difference in the total charge as a result of the biosensing, and C is the capacitance

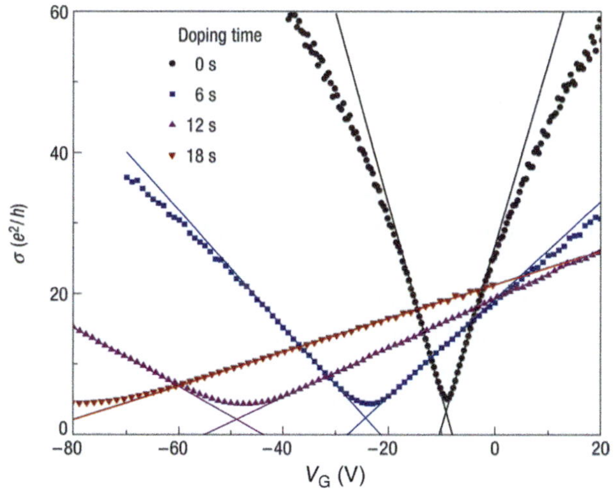

Figure 8.3 Conductivity as a function of the gate voltage measured at $T = 20$ K in ultrahigh vacuum conditions. Black circles, blue squares, purple triangles, and red triangles are the transfer curves of pristine graphene and graphene with three potassium doping concentrations, respectively. Source: Reproduced with permission from Chen et al. [34]/ Springer Nature.

of the dielectric layer between graphene and silicon or the electrical double layer between graphene and adsorbates in an ionic liquid environment. Thus, the Dirac point is an extraordinary sensing parameter that can detect a single atom adsorbed per 1000 carbon atoms in a graphene lattice [35].

The transconductance g_m ($=dI/dV_G$), a linear fit of the I–V_G curve far from the Dirac point [34–36], is another important sensing parameter. This parameter is derived from charge carrier mobility μ given by two different representations, a differential form $\mu = 1/e \times d\sigma/dn \approx dI/dV_G$ [36] and a linear relation $\mu = \sigma/en \approx I/V_G$ [38]. Therefore, g_m serves as a V_G independent sensing parameter if either μ is constant or the I–V_G curve shows a linear relationship.

In fact, the linear I–V_G relationship, which is widely used in graphene biosensing, is not an intrinsic characteristic of graphene but rather it is sublinear [7]. However, the I–V_G relation becomes linear when Coulomb scattering is the dominant scattering interaction between the charge carriers and the charged impurities in graphene. In other words, the linear relationship is only valid when the movement of electrons or holes in graphene is impeded by Coulomb scattering induced by impurities near graphene, which has been proved by the I–V_G measurements using the charged impurity decorated on graphene [34–36]. For example, Figure 8.3 shows the dependence of the I–V_G curves on the potassium atom decoration. Pristine graphene reveals the sublinear I–V_G curve far from the Dirac point (black circles). As the potassium concentration on the graphene surface increases, the I–V_G curves far from the Dirac point become more linear (blue, purple, and red markers). These results indicate that g_m is an accurate and reliable sensing parameter only if charged impurities (Coulomb scatterers) are near graphene [32, 34, 35]. It has been proved that the mobility of the charge carriers in graphene is constant and inversely proportional to the concentration of charged adsorbates ($\mu \sim 1/n_{imp}$, where n_{imp} is the charged adsorbates' concentration), further confirming that g_m is the conditional sensing parameter [32, 34].

8.4 Disordered Bilayer Graphene Field-Effect Biosensors Functionalized with Pyrene Derivatives

Despite the unprecedented sensitivity of graphene to atomic and molecular adsorption, the main drawback of graphene-based biosensing is that graphene adsorbs biomolecules nonspecifically. Such nonspecific adsorption deteriorates the detection accuracy of the target biomarkers and limits the overall biosensing capacity [39–41]. To overcome the nonspecificity problem of pristine graphene, the chemical modification of graphene has been suggested. The conventional method is to functionalize graphene with bifunctional linker molecules. One end of the linker includes an aromatic compound that sticks to graphene via π–π stacking and the other end has a functional compound that specifically binds to the target residue of biomolecules [40–43].

For example, pyrene-maleimide has been used to functionalize graphene using pyrene and capture a cysteine residue of target proteins using maleimide [40, 44, 45].

On the other hand, the chemical functionalization could lead to a change in the electronic properties of pristine graphene since the staking of pyrene acts like an additional layer of graphene. Structurally, pyrene consists of four fused benzene rings, i.e. 16 carbon atoms constitute a honeycomb structure terminated by hydrogen atoms on the outside dangling bonds, which is the identical crystal structure of graphene, except for the hydrogen terminations. Therefore, it is reasonable to assume that graphene functionalized with pyrene-based linker molecules could evolve from monolayer to BLG.

Raman spectroscopy is a powerful tool to identify monolayer graphene and BLG [12, 42]. In particular, the Raman 2D peak around 2700 cm^{-1} is one of the specific identifiers to distinguish monolayer graphene from BLG. The 2D peak signal of monolayer graphene exhibits a single Lorentzian peak with its full width half maximum (FWHM) as narrow as 30 cm^{-1}, whereas BLG reveals the broadened peak of the 2D peak, which can be deconvoluted into four subpeaks. This Raman signature of monolayer and BLG is attributed to the double resonance (DR) process [46]. In contrast to other Raman signals that arise from carbon atom vibrations within graphene, the 2D band is generated by two consecutive electronic processes. Initially, an electron is excited from the valence band to the conduction band upon absorption of the incident laser, and then interacts with a phonon to transit between the K and K' valleys in reciprocal space. Therefore, the transition between these two valleys is responsible for the appearance of the Raman 2D peak. During the intraband transition (Figure 8.4), the electron interacts with the phonon to emit or adsorb momentum, which causes the 2D peak. According to the DR process, the number of allowed transitions between K and K' is determined by the graphene layers. For example, only one transition of the electron with the help of a phonon wave vector q is possible in monolayer graphene (Figure 8.4a). For the BLG case, four possible transitions (q_{1A}, q_{1B}, q_{2A}, and q_{2B}) are allowed (Figure 8.4b). Thus, the overlapping of the 2D peak of each transition results in the broadening of the overall 2D peak of BLG.

Considering the crystallographic similarity of graphene and pyrene, graphene functionalized with the pyrene linker molecules could be regarded as BLG. To probe BLG formation on functionalized graphene, the 2D Raman peak was examined (Figure 8.5). As expected, pristine graphene revealed a representative Raman 2D spectrum around the Raman shift of 2675 cm^{-1} (Figure 8.5a). The experimental data (black circles) fit the Lorentzian distribution (red curve) having an FWHM of 29.9 cm^{-1}, which is a fingerprint of monolayer graphene [46]. On the other hand, the pyrene-functionalized graphene exhibits the broadened Raman 2D peak (red curve in Figure 8.5b), which fits the constituent of four Lorentzian peaks (blue curves in Figure 8.5b) rather than a single Lorentzian peak. This result directly proves that the chemical functionalization with pyrene linkers on graphene transforms into BLG [12, 37, 46].

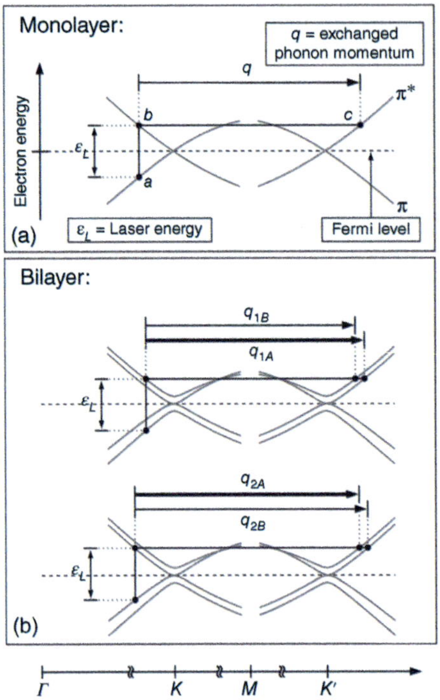

Figure 8.4 Energy band structure and double resonance processes in (a) monolayer graphene and (b) bilayer graphene. Source: Reproduced with permission from Ferrari et al. [46]/American Physical Society.

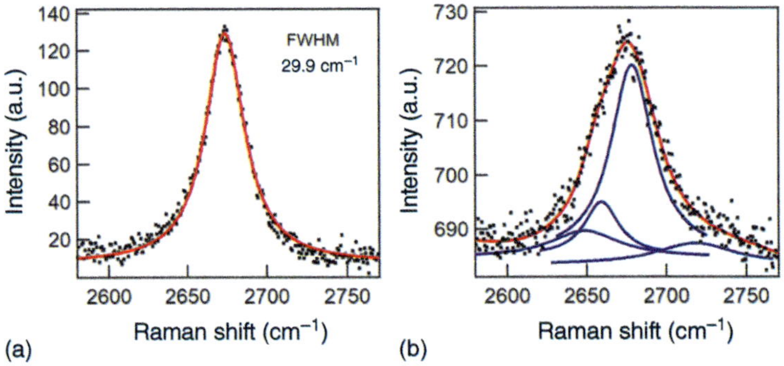

Figure 8.5 The Raman 2D peak of graphene-based biosensors. (a) Black circles are experimental data obtained from pristine graphene, and the red curve is the Lorentzian fit. The full width at half maximum of the fit is 29.9 cm^{-1}. (b) The Raman 2D peak from the pyrene-functionalized graphene. Black circles represent experimental data, blue curves are four deconvoluted Lorentzian fits, and the red line is a constituent by summation of the blue curves. Source: Reproduced with permission from Ferrari et al. [46]/American Chemical Society.

8.5 Quadratic Fit Analysis of the Nonlinear Transconductance of Disordered Bilayer Graphene Field-Effect Biosensors

Due to the BLG formation of the pyrene-functionalized graphene biosensors, the conventional sensing parameters used to analyze sensing signals for the monolayer graphene biosensors are no longer applicable. In the energy band structure, the electronic properties of BLG at low energies near the charge neutrality point (V_{CNP}) are described by massive particles, along with the quadratic dispersion relation $E = \hbar^2k^2/2m^{*2}$, whereas those of graphene are determined by massless Dirac fermions, following the linear dispersion relationship $E = \hbar v_F k$. Here, $m^* = \gamma_1/2v_F^2$ is the effective mass of electrons in BLG, γ_1 is the interlayer hopping strength of electrons, $\hbar$ is the reduced Planck constant, k is the wave vector of electrons, and v_F is the Fermi velocity of graphene. Thus, the difference in the energy band structure of graphene and BLG requires alternative sensing parameters that fit the electronic properties of BLG.

Unlike V_D which is a special nomenclature of the charge neutrality point for the massless relativistic fermions, V_{CNP} is not as accurate as V_D for measuring the external charge doping because of the broadening of the V_{CNP} in BLG [18, 35, 37]. Furthermore, another conventional sensing parameter g_m, is no longer reliable parameter to determine the adsorption of target biomolecules because the electronic properties of BLG are completely different from that of graphene. For instance, the I–V_G relationship in BLG becomes nonlinear, even though charged impurity scattering is the dominant mechanism for charge carrier transport in BLG [37, 47, 48].

In BLG, the source–drain current follows the gate voltage-dependent power law behavior $I \propto V_G\alpha$, which relies on the concentration of the biomolecules adsorbed on the surface. Here, α is a constant between 1 and 1.5, which is determined by both charged impurity scattering and charge screening induced by adsorbates [37, 47, 48]. The superlinearity of the I–V_G relationship in BLG suggests that a new transconductance form is needed to analyze the electrical signal of the BLG biosensors. In addition, some structural discrepancies between pyrene and graphene lead to the imperfect crystal structure of pyrene-functionalized BLG, which should be considered to precisely describe the electronic properties of the BLG biosensors. Furthermore, the degree of the structural disorder increases as the adsorption of the target biomolecules increases on the sensing surface. Therefore, a new sensing parameter for the disordered BLG should reflect the strength of the disorder induced by the external adsorbates.

The electronic properties of the pyrene-functionalized graphene biosensors could be described as a disordered BLG model [49]. In the model, the conductivity σ of BLG at the Fermi energy is given by

$$\sigma \approx g_v g_s \frac{e^2}{2\pi^2\hbar}\left[1+\left(\frac{|\varepsilon|}{\Gamma}+\frac{\Gamma}{|\varepsilon|}\right)\arctan\left(\frac{|\varepsilon|}{\Gamma}\right)+\frac{4\pi\varepsilon_{\text{trig}}}{\Gamma}\right] \tag{8.1}$$

where ε is the Fermi energy, g_V and g_s are the valleys and spin degeneracies, e is the elementary charge, ε_{trig} is the trigonal wrapping energy [50], and Γ is the strength of disorder in BLG. If the conductivity is near V_{CNP} and the strength of disorder in BLG is high, then $\tan^{-1}(|\varepsilon|/\Gamma)$ is approximately equal to $|\varepsilon|/\Gamma$. Then, the conductivity in Eq. (8.1) can be simplified as

$$\sigma \approx g_V g_s \frac{e^2}{2\pi^2\hbar}\left[\left(2+\frac{4\pi\varepsilon_{trig}}{\Gamma}\right)+\left(\frac{|\varepsilon|}{\Gamma}\right)^2\right] \tag{8.2}$$

Plugging Eq. (8.2) into Ohm's law, the conductivity can be expressed as electrical current I of the graphene biosensors,

$$I = \frac{W}{L}\sigma V_{sd} \tag{8.3}$$

where V_{sd} is the source–drain bias and W and L are the width and length of the graphene channel, respectively. Equation (8.3) can be divided into two terms as

$$I \approx g_V g_s \frac{e^2}{2\pi^2\hbar}\frac{W}{L}V_{sd}\left(2+\frac{4\pi\varepsilon_{trig}}{\Gamma}\right)+g_V g_s \frac{e^2}{2\pi^2\hbar}\frac{W}{L}V_{sd}\left(\frac{\varepsilon}{\Gamma}\right)^2 \tag{8.4}$$

Using the energy-gate voltage relation $\varepsilon = \hbar^2k^2/2m^{*2} \approx n \approx V_G$, the current can be expressed as a function of the gate voltage,

$$I \approx a + bV_G^{\,2} \tag{8.5}$$

where both a and b are constants related to the disorder of BLG. The coefficient a is strongly influenced by the measurement conditions, such as contact resistance, whereas the coefficient b is independent of the external experimental conditions. Furthermore, b is inversely proportional to the square of the strength of disorder Γ induced by the adsorption of the target biomolecules during biosensing operations. Equation (8.5) indicates that the electrical current in disordered BLG is a quadratic function of the gate voltage near the charge neutrality point. Therefore, coefficient b serves as an alternative sensing parameter for the pyrene-functionalized BLG biosensors. Since b is analogous to the transconductance, it could be considered as a second-order transconductance $g_m^{(2)}$.

To compare g_m and $g_m^{(2)}$, a set of transfer curves was obtained from pristine graphene, pyrene-functionalized graphene, and the pyrene-functionalized graphene biosensor in the presence of target biomolecule lysozymes (Figure 8.6). First, the linear fit to the experimental data was performed by varying an interval of the gate voltage in order to determine g_m. In Figure 8.6a, the I–V_G curve obtained from pristine graphene shows a nonlinear behavior around V_{CNP} and almost linear behavior for $V_G > 0.15$ V. The linear regime fits well with the experimental data on both gate voltage intervals of $0.15\,\text{V} < V_G < 0.3\,\text{V}$ (blue) and $0.25\,\text{V} < V_G < 0.4\,\text{V}$ (cyan). However, the functionalization of pyrene maleimide on graphene changes the I–V_G curve, resulting in a significant deviation from each fit of two different gate intervals

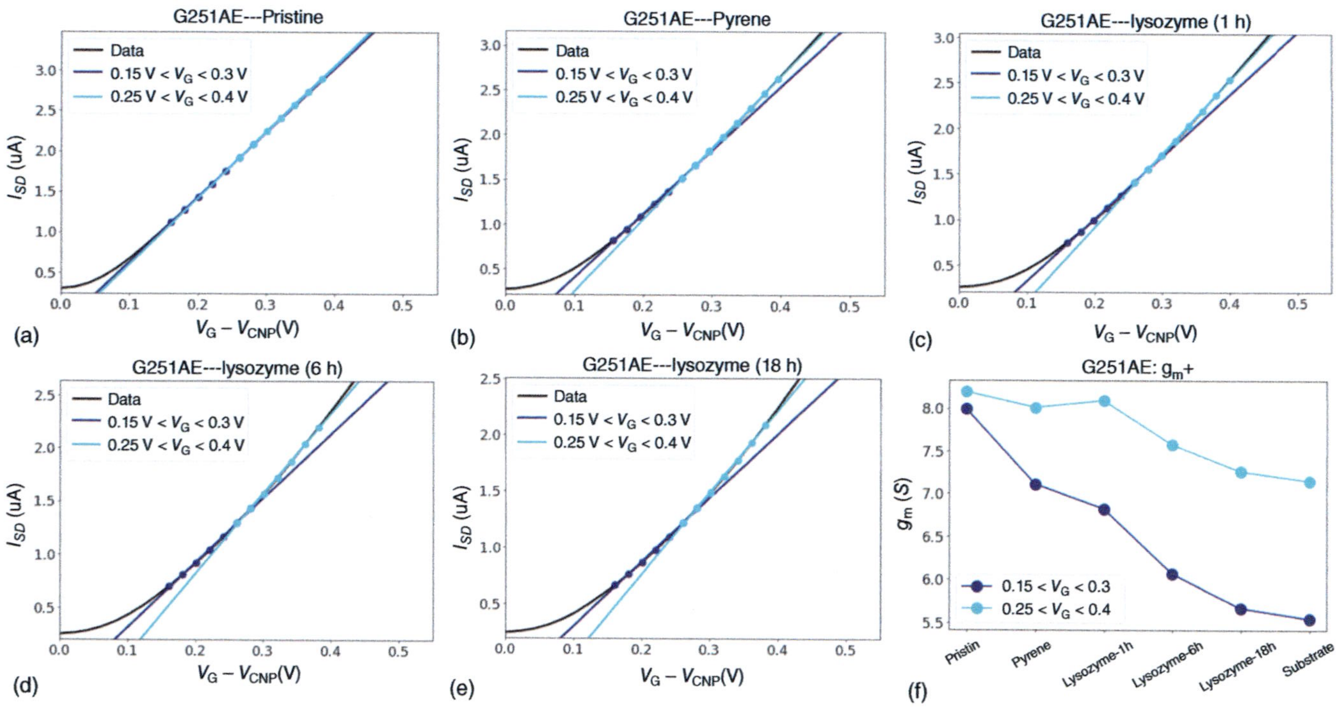

Figure 8.6 The $I-V_g$ curves of pristine graphene and functionalized graphene biosensors in the absence and presence of target proteins. Solid circles are data points used for the linear fit analysis (solid lines). Blue and cyan color correspond to the interval of $0.15\,V < V_G < 0.3\,V$ and $0.25\,V < V_G < 0.4\,V$, respectively. (a) Pristine graphene, (b) graphene functionalized with pyrene-maleimide, and bioconjugation of the functionalized graphene with lysozyme for (c) 1 hour, (d) 6 hour, and (e) 18 hour. (f) The comparisons of transconductance g_m acquired from the two different gate voltage intervals. Source: Reproduced with permission from Ferrari et al. [46]/American Chemical Society.

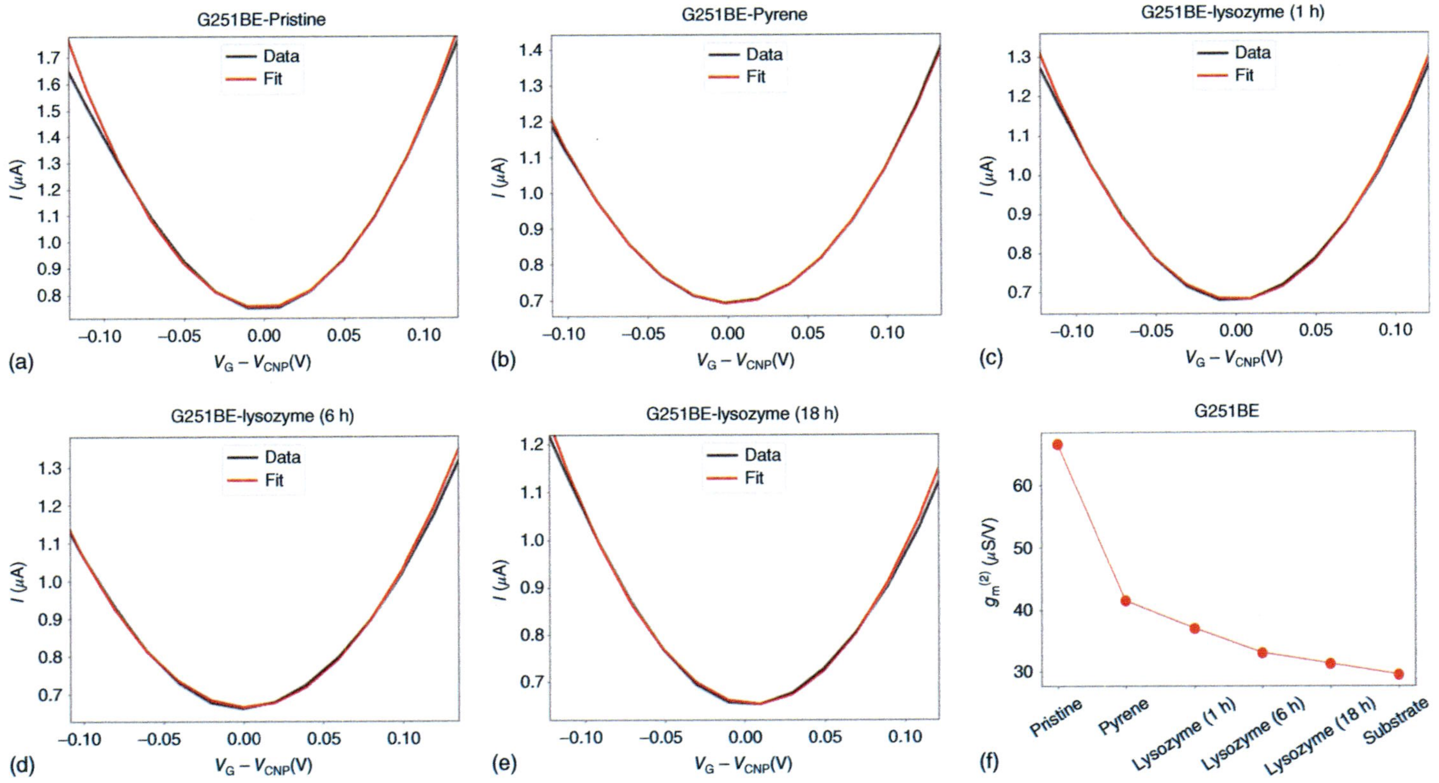

Figure 8.7 The analysis of the graphene biosensor signals with the disordered BLG model. The $I - V_G$ curves of (a) pristine graphene, (b) graphene functionalized with pyrene-maleimide, and pyrene-functionalized graphene biosensor with lysozyme bioconjugation for (c) 1 hour, (d) 6 hour, and (e) 18 hour. (f) The trends of the second-order transconductance $g_m^{(2)}$ for each condition. Black and red curves in (a)–(e) respectively, and are experimental data that fit with Eq. (8.5). Source: Reproduced with permission from Ferrari et al. [46]/American Chemical Society.

(Figure 8.6b). Such discrepancy between two fitting lines is further increased by increasing the incubation time of the target protein (Figure 8.6c–e). Figure 8.6f shows the summary of the g_m values according to each graphene condition. The more the target biomolecule's attachment increases, the more the difference between two g_m values becomes significant. The g_m values vary widely by the selected gate intervals, which makes g_m unreliable for the sensing parameter.

On the other hand, the second-order transconductance $g_m^{(2)}$ extracted from the fit to the quadratic Eq. (8.5) reveals reliable behavior with increasing the incubation time of lysozyme, as shown in Figure 8.7. Not surprisingly, the I–V_G curve obtained from pristine graphene is slightly off from the fit of Eq. (8.5) (Figure 8.7a). In contrast, the pyrene-functionalized graphene biosensor reveals the perfect match between the experimental data and the fit with the quadratic Eq. (8.5), which implies the crystallographic transformation of graphene into BLG, along with the evidence of the Raman 2D peak broadening (Figure 8.7b). As the incubation time of lysozyme increases, the I–V_G curve near V_{CNP} becomes wide, resulting in the decrease of coefficient b in the Eq. (8.5). The disordered BLG model predicts that $g_m^{(2)}$ decreases when the strength of the disorder increases. In other words, the $g_m^{(2)}$ values decrease as the amount of the captured target biomolecule lysozyme on the graphene biosensor increases (Figure 8.7). It is noteworthy to point out that a significant change in $g_m^{(2)}$ between pristine graphene and pyrene functionalization on graphene (Figure 8.7a,b) implies the evolution of graphene to BLG. Based on the nearly perfect fit to Eq. (8.5) with the experimental data and the distinguishable changes in $g_m^{(2)}$ with increasing the disorder in BLG, the newly introduced sensing parameter $g_m^{(2)}$ is reliable and accurate to analyze the electronic signals obtained from the pyrene-functionalized graphene biosensors.

8.6 Conclusion

The formation of BLG from monolayer graphene with pyrene-based functionalization is discussed based on the theory of the DR process in Raman spectroscopy analysis. Because of the different band structures of bilayer graphene, the traditional transconductance g_m is no longer useful to analyze the electrical signals of functionalized graphene sensors. To mitigate this issue, a new transconductance $g_m^{(2)}$ is proposed using the electronic transport theory of disordered bilayer graphene. Compared to the g_m analysis, $g_m^{(2)}$ shows an excellent fit with all experimental sensing data, suggesting it as an alternative and reliable sensing parameter.

Acknowledgment

The work has been supported financially by the NIH/NIGMS (P30GM10332).

References

1 Novoselov, K.S., Geim, A.K., Morozov, S.V. et al. (2004). Electric field effect in atomically thin carbon films. *Science* 306: 666–669.

2 Novoselov, K.S., Geim, A.K., Morozov, S.V. et al. (2005). Two-dimensional gas of massless Dirac fermions in graphene. *Nature* 438: 197–200.

3 Geim, A.K. (2009). Graphene: status and prospects. *Science* 324: 1530.

4 Katsnelson, M.I., Novoselov, K.S., and Geim, A.K. (2006). Chiral tunnelling and the Klein paradox in graphene. *Nat. Phys.* 2: 620–625.

5 Xia, F., Mueller, T., Lin, Y.-m. et al. (2009). Ultrafast graphene photodetector. *Nat. Nanotechnol.* 4: 839–843.

6 Bolotin, K.I., Sikes, K.J., Jiang, Z. et al. (2008). Ultrahigh electron mobility in suspended graphene. *Solid State Commun.* 146: 351–355.

7 Bolotin, K.I., Sikes, K.J., Hone, J. et al. (2008). Temperature-dependent transport in suspended graphene. *Phys. Rev. Lett.* 101: 096802.

8 Nair, R.R., Blake, P., Grigorenko, A.N. et al. (2008). Fine structure constant defines visual transparency of graphene. *Science* 320: 1308–1308.

9 Blake, P., Hill, E.W., Neto, A.H.C. et al. (2007). Making graphene visible. *Appl. Phys. Lett.* 91: 063124.

10 Woo, S.O. and Teizer, W. (2013). The effect of electron induced hydrogenation of graphene on its electrical transport properties. *Appl. Phys. Lett.* 103: 041603.

11 Choi, H., Choi, J., Kim, J. et al. (2014). Flexible and transparent gas molecule sensor integrated with sensing and heating graphene layers. *Small* 10: 3685–3691.

12 Woo, S.O., Froberg, J., Pan, Y. et al. (2020). Protein detection using quadratic fit analysis near the Dirac point of graphene field-effect biosensors. *ACS Appl. Electron. Mater.* 2: 913–919.

13 Liao, L., Lin, Y., Bao, M. et al. (2010). High-speed graphene transistors with a self-aligned nanowire gate. *Nature* 467: 305–308.

14 Bae, S., Kim, H., Lee, Y. et al. (2010). Roll-to-roll production of 30-inch graphene films for transparent electrodes. *Nat. Nanotechnol.* 5: 574–578.

15 Abraham, J., Vasu, K.S., Williams, C.D. et al. (2017). Tunable sieving of ions using graphene oxide membranes. *Nat. Nanotechnol.* 12: 546–550.

16 Heerema, S.J. and Dekker, C. (2016). Graphene nanodevices for DNA sequencing. *Nat. Nanotechnol.* 11: 127–136.

17 Yavari, F. and Koratkar, N. (2012). Graphene-based chemical sensors. *J. Phys. Chem. Lett.* 3: 1746–1753.

18 McCann, E. and Koshino, M. (2013). The electronic properties of bilayer graphene. *Rep. Prog. Phys.* 76: 056503.

19 Georgakilas, V., Otyepka, M., Bourlinos, A.B. et al. (2012). Functionalization of graphene: covalent and non-covalent approaches, derivatives and applications. *Chem. Rev.* 112: 6156–6214.

20 Kuila, T., Bose, S., Mishra, A.K. et al. (2012). Chemical functionalization of graphene and its applications. *Prog. Mater. Sci.* 57: 1061–1105.

21 Zhang, X., Liu, T., Boyle, A. et al. (2022). Dielectric-modulated biosensing with ultrahigh-frequency-operated graphene field-effect transistors. *Adv. Mater.* 34: 2106666.

22 Han, Q., Pang, J., Li, Y. et al. (2021). Graphene biodevices for early disease diagnosis based on biomarker detection. *ACS Sens.* 6: 3841–3881.

23 Tetlow, H., de Boer, J. .P., Ford, I.J. et al. (2014). Growth of epitaxial graphene: theory and experiment. *Phys. Rep.* 542: 195–295.

24 Yan, K., Fu, L., Peng, H., and Liu, Z. (2013). Designed CVD growth of graphene via process engineering. *Acc. Chem. Res.* 46: 2263–2274.

25 Zhang, Y., Zhang, L., and Zhou, C. (2013). Review of chemical vapor deposition of graphene and related applications. *Acc. Chem. Res.* 46: 2329–2339.

26 Geng, D., Wang, H., and Yu, G. (2015). Graphene single crystals: size and morphology engineering. *Adv. Mater.* 27: 2821–2837.

27 Li, M., Liu, D., Wei, D. et al. (2016). Controllable synthesis of graphene by plasma-enhanced chemical vapor deposition and its related applications. *Adv. Sci.* 3: 1600003.

28 Deokar, G., Avila, J., Razado-Colambo, I. et al. (2015). Towards high quality CVD graphene growth and transfer. *Carbon* 89: 82–92.

29 Tai, L., Zhu, D., Liu, X. et al. (2017). Direct growth of graphene on silicon by metal-free chemical vapor deposition. *Nano-Micro Lett.* 10: 20.

30 Losurdo, M., Giangregorio, M.M., Capezzuto, P., and Bruno, G. (2011). Graphene CVD growth on copper and nickel: role of hydrogen in kinetics and structure. *Phys. Chem. Chem. Phys.* 13: 20836–20843.

31 Yang, Q., Zhang, Z., Zhu, W., and Wang, G. (2019). Growth of large-area high-quality graphene on different types of copper foil preannealed under positive pressure H_2 ambience. *ACS Omega* 4: 5165–5171.

32 Adam, S., Hwang, E.H., Galitski, V.M., and Das Sarma, S. (2007). A self-consistent theory for graphene transport. *Proc. Natl. Acad. Sci. U.S.A.* 104: 18392–18397.

33 Adam, S., Jung, S., Klimov, N.N. et al. (2011). Mechanism for puddle formation in graphene. *Phys. Rev. B* 84: 235421.

34 Chen, J.H., Jang, C., Adam, S. et al. (2008). Charged-impurity scattering in graphene. *Nat. Phys.* 4: 377–381.

35 Woo, S., Hemmatiyan, S., Morrison, T.D. et al. (2017). Temperature-dependent transport properties of graphene decorated by alkali metal adatoms (Li, K). *Appl. Phys. Lett.* 111: 263502.

36 Woo, S.O. and Teizer, W. (2015). Effects of electron beam induced Redox processes on the electronic transport in graphene field effect transistors. *Carbon* 93: 693–701.

37 Woo, S.O., Yudhistira, I., Hemmatiyan, S. et al. (2019). Transport properties of bilayer graphene decorated by K adatoms in the framework of Thomas-Fermi screening. *Phys. Rev. B* 99: 085416.

38 Ashcroft, N.W. and Mermin, N.D. (1976). *Solid State Physics*, 1e, 826. USA: Saunders College Publishing.

39 Jiang, S., Cheng, R., Wang, X. et al. (2013). Real-time electrical detection of nitric oxide in biological systems with sub-nanomolar sensitivity. *Nat. Commun.* 4: 2225.

40 Choi, Y.K., Moody, I.S., Sims, P.C. et al. (2012). Single-molecule lysozyme dynamics monitored by an electronic circuit. *Science* 335: 319–324.

41 Georgakilas, V., Tiwari, J.N., Kemp, K.C. et al. (2016). Noncovalent functionalization of graphene and graphene oxide for energy materials, biosensing, catalytic, and biomedical applications. *Chem. Rev.* 116: 5464–5519.

42 Gao, N., Gao, T., Yang, X. et al. (2016). Specific detection of biomolecules in physiological solutions using graphene transistor biosensors. *Proc. Natl. Acad. Sci. U.S.A.* 113: 14633–14638.

43 Wu, G., Dai, Z., Tang, X. et al. (2017). Graphene field-effect transistors for the sensitive and selective detection of *Escherichia coli* using pyrene-tagged DNA aptamer. *Adv. Healthcare Mater.* 6: 1700736.

44 Choi, Y., Moody, I.S., Sims, P.C. et al. (2012). Single-molecule dynamics of lysozyme processing distinguishes linear and cross-linked peptidoglycan substrates. *J. Am. Chem. Soc.* 134: 2032–2035.

45 Choi, Y., Olsen, T.J., Sims, P.C. et al. (2013). Dissecting single-molecule signal transduction in carbon nanotube circuits with protein engineering. *Nano Lett.* 13: 625–631.

46 Ferrari, A.C., Meyer, J.C., Scardaci, V. et al. (2006). Raman spectrum of graphene and graphene layers. *Phys. Rev. Lett.* 97: 187401.

47 Das Sarma, S., Hwang, E.H., and Rossi, E. (2010). Theory of carrier transport in bilayer graphene. *Phys. Rev. B* 81: 161407.

48 Adam, S. and Das Sarma, S. (2008). Boltzmann transport and residual conductivity in bilayer graphene. *Phys. Rev. B* 77: 115436.

49 Koshino, M. (2009). Electronic transport in bilayer graphene. *N. J. Phys.* 11: 095010.

50 Castro Neto, A.H., Guinea, F., Peres, N.M.R. et al. (2009). The electronic properties of graphene. *Rev. Mod. Phys.* 81: 109–162.

9

Theoretical and Experimental Characterization of Molecular Self-Assembly on Graphene Films

Kishan Thodkar[1], Pierre Cazade[2], and Damien Thompson[2]

[1]*ETH Zurich, Micro- & Nanosystems, Department of Mechanical & Process Engineering, Tannenstrasse 3, 8092 Zurich, Switzerland*

[2]*University of Limerick, Bernal Institute, Department of Physics, Limerick V94 T9PX, Ireland*

9.1 Introduction

Graphene [1] has emerged as an attractive substrate for field-effect transistors (FETs), spurred by Schwierz's vision of graphene transistors as a key enabler and driver of post-silicon electronics [2]. Beyond the excellent inherent electronic mobility of the π-conjugated graphene two-dimensional (2D) sheet, nanofabrication of extremely thin channels in graphene FETs provides miniaturized high-performance electronics with excellent gate control over the channel barrier. The engineering of ultrathin channels in graphene FETs also avoids the short-channel effects (SCEs) that hamper device integration of silicon-based metal-oxide-semiconductor field-effect transistors (MOSFETs) [3], in which the channel length is comparable to the depletion layer widths of the source and drain junctions. In addition, graphene exhibits high thermal conductivity and optical transparency.

FETs control the flow of current in a semiconductor by altering the conductivity between the electrical source and drain of the device when voltage is applied. This means that the FET performance depends critically on the gate material, with electrons or holes acting as charge carriers. The archetypal FET is the MOSFET. The potential for graphene-based FETs as bioanalytical sensors has been comprehensively reviewed – for the most recent authoritative reviews published at the time of writing in early 2022, we refer the interested reader to [4–7]. Here, we focus specifically on how molecular self-assembly, as opposed to conventional chemical deposition, bonding, and film growth processes, can potentially tune the electronic properties of electrolyte-gated graphene FETs.

Self-assembly is advantageous because, compared with traditional chemical synthesis involving the catalytic promotion of covalent bond forming/breaking, self-assembly uses individually weak, reversible noncovalent interactions, including

Graphene Field-Effect Transistors: Advanced Bioelectronic Devices for Sensing Applications,
First Edition. Edited by Omar Azzaroni and Wolfgang Knoll.

metal–π, electrostatic, H-bonding, π–π, and van der Waals's binding, in the spontaneous emergence of nanostructured films. These soft, bioinspired assembly strategies can be less expensive, more eco-friendly, and more versatile in terms of both building blocks and possible end products than traditional chemical approaches.

The self-assembly of (bio)molecular films on 2D platforms such as graphene opens the door to fully integrating bio-nano sensing devices with potential applications as inexpensive, portable electrochemical sensors based on a programmable change in impedance due to change in dielectric constant resulting from complexation of analyte molecules to the coated sensor surface. Better understanding and control of functionalization also paves the way for broader applications of molecularly functionalized graphene in industrial polymer nanocomposites, energy storage or conversion (batteries, fuel cells, solar cells, and supercapacitors), electrochemical sensors, and optical devices [8–10]. Sustainable materials such as peptide films are crucial for next-generation technologies to replace polluting or non-biocompatible materials, which may be harmful due to the metal-based catalysts or strong acid solvents used in their production, high temperatures produced, or toxic side products. The materials themselves may also contain harmful ingredients, e.g. lead zirconate titanate PZT used in televisions and other products, or conducting polymers such as polyaniline, polypyrrole, and PDOT used in environmental/biomedical sensing, which have low but not insignificant cytotoxicity [11, 12].

The challenges include the development of low-cost, high-yield methods of creating large surface areas of low-defect graphene sheets, effective electroding to make contact with CMOS device architectures, identification/engineering of high-performance materials to coat the graphene with the required adhesion to graphene, ability to pack into well-formed films, molecular recognition of analyte, and a significant, analyte-specific binding-induced change in electrical properties that can be simply and reliably detected by the graphene-device circuit. Below we describe recent efforts from ourselves and others to characterize and exploit molecular films on graphene for device applications, guided by fundamental science and supported by atomic-level modeling of the structure, dynamics, and electrical properties of the graphene–film interface.

9.2 Experimental Tools to Characterize Molecular Functionalization of Graphene

Graphene-based field-effect transistors can be functionalized using a diverse, wide range of organic and inorganic molecules. Such a functional layer in design can consist of molecules with a complex structure, size, and net electronic charge, positioned directly on the graphene surface (see Figure 9.1).

The unique chemical structure of graphene makes it highly suitable for the non-covalent linking of molecules on the graphene surface. Researchers have used different techniques to adsorb molecules on graphene, namely: (i) dispensing, (ii) evaporation, and (iii) synthesis. Dispensing technique can be performed using a

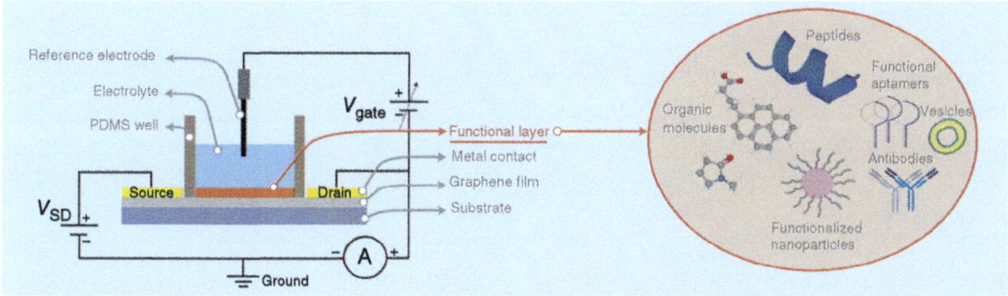

Figure 9.1 A schematic of a graphene-based electrolyte-gated field-effect transistor with a diverse functional layer.

direct dispensing approach using a "target" molecule diluted in solvents or by using a spray coating method. Commercially available, microliter-dispensing tools are routinely used in the pharmaceutical industry to create diagnostic test kits. Spray coating, for instance, has been widely used to study molecular self-assembly on graphene using scanning tunneling microscopy. Spin coating can also be combined after dispersion. The evaporation technique is typically used in the presence of a carrier gas in the background with the "target" molecule either in solid or liquid form. This technique is particularly advantageous when the combination of nanoscale sticking and friction – so-called "stiction" – limits the usage of the dispensing approach. Alternatively, the two techniques can be combined in a complex, multistep chemical synthesis of the desired functional layer on the graphene surface. However, it is essential to note that achieving an understanding of the functional layer design requirements in terms of its morphology, influence on the device performance, and reproducibility is a prerequisite for the development of robust fabrication process chains to facilitate the production of state-of-the-art 2D material-based biosensing platforms.

9.2.1 Considering the Three Distinct Techniques Available for Functionalizing Graphene Are the Outcomes of the Three Functionalization Techniques Consistent, Similar, Reproducible Across all Three Techniques?

It is unlikely to obtain consistent surface functionalization morphology across all three techniques. For example, dispensing and microfluidic approaches are fundamentally different, employing different dominant principles, and kinetics and yielding different functionalization morphologies for a given concentration of a molecular solution (see Figure 9.2). Secondary effects of molecular doublet and triplet formation may also occur, leading to different preferred stacking behavior. Suitable solvents need to be used in such cases to minimize cluster or aggregate formation during the surface functionalization. Note that the advancements in microfluidics make it possible to design and influence the flow dynamics during the functionalization process. However, characterizing the nanoscale molecular functionalization morphology on the sensor surface is a key challenge.

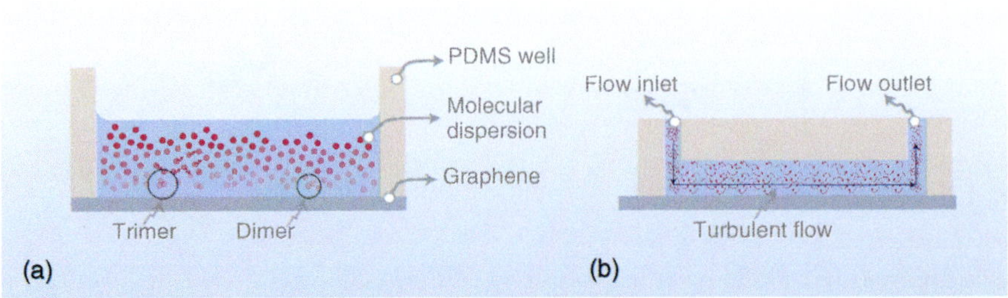

Figure 9.2 A simplified schematic of dispensing and microfluidic approach used to functionalize graphene films. (a) Dispensing approach and (b) microfluidic approach.

9.2.2 What Tools and Methods Are Available to Perform Such a Characterization of Molecular Self-Assembly Across the Nano to Macro Scale?

Scanning tunneling microscopy (STM) provides state-of-the-art resolution to characterize features at the atomic scale. STM has been widely used to characterize adsorbed molecules on graphite and other 2D materials (e.g. MoS_2) [13, 14]. Graphite flake consists of stacked, single-layer graphene films, weakly bound by π–π interactions. Alternatively, single-layer graphene films on metal substrates [15, 16] and epitaxial graphene [17] can be utilized to study molecular interaction on pristine, single-layer graphene films [18]. Pathipati et al. utilized a spin coating approach to deposit monolayer of pyrene butyric acid (PBA) molecules ($2\,\mu g\,ml^{-1}$ of PBA in methanol) on reduced graphene oxide (rGO) flakes [19]. The spin coating approach offered a convenient route to disperse a single layer of PBA molecules on graphene and to resolve molecular adsorption configuration and stacking tendency using STM (see Figure 9.3).

The study from Pathipati et al. [19] reveals hydrogen bonding dominated PBA–rGO binding interaction and supramolecular packing as dimer units ($1.08\ nm \times 2.31\,nm$) on the rGO surface. Furthermore, calculations confirm the noncovalent yet robust nature of the interface with a weak charge transfer (0.06 eV/dimer) and a large H-bond-driven binding energy of −2.6 eV/dimer at the PBA-rGO interface. Such a characterization using the STM offers extensive insight into the molecular adsorption at the nanoscale.

To characterize the influence of functionalization at a micro to macro scale, electrical measurements such as field effect (FET) and electrochemical characterization are widely used. In Figure 9.3e, conductivity vs. gate voltage sweep characteristics of bare and PBA-functionalized rGO sample is shown. We can observe two changes in the FET characteristics on PBA deposition, namely (i) drop in conductivity and (ii) shift of the Dirac potential toward higher gate voltage.

Adsorption of PBA molecules on rGO leads to p-type doping of the sample, with an estimate of $1.3 \times 10^{12}\ cm^{-2}$ of charge density calculated from the Dirac voltage shift of 8 V [19]. Such observations are essential to characterize the influence of the functional layer on electrical properties in the development of graphene-based

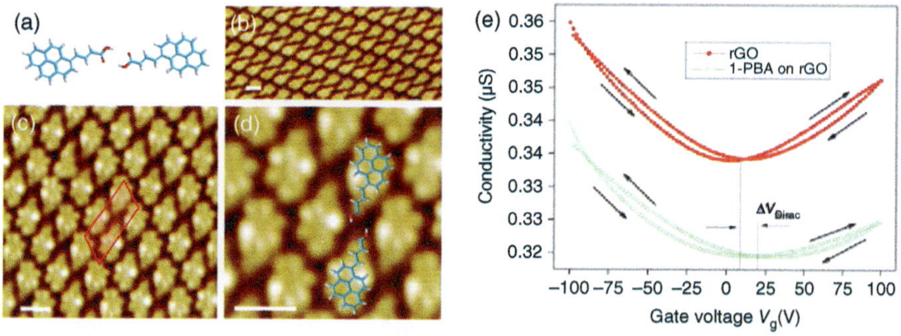

Figure 9.3 PBA dimer formation on rGO, observed using STM. (a) PBA dimer arrangement, (b–d) topography images obtained at tunneling current of 0.3 and 0.5 nA (sample bias, V_s = 0.5 V), scale (estimate) 1 nm. (e) Conductivity vs. gate voltage sweep of the bare and PBA-functionalized rGO transistor. The black arrow on the curves indicates the direction of the gate sweep. Note the shift of the Dirac voltage after PBA functionalization. Source: Reproduced with permission from Pathipati et al. [19]/Elsevier.

biosensing applications. There are several methods to characterize functionalized graphene films at micro and macro scales, namely UV–visible absorption, Raman spectroscopy, fluorescence microscopy, nuclear magnetic resonance (NMR), X-ray photoelectron spectroscopy (XPS), terahertz imaging, electron microscopy techniques (such as SEM and TEM), and surface probe microscopy (such as AFM and STM), and more. Raman spectroscopy has been widely used to characterize carbon nanomaterials [20, 21]. Using large-area Raman mapping, many characteristics such as doping, number of layers, type of defects [22, 23], grain boundaries, and strain [24] can be measured in graphene films. Characterizing the changes in the Raman peak shape and position offers extensive insight into the binding and stacking behavior of the molecules on a graphene surface [25]. For instance, an AB stacking of PBA molecules on graphene can lead to the splitting of the 2D Raman peak [26]. Hinnemo et al. studied the influence of increasing PBA concentration on R_{sp^2} ratio in PBA functionalized graphene films. Estimating the R_{sp^2} ratio provides insight into the PBA monolayer formation and stacking behavior on CVD graphene films. In this study, a droplet-dispensing approach (PBA diluted in dimethylformamide for one hour) was followed during the PBA functionalization of CVD graphene films.

The R_{sp^2} was estimated by considering the ratio of $[A_{(SLG+PBA)} - A_{(SLG)}]/A_{(SLG)}$, wherein "$A$" represents the area of the respective Raman G peak. By plotting the R_{sp^2} at respective PBA concentrations, Langmuir Isotherm behavior was observed (see Figure 9.4a). In Figure 9.4b, the intensity count of the R_{sp^2} peak vs. the PBA concentration is shown. The green dotted lines are the fitting results with $R_{sat} = 2.2$ and adsorption equilibrium constant $K = 4.5 \times 10^{-2}$ mM^{-1}. The authors estimate a PBA surface density of 5.3 PBA molecules per nm^2 under saturation conditions.

Fluorescent dyes are extensively used to investigate biomolecular interactions and ligand–receptor binding kinetics. In the case of weakly fluorescing functional molecules (e.g. PBA or graphene-binding peptides), additional fluorescent dyes or tags are useful in facilitating the indirect investigation of binding kinetics.

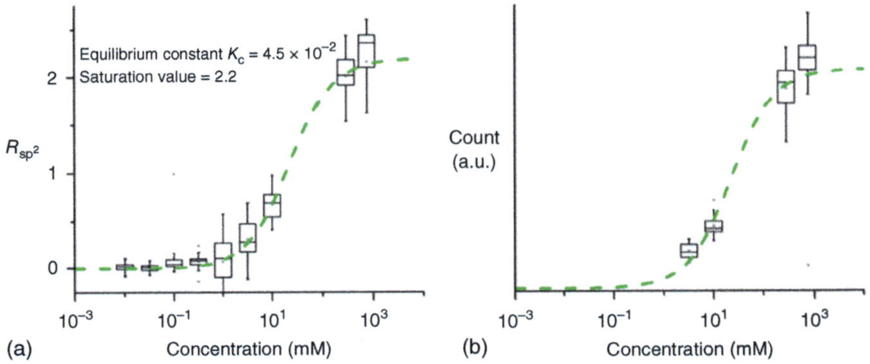

Figure 9.4 (a) Ratio of carbon peaks of PBA and graphene, characterized using Raman spectroscopy vs. PBA concentration. (b) D' Raman peak intensity vs. PBA concentration. The green dotted line is the fit to the Langmuir model ($K = 4.5 \times 10^{-2}$ mM^{-1}). Source: Reproduced with permission from Hinnemo et al. [26]/American Chemical Society.

Thodkar et al. utilized a fluorescence imaging technique to characterize up to $1 \times 1\,mm^2$ regions of pyrene- and peptide-functionalized graphene films. Fluorescent tag molecules such as FITC (fluorescein-5-isothiocyanate) in combination with linker chemistry were used to covalently bind FITC molecules to PBA (see Figure 9.5). A droplet dispensing approach was utilized to functionalize the graphene films in the study.

The surface morphology over large areas was studied using a standard fluorescent microscope, shown in Figure 9.5. Distinct features are observable within the images, with globular structures dominant in the pyrene-functionalized sample (Figure 9.4a) and thread-like structures present in the peptide-functionalized

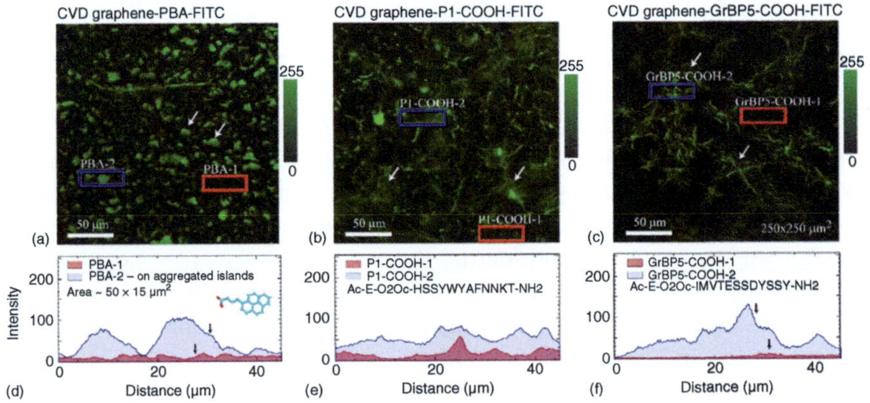

Figure 9.5 Fluorescence microscopy images ($250 \times 250\,\mu m^2$) of (a) FITC-tagged pyrene, (b, c) peptide-functionalized CVD graphene films. The vertical color bar is representative of the gray value intensity of the images. The average intensity vs. distance of the areas highlighted within blue and red regions ($50 \times 15\,\mu m^2$) are presented in figures (d–f). Inset: Respective chemical structure of the pyrene and peptide molecules, scale bar is $50\,\mu m$. Source: Reproduced with permission from Thodkar et al. [27]/American Chemical Society.

sample (Figure 9.5b,c). The blue region in Figure 9.4a represents regions with strong fluorescence, which can be visualized by extracting the average gray value intensity within the selected region as presented in Figure 9.5d. The red region within Figure 9.5a offers a relatively lower fluorescent background. Fluorescent labeling offers an indirect, easier, and quicker means to image and characterize large regions of functionalized graphene films in a reliable manner. Fluorescence quenching on direct adsorption of FITC molecules on graphene films has been reported by Lin et al. [28]. Bright features in the images can occur due to "free" FITC molecules not being in direct contact with the underlying graphene layer. The higher stacking of the molecular linker present between the FITC and graphene film can cause such a limitation in fluorescence-quenching via graphene layer.

Spectroscopic ellipsometry and contact angle measurements offer alternative means to characterize functionalized graphene films. Hinnemo et al. [26] characterized monolayer self-assembly of PBA molecules on graphene films using the two techniques. The effective thickness of adsorbed PBA increases with the increase in molar concentration of PBA, with a saturation thickness of 0.7 nm, as shown in Figure 9.6. They reported horizontal self-assembly at a low molar concentration of PBA and vertical self-assembly of PBA molecules at high molar concentrations. This finding provides unique insight into the role of concentration during the functionalization process.

Interestingly, the inherent hydrophobic nature of the bare graphene surface can be altered with the increase in PBA concentration. A gradual decrease in contact angle with an increase in PBA concentration can be observed in Figure 9.5b. Both these methods offer a macroscale approach to characterizing molecular stacking behavior on graphene films.

Besides the great potential of pyrene-based molecular linkers, there are a variety of functional molecules used for graphene functionalization, such as peptides, aptamers, and nanoparticles. Several peptide sequences have been explored for their ability to self-assemble into well-formed films on graphene [27, 29–32]. The peptide sequence of commonly used graphene binding peptides (GrBP) is IMVTESSDY-SSY. The sequence YSSY, consisting of Tyrosine amino acid, offers the possibility to

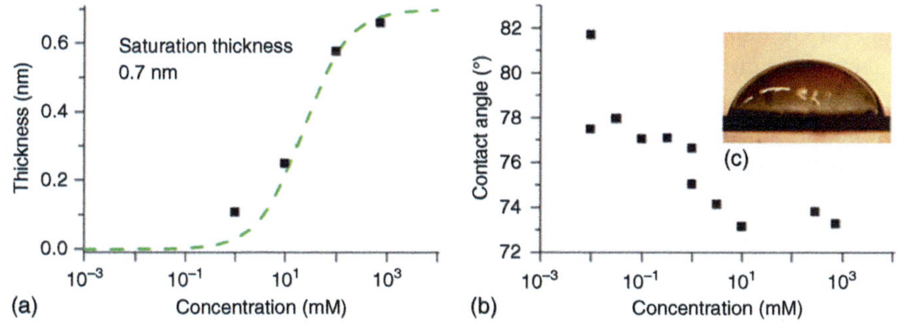

Figure 9.6 (a) Characterization of effective thickness of the self-assembled PBA layer on graphene surface and (b) Contact angle measurements vs. PBA concentration. Inset: (c) Optical image of the contact angle characterization. Source: Reproduced with permission from Hinnemo et al. [26]/American Chemical Society.

bind to graphene via π–π stacking. The terminals of the peptide can be capped with functional chains or passive molecules. Hayamizu et al. reengineered the GrBP terminal and measured the influence of the sequence on surface adsorption (see Figure 9.7) [30]. Performing peptide side-chain engineering is a simple and highly effective means of changing both the overall net charge and the local features of the electrostatic potential in graphene-bound peptide layers (Figure 9.7).

The surface binding process was performed by droplet dispensing of the peptide solution on the nanomaterial surface. Interestingly, a strong response to the respective GrBP surface coverage state can be monitored using field-effect measurements (see Figure 9.8). A gradual shift of the charge neutrality point, indicating hole doping during the liquid-gate measurement, is observable. The occurrence of the double peak in the conductivity during gate voltage sweep depicts the localized doping influence of the GrBP adsorption. However, at higher surface coverage, double peak features are not observable. The double peak resistance feature can be modeled by considering two independent graphene regions with distinct charge carrier densities.

In a more recent study, the pyrene and peptide functionalization of graphene enabled via dispensing approach was studied using electrolyte gating. Thodkar et al. compared the distinct features such as changes in the relative slope, charge neutrality point, mobility of pyrene, and peptide-functionalized GFET characteristics. Two types of peptides were considered during the study, namely P1 (peptide sequence: HSSYWYAFNNKT) and GrBP, with the ends of the peptides caped with either carboxylic acid or maleimide groups.

Consistent changes in functionalization at multiple electrodes were observed in the measurements. A decrease of 5% relative slope was observed in pyrene functionalized GFETs, whereas a larger decrease (35%) in slope was observed after peptide functionalization. A relative charge neutrality point shift of up to −64% in pyrene functionalized GFETs, and −1% to +3.5% change was observed in peptide-functionalized GFETs (see Figures 9.9 and 9.10).

The net shift of the charge neutrality point can be understood by considering the intrinsic electron withdrawing or donating properties of the pyrene and peptide molecules. A similar observation of hole doping was identified in rGO samples by Pathipati et al. [19]. The significant change in field-effect mobility measured in the peptide-functionalized samples confirms the large role of charge impurity scattering effects at the graphene–peptide interface. The weak shift of the charge neutrality point in peptide-functionalized samples is likely due to the relatively weaker π–π interaction at the graphene–peptide interface. Note the difference in the size of the molecular linkers. The length of the PBA molecule is 1.2 nm, which is three times shorter than the peptide sequences (5 nm). Such physical conditions of the molecular linkers, together with their chemical and electronic properties, must be understood in order to gauge their impact on the sensor characteristics.

Primarily, the stacking and aggregation tendencies of the molecular linkers under sample functionalization conditions have to be thoroughly mapped using complementary characterization methods. This step is crucial to assess the linker suitability and the anticipated detection limitation during the biosensing applications, as sketched in Figure 9.11.

Name	Sequence												Net charge
	1	2	3	4	5	6	7	8	9	10	11	12	
GrBP5-WT	I	M	V	T	E	S	S	D	Y	S	S	Y	−2
GrBP5-M2	I	M	V	T	E	S	S	D	W	S	S	W	−2
GrBP5-M6	I	M	V	T	A	S	S	A	Y	D	D	Y	−2
GrBP5-M8	I	M	V	T	A	S	S	A	Y	R	R	Y	+2

(a)

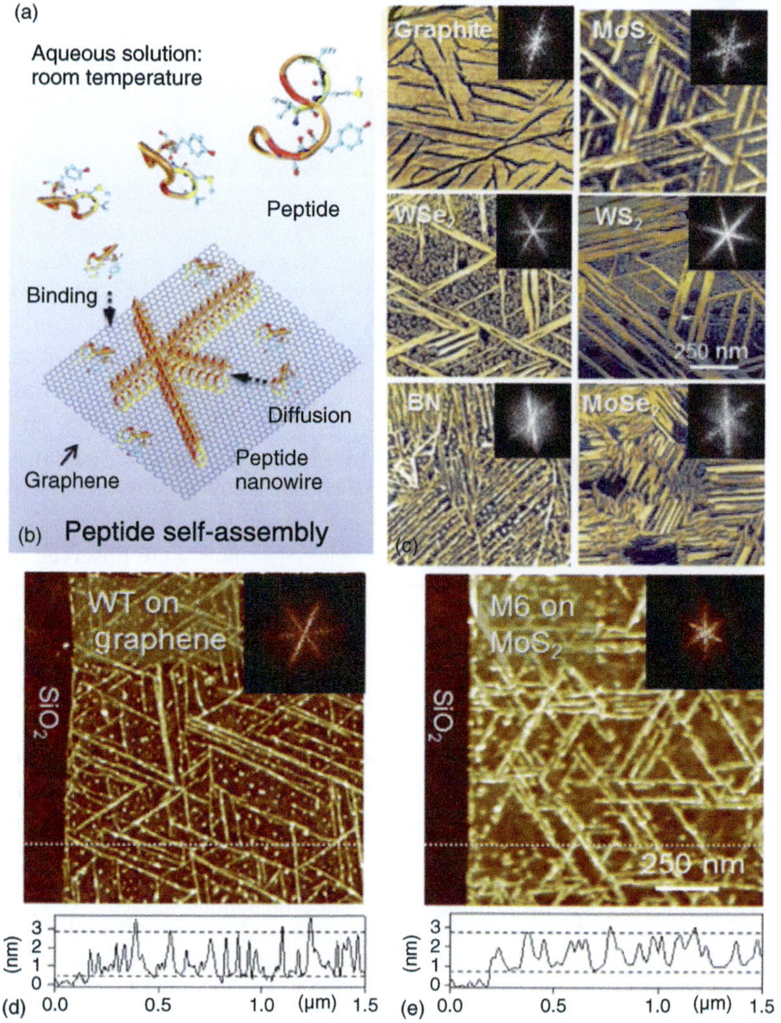

Figure 9.7 Overview of the GrBP peptide surface adsorption study on various 2D materials. (a) Table of GrBP sequences with their respective net charge. (b) Schematic of the various peptide binding and organization processes. (c) AFM characterization of peptide arrangement on graphite, MoS_2, WSe_2, WS_2, hBN, and $MoSe_2$. (d) GrBP wild type peptide sequence arrangement on graphene and (e) GrBP M6 variant arrangement on MoS_2, with respective AFM line scan. Inset: FFT of the respective AFM images. Source: Reproduced from Hayamizu et al. [30], © 2016, Springer Nature/CC BY 4.0.

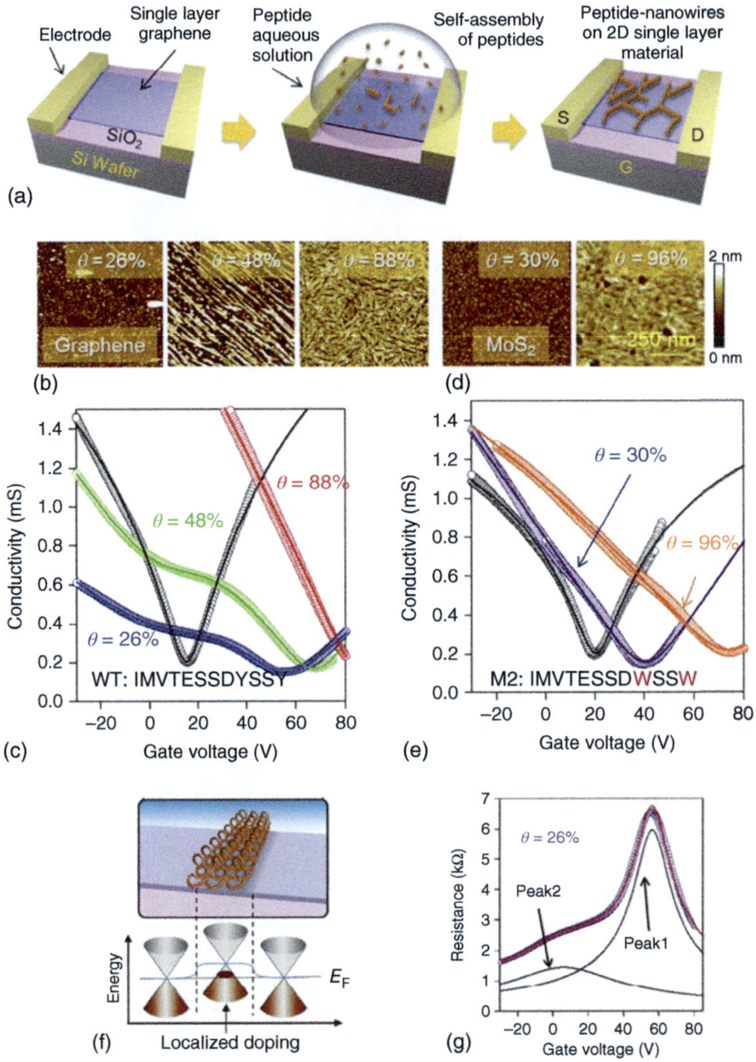

Figure 9.8 Overview of the field-effect characterization of GrBP peptide surface adsorption on graphene and MoS_2. (a) Schematic of the peptide self-assembly toward nanowire formation. (b, d) AFM characterization of the peptide-functionalized graphene (MoS_2) at surface coverage of 26%, 48% (30%), and 88% (96%). (c, e) Conductivity vs. gate voltage characteristics of the wild type GrBP (GrBP-M2) sequence at 26, 48 (30%), and 88% (96%) surface coverage ratio. (f) Schematic of the doping profile of GrBP binding on graphene. (g) Peak fit of the resistance vs. gate voltage measurement of the GrBP-WT (26% surface coverage). Source: Reproduced from Hayamizu et al. [30], © 2016, Springer Nature/ CC BY 4.0.

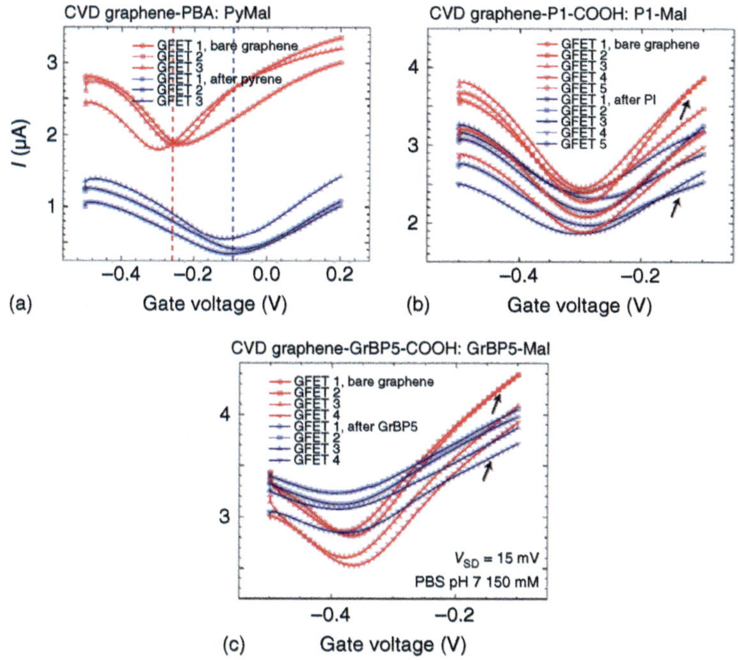

Figure 9.9 Overview of the electrolyte-gated field-effect characterization of pyrene and peptide-functionalized GFETs. Current vs. gate voltage characteristics of bare (red) and after functionalization (blue), of pyrene (a, three GFETs), and peptide P1 (b, five GFETs) and GrBP (c, four GFETs) functionalized GFETs. A PBS buffer with pH 7, 150 mM was used as an electrolyte during these experiments. Source: Reproduced with permission from Thodkar et al. [27]/American Chemical Society.

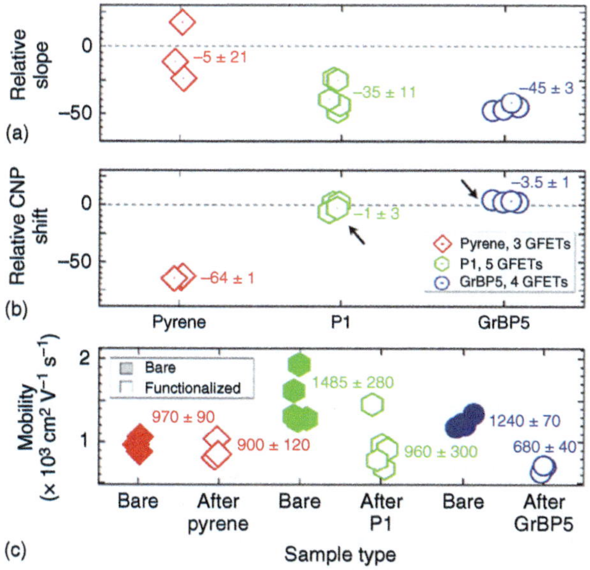

Figure 9.10 The relative change in (a) slope, (b) charge neutrality point and (c) field-effect mobility of bare and pyrene (red) and peptides P1 (green) and GrBP (blue) functionalized GFETs. Source: Reproduced with permission from Thodkar et al. [27]/American Chemical Society.

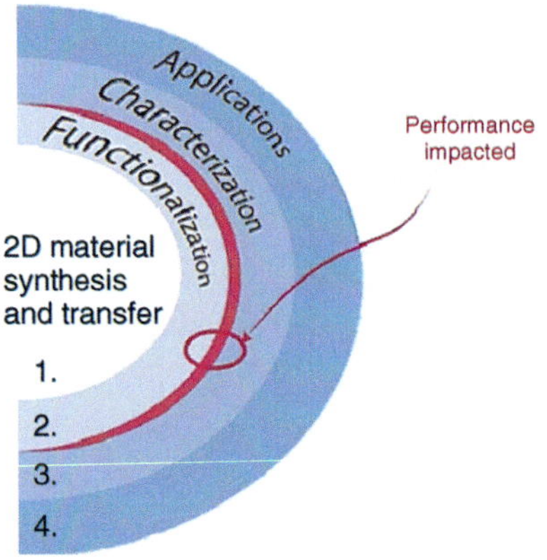

Figure 9.11 A schematic of the anticipated limitation in sensor performance during sensor surface functionalization.

9.3 Atomistic Insights to Guide Molecular Functionalization of Graphene

Theory and modeling guide the rational design of self-assembled films on graphene FETs. In this second part of this chapter, following the extensive discussion above of device architecture and identification of the important features for technology applications, we provide here a detailed tutorial-style presentation of predictive modeling to guide future experiments. We outline how to build large atomistic models and efficiently run molecular dynamics (MD) simulations on high-performance computing platforms of molecular assemblies on flat electronic substrates such as graphene.

We describe efficient data analytics and visualization methods to predict the structure, dynamics and energetics of SAM-functionalized graphene FETs. The methodology is readily adaptable and transferable to other types of molecular self-assembly to tune a wide variety of potential electrolyte-gated graphene FETs as well as assemblies on other electronic substrates, including other 2Ds, metals, and metal oxides, requiring only gathering and/or generation of force field data to describe the potential energy surface of the substrate. These parameters are generally obtained by *ab initio* or density functional theory (DFT) quantum mechanical calculations on representative fragments, e.g. methylamine above pyrene, to describe the anchoring of alkylamines on graphene using a very high level of theory [33]. The QM data may be benchmarked or further fit to known experimental data, where available. Note that for covalently bonded films, e.g. films of PEG on silica bonded by O–Si, which we do not discuss in detail here due to lack of self-healing, preliminary quantum mechanical calculations are also required to describe the length, strength, and polarity of the tethering pairwise covalent bond. Metal-thiolate bonds (Figure 9.12) are intermediate in strength between strong covalent bonds and physisorbed complexes, −1.7 eV per bond, which, coupled with weak metal–metal bonds, particularly in gold substrates, allows self-healing and self-correction during SAM formation.

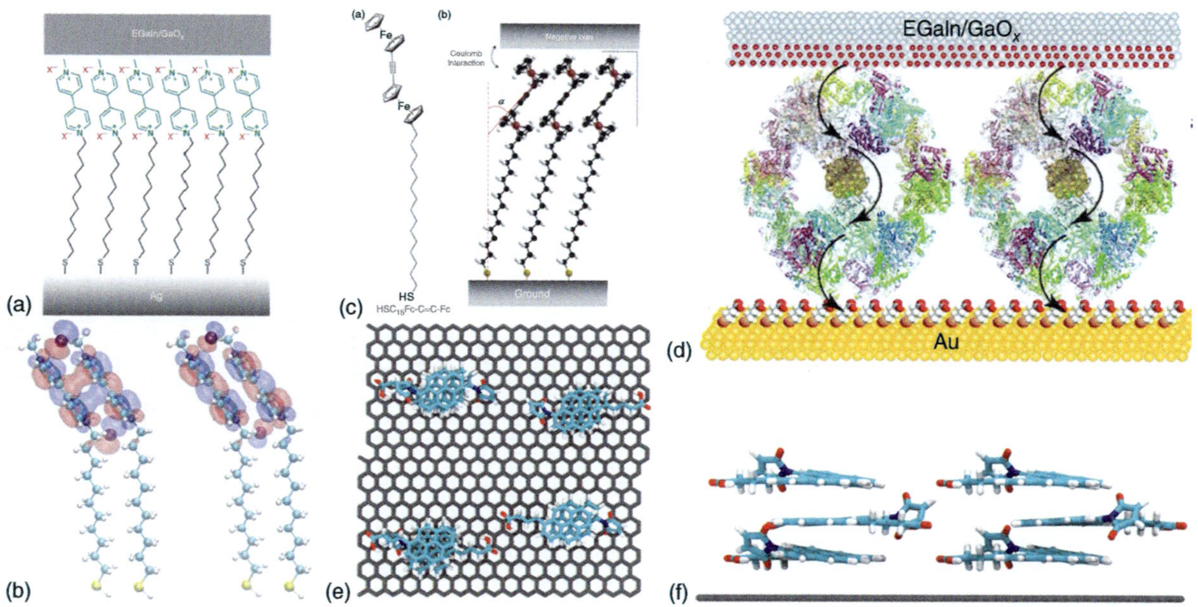

Figure 9.12 Example of various SAM-based substrate-bound complexes and electrical junctions, modeled by means of DFT and/or MD simulations (a–c, e, and f). (a, b) show an anchored monolayer of viologen units with an aromatic head group tethered to the substrate via an alkyl chain anchored through a thiolate-gold bond. Source: Han et al. [34]/American Chemical Society. (c) shows a diferrocene-based redox-active SAM. Source: Chen et al. [35]/Springer Nature. (d) shows a ferritin protein-based SAM, the size and slow dynamics of which present a challenge for molecular simulations. Source: Nipun et al. [36]/Royal Society of Chemistry. (e, f) show stacking of simple aromatic molecules on a graphene sheet. Source: Adapted from Thodkar et al. [27]. (a–d) illustrates functionalization involving the formation of a chemical bond with the substrate, while (e, f) show functionalization via reversible physisorption on the substrate.

Only force fields that have been properly parametrized using *ab initio* quantum mechanics or DFT provide a sound theoretical framework for classical MD simulations. Once carefully parametrized and benchmarked as appropriate, the MD models contain the correct molecular geometries and atomic charge distributions at the interface, and the simulation cell can then be energy minimized, thermalized, and equilibrated. It is essential to efficiently and accurately determine the endpoints of each of these stages of simulation to avoid making predictions based on poorly equilibrated and/or too-short sampling times. Figure 9.13 shows timelines of properties typically used to determine at what point systems become well-equilibrated for a small lysozyme crystal (a, b, and c) [37] and for a large protein complex (cellulosome, d–f). In both cases, the temperature is already equilibrated from the heating phase; the total energy (potential and kinetic), however, is not and takes from a few ns (small system) to half a μs (for a large system) to converge to a stable average. The root-mean-square deviation (RMSD) is another property used to determine when a surface-adsorbed protein reaches a stable conformation, which is easily identified for lysozyme. The cellulosome complex on the other hand reaches a form of dynamic equilibrium where it interchanges between the same set of conformations. This is expected for large flexible systems and is considered equilibrated. Once carefully equilibrated, production runs at ambient room temperature can provide physically meaningful predictions based on large numbers of statistically independent structural snapshots.

For the examples above, the production phase would start as soon as the three monitored properties plateau. The predictive models then can be used to guide experiments by identifying the most likely molecular stacking modes on the substrate. This also allows us predict the molecule binding energies to graphene, and supramolecular packing energies of the SAM molecules, for various potential candidate molecules in the FET gate material, e.g. using robust statistical analyses to assess the quality and predict the device performance of SAM-coated graphene FETs made with materials such as pyrene and peptide building blocks, which can complement experiments characterizing the physicochemical and electrical properties of the device. A summary of these data was presented as part of [27], but here we present a more detailed analysis of our model choice and data.

Atomistic modeling drives further experimental exploration of materials selection and device optimization, in particular for biosensing, which requires a mechanically stable and electrically coupled interface between the film and substrate. Here, we review efforts to measure and control charge transfer, doping, the role of defects/residues, type of molecules (pyrene, peptides, biomolecules, e.g. aptamers and antibodies), and interfacial capacitance toward development of effective, inexpensive bioanalytical sensors for rapid, reliable medical diagnostics, and health monitoring. The new science that is emerging is also seeding applications of gated SAM-coated graphene FETs in environmental monitoring, among other potential future applications of these promising materials.

Functionalized graphene entails a wide range of molecules, from small aromatic molecules (e.g. pyrene) to large proteins such as antibodies. The diversity of size and properties of these molecules necessitates the development and efficient utilization of equally diverse modeling techniques on high-performance computing platforms.

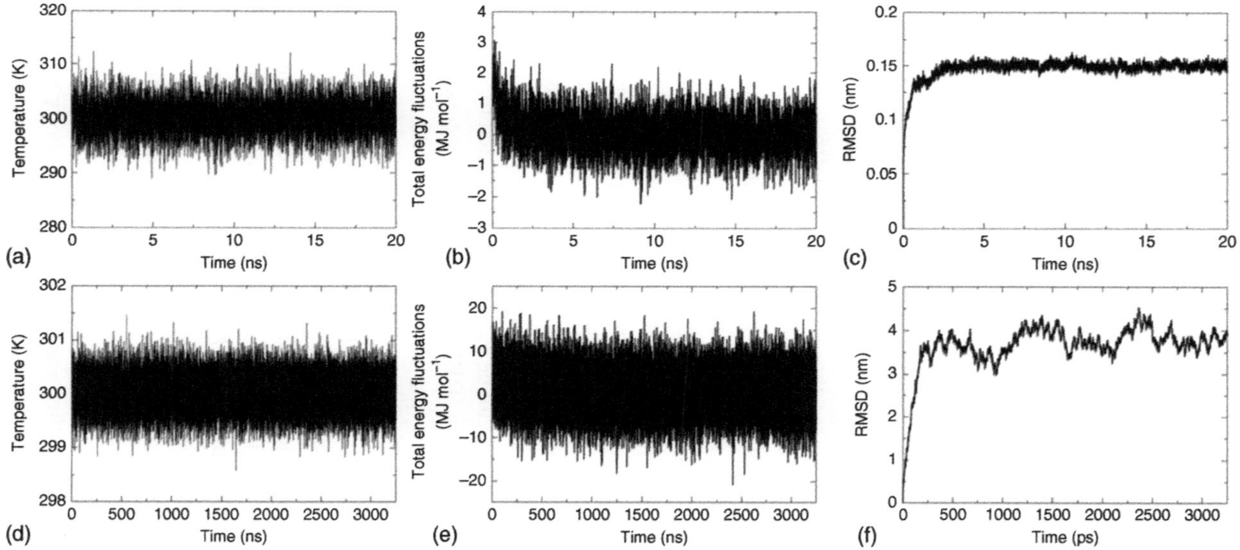

Figure 9.13 Examples of various properties used to evaluate equilibration of molecular dynamics models. (a–c) show the timelines of temperature, total energy fluctuations (potential plus kinetic), and the root-mean-square deviation (RMSD) of atoms in a lysozyme crystal (11 000 atoms). Source: Adapted with O'Donnell et al. [37]. (d–f) show the same properties for a much larger protein complex (cellulosome, almost two million atoms).

SAMs based on small aromatic molecules can be directly modeled by means of QM methods such as DFT, even for systems up to a few thousand atoms using the most efficient algorithms, e.g. linear scaling DFT. The rigidity of aromatic molecules limits their number of degrees of freedom, reducing the need for extensive sampling by Monte Carlo or MD methods. Most DFT calculations are static and performed at 0 K, making it therefore crucial to consider multiple configurations to identify which is most energetically favorable. This is particularly important for heterogeneous SAMs based on aromatic molecules as the latter tend to stack [27]. Not only may the molecules stack in different orders but also with different geometries. Even a single phenyl molecule can have different binding modes onto the graphene layer, side-on, tilted, and flat with small energy barriers separating their interconversion. The various stacking combinations and binding patterns must then be thoroughly studied to identify the ones most representative of the experimental system. This is typically achieved by comparing the enthalpy of each configuration. This approach enables the evaluation of the energetic gain of adding more layers to the stack and, therefore, the estimation of its upper limit. Note the energy here corresponds to the internal energy and not to Helmholtz free energy and neglects entropic effects. Therefore, we generally assess energy differences between different complexation geometries or poses, assuming negligible differences in configurational entropy between the alternative models, rather than rely on absolute values.

For periodic systems, i.e. simulation of 3D crystals or adsorption of ordered, large-area SAMs on 2D surfaces using Bloch's theorem, estimation of the free energy is relatively straightforward as the main source of entropy is the phonons. Tools like PHONOPY [38] provide a fast and efficient approach to computing phonons. However, for systems with a few hundred atoms and no internal symmetry, such an approach can be prohibitive, as more than a thousand single-point calculations are required. Only electronic entropy can be reasonably accounted for. Aside from energetics and geometric configurations, DFT calculations give access to the band structure density of states (DOS) and partial or projected-to-atoms density of states (PDOS), which identify the atoms and orbitals contributing the most to charge transport around the Fermi level. The work function can also be obtained by mapping the electrostatic potential across the interface, and the transport mechanism can be modeled through the Landauer formula. Other properties, including the UV–vis spectrum (electronic transitions through TDDFT), may also be predicted from the computed electronic structures. As a final essential ingredient in the computational input, dispersion interactions are critical to model van der Waals interactions in QM calculations. A correction method should be used as methods providing an *ab initio* description of the electronic correlation are computationally expensive (MP2, coupled cluster, etc.). A popular set of corrections are the so-called D3 corrections by Grimme et al. [39]. Another approach uses many-body dispersion (MBD) corrections [40–42] which aims at departing from the pairwise interaction model to consider higher orders such as three-body (note that D3 corrections already include a three-body term) and four-body interactions, up to n-body. A new Hamiltonian can be built and diagonalized. These methods are typically combined with the PBE functional, which is one of the most widely used functional for periodic systems.

An example of calculations for the SAM-functionalized graphene junctions described here would be for systems up to 500 atoms, the use of VASP [43] code to carry out the DFT calculations using the PBE functional [44] and projector augmented wave (PAW) pseudopotentials [45], Gaussian smearing and a width of 0.05 eV, and a plane wave cut-off of 800 eV.

Sufficiently dense, typically $3 \times 3 \times 1$, k-point sampling meshes are generated via an automatic Monkhorst–Pack scheme with convergence assessed by finding the point at which the computed structures stop changing significantly with respect to mesh density. For metal-containing systems, meshes of $8 \times 8 \times 1$ or higher may be required to obtain sufficiently accurate band structures, necessitating the use of highly-parallelized supercomputing to obtain solutions within a reasonable time. In general, the suitability of all the parameters is assessed by systematically increasing the detail (varying the cutoff, increasing the number of k-points, possibly also increasing the unit cell size, including the width of the vacuum layer for molecule-on-surface calculations) until the computed structures and energies stop varying. For larger systems up to a few thousand atoms, CP2K software [46] may provide a computationally expedient means of preserving full DFT accuracy, used with the PBE functional [44] Grimme D3 corrections [39], and PAW pseudopotentials [45]. CP2K uses a mix of Gaussian basis set and plane waves [47] to describe the wave function. For adsorption of purely organic molecules on graphene, double zeta basis set used together with a direct cutoff of 900 Ry and a relative cutoff of 60 Ry is generally sufficient to generate predictions that are converged with respect to the level of theory. Depending on the properties, a k-point-based approach is used, or a supercell can be used for sampling. Again, the suitability of these parameters is assessed by systematically increasing the detail (varying the cutoff, increasing the number of k-points, supercell size) until the computed structures and energies stop varying. CP2K makes it also possible to obtain the crystal dipole moment of the system, as opposed to just the intrinsic free-molecule dipole moment, which may play a role in directing the self-assembly process along particular crystallographic planes, which in turn influences the morphology and electronic properties of the material.

For larger and more complex systems such as those involving long peptides, a more extensive sampling method is required, like classical Monte Carlo (MC) or classical MD. For molecules with few internal degrees of freedom (water, aromatic molecules, planar molecules, etc.), MC simulations are an efficient approach to obtaining reliable and experimentally relevant configurations. Furthermore, provided the chemical potential of these molecules is known or can be accurately estimated, the grand canonical Monte Carlo (GCMC) method makes it possible to model and predict the adsorption isotherms. For molecules with a larger number of internal degrees of freedom (such as floppy peptides and alkyl chains with free rotations of bond and torsion angles) or for which a time trajectory is required, MD is the method of choice. The main benefit of MD is the access to timeline properties and time correlation functions, which provide insight into the timescales of the processes taking place in the system, e.g. adhesion, stacking, and nucleation. Correlation functions can also be directly compared to experimental data such as NMR. By construction, MC and MD are designed to describe the states of lower energy, which contribute the most to the components of the partition function used to describe equilibrium properties. Therefore, both are ill-suited to describe entropy and free energies. Like DFT calculations, energetic comparisons involve enthalpies, in particular variations of enthalpy between different configurations and/or binding mechanisms. Although several methods for estimating free energy have been developed, they are computationally demanding and are only used on promising configurations to reduce overall time to solution. Unlike DFT, MC, and MD do not require systematic starting configurations as more stable ones will be generated during the simulation and identified based on statistical occurrence, provided the density and ratio of each component match experimental data. Chemically bound SAMs are somewhat an exception as the molecular chains must be oriented so that the atom forming the bond with the substrate is in close proximity to the surface. A full simulation is performed with the molecules not directly tethered to the surface in pairwise covalent bonds, but with the binding atom constrained to stay within the plane at a binding distance from the substrate. The molecules are free to move in the plane of the substrate but not to completely dissociate from it. This enables the SAM to sample a wide variety of physically realistic surface structures. After this initial simulation to form a converged surface coverage, the SAM is bound via pairwise interactions to the surface before the production phase for chemically bound SAMs, e.g. strongly chemisorbed O–Si or weakly chemisorbed S–Au with bond strength of −1.7 eV conferring stability but also allowing self-correction due to weak Au–Au bonds enabling extensive adatom migration in the creation of well-packed films. Representative configurations from MD and MC simulations can then be used as starting geometries for DFT calculations for further characterization. For systems such as alkylamines on graphene, the molecules naturally remain weakly associated to the surface through van der Waals interactions and do not require any constraints/tethering.

Classical MD and MC depend on the accuracy of the force field used for the model. For peptides, CHARMM 36 m [48] and AMBER [49, 50] are the most widely used, and they come each with means to parametrize new molecules that are not part of the existing standard force field: for CHARMM, ParamChem [51, 52], which provides parameters compatible with the CHARMM general force field (CGenFF) [53], and for AMBER, AmberTools [54] which provides parameters for GAFF [55, 56]. Additionally, the partial charges for the atoms of these molecules are derived using the restrained electrostatic potential (RESP) [57] scheme based on quantum mechanical calculations with Gaussian [58], followed by charge fitting with Antechamber [56]. Typically, MD simulations are carried out using software like Gromacs 2018. Four [59] packages with a time step of 2 fs using the Leapfrog integrator [60]. Bond lengths to hydrogen are constrained using the LINCS [61] or SHAKE [62, 63] algorithm. Long-range electrostatics are treated by the particle mesh Ewald (PME) method [64]. Graphene and SAM molecules are coupled separately to an external heat bath (300 K) with the coupling time constant of 1 ps using the velocity rescaling method [65]. All systems are minimized for 100 ps and equilibrated for 500 ps in constant volume NVT ensemble. Production phases from 100 ns to 1 μs (and beyond where computationally feasible and required to capture slow dynamics, e.g. lateral protein–protein interface formation in a graphene-bound protein SAM) are then carried out in the constant-volume NVT ensemble or constant-pressure NPT ensemble to obtain statistical mechanics properties for the surface-bound complexes. The length of the production phase is based on the decorrelation time of the SAM structure. Structures are saved every 10 ps to obtain a large set of statistically independent snapshots of the system.

It is clear then that with recent advances in hardware and software capabilities, simulations are providing a valuable ally to experiments, predicting and exploring nanoscale phenomena that are very difficult to capture using experiments alone. We expect that integrated simulation/experiment codesign strategies will become commonplace in the search for new functionalized 2D materials and their rational atomic-scale engineering for technology applications at the nanobio interface.

References

1 Novoselov, K.S., Geim, A.K., Morozov, S.V. et al. (2004). Electric field effect in atomically thin carbon films. *Science (New York, N.Y.)* **306** (5696): 666–669.

2 Schwierz, F. (2010). Graphene transistors. *Nat. Nanotechnol.* **5** (7): 487–496.

3 Chaudhry, A. and Kumar, M.J. (2004). Controlling short-channel effects in deep-submicron SOI MOSFETs for improved reliability: a review. *IEEE Trans. Device Mater. Reliab.* **4** (1): 99–109.

4 Li, H., Shi, W., Song, J. et al. (2019). Chemical and biomolecule sensing with organic field-effect transistors. *Chem. Rev.* **119** (1): 3–35.

5 Zhang, A. and Lieber, C.M. (2016). Nano-bioelectronics. *Chem. Rev.* **116** (1): 215–257.

6 Huang, H., Feng, W., and Chen, Y. (2021). Two-dimensional biomaterials: material science, biological effect and biomedical engineering applications. *Chem. Soc. Rev.* **50** (20): 11381–11485.

7 Usha, S.P., Manoharan, H., Deshmukh, R. et al. (2021). Attomolar analyte sensing techniques (AttoSens): a review on a decade of progress on chemical and biosensing nanoplatforms. *Chem. Soc. Rev.* **50** (23): 13012–13089.

8 Xu, M., Liang, T., Shi, M., and Chen, H. (2013). Graphene-like two-dimensional materials. *Chem. Rev.* **113** (5): 3766–3798.

9 Ioniţă, M., Vlăsceanu, G.M., Watzlawek, A.A. et al. (2017). Graphene and functionalized graphene: extraordinary prospects for nanobiocomposite materials. *Composites Part B* **121**: 34–57.

10 Singh, V., Joung, D., Zhai, L. et al. (2011). Graphene based materials: past, present and future. *Prog. Mater Sci.* **56** (8): 1178–1271.

11 Editorial (2020). Conducting polymers forward. *Nat. Mater.* **19** (9): 921.

12 Ibanez, J.G., Rincón, M.E., Gutierrez-Granados, S. et al. (2018). Conducting polymers in the fields of energy, environmental remediation, and chemical–chiral sensors. *Chem. Rev.* **118** (9): 4731–4816.

13 Ludwig, C., Gompf, B., Petersen, J. et al. (1994). STM investigations of PTCDA and PTCDI on graphite and MoS_2. A systematic study of epitaxy and STM image contrast. *Z. Phys. B: Condens. Matter* **93** (3): 365–373.

14 Zimmermann, U. and Karl, N. (1992). Epitaxial growth of coronene and hexa-peri-benzocoronene on MoS2(0001) and graphite (0001): a LEED study of molecular size effects. *Surf. Sci.* **268** (1): 296–306.

15 Coraux, J., N'Diaye, A.T., Busse, C., and Michely, T. (2008). Structural coherency of graphene on Ir(111). *Nano Lett.* **8** (2): 565–570.

16 Martoccia, D., Willmott, P.R., Brugger, T. et al. (2008). Graphene on Ru(0001): a 25 × 25 supercell. *Phys. Rev. Lett.* **101** (12): 126102.

17 Rutter, G.M., Crain, J.N., Guisinger, N.P. et al. (2007). Scattering and interference in epitaxial graphene. *Science (New York, N.Y.)* **317** (5835): 219–222.

18 Ogawa, Y., Niu, T., Wong, S.L. et al. (2013). Self-assembly of polar phthalocyanine molecules on graphene grown by chemical vapor deposition. *J. Phys. Chem. C* **117** (42): 21849–21855.

19 Pathipati, S.R., Pavlica, E., Treossi, E. et al. (2013). Modulation of charge transport properties of reduced graphene oxide by submonolayer physisorption of an organic dye. *Org. Electron.* **14** (7): 1787–1792.

20 Malard, L.M., Pimenta, M.A., Dresselhaus, G., and Dresselhaus, M.S. (2009). Raman spectroscopy in graphene. *Phys. Rep.* **473** (5): 51–87.

21 Ferrari, A.C. and Basko, D.M. (2013). Raman spectroscopy as a versatile tool for studying the properties of graphene. *Nat. Nanotechnol.* **8** (4): 235–246.

22 Eckmann, A., Felten, A., Mishchenko, A. et al. (2012). Probing the nature of defects in graphene by Raman spectroscopy. *Nano Lett.* **12** (8): 3925–3930.

23 Cançado, L.G., Jorio, A., Ferreira, E.H.M. et al. (2011). Quantifying defects in graphene via Raman spectroscopy at different excitation energies. *Nano Lett.* **11** (8): 3190–3196.

24 Robinson, J.A., Puls, C.P., Staley, N.E. et al. (2009). Raman topography and strain uniformity of large-area epitaxial graphene. *Nano Lett.* **9** (3): 964–968.

25 Huang, S., Ling, X., Liang, L. et al. (2015). Molecular selectivity of graphene-enhanced Raman scattering. *Nano Lett.* **15** (5): 2892–2901.

26 Hinnemo, M., Zhao, J., Ahlberg, P. et al. (2017). On monolayer formation of pyrenebutyric acid on graphene. *Langmuir* **33** (15): 3588–3593.

27 Thodkar, K., Cazade, P.-A., Bergmann, F. et al. (2021). Self-assembled pyrene stacks and peptide monolayers tune the electronic properties of functionalized electrolyte-gated graphene field-effect transistors. *ACS Appl. Mater. Interfaces* **13** (7): 9134–9142.

28 Lin, W., Tian, B., Zhuang, P. et al. (2016). Graphene-based fluorescence-quenching-related Fermi level elevation and electron-concentration surge. *Nano Lett.* **16** (9): 5737–5741.

29 Kim, S.N., Kuang, Z., Slocik, J.M. et al. (2011). Preferential binding of peptides to graphene edges and planes. *J. Am. Chem. Soc.* **133** (37): 14480–14483.

30 Hayamizu, Y., So, C.R., Dag, S. et al. (2016). Bioelectronic interfaces by spontaneously organized peptides on 2D atomic single layer materials. *Sci. Rep.* **6** (1): 1–9.

31 Katoch, J., Kim, S.N., Kuang, Z. et al. (2012). Structure of a peptide adsorbed on graphene and graphite. *Nano Lett.* **12** (5): 2342–2346.

32 Kim, S.S., Kuang, Z., Ngo, Y.H. et al. (2015). Biotic–abiotic interactions: factors that influence peptide–graphene interactions. *ACS Appl. Mater. Interfaces* **7** (36): 20447–20453.

33 Song, P., Suchand Sangeeth, C.S., Thompson, D. et al. (2016). Noncovalent self-assembled monolayers on graphene as a highly stable platform for molecular tunnel junctions. *Adv. Mater.* **28** (4): 631–639.

34 Han, Y., Nickle, C., Zhang, Z. et al. (2020). Electric-field-driven dual-functional molecular switches in tunnel junctions. *Nat. Mater.* **19** (8): 843–848.

35 Chen, X., Roemer, M., Yuan, L. et al. (2017). Molecular diodes with rectification ratios exceeding 105 driven by electrostatic interactions. *Nat. Nanotechnol.* **12** (8): 797–803.

36 Gupta, N.K., Pasula, R.R., Karuppannan, S.K. et al. (2021). Switching of the mechanism of charge transport induced by phase transitions in tunnel junctions with large biomolecular cages. *J. Mater. Chem. C* **9** (33): 10768–10776.

37 O'Donnell, J., Cazade, P.-A., Guerin, S. et al. (2021). Piezoelectricity of the transmembrane protein ba_3 Cytochrome *c* oxidase. *Adv. Funct. Mater.* **31** (28): 2100884.

38 Togo, A. and Tanaka, I. (2015). First principles phonon calculations in materials science. *Scr. Mater.* **108**: 1–5.

39 Grimme, S., Antony, J., Ehrlich, S., and Krieg, H. (2010). A consistent and accurate ab initio parametrization of density functional dispersion correction (DFT-D) for the 94 elements H-Pu. *J. Chem. Phys.* **132** (15): 154104.

40 Bučko, T., Lebègue, S., Gould, T., and Ángyán, J.G. (2016). Many-body dispersion corrections for periodic systems: an efficient reciprocal space implementation. *J. Phys. Condens. Matter* **28** (4): 45201.

41 Tkatchenko, A., DiStasio, R.A. Jr., Car, R., and Scheffler, M. (2012). Accurate and efficient method for many-body van der Waals interactions. *Phys. Rev. Lett.* **108** (23): 236402.

42 Ambrosetti, A., Reilly, A.M., DiStasio, R.A. Jr., and Tkatchenko, A. (2014). Long-range correlation energy calculated from coupled atomic response functions. *J. Chem. Phys.* **140** (18): 18A508.

43 Hafner, J. (2007). Materials simulations using VASP – a quantum perspective to materials science. *Comput. Phys. Commun.* **177** (1): 6–13.

44 Perdew, J.P., Chevary, J.A., Vosko, S.H. et al. (1992). Atoms, molecules, solids, and surfaces: applications of the generalized gradient approximation for exchange and correlation. *Phys. Rev. B: Condens. Matter* **46** (11): 6671.

45 Kresse, G. and Joubert, D. (1999). From ultrasoft pseudopotentials to the projector augmented-wave method. *Phys. Rev. B: Condens. Matter* **59** (3): 1758.

46 Hutter, J., Iannuzzi, M., Schiffmann, F., and VandeVondele, J. (2014). cp2k: atomistic simulations of condensed matter systems. *Wiley Interdiscip. Rev.: Comput. Mol. Sci.* **4** (1): 15–25.

47 Lippert, B.G., Hutter, J., and Parrinello, M. (1997). A hybrid Gaussian and plane wave density functional scheme. *Mol. Phys.* **92** (3): 477–488.

48 Huang, J., Rauscher, S., Nawrocki, G. et al. (2017). CHARMM36m: an improved force field for folded and intrinsically disordered proteins. *Nat. Methods* **14** (1): 71–73.

49 Tian, C., Kasavajhala, K., Belfon, K.A.A. et al. (2020). ff19SB: amino-acid-specific protein backbone parameters trained against quantum mechanics energy surfaces in solution. *J. Chem. Theory Comput.* **16** (1): 528–552.

50 Maier, J.A., Martinez, C., Kasavajhala, K. et al. (2015). ff14SB: improving the accuracy of protein side chain and backbone parameters from ff99SB. *J. Chem. Theory Comput.* **11** (8): 3696–3713.

51 Vanommeslaeghe, K., Raman, E.P., and MacKerell, A.D. (2012). Automation of the CHARMM general force field (CGenFF) II: assignment of bonded parameters and partial atomic charges. *J. Chem. Inf. Model.* **52** (12): 3155–3168.

52 Vanommeslaeghe, K. and MacKerell, A.D. (2012). Automation of the CHARMM general force field (CGenFF) I: bond perception and atom typing. *J. Chem. Inf. Model.* **52** (12): 3144–3154.

53 Vanommeslaeghe, K., Hatcher, E., Acharya, C. et al. (2010). CHARMM general force field: a force field for drug-like molecules compatible with the CHARMM all-atom additive biological force fields. *J. Comput. Chem.* **31** (4): 671–690.

54 Case, D.A., Aktulga, H.M., Belfon, K. et al. (2021). *Amber 2021*. San Francisco: *University of California*.

55 Wang, J., Wolf, R.M., Caldwell, J.W. et al. (2004). Development and testing of a general amber force field. *J. Comput. Chem.* **25** (9): 1157–1174.

56 Wang, J., Wang, W., Kollman, P.A., and Case, D.A. (2006). Automatic atom type and bond type perception in molecular mechanical calculations. *J. Mol. Graphics Modell.* **25** (2): 247–260.

57 Bayly, C.I., Cieplak, P., Cornell, W., and Kollman, P.A. (1993). A well-behaved electrostatic potential based method using charge restraints for deriving atomic charges: the RESP model. *J. Phys. Chem.* **97** (40): 10269–10280.

58 Frisch, M.J., Trucks, G.W., Schlegel, H.B., et al. (2016). Gaussian 16 Rev. C. 01. *Wallingford CT*.

59 Van Der Spoel, D., Lindahl, E., Hess, B. et al. (2005). GROMACS: fast, flexible, and free. *J. Comput. Chem.* **26** (16): 1701–1718.

60 Hockney, R.W. (1970). The potential calculation and some applications. *Methods Comput. Phys.* 9: 136.

61 Hess, B., Bekker, H., Berendsen, H.J.C., and Fraaije, J.G.E.M. (1997). LINCS: a linear constraint solver for molecular simulations. *J. Comput. Chem.* **18** (12): 1463–1472.

62 Miyamoto, S. and Kollman, P.A. (1992). Settle: an analytical version of the SHAKE and RATTLE algorithm for rigid water models. *J. Comput. Chem.* **13** (8): 952–962.

63 Ryckaert, J.-P., Ciccotti, G., and Berendsen, H.J.C. (1977). Numerical integration of the Cartesian equations of motion of a system with constraints: molecular dynamics of n-alkanes. *J. Comput. Phys.* **23** (3): 327–341.

64 Darden, T., York, D., and Pedersen, L. (1998). Particle mesh Ewald: an $N \log(N)$ method for Ewald sums in large systems. *J. Chem. Phys.* **98** (12): 10089.

65 Bussi, G., Donadio, D., and Parrinello, M. (2007). Canonical sampling through velocity rescaling. *J. Chem. Phys.* **126** (1): 14101.

10

The Holy Grail of Surface Chemistry of CVD Graphene: Effect on Sensing of cTNI as Model Analyte

Adrien Hugo[1], Teresa Rodrigues[2], Marie-Helen Polte[2], Yann R. Leroux[3], Rabah Boukherroub[1], Wolfgang Knoll[4], and Sabine Szunerits[1,4]

[1] *University of Lille, CNRS, Centrale Lille, Universite Polytechnique Hauts-de-France, UMR 8520 - IEMN, Lille 59000, France*
[2] *AIT Austrian Institute of Technology GmbH, Biosensor Technologies, Tulln 3430, Austria*
[3] *University of Rennes, CNRS, ISCR - UMR 6226, Campus de Beaulieu, Rennes 35000, France*
[4] *Danube Private University, Laboratory for Life Sciences and Technology (LiST), Faculty of Medicine and Dentistry, 3500 Krems, Austria*

10.1 Introduction

The past few years have witnessed an overwhelming research interest in graphene for biosensing-related applications [1–7]. Researchers around the world are engaging themselves in benchmarking chemical and physical-derived graphene nanomaterials as transducer elements, with chemical vapor deposition (CVD) graphene in the form of mono, bi and multiple layers are particularly appealing for field-effect based biosensing. Field-effect transistor (FET)-based sensing is highly adapted in the context of portable electronic devices, but also as the field-effect is capacitance-related and known to be very sensitive to surface changes. Upon the occurrence of receptor/analyte binding over either the gate/dielectric or dielectric/semiconductor interface, a shift in the conductance of the semiconductor is often proportional to the analyte concentration, as required for any reliable sensor architecture. The main drawback with conventional silicon-based transistors is the high cost of silicon lithography operated in clean room, which is not in line with the construction of disposable sensors. Nevertheless, their sensitivities are high, with subthreshold swings of c. $70\,mV\,dec^{-1}$ for conventional inorganic MOSFETs and significantly lower values for nonconventional FETs such as tunneling FETs ($10–20\,mV\,dec^{-1}$) [8]. Considering low-cost alternative such as organic thin-film field effect transistors (OFETs) and electrolyte-gated OFETs are options [9]. While it remains challenging for OFETs to compete with silicon-based analogues in terms of carrier mobility and operating frequency, electrolyte-gated OFETs, where the organic semiconducting layer is in contact with the electrolyte, can operate using bias voltages below 1 V and

Graphene Field-Effect Transistors: Advanced Bioelectronic Devices for Sensing Applications,
First Edition. Edited by Omar Azzaroni and Wolfgang Knoll.

emerge as biosensing platforms [10, 11]. In an attempt to understand the influence of the Debye' length on the distance at which binding events can be detected, a successful analysis until 30× the Debye's length was reported [11].

The high carrier mobility of CVD graphene with large changes in conductivity upon target molecule binding is attracting high interest to overcome some of the limitations of silicon-based FETs (Figure 10.1a) [12]. The chemical inertness of graphene in aqueous solutions along with its wide potential window are additional important features of graphene-based field effect transmittor (GFET) sensing. Debye screening is particularly significant in GFET, where no insulation exists and zero distance can be achieved between the graphene and the analyte target (Figure 10.1b). GFET biosensing under physiological conditions free from the limitations of Debye screening is easily awaited [13, 14]. If the target generates products that freely diffuse to the graphene transducer, analyte detection is independent of Debye screening. This concept has been proven by Ono et al. via the immobilization of urease enzymes and measuring their enzymatic kinetics [15]. In this case, conversion of urea into ammonia, in femtoliter microdroplets results in a local increase in pH, an increase in graphene hole carrier density, and an increase in the GFET hole current [15].

On the way to achieve sensitive and specific GFET-based sensing, two critical challenges arise: first, integration of molecular receptors onto graphene transducers with molecular specificity toward target analyte; and second, structure-functional relationships should be based on robust surface chemistry characterization. The first challenge requires an attachment methodology that is both synthetically accessible and promotes changes in the electrical characteristics of the graphene channel when exposed to different environments and upon target analyte binding. Traditionally, receptor molecules are immobilized onto graphene transducers using two different methods (Figure 10.1c): covalent and noncovalent schemes. Functionalization of graphene by covalent grafting presents beneficial shelf-life stability often required in sensor applications [16–18]. This strategy causes important change in the electric band structure of graphene due to the rehybridization of sp^2 carbons to sp^3 carbons. The modification of the conjugation length of the delocalized carbon lattice introduces a band gap of 0.4 eV into the graphene network [19]. In most cases, the covalent functionalization of graphene is obtained using aryldiazonium chemistry, where the functionalization is not restricted to graphene edges but also happens on the basal plane of the graphene lattice. The formation of disordered polyaryl multilayers induced by the high reactivity of the generated aryl radical species during aryldiazonium chemistry represents a marked inconvenience and needs to be overcome to fully exploit this strategy. The other approach, noncovalent, is to affix the receptor to an inert aromatic ligand and utilize strong π–π stacking interactions. Relying on π–π stacking interactions of aromatic ligands with graphene has the obvious benefit that this approach retains the structures and hence the electrical properties of the graphene lattice. Pyrene-based surface ligands are the most widely employed, as the van der Waals forces between the graphene and the pyrene backbone ensure their tight binding via π–π stacking [20–22]. Pyrene molecules can be easily functionalized and are stable to a large range of reactive functional groups. Furthermore, other aromatic

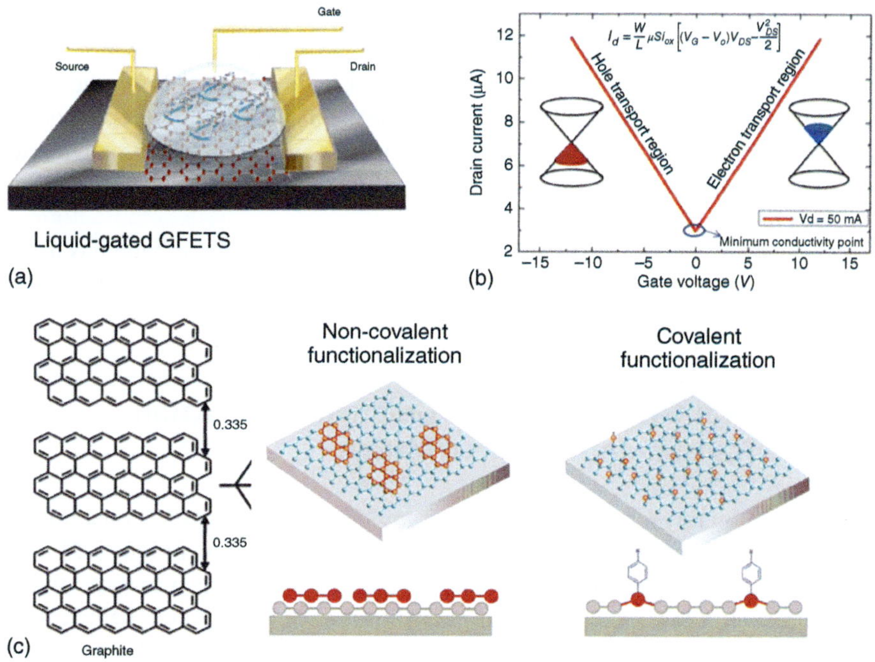

Figure 10.1 Graphene-based field effect transistor sensing. (a) Schematic of GFETs polarized in common liquid-gated modes by graphene beneath the source–drain gold contact. (b) Tuning of chemical potential in graphene by liquid gating. (c) Distance between graphene sheets and surface chemistry approaches for bioreceptor integration.

ligands can be envisaged. For instance, the adsorption of dopamine (DA) on CVD graphene theoretically studied in the framework of density functional theory [23], concluded that DA molecule parallel or quasi parallel to the surface of graphene are the most stable configurations. The adsorption of DA onto graphene implies attractive interactions of dispersive nature that compete with a Pauli repulsion between the occupied π orbitals of DA ring and π orbitals of graphene [23]. As DA is an interesting analyte to sense itself, graphene modification with DA ligands has been limited to the decoration of chemically derived graphene oxide nanosheets [24].

The second aspect, the structure-functional relationships, is related to the use of a set of surface characterization tools to allow assessment of the surface modification aspects. Among the analytical methods for surface chemistry (scanning electron microscopy [SEM], Raman spectroscopy, X-ray photoelectron spectroscopy [XPS], and atomic force microscopy [AFM]), SEM is a good indicative tool of folds and wrinkles as well as general observation of graphene's uniformity, especially after its transfer to the desired substrate, but it is not so efficient to evaluate graphene's composition and chemical purity. Raman spectroscopy can semi-quantitatively provide information on the number of layers and defects present on a graphene sheet by looking at the three main bands that characterize this material. It is also possible to evaluate graphene's doping effect, providing valuable information about graphene's quality. XPS is a valuable tool to investigate graphene's chemical composition, which in turn will be a good indication of the presence of residues and contaminations on the surface. Furthermore, it is used to confirm graphene's functionalization by the presence of certain specific peaks. Finally, AFM is a straightforward and fast technique to evaluate graphene's thickness, and thus, it gives information on the number of layers as well as the presence or absence of polymeric residues. Due to the instrument's resolution, it is not a suitable technique to analyze graphene's functionalization.

Independent of the surface modification strategy used, CVD graphene has to be first integrated into FET-based sensors. Some details about CVD graphene are given in the following.

10.2 General Overview of CVD Graphene Production, Substrate Transfer and Characterization

To obtain high-quality large-area monolayer graphene, growth by CVD on a transition metal surface, such as copper or nickel foil, is the preferred technique [25]. The CVD process is based on gaseous molecules (nitrogen and hydrogen as respectively inert and reducing gases with other carbon sources, such as methane) being introduced into a closed chamber at high temperature and pressure. The flow rates of vapors on the dosing tube are varied to produce different fluxes of molecules onto the surface. Typical carbon precursors are ethene and methane, but it is also possible to grow graphene using a variety of different larger carbon molecules [26]. Generally, the formation of graphene on transition metals through CVD follows a two-step mechanism: (i) dilution or incorporation of carbon into the metal, which is better known as dissolution, and (ii) formation of graphene through rapid cooling, also known as segregation (Figure 10.2a) [28]. Higher temperatures during the growth of graphene sheets are used to ensure increased homogeneity of nucleation events due to the enhanced diffusion of active species. In addition, the nucleation density is often increased by improving the surface of the metal with careful polishing and by lowering the proportion of carbon precursor with respect to that of the inert gas and etching gas (H_2), allowing the production of single-crystal graphene grains with sizes up to a few millimetres [26].

The substrate on which graphene is grown is not suitable for most applications. Therefore, after growth on a metal surface, the single-layer graphene must be transferred onto the desired substrate (e.g. SiO_2 wafer and gold interdigitated electrodes). This is most commonly achieved using metal etching/polymer-assisted techniques (Figure 10.2a) [29]. The transfer process is usually prone to leaving graphene with cracks, wrinkles, residues, and contaminations, and is the first source of decreasing graphene's performance. These impurities will create undesired doping that alters the electronic performance and surface composition by creating charge-scattering centers and charge gradients [30].

Therefore, the cleanness of the graphene film is of high importance to achieve homogenous functionalization of the surface and is an essential part of GFET technology. Despite the solubility of polymethyl methacrylate (PMMA), it is difficult to remove this polymer layer using common solvents owing to the strong chemical reactivity of the graphene surface, which in turn can hinder graphene performance. Thus, the removal of PMMA residues is most efficiently by thermal annealing treatments [31] as well as plasma and ion beam treatments. Furthermore, it is crucial to characterize the samples before and after removal of the polymer, ensuring proper cleaning [32]. It is important to take into account that increasing the annealing temperature might produce amorphous carbon on graphene without necessarily removing the polymer residues, and it has already been investigated the effective removal

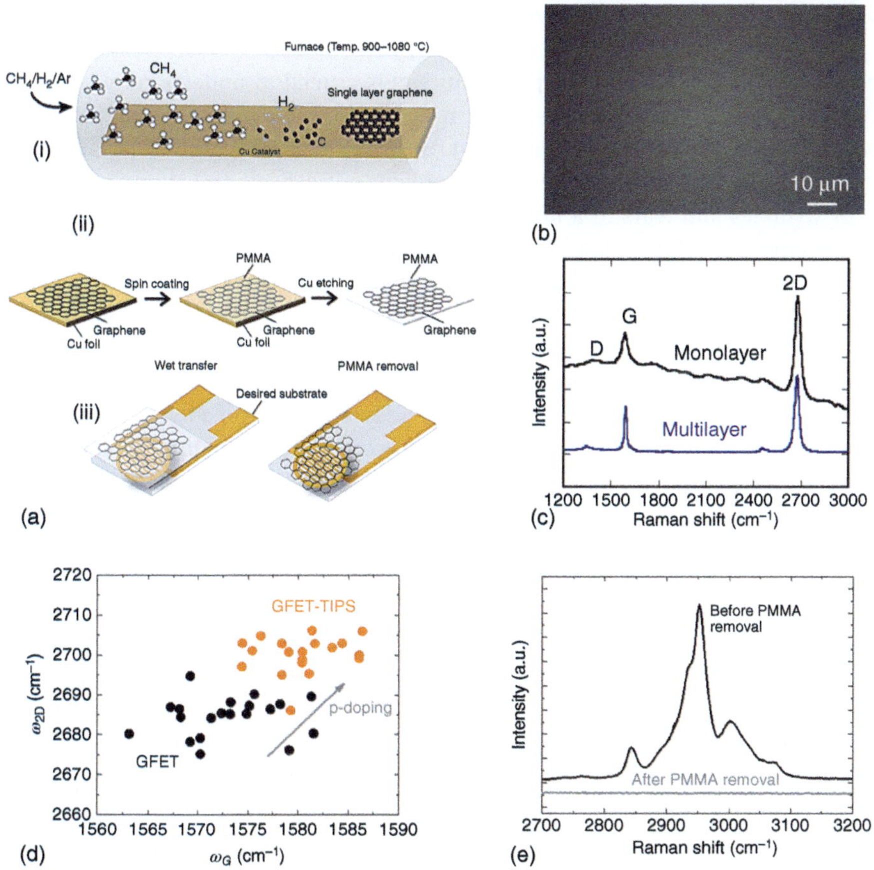

Figure 10.2 CVD graphene and its characteristics. (a) Steps in the CVD of graphene on a copper substrate and its transfer: (i) CVD process; (ii) growth mechanism of graphene via deposition; and (iii) schematic illustration of the wet transfer process of graphene via etching copper and dry transfer process of graphene via lamination/delamination. (b) SEM images of graphene grown on Cu foil using H_2/CH_4 : 10/1 sccm at 1070 °C (Si/SiO_2 300 nm). (c) Raman measurements taken at dark and bright spots of SEM. (d) Scatter plot of 2D peak position vs. G peak position for GFET (black) and chemically modified GFET. Source: Adapted from Mishyn et al. [27]. (e) Raman spectrum of PMMA coated GFET before (black) and after exposing the graphene/PMMA film to UV/ozone, followed by hot acetone washing for 30 minutes (red). Source: Reproduced with permission from Mishyn et al. [27]/Royal Society of Chemistry.

of polymer residues at different annealing temperatures and ambient conditions (e.g. H_2, Ar, mixed gases, and low-level vacuum). More recently, increased efforts have focused on achieving polymer-free transfer methods prior to graphene growth by CVD to reduce the impurities as much as possible [29]. A few examples are roll-to-roll transfer, a low-cost technique with short processing but limited to flexible substrates [33, 34] and electrochemical bubble transfer approaches have also been reported, but is restricted to conductive substrates and require voltage optimization [29].

10.3 Evaluation of Graphene Topographical Quality

Evaluation of graphene's topographical quality is crucial to understanding how it will behave as a channel in FETs – concerning its electronic properties as well as its capacity to withstand the immobilization of probes and receptors. SEM is widely employed for *ex situ* characterizations of graphene nanosheets and allows visualization of wrinkles and tears in a nondestructive manner. In addition to being a fast and convenient tool for evaluating micro- and nano-scale features, it allows for differentiate of the presence of single-layer or few-layer graphene (Figure 10.2b). Dark spots are referred to carbon clusters that are known to be nucleation centers forming in the initial stages of graphene growth. These nucleation points are attracted to impurities on Cu foil in the early stage of graphene growth due to their much lower nucleation barriers. While pure Cu does not adsorb carbon precursors, impurities on Cu foil have high saturation of carbon precursors, resulting in multi-layered carbon cluster formation. As the number of graphene layers increases, more electrons generated from the substrate will be blocked, leading to lower brightness in SEM images. Hiura et al. found a linear relationship between the SEM contrast of graphene and the number of layers in the sample, demonstrating that SEM can be easily used to identify the number of layers on a graphene substrate by calibrating the brightness [35]. Rather than the brightness, the contrast highly depends on the SEM detector used – where folding lines, for example, are invisible on a backscattering electron SEM image, they become visible when using secondary electron SEM images collected by ETD and especially through the in-lens detector (TLD) [36].

While SEM is a very useful technique to evaluate graphene topographical quality, it is not indicative of the graphene's chemical nature, and it is not adapted to probe the possible polymer residues on its surface. Raman spectroscopy can quantitatively reveal graphene's fingerprint for single- and few-layer graphene, allowing unambiguous and nondestructive analysis of graphene substrates [37, 38]. Moreover, this technique is capable to provide information on the graphene lateral size and electronic doping when analyzing its prominent bands: the D band at $1350\,cm^{-1}$, the G band at $1580\,cm^{-1}$ and the two-dimensional (2D) band (also known as G′) at $2700\,cm^{-1}$ (Figure 10.2c). The G band is due to the in-plane vibrational mode involving the sp^2-hybridized carbon atoms in the graphene lattice, corresponding to the

E_{2g} phonon at the Brillouin zone center (Γ). It can, thus, be used to estimate the number of layers presents in a graphene sample – the peak position is expected to downshift to $\approx$1581 cm^{-1} in few-layer graphene when compared to $\approx$1587 cm^{-1} in monolayer graphene (Figure 10.2c). On the other hand, the 2D band is the second order of zone-boundary phonons due to two phonon lattice vibrational processes. Thus, the 2D peak position and shape, and the I_{2D}/I_G ratio (where I refers to intensity), are commonly used to estimate the number of layers and doping effects (Figure 10.2d) in graphene samples. The D peak is due to the breathing modes of six-atom rings and requires a defect for its activation by a double resonance process. Considering all the information that Raman spectroscopy can bring, it is also used as an indicative tool of the presence of impurities on graphene, where it was already used to characterize CVD-grown graphene covered and uncovered with PMMA and after thermal annealing treatments (Figure 10.2e).

In addition to Raman spectroscopy, XPS is the most adapted approach to provide information about the chemical nature of graphene by identifying the chemical states on a surface as well as the elemental composition, empirical formula, and electronic states of the elements present. The relevant bands in the XPS spectra of graphene are the C 1s and O 1s. The C 1s main peak at $\approx$284 eV is related to the carbon bonds with graphitic contribution of C–C sp^2 (284.6 eV), next to impurities or defects such as C–OH (285.7 eV), C–O (286.6 eV), C=O (287.5 eV), C–C=O (288.7) and O–C=O (290 eV) (Figure 10.3a). The presence of bands at around 285.0 eV is related to sp^3-hybridized carbon atoms. XPS is a powerful tool to not only evaluate graphene's chemical structure but also indicate the effectiveness of PMMA removal. In this line, XPS measurements were performed on pristine graphene and on graphene transferred onto SiO_2/Si substrates before thermal annealing, but after attempting to remove PMMA with acetone [31]. The sp^2-hybridized carbon peak at $\approx$284.6 eV and the sp^3-hybridized carbon peak at $\approx$285.3 eV are clearly visible in all cases (Figure 10.3a). The transferred graphene shows additional peaks associated with PMMA, notably the carbon-based polymer backbone at $\approx$285.7 eV, as well as carboxyl groups at 289.0 eV and methoxy functional group at $\approx$286.8 eV. Thermal annealing is crucial for the effective removal of PMMA residues as seen in the reduction in the carbon bonds corresponding to PMMA after annealing at 400 °C and 600 °C. H_2-annealing is clearly more effective than Ar-annealing at 400 °C with PMMA free graphene observed in both cases after 600 °C annealing.

AFM has also been part of graphene research due to its ability to probe individual graphene flakes and provide nanoscale details at the atomic level. It is mostly employed to check the uniformity of graphene and distinguish few- from single-layer graphene by evaluation of the film thickness. Topographic maps of graphene flakes prepared on silicon dioxide can reveal the expected 300 pm graphene steps between successive layers (Figure 10.3b). AFM is also adapted to indicate if the removal of PMMA, commonly used as supporting substrate during the polymer transfer process, has been efficient (Figure 10.3b) [30].

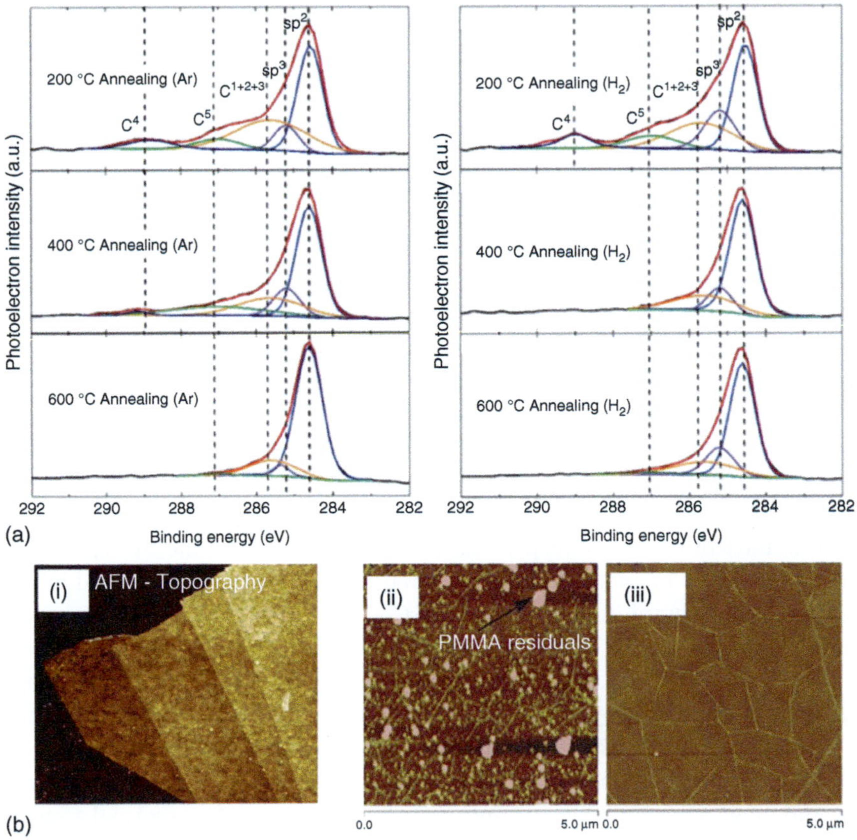

Figure 10.3 XPS and AFM characterization of graphene nanosheets. (a) XPS C 1s analysis of transferred graphene after argon and H_2 gas annealing. Source: Reprint permission form Ahn et al. [31]. (b) AFM images of the graphene surface: (i) mono, bi, and multilayer graphene (https://www.bruker.com/fr/products-and-solutions/microscopes/materials-afm/afm-for-graphene.html). (ii) AFM of graphene treated with acetone for two hours to remove PMMA. (iii) Graphene treated with acetone vapor, two minutes acetone immersion, and three hours of annealing. Source: Reproduced from Chen et al. [30], John Wiley & Sons/ CC BY 4.0.

10.4 CVD Graphene for FET-Based Sensing

Carbon nanotube (CNT) FETs have been considered as the most promising candidate for label-free biosensing applications around 2008. Indeed, these nanometric pseudo-1D cylinders made of sp^2-hybridized carbon exhibit exceptional properties, including the ability to carry a current density up to $10^9\,A\,cm^{-2}$ [39]. Using CNTs with their excellent electrical conductivity and high-specific surface area (surface/volume ratio), as a transducing material was a large step further for highly sensitive and label-free detection of biological species, since charged species binding at the CNT surface can induce large electrical carrier density modulation. However, the chirality and length heterogeneity inherent to CNT fabrication and the subsequent variability of their electronic properties have been the main limitations for the mass production of robust CNT-based FET devices [40].

Graphene, as a 2D counterpart of CNT, quickly attracted attention in the same period due to extremely high charge carrier mobility reaching $10^4 cm^2\ V^{-1}\ s^{-1}$ at room temperature [41]. This planar carbon allotrope can be synthetized over large area by CVD and further transferred onto any desired substrate [42], enabling the fabrication of a variety of devices, such as FETs, using well known large-scale micro/nano-fabrication techniques.

Back-gated devices are the simplest device configuration, allowing to electrostatically modulate the charge carrier density in graphene. Graphene constitutes the transistor channel in this configuration and is contacted by two metallic contacts, namely drain and source. Usually, a Si(p+++)/SiO_2 is used as substrate, the doped Si bulk serving as the transistor's gate, separated from the graphene channel by a SiO_2 dielectric layer (Figure 10.4a). GFETs present a unique V-shaped current-voltage transfer characteristic (Figure 10.4b), resulting from its unique electronic band structure: [46] there is no bandgap between the conduction and valence bands whose intersection also coincides with a zero density of states. Consequently, when a large positive (negative) gate voltage is applied, the Fermi level in graphene is

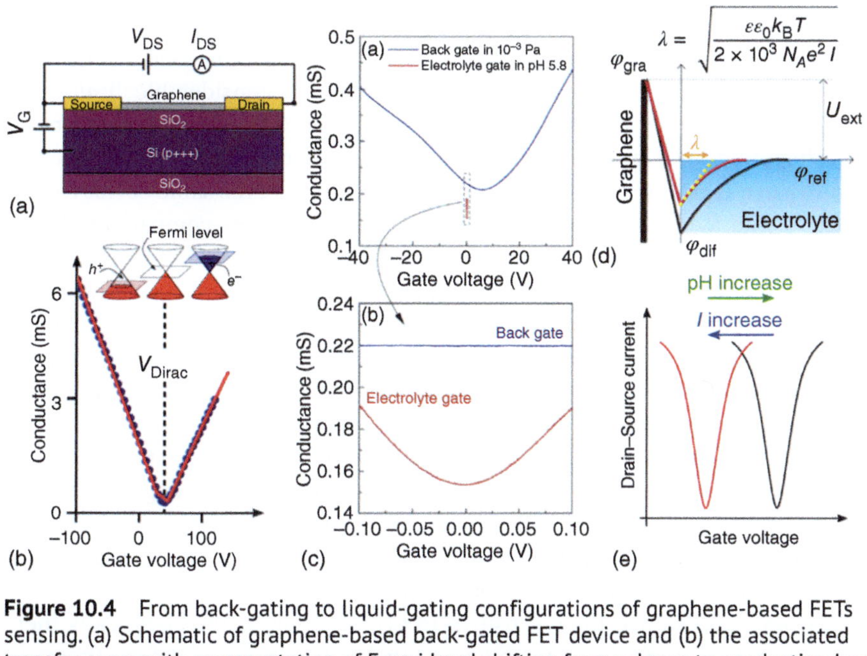

Figure 10.4 From back-gating to liquid-gating configurations of graphene-based FETs sensing. (a) Schematic of graphene-based back-gated FET device and (b) the associated transfer curve with representation of Fermi level shifting from valence to conduction bands. Source: Adapted from Novoselov et al. [43]. (c) Comparison of the transfer curves of back-gated (top) and liquid-gated GFET (bottom). Source: Reproduced with permission from Ohno et al. [44]/American Chemical Society. (d) Schematic of the electrical potential drop at the graphene-electrolyte interface in the liquid-gated configuration and the associated Debye length λ. Source: Adapted from Hess et al. [45]. (e) Schematic of ionic strength and pH induced shift of the transfer curve.

shifted into the conduction (valence) band and the electrical conductivity in the graphene channel is governed by electrons (holes), the point of minimal conductivity corresponding to the so-called "Dirac point." The potential needed to reach the Dirac point (V_{Dirac}) is typically used to monitor graphene chemical doping by its environment, while the slope of each branch of the I/V curve is reflecting the mobility of the associated predominant charge carrier. Mobility ultimately serves as the final device performance metric since mobility can be drastically reduced during the fabrication process due to polymer residues [47], charges impurities trapped at the graphene/substrate interface [48], substrate roughness, mechanical strain, and graphene lattice defects [49].

Since graphene is atomically thin, bulk charge distribution effects that might screen any alteration of the π-electron cloud at the surface are absent, making graphene extremely sensitive to its physico-chemical environment. Combined with exceptional charge carrier mobility, graphene naturally became a great candidate as a superior transducing material, in particular for biosensing applications. The detection principle lies in charged biological species specifically binding to receptors immobilized on the graphene surface and inducing a shift of V_{Dirac} or a modulation of the drain–source current. However, back-gated FET devices require application of a large electrical potential (up to 100 V) to the gate in order to obtain a sufficient modulation of the drain–source current and monitoring V_{Dirac}, due to the capacitance of the dielectric used (typically, for a 290 nm layer of thermally oxidized SiO_2, the capacitance is in the order of $10\,nF\,cm^{-2}$). For power consumption and integration purposes, this operation condition is therefore a considerable limitation. Instead of using thinner dielectric materials, with higher the risk of lower breakdown voltage and complex fabrication processes, one can turn to "liquid/electrolyte/solution"-gated GFETs (Figure 10.1a). In this configuration, the same device can be operated within a considerably narrower gate voltage window compared to back-gated devices (<1 V) (Figure 10.4c) by using a reference electrode immersed in an electrolyte in contact with the graphene channel. As a response to the excess of charges at the graphene surface, ions in the electrolyte accumulate at the graphene–electrolyte interface to maintain the system electroneutrality, creating a highly capacitive "electrochemical double-layer" (EDL) (typically in the order of $\mu F\,cm^{-2}$). The large electrical potential gradient in the EDL is at the core of the gating mechanism and the source of the enhanced electrical sensitivity of liquid-gated GFETs. In liquid environment, GFETs have shown to be sensitive to ionic strength, as higher ionic concentrations lead to smaller potential drop at the graphene–electrolyte interface [45]. Thus a lower gate potential is necessary to reach V_{Dirac} and the I/V curve is shifted toward negative potential (Figure 10.4d,e).

GFETs have also shown to be sensitive to pH variation as the chemically inert carbon network of graphene inevitably presents locally structural defects (point or line defects, graphene edges) [50] rich in (de)protonable oxygen groups [51]. The excess of charges from the presence of O^-/OH_2^+ groups at the graphene surface at high/low pH typically induces graphene n-doping (p-doping) (Figure 10.4e). Such pH sensitivity is therefore largely dependent on graphene source, but also on the quality of the device fabrication process. Both ionic strength and pH factors need to be controlled in biosensing experiments to avoid any parasitic signals unrelated to the detection of the target analyte. The role of ionic strength is particularly important as it governs the depth of the electrical potential exponential decay away from the surface through a decaying constant called the Debye-length. As the EDL thickness varies inversely to the square root of the ionic strength, the higher the ionic strength is, the shorter the Debye-length. Any charge located further than a few Debye-lengths away from the sensor surface will be screened, making the label-free and real-time detection of biological species in physiological media such as 1X phosphate-buffered saline (PBS), typically showing a screening length below 1 nm, challenging. Hence the functionalization of graphene with receptors specifically targeting and immobilizing the analyte of interest close to the surface in a robust manner is essential for the development of highly sensitive GFET biosensors.

10.4.1 Diazonium Chemistry on CVD Graphene

Functionalization of graphene with organic molecules has boomed over the past few years. Diazonium salt chemistry being a valuable approach as it allows covalent attachment under mild conditions of different surface functional groups [52–55]. Aryl diazonium salt chemistry (Figure 10.5a) dates back to the work of Pinson and coworkers in 1992 [56] and has since then been widely reported due to the ease of preparation, rapid reaction, and strong aryl-substrate covalent binding. Functionalization with aryl diazonium salts results also in the opening of the bandgap and shifting of Fermi level, both desirable in fabrication of electronic devices [16]. Another advantage of the grafting of aryl diazonium moieties is that they are able to carry a wide variety of functional groups and thus capable to improve the surface chemistry [57]. Aryl diazonium molecules react with graphene via an electron transfer mechanism, where the delocalized electrons of the graphene lattices are transferred to diazonium cations ($^{+}N_2$–C), resulting in the formation of aryl radicals and the release or N_2 (Figure 10.4a). The formed aryl radicals can react with a carbon atom of the graphene sheet and gives rise to a covalent bond. As the reaction proceeds, the radicals can also bind to the aryl groups already immobilized on the surface, forming multilayers. This concept is widely discussed in the literature for the integration of functional groups onto GFETs [18, 27, 55, 58]. It has been reported by the team of Strano and coworkers that aryl diazonium chemistry on graphene is controlled by the underlying substrate with a stark difference in rate of electron-transfer reactions proceeding rapidly for graphene supported on SiO_2 and Al_2O_3, but negligible on alkyl-terminated and hexagonal boron nitride (hBN) surfaces, as shown by Raman spectroscopy [55]. The effect of time (0–5–10–30–185 minutes) on chemical functionalization of GFET with 4-nitrobphenyl groups was determined by Tour and coworkers, and revealed a clear decrease in conductivity after functionalization, with a drop in current twofold current drop within the first five minutes of reaction (Figure 10.5b) [18].

Electrochemical functionalization can be used to overcome substrate related limitations with diazonium chemistry [59]. For instance, the Strano group grounded the source and drain electrodes of a GFET electrically and applied a positive voltage to

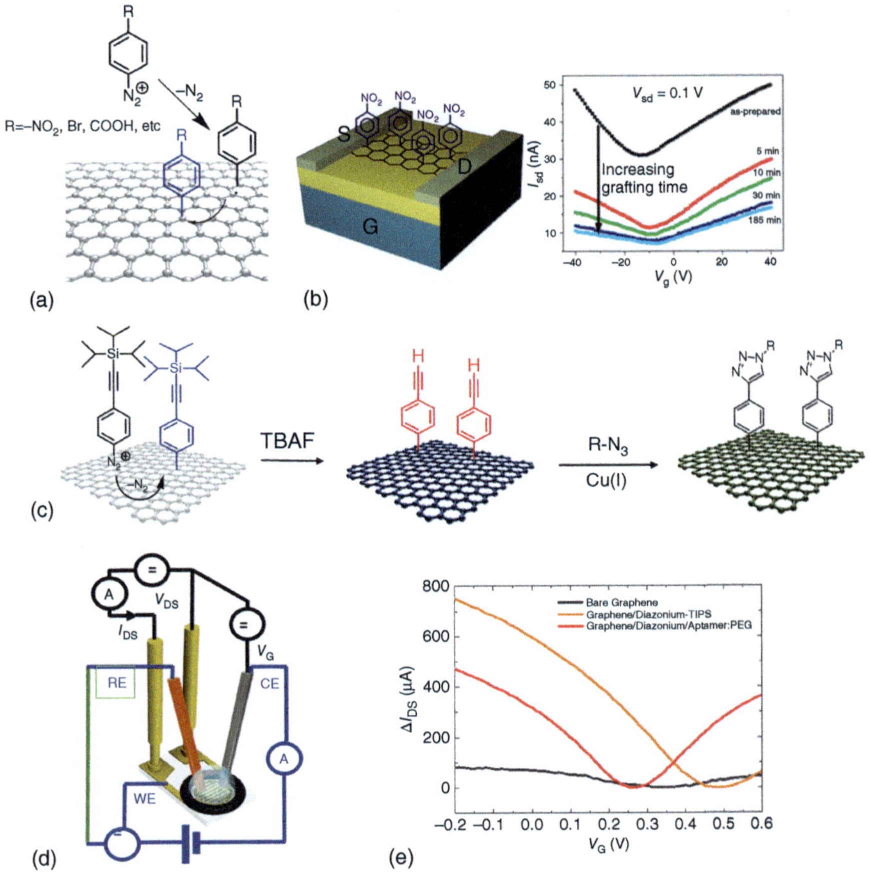

Figure 10.5 Covalent graphene functionalization using diazonium chemistry. (a) Schematic of covalent modification of graphene with aryl diazonium molecules. (b) Chemical functionalization by diazonium salts of graphene nanoribbons. The kinetics of the functionalization can be monitored by probing the electrical properties of GNRs, with a reaction with 4-nitrobenzene diazonium tetrafluoroborate is reasonably fast, such that >60% of the maximum change in the electrical properties was observed after less than five minutes of grafting at room temperature. Source: Reprint with permission from Sinitskii et al. [18]. (c) Formation of ethynyl-terminated GFET via deprotection of GFET-TIPS followed by immobilization of azide-terminated receptors through the Cu(I) "click" chemistry approach. (d) Set-up allowing electrochemical grating and electrical measurements on same device using graphene-coated interdigitated electrodes as support. (e) $I_{DS}V_G$ of GFET before and after TIPS anchoring and deprotection.

the solution using a tungsten probe for 30 seconds [59]. The applied electrical field creates an electrical double layer close to the graphene surface. This raises the Fermi level of graphene and increases the rate of electron transfer. The positive applied voltage from the tungsten wire concentrates the positively charged diazonium cations within the double layer, which accelerate the grafting reaction. The calculated field-effect hole mobilities decreased while an increase in the Raman I_D/I_G values evidencing clearly that chemical functionalization happened.

We showed lately the interest of electrochemically driven diazonium chemistry on GFET devices for sensing [60]. We opted for the electrochemical grafting of a triisopropylsilyl (TIPS) protected ethynylphenyl diazonium salt (Figure 10.5c), 4-[(triisopropylsilyl)ethynyl]benzene diazonium tetrafluoroborate (TIPS-Eth-ArN_2^+), known for its surface limiting grafting capabilities due to the presence of the bulky TIPS group [61, 62]. Electrochemical grafting was achieved by using the graphene channel of the GFET as a working electrode thanks to the liquid-gated configuration of our system (Figure 10.5d) and performing a cyclic voltammetric scan. Electrochemical reduction of TIPS-Eth-ArN_2^+ occurs at −0.2 V vs. Ag/AgCl with film grafting completed in one cycle. Different from other aryl diazonium salts, the bulky TIPS-Eth-ArN_2^+ precursor limits the surface modification to a functional monolayer and prevents the formation of multilayer branched structures, as observed for 4-nitrophenyl diazonium salts [58]. This controlled covalent functionalization of the graphene channel results in a charge mobility of the GFET of 1739 ± 376 and $1698 \pm 536\,cm^2\,V^{-1}\,s^{-1}$ for the holes and electrons, respectively, as determined from 10 different devices. A positive shift of the Dirac point is observed and correlated with p-doping effects. The preservation of such high charge mobility is exceptional and in contrast to other reports where charge mobilities of graphene around $600\,cm^2\,V^{-1}\,s^{-1}$ were reported [63]. We believe this observation is device-specific and related to the diazonium salt used. Comparable experiments with 4-nitrobenzene diazonium tetrafluoroborate as well as other diazonium ligands did not result in comparable high charge mobilities [27]. The high charge mobilities were even preserved after TIPS deprotection and integration of cardiac troponin I (cTnI) aptamers via click chemistry (Figure 10.5e). Independent of the underlying mechanism for the enhanced electrical performance, the aptamer-modified GFET was investigated in the following for its possibility to sense cTNI. This opens the field for receptor-based GFET sensing, as exemplified by electrical sensing of biomarkers such as cTnI, via integration of a cTnI-specific aptamer onto GFET using the described TIPS-based surface modification strategy (Figure 10.5c). To insure working with an acceptable anti-fouling device, a poly(ethylene glycol) (PEG) function was clicked onto the ethynyl-terminated GFET at the same time as the cTnI-aptamer (Figure 10.6a) [60].

From Figure 10.6b, one can clearly see that the cTnI aptamer-modified GFET can convert the binding interaction of different cTnI concentrations into an electrical signal change. By performing a linear fit to the measurement data in the linear range, a limit of detection (LoD) equal to $3.34\,pg\,ml^{-1}$ ($S/N = 3$) was determined, with a dynamic response covering the concentration ranges relevant for revealing the magnitude of eventual myocardial injury in serum (Figure 10.6b). The aptamer GFET clearly distinguishes between low ($<15\,pg\,ml^{-1}$), medium ($15\,pg\,ml^{-1} >$ c-TnI $< 500\,ml^{-1}$), and high ($>500\,ml^{-1}$) cTnI levels in serum of patients and is one of the first examples of a GFET sensor with real clinical potential application. This approach results in addition of a GFET selective for cTnI (Figure 10.6c) as current changes recorded using another cardiac troponin protein, BNP-32, are in the opposite direction compared to cTnI. This different behavior might give an indication of the way cTnI interacts with the aptamer. A current decrease is consistent with aptamer

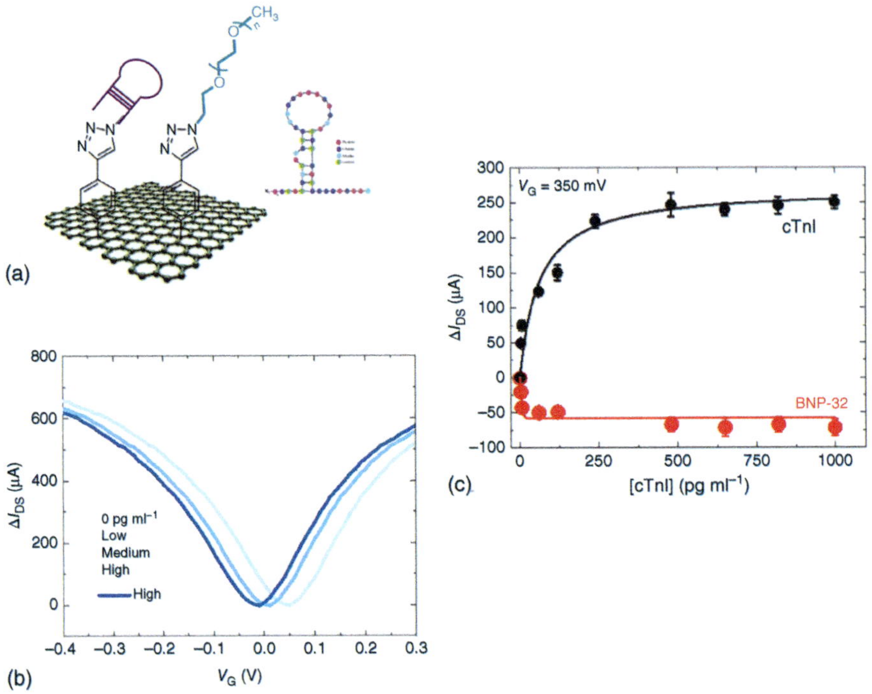

Figure 10.6 Cardiac troponin level sensing using GFET modified with cTnI aptamer using diazonium based chemistry. (a) GFET-based sensing of cTnI using an aptamer-modified GFET sensor. (b) Graphene transfer characteristics after exposure to patient samples with different cTnI levels: low (cTnI <15 pg ml^{-1}), medium (15 pg ml^{-1} > c-TnI < 500 ml^{-1}), high troponin concentration (cTnI >500 pg ml^{-1}), (c) ΔI_{DS} for different cTnI concentrations (black) as well as BNP-32 (red) at a fixed gate voltage V_G = 350 mV.

reorientation toward the sensing channel, as the increase of negatively charged backbones is depleting the graphene channel electrostatically, while aptamer stem-loop reorientation away from the graphene channel leads to current increase. How does this sensor compare to a GFET sensor where the same cTnI aptamer is integrated via a noncovalent surface modification strategy?

10.4.2 Pyrene Chemistry on CVD Graphene

Commonly used as florescent probe within the context of spectroscopically based applications, pyrene and its differently terminated variations (Figure 10.7a) are becoming increasingly important as a straightforward way to modify graphene in a noncovalent manner. Relying on π–π stacking, rather than rehybridization of the carbon atoms, this approach also takes advantage of the aromaticity of the pyrene units, but instead of utilizing the fluorescent properties of the molecule, it exploits the interaction of the π-orbital system of the pyrene molecule core with graphene. This mimics the same principle that also naturally occurs for the multilayer graphene sheets that form graphite and are solely stable due to the accumulative effect of van der Waals forces. Based on this, pyrene moieties can be seen as small elements employed to conveniently modify the graphene surface while at the same time preserving the structural and electrical properties of graphene [2, 20, 21]. Furthermore, tripodal pyrene binders [64, 65] have been proposed (Figure 10.7a) with the belief that they can provide higher stability over time. The seemingly most popular pyrene derivative is the 1-pyrenebutanoic acid succinimidyl ester (PBASE) [7]. This pyrene linker features an *N*-hydroxysuccinimide (NHS) ester that is conveniently reactive with primary amines, forming stable amide bonds. This is of outmost practicability considering that primary amines are one of the most abundant functional groups within proteins, peptides, and many other macromolecules. As lately stated by Wang et al. [66] the conditions used for modification of graphene with PBASE are crucial for maximizing device performance. They optimized this process by incubating graphene in dry nitrophenyl dimethylformamide (DMF) solutions of 50 mM PBASE for one to eight hours at room temperature. A time of four hours seemed optimal, as longer time resulted in aggregation of pyrene groups onto the monolayer graphene forming a structure similar to that of bilayer graphene.

The downside of PBASE is the fast occurrence of hydrolysis of its ester group in aqueous media, which might limit its conjugation with bioreceptors. Indeed, we have demonstrated lately the limitations of PBASE for achieving high ligand loading [22]. According to AFM images of PBASE-modified GFETs for increased time spans (Figure 10.7b); the deposition of PBASE molecules is clearly visible forming a dense pyrene layer after 1 hour, and a multilayer after 24 hours. To estimate the amount of receptor ligands which can be integrated into PBASE modified GFET for different incubation times, posttreatment of the ester groups with aminoferrocene was performed readily after (Figure 10.7c). From the cyclic voltammogram of the ferrocene-modified GFET, the amount of surface linked ferrocene was evaluated using $\Gamma = Q/nFA$, where Q is the passed charge, n the number of exchanged electrons ($n = 1$), F the Faraday constant, and A the electroactive surface area of the electrode. In the case of 4 hours incubation, a surface coverage value of $6.44 \pm 1.52 \times 10^{-11}$ mol cm^{-2} was extracted while no ferrocene ligands were detected after 24 hours of incubation. Considering the maximum theoretical surface coverage density of PBASE as 4.76×10^{-10} mol cm^{-2} [67], the surface coverage of ferrocene corresponds to less than a quarter after 4 hours, and nothing after 24 hours due to the labile character of the ester group. Using PBA as surface ligand and linking aminoferrocene to the modified GFET/PBA surface shows that the surface coverage of ferrocene was largely increased to $4.14 \pm 0.34 \times 10^{-10}$ mol cm^{-2} after 2 hours, reaching $4.17 \pm 0.51 \times 10^{-10}$ mol cm^{-2} after 12 hours. Our group used 1-PBA recently for the covalent grafting of an aptamer specific to the human papillomavirus, HPV-16 E7 protein [20].

Different to PBA and PBASE, modification of the GFET with PMAL (Figure 10.7a) results in a chemical linkage of thiolated ferrocene derivative with coverage as high as $4.18 \pm 0.25 \times 10^{-10}$ mol cm^{-2} after only two hours. While longer PMAL incubation times did not show increased ferrocene loading, nm-thick PMAL films were evidenced by AFM (Figure 10.7d), which can partially block electron transfer.

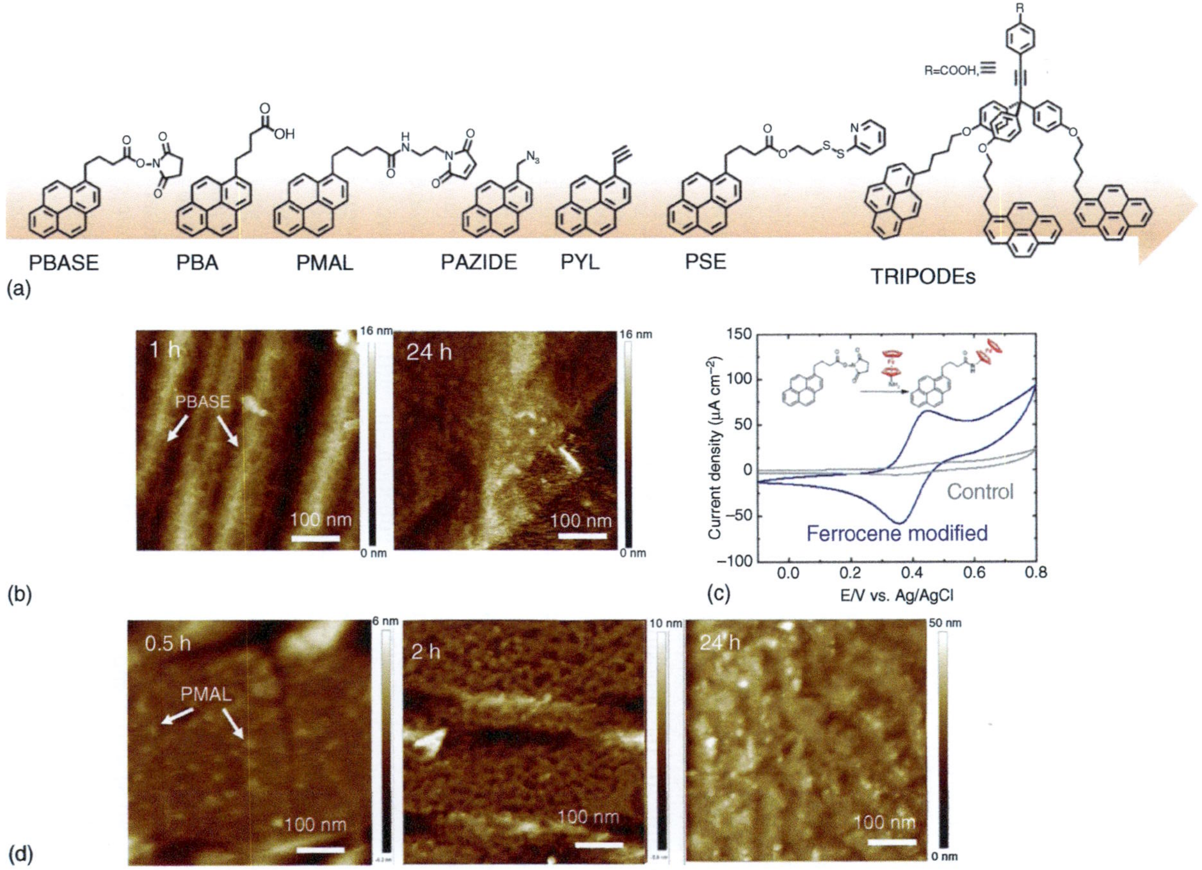

Figure 10.7 Pyrene as a versatile ligand for GFET. (a) Different pyrene ligands reported: 1-pyrenebutanoic acid succinimidyl ester (PBASE), 1-pyrenebutyric acid (PBA), pyrene-propanoic amide ethylmaleimide (PMAL), 1-(azidomethyl)-pyrene (PAZIDE), 1-ethynylpyrene (PYL), 1-pyrene butanoic acid pyridyl disulphide ethyl ester (PSE), pyrene tripode (TRIPODE-YL). (b) AFM images of GFETs immersed for 1 and 24 hours into PBASE (5 mM, DMF) solution. (c) Cyclic voltammogram of ferrocene modified GFET (four hours) in acetonitrile/NBu_4PF_6 (0.1 M) (blue), control (gray): unmodified GFET interface immersed in aminoferrocene, scan rate = 100 mV s^{-1}; (inset: reaction of PBASE with aminoferrocene). (d) AFM images of GFETs immersed for 0.5–2–24 hours in PMAL (5 mM, DMF) solutions.

The importance of the pyrene immersion time on GFET device operation for sensing related aspects was demonstrated using cTnI as model biomarker. Integration of thiol-modified cTnI aptamers was achieved using thiol-maleimide chemistry (Figure 10.8a). The transfer characteristics (Figure 10.8b) of the cTnI aptamer-modified GFET during exposure to increasing cTnI concentrations in 0.01×PBS/1 mM $MgCl_2$ buffer showed a sharp increase of the Dirac point shift toward negative gate voltage with increasing cTnI (Figure 10.8c) and a LoD = 22±1 pg ml^{-1} was determined. Compared to the GFET discussed above (covalent functionalization using diazonium chemistry (Figure 10.6), the LoD increased by seven times but still comparable with the additional advantage of using considerably simpler fabrication processes, without the need of complex chemistry methods for the functionalization of the sensing area. The sensor remains adequate for differentiating between low-medium- and high levels of cTnI in serum of patients and to give guidelines of heart attack related issues. With a view to showing how the optimization of the density of active functionality through PMAL incubation time can affect the biosensor's final performance, a GFET device was functionalized with PMAL for overnight (17 hours). As seen in Figure 10.8b, the device response for a sensor modified for 17 h falls below the signal obtained for the GFET device incubated for two hours with PMAL, demonstrating that the linker incubation conditions are of primary importance to promote a robust and functional biorecognition layer on GFET for optimizing the sensor performance.

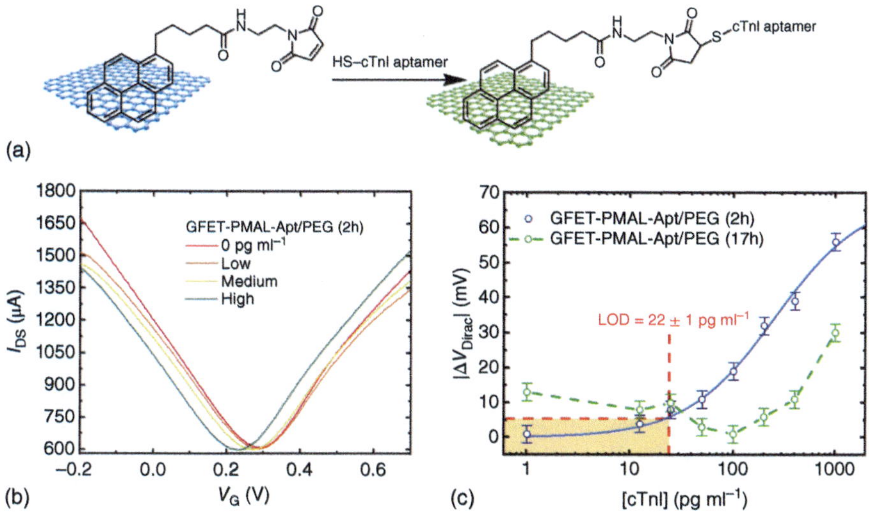

Figure 10.8 Cardiac troponin level sensing using GFET modified with cTnI aptamer using PMAL as surface linker. (a) Linkage of thiol-terminated cTnI aptamer to PMAL modified GFET interface. (b) Absolute Dirac point variations ($\Delta V_{Dirac} = V_{Dirac}$ ([cTnI]) − V_{Dirac} (Buffer)) as a function of cTnI concentration using logarithmic scale for 2 hours GFET modified 2 and 17 hours with PMAL. (c) Transfer characteristics after stabilization for low (cTnI <15 pg ml^{-1}), medium (15 pg ml^{-1} > cTnI < 500 pg ml^{-1}), and high troponin concentration (cTnI > 500 pg ml^{-1}) in 0.01× PBS/$MgCl_2$ (1 mM) (pH 7.4) without washing steps, applied V_{DS} = 50 mV.

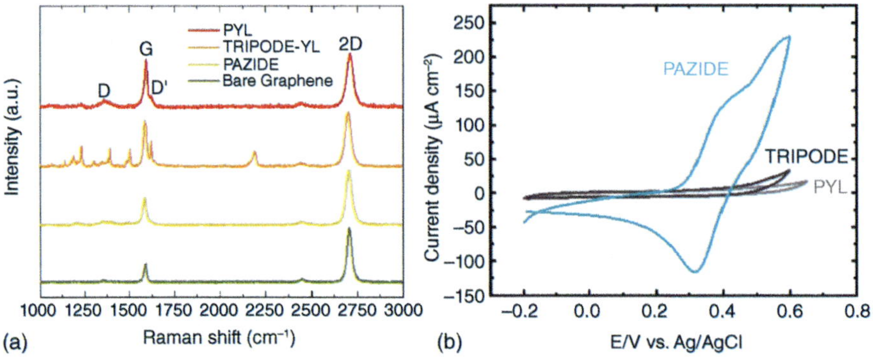

Figure 10.9 Ethynyl- and azide-terminated pyrene surface ligands on GFETs: (a) Raman spectra of GFET interfaces modified for two hours with different pyrene ligands: 1-(azidomethyl)-pyrene (PAZIDE), 1-ethynylpyrene (PYL), and TRIPODE-ethynyl ligand (5 mM, DMF). (b) Cyclic voltammograms of ferrocene modified GFET in acetonitrile/NBu_4PF_6 (0.1 M): PAZIDE modified with (tetradec-13-ynyl)ferrocene (blue), PYL and TRIPODE-YL modified with N_3-ferrocene (grey and black), scan rate = $100\,mV\,s^{-1}$.

While PMAL modification gave well-performing GFET sensors, the last question we asked ourselves was whether using a pyrene ligand with an alkynyl group (Figure 10.7a) would allow to "click" azido-terminated cTnI aptamer in a similar manner to that performed for GFET modified by TIPS-Eth-ArN_2^+ (Figure 10.6a)? The Raman plot of 1-ethynylpyrene modified graphene (PYL) (two hours, 5 mM DMF) (Figure 10.9a) revealed several Raman bands of different intensities between 1140 and 1590 cm^{-1} and splitting of the G with a G′ component. Using an ethynyl-terminated tripode, the Raman signals between 1140 and 1590 cm^{-1} are less pronounced than for PYL but still visible. Clicking on azido-terminated ferrocene derivatives at these modified interfaces shows no integration of the ferrocene unit (Figure 10.9b), even though the Raman spectra show the integration of the ligands. Even if the reasons are still not clear, this might be linked to stacking of the triple bond within the graphene backbone or due to chemical side reactions between the free ethynyl moieties and the graphene surface. It has already been demonstrated that graphene can act as a diene or a dienophile in a Diels–Alder reaction [68]. Alkynes derivatives are also well-known molecules that can act as dienophiles, even if they are not activated, and the assumption can be made that such reaction can occur on our GFET surface, thus limiting the availability of alkynes for further "click" reactions. Integrating 1-(azidomethyl) pyrene on the GFET (two hours, 5 mM DMF) shows comparable Raman characteristics underlying the integration of the ligand to the GFET surface. Clicking on a rather long ethynylalkyl ferrocene derivative (tetradec-13-ynyl)ferrocene) [69], resulted in a ferrocene coverage of $3.06\times10^{-10}\,mol\,cm^{-2}$ ferrocene units (Figure 10.9b).

10.5 Conclusion

Improvement of the charge transfer characteristics of GFETs is one of the essential steps for achieving highly sensitive GFET-based sensors. If a chemical functionality can be integrated with concomitant improvement of GFET characteristics, one is close to a workable GFET sensor for a wide range of applications. The use of the 4-[(triisopropylsilyl)ethynyl]benzene diazonium tetrafluoroborate (TIPS-Eth-ArN_2^+) salt is proposed here as one of the possibilities to reach this double goal on a GFET device based on graphene coated interdigitated electrodes, which showed initial low charge mobility. One has to keep in mind that the transistor's channel width (W) and length (L) directly affect the electrical mobility of the device, with the mobility being inversely proportional to the aspect ratio W/L. Our device architecture based on graphene coated interdigitated electrodes exhibited a high W/L ratio (4.9×10^4) compared to other reported devices with W/L ratios of 1 or lower and the initial low mobility was mainly due to the small length and large width attributed to the presence of interdigitated electrodes. The large width will increase the channel surface area defining the number of immobilized bioreceptors at the sensor surface, which directly influences the target binding, although it might compromise the device electronic performance. This emphasizes the importance of incorporating diazonium-based chemistry on devices that had low mobility at start, which in turn lowers device sensitivity as a biosensor. In the case of GFETs with acceptable charge mobility, pyrene-containing molecules are valuable components for the functionalization of GFET sensors due to their electronic extended rigid structure and the large range of functional groups possible at the other end of the pyrene ligand. Some of them have been more closely analyzed in this work for the ability to form mono and bilayers on the GFET surface and their post-functionalization with bioreceptors. Using ferrocene as electrochemical probe revealed maleimide-modified pyrene to be outperforming carboxylic acid- and ester-terminated pyrene anchors. 1-Ethynylpyrene proved to integrate well into the graphene nanosheets; post-functionalization was prevented however due most likely to chemical interactions with the graphene network. The same was true for a more complex ethynyl-terminated pyrene TRIPODE, preventing click chemistry with azido-terminated bioreceptors, such as N_3-cTnI aptamers as exemplified using TIPS modified GFET chips. Click chemistry might be considered using 1-(azidomethyl)-pyrene as surface anchor for which clicking of (tetradec-13-ynyl)ferrocene resulted in surface coverage of $3.06 \pm 0.92 \times 10^{-10}$ mol cm^{-2}, comparable to that of PMAL modified GFET ($4.18 \pm 0.25 \times 10^{-10}$ mol cm^{-2}). The efficiency to click ethynyl-terminated bioreceptors still has to be confirmed in the future. Comparing TIPS and PMAL modified GFET and integrated bioreceptors based on N_3-cTnI or thiol-cTnI aptamer, LoD values of 3.34 ± 0.9 pg ml^{-1} (TIPS-based GFET) and LoD 22 ± 1 pg ml^{-1} (PMAL) were extracted for optimized sensing structures, allowing in both cases the differentiation between healthy, and people with low and high risk for myocardial infraction (AMI). The ease of the surface modification with pyrene makes it better adapted for upscaling and commercialization in the long run.

References

1 Fu, W., Jiang, L., van Geest, E.P. et al. (2017). Sensing at the surface of graphene field-effect transistors. *Adv. Mater.* 29: 1603610.

2 Georgakilas, V., Toiwari, J.N., Kemp, C. et al. (2016). Noncovalent functionalization of graphene and graphene oxide for energy materials, biosensing, catalytic, and biomedical applications. *Chem. Rev.* 9: 5464–5519.

3 Kotlowski, C., Aspermair, P., Khan, H.U. et al. (2018). Electronic biosensing with flexible organic transistor devices. *Flex. Print. Electron.* 3: 034003.

4 Lopez, A. and Liu, J. (2020). Covalent and noncovalent functionalizaiton of grpahene oxide with DNA for smart sensing. *Adv. Intell. Syst.* 2: 2000123.

5 Reiner-Rozman, C., Kotlowski, C., and Knoll, W. (2016). Electronic biosensing with functionalized rGO FETs. *Biosensors* 6 (6): 17.

6 Reiner-Rozman, C., Larisika, M., Nowak, C., and Knoll, W. (2015). Graphene-based liquid-gated field effect transistor for biosensing: theory and experiments. *Biosens. Bioelectron.* 70: 21–27.

7 Tsang, D.K.H., Lieberthal, T.J., Watts, C. et al. (2019). Chemically functionalised graphene FET biosensor for the label-free sensing of exosomes. *Sci. Rep.* 9: 13946.

8 Sarkar, D. and Banerjee, K. (2012). Proposal for tunnel-field-effect-transistor as ultra-sensitive and label-free biosensors. *Appl. Phys. Lett.* 100: 143108.

9 Wang, D., Noël, V., and Piro, B. (2016). Electrolytic gated organic field-effect transistors for application in biosensors – a review. *Electronics* 5: 9.

10 Buth, F., Donner, A., Sachsenhauser, M. et al. (2012). Biofunctional electrolyte-gated organic field-effect transistors. *Adv. Mater.* 24: 4511–4517.

11 Palazzo, G., De Tullio, D., Magliulo, M. et al. (2014). Detection beyond Debye's length with an electrolyte-gated organic field-effect transistor. *Adv. Mater.* 27: 911–916.

12 Hwang, E.H., Adam, S., and Das Sarma, S. (2007). Carrier transport in two-dimensional graphene layers. *Phys. Rev. Lett.* 98: 186806.

13 Gao, N., Gao, T., Yang, X. et al. (2016). Specific detection of biomolecules in physiological solutions using graphene transistor biosensors. *Proc. Natl. Acad. Sci. U.S.A.* 113: 14633–14638.

14 Nakatsuka, N., Yang, K.-A., Abendroth, J.M. et al. (2018). Aptamer–field-effect transistors overcome Debye length limitations for small-molecule sensing. *Science* 362: 319–324.

15 Ono, T., Kani, Y., Inou, K. et al. (2019). Electrical biosensing at physiological ionic strength using graphene field-effect transistor in femtoliter microdroplet. *Nano Lett.* 19: 4004–4009.

16 Paulus, G.L.C., Wang, Q.H., and Strano, M.S. (2012). Covalent electron transfer chemistry of graphene with diazonium salts. *Acc. Chem. Res.* 46: 160–170.

17 Park, J. and Yan, M. (2013). Covalent functionalization of graphene with reactive intermediates. *Acc. Chem. Res.* 46: 181.

18 Sinitskii, A., Dimiev, A., Corley, D.A. et al. (2010). Kinetics of diazonium functionalization of chemically converted graphene nanoribbons. *ACS Nano* 4: 1949–1954.

19 Niyogi, S., Bekyarova, E., Itkis, M.E. et al. (2010). Spectroscopy of covalently functionalized graphene. *Nano Lett.* 10: 4061–4066.

20 Aspermair, P., Mishyn, V., Bintinger, J. et al. (2021). Reduced graphene oxide-based field effect transistors for the detection of E7 protein of human papillomavirus in saliva. *Anal. Bioanal. Chem.* 413: 779–787.

21 Ingrosso, C., Bianco, G.V., Corricelli, R. et al. (2015). Photoactive hybrid material based on pyrene functionalized PbS nanocrystals decorating CVD monolayer graphene. *ACS Appl. Mater. Interfaces* 7: 4151–4159.

22 Mishyn, V., Hugo, A., Rodrigues, T. et al. (2022). The holy grail of pyrene-based surface ligands on the sensitivity of graphene-based field effect transistors. *Sens. Diagn.* 1: 235–244.

23 Rossi Fernández, A.C. and Castellani, N.J. (2017). Noncovalent interactions between dopamine and regular and defective graphene. *ChemPhysChem* 18: 2065–2080.

24 Kaminska, I., Das, M.R., Coffinier, Y. et al. (2021). Reduction and functionalization of graphene oxide sheets using biomimetic dopamine derivatives in one step. *ACS Appl. Mater. Interfaces* 4: 1016–1020.

25 Al-Hilfi, S.H., Derby, B., Martinc, P.A., and Whitehead, J.C. (2020). Chemical vapour deposition of graphene on copper–nickel alloys: the simulation of a thermodynamic and kinetic approach. *Nanoscale* 12: 15283–15294.

26 Tetlow, H., Posthuma de Boer, J., Ford, I.J. et al. (2014). Growth of epitaxial graphene: theory and experiment. *Phys. Rep.* 542: 195–295.

27 Mishyn, V., Rodrigez, T., Leroux, Y.R. et al. (2021). Controlled covalent functionalization of a graphene-channel of a field effect transistor as an ideal platform for (bio)sensing applications. *Nanoscale Horiz.* 6: 819–829.

28 Liu, P., Cottrill, A.L., Kozawa, D. et al. (2018). Emerging trends in 2D nanotechnology that are redefining our understanding of "Nanocomposites". *NanoToday* 21: 18–40.

29 Ullah, S., Yang, X., Ta, H.Q. et al. (2021). Graphene transfer methods: a review. *Nano Res.* 14: 3756–3772.

30 Chen, Y., Gong, X.-L., and Gai, J.-G. (2016). Progress and challenges in transfer of large-area graphene films. *Adv. Sci.* 3: 1500343.

31 Ahn, Y., Kim, J., Ganorkar, S. et al. (2016). Thermal annealing of graphene to remove polymer residues. *Mater. Exp.* 6: 69–76.

32 Lin, Y.-C., Lu, C.-C., Yeh, C.-H. et al. (2012). Graphene annealing: how clean can it be? *Nano Lett.* 12: 414–419.

33 Bae, S., Kim, H., Lee, Y. et al. (2010). Roll-to-roll production of 30-inch graphene films for transparent electrodes. *Nat. Nanotech.* 5: 574–578.

34 Chandrashekar, B.N., Deng, B., Shankaregowda Smitha, A.S. et al. (2015). Roll-to-roll green transfer of CVD graphene onto plastic for a transparent and flexible triboelectric nanogenerator. *Adv. Mater.* 27: 5210–5216.

35 Hiura, H., Miyazaki, H., and Tsukagoshi, K. (2010). Determination of the number of graphene layers: discrete distribution of the secondary electron intensity stemming from individual graphene layers. *Appl. Phys. Exp.* 3: 095101.

36 Lee, J., Zheng, X., Robert, R.C., and Feng, P.X.-L. (2015). Scanning electron microscopy characterization of structural features in suspended and non-suspended graphene by customized CVD growth. *Diam. Relat. Mater.* 54: 64–73.

37 Ferrari, A.C., Meyer, J.C., Scardaci, V. et al. (2006). Raman spectrum of graphene and graphene layers. *Phys. Rev. Lett.* 97: 187401.

38 Ferrari, A.C. and Basko, D.M. (2013). Raman spectroscopy as a versatile tool for studying the properties of graphene. *Nat. Nanotechnol.* 8: 235–246.

39 De Volder, M.F.L., Tawfick, S.H., Baughman, R.H., and Hart, A.J. (2013). Carbon nanotubes: present and future commercial applications. *Science* 339: 535–539.

40 Wang, C., Takei, K., Takahashi, T., and Javey, A. (2013). Carbon nanotube electronics – moving forward. *Chem. Soc. Rev.* 42: 2592–2609.

41 Li, X., Magnuson, C.W., Venugopal, A. et al. (2010). Graphene films with large domain size by a two-step chemical vapor deposition process. *Nano Lett.* 10: 4328–4334.

42 Li, X., Cai, W., An, J. et al. (2009). Large-area synthesis of high-quality and uniform graphene films on copper foils. *Science* 324: 1312–1314.

43 Novoselov, K.S., Geim, A.K., Morozov, S.V. et al. (2004). Electric field effect in atomically thin carbon films. *Science* 306: 666–669.

44 Ohno, Y., Maehashi, K., Yamashiro, Y., and Matsumoto, K. (2009). Electrolyte-gated graphene field-effect transistors for detecting pH and protein adsorption. *Nano Lett.* 9: 3318–3322.

45 Hess, L.H., Seifert, M., and Garrido, J.A. (2013). Graphene transistors for bioelectronics. *Proc. IEEE* 101: 1780–1792.

46 Castro Neto, A.H., Guinea, F., Peres, N.M.R. et al. (2009). The electronic properties of graphene. *Rev. Mod. Phys.* 81: 109.

47 Gammelgaard, L., Caridad, J.M., Cagliani, A. et al. (2014). Graphene transport properties upon exposure to PMMA processing and heat treatments. *2D Mater.* 1: 035005.

48 Nagashioa, K., Yamashita, T., Nishimura, T. et al. (2011). Eectrical transport properties of graphene on SiO_2 with specific surface structures. *J. Appl. Phys.* 110: 024513.

49 Sarma, S.D., Adam, S., Hwang, E.H., and Rossi, E. (2011). Electronic transport in two-dimensional graphene. *Rev. Mod. Phys.* 83: 407.

50 Banhart, F., Kotakoski, J., and Krasheninnikov, A.V. (2011). Strucural defects in graphene. *ACS Nano* 5: 26–41.

51 Tan, X., Chuang, H.-J., Lin, M.-W. et al. (2013). Edge effects on the pH response of graphene nanoribbon field effect transistors. *J. Phys. Chem. C* 51: 27155–27160.

52 Gautier, C., Lopez, I., and Breton, T. (2021). A post-functionalization toolbox for diazonium (electro)-grafted surfaces: review of the coupling methods. *Mater. Adv.* 2: 2773–2810.

53 Xia, Y., Martin, C., Seibel, J. et al. (2020). Iodide mediated reductive decomposition of diazonium salts: towards mild and efficient covalent functionalization of surface-supported graphene. *Nanoscale* 12: 11916–11926.

54 Ambrosio, G., Brown, A., Daukiya, L. et al. (2020). Impact of covalent functionalization by diazonium chemistry on the electronic properties of graphene on SiC. *Nanoscale* 12: 9032–9037.

55 Wang, Q.H., Jin, Z., Kim, K.K. et al. (2012). Understanding and controlling the substrate effect on graphene electron-transfer chemistry via reactivity imprint lithography. *Nat. Chem.* 4: 724–732.

56 Delamar, M., Hitmi, R., Pinson, J., and Saveant, J.-M. (1992). Covalent modification of carbon surfaces by grafting of functionalized aryl radicals produced from electrochemical reduction of diazonium salts. *J. Am. Chem. Soc.* 114: 5883.

57 Pinson, J. and Podvorica, F. (2005). Attachment of organic layers to conductive or semiconductive surfaces by reduction of diazonium salts. *Chem. Soc. Rev.* 34: 429.

58 Fan, X.-Y., Nouchi, R., Yin, L.-C., and Tanigaki, K. (2010). Effects of electron-transfer chemical modification on the electrical characteristics of graphene. *Nanotechnology* 21: 475208.

59 Shih, C.-J., Wang, Q.H., Jin, Z. et al. (2013). Disorder imposed limits of mono- and bilayer graphene electronic modification using covalent chemistry. *Nano Lett.* 13: 809.

60 Rodrigues, T., Mishyn, V., Leroux, Y.R. et al. (2021). Highly performing graphene-based field effect transistor for the differentiation between mild-moderate-severe myocardial injury. *Nano Today* 43: 101391.

61 Leroux, Y.R. and Hapiot, P. (2013). Nanostructured monolayers on carbon substrates prepared by electrografting of protected aryldiazonium salts. *Chem. Mater.* 25: 489–495.

62 Leroux, Y.R., Hui, F., Noel, J.-M. et al. (2010). Efficient covalent modification of a carbon surface: use of a silyl protecting group to form an active monolayer. *J. Am. Chem. Soc.* 132: 14039–14041.

63 Huang, P., Zu, H., Jiang, L. et al. (2011). Graphene covalently binding aryl groups: conductivity increases rather than decreases. *ACS Nano* 5: 7945–7949.

64 Mann, J.A., Alava, T., Craighead, H.G., and Dichtel, W.R. (2013). Preservation of antibody selectivity on graphene by conjugation to a tripod monolayer. *Angew. Chem. Int. Ed.* 52: 3177–3180.

65 Mann, J.A., Rodríguez-López, J., Abruña, H.D., and Dichtel, W.R. (2011). Multivalent binding motifs for the noncovalent functionalization of graphene. *J. Am. Chem. Soc.* 13: 17614–17617.

66 Wang, S., Hossain, M.Z., Shinozuka, K. et al. (2020). Graphene field-effect transistor biosensor for detection of biotin with ultrahigh sensitivity and specificity. *Biosens. Bioelectron.* 165: 112363.

67 Cao, M., Fu, A., Wang, Z. et al. (2014). Electrochemical and theoretical study of p-p stacking interactions between graphitic surfaces and pyrene derivatiges. *J. Phys. Chem. B* 118: 2650–2659.

68 Sarkar, S., Bekyarova, E., Niyogi, S., and Haddon, R.C. (2011). Diels–Alder chemistry of graphite and graphene: graphene as Diene and Dienophile. *J. Am. Chem. Soc.* 133: 3324–3327.

69 Zhong, D., Blömker, T., Wedeking, K. et al. (2009). Surface-mounted molecular rotors with variable functional groups and rotation radii. *Nano Lett.* 9: 4387–4391.

11

Sensing Mechanisms in Graphene Field-Effect Transistors Operating in Liquid

Tilmann J. Neubert and Kannan Balasubramanian

Humboldt-Universität zu Berlin, School of Analytical Sciences Adlershof (SALSA), Department of Chemistry & IRIS Adlershof, Unter den Linden 6, 10099, Berlin, Germany

11.1 Introduction

The use of graphene field-effect transistors (GFETs) for label-free (bio)sensing has been demonstrated in many studies and reviewed extensively over the past years [1–3]. For sensing applications, especially in the case of biologically relevant analytes, operation in liquids such as physiological fluids is usually a prerequisite [4–6]. Gaining a good understanding of the solid–liquid interface between graphene and the analyte solution is of huge importance to comprehending the mechanisms behind the signal generation in GFETs operated in liquid. A clear understanding of the underlying mechanisms of field-effect transistor (FET) operation is also fundamentally essential for pushing forward the widespread use of GFETs in sensing applications. For example, in medical diagnostics, very high reliability and reproducibility are essential [7]. With unclear mechanistic understanding or with the interplay of several complicated mechanisms, it is very difficult to achieve this target.

Here, we start with the well-established picture of FET operation in air and draw comparisons to its working principle in liquids in general. Following this, we present the key aspects of using graphene as an FET channel and important caveats when using graphene FETs in liquid. Next, we focus on possible sensing mechanisms from a theoretical point of view. Since a graphene sheet immersed in liquid behaves as an electrode in solution, many concepts can be explained from an electrochemistry perspective, which is the topic of Section 11.7. In presenting the fundamental aspects, we will draw several parallels to results (both theory and experiments) available on carbon nanotubes (CNTs), due to many similarities in FET operation using CNTs and graphene. Finally, we show selected graphene-FET-based sensor examples from the literature and highlight the situations where the mechanistic explanation is quite clear.

Graphene Field-Effect Transistors: Advanced Bioelectronic Devices for Sensing Applications,
First Edition. Edited by Omar Azzaroni and Wolfgang Knoll.

11.2 Field-Effect Operation in Liquid Compared to Operation in Air

There are basically two device configuration of FETs operating in air, namely the junction FET and metal-oxide-semiconductor field-effect transistor (MOSFET) [8]. Most of the one-dimensional (1D) and two-dimensional (2D) nanomaterial-based FETs operate as MOSFETs in air. In such devices, the key component responsible for operation is the metal-oxide-semiconductor (MOS) capacitor, wherein a semiconductor channel is separated by a metal (or a highly doped semiconductor), typically through an insulating oxide layer (Figure 11.1a). The metallic layer acts as a gate electrode, and the insulating layer is referred to as gate oxide or gate insulator. Depending on whether the gate insulator is below or above the semiconductor channel, such devices are referred to as back-gated or top-gated FETs, respectively. By varying the voltage applied at the gate electrode (gate voltage, V_G), the electric field across the gate insulator is varied, which in turn electrostatically modulates the conductivity of the semiconductor channel. By using two probes on the semiconductor, referred to as source and drain electrodes, the conductance of the channel (G_{sd}) or the current through the channel (I_{SD}) is measured using DC or AC measurements. A typical FET transfer characteristic (also referred to as gate voltage dependence) for a p-type semiconducting CNT is shown in Figure 11.1c, where it can be seen that the conductance can be modulated over several orders of magnitude, characterized by a high ON/OFF (current) ratio. Another related parameter for an FET is transconductance (g_m), which indicates the change in drain current attainable for a given change in gate voltage: $g_m = \Delta I_{sd}/\Delta V_G$. For sensors, g_m is an important parameter. A higher g_m value indicates a better measurement sensitivity to detect changes in gate voltage via a corresponding change in channel current.

When operating in liquid, an electrical double layer (EDL) is formed at the interface between the transistor channel and the liquid (Figure 11.1b) [9, 10]. The EDL forms a capacitor, which is analogous to the gate insulator of an FET operating in air. The voltage at the EDL capacitor can be modulated by using an electrode in contact with the liquid, whose electrode potential needs to be constant and does not vary with the composition of the liquid. This can be achieved optimally using a reference electrode [7, 11]. The formation of an EDL requires dissociated ions in solution, which are delivered by an electrolyte. Even in deionized/distilled water under ambient conditions, there are dissociated ions, e.g. due to trace impurities or dissolved carbon dioxide. Hence, a capacitance builds up at the nanomaterial–liquid interface, even in the absence of deliberately added salt. With increasing ionic strength (brought about by adding electrolytes), the capacitance increases due to a decreasing Debye length [12].

In most cases, this EDL capacitance is much higher in comparison to the capacitance of a gate insulator, due to the smaller EDL thickness and the relatively high dielectric constant of the liquid in comparison to that of the gate oxide. As a result, the same ON/OFF ratio can be achieved in a smaller gate voltage range thereby yielding a comparatively higher g_m (see Figure 11.1c). FETs operating in liquids have been referred to as liquid-gated, solution-gated, or electrolyte-gated FETs.

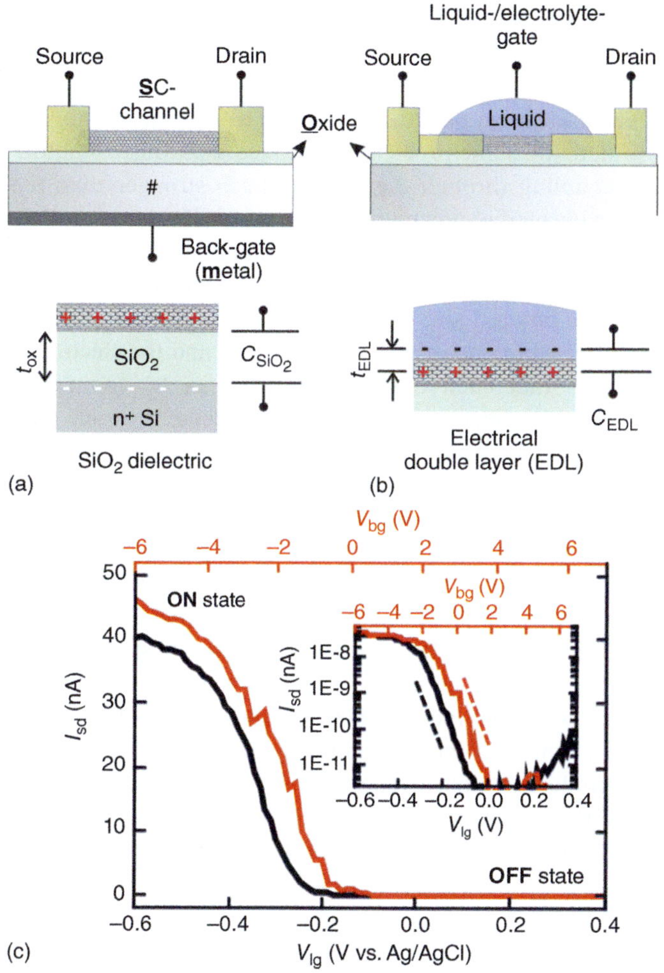

Figure 11.1 Schematic of (a) back-gated (b) and liquid-gated FET setup with a semiconductor (sc) channel, showing the different capacitances formed at the interface. # refers to highly doped semiconductor (e.g. n^+-Si). (c) Transfer curves obtained for an individual semiconducting carbon nanotube under back-gated (top curve) and liquid-gated (bottom curve) operation. V_{bg}: back-gate voltage, V_{lg}: (liquid) gate voltage at the reference electrode. Source: Adapted from Heller et al. [9].

However, a more general term is electrochemical FET, which will be introduced later. Since all that is needed is an EDL at the interface, the medium does not just have to be water (with or without salt). Even, organic solvents, ionic liquids, or even polymers can be used as the gating medium. In all cases, some form of electrolyte must be present for the build-up of EDL, whose capacitive nature will induce an electric field-effect to modulate the channel conductance. One example is a CNT FET in a polymer matrix with embedded salts, which is referred to as solid polymer-electrolyte gated FET [10].

11.3 Caveats When Operating FETs in Liquid

There are some demonstrations of using back-gated FETs immersed in a liquid, without the use of a gate electrode in liquid [13, 14]. Although one observes an FET response in such devices, there are some problems associated with this configuration. Since the capacitive coupling through the EDL is much stronger than the capacitance due to the gate oxide on the substrate, interactions at the nanomaterial–liquid interface have a dominant effect on the operating point of the semiconductor channel. Furthermore, this results in serious problems of hysteresis in the FET transfer characteristics, i.e. the forward and backward gate voltage sweeps are no longer identical. Use of a proper reference electrode as the gate and the electrolyte as a gating medium ensures that such hysteresis effects can be completely avoided [10]. An Ag/AgCl reference electrode is the gate electrode of choice used widely in liquid-gated graphene FETs. Another alternative, reported rarely, involves the use of a metallic wire or metal film as gate electrode [15, 16]. This approach has severe shortcomings. Such metallic wires are polarizable, although often not ideally polarizable [12], and hence do not provide for a stable reference potential.

Figure 11.2 shows current–voltage (*I–V*) curves measured in an electrolyte solution using two Ag/AgCl electrodes (RE vs. RE) or using two Pt electrodes. The *I–V* curves at the Ag/AgCl electrodes show a linear behavior with no hysteresis independent of the scan rate with the slope giving a measure of the electrolyte resistance. This is indicative of both electrodes exhibiting ideal non-polarizability behavior allowing for precise control of the applied potential. On the contrary, the *I–V* curves at the two Pt electrodes are capacitive, signifying that these electrodes are polarizable. The charging of the EDL capacitance at both electrodes competes with the precise control of the potential. Furthermore, the latter behavior indicates that

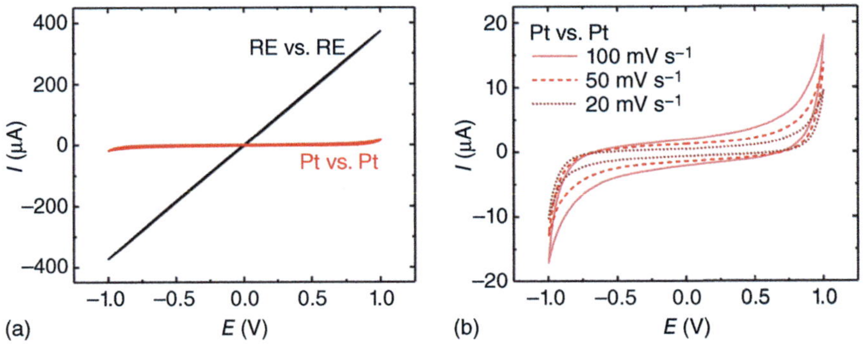

Figure 11.2 (a) Current–voltage (*I*–*V*) curve measured in 0.1 M KCl between two Ag/AgCl electrodes (RE vs. RE) in comparison to measurement using two Pt wires (Pt vs. Pt). (b) A zoomed-in view of the *I*–*V* curve measured using two Pt wires. While the Ag/AgCl electrodes show low resistance with no hysteresis at any scan rate, the Pt-wires show capacitive behavior with a large difference between forward and backward currents, which furthermore vary according to the scan rate used. The exponential increase in current at the extreme voltages in (b) is due to hydrogen evolution.

the actual voltage is dependent on the direction of the voltage sweep. While FET operation is still possible using simple metal electrodes, the transfer curves are very much prone to hysteresis and instability, and do not allow for reliable and reproducible operation, such as required in sensor applications.

11.4 Graphene FETs in Liquid

Graphene as the charge carrier channel in an FET is a special case since graphene is neither a semiconductor, nor a metal [17]. The electronic structure resembles that of a semiconductor, however, with a zero bandgap at zero temperature. Hence, in contrast to a semiconductor, there is a finite density of states at the Fermi level at room temperature. In other words, it is not possible to switch OFF the conductance in the graphene channel completely. As a result, the ON/OFF ratio for a graphene FET either in air or in liquid are rather low, typically less than 3. There is another difference with respect to bulk semiconductor FETs, arising due to quantum effects. As we had seen earlier, the key component of an FET is the MOS capacitor. In the presence of an electrolyte, the EDL capacitance enables the possibility of a field effect. However, for low-dimensional systems, especially at the nanoscale, where quantization effects become more important and the energy levels become discretized, the quantum capacitance occurring in series with the EDL capacitance starts to contribute significantly to the net capacitance [18]. The quantum capacitance occurs due to the unique electronic structure of nanostructures [10, 18–20]. In typical bulk semiconductors, the quantum capacitance is higher than the EDL due to the high density of states and continuum nature of the conduction and valence bands. In contrast, in graphene, the low density of states at the Fermi level leads to an influence of this effect on the FET characteristics. The density of states in graphene varies with the applied gate voltage, and hence the net capacitance at the graphene–liquid interface (GLI) is a function of the gate voltage [19]. Moreover, since the net capacitance is dependent on the active graphene area in the FET channel, the effect of quantum capacitance decreases significantly as we go to large area devices, where the interfacial capacitance is mainly dictated by the EDL.

Another important aspect while operating graphene FETs in liquid concerns the design of the source and drain contacts. In the ideal case, we are mainly interested in using the graphene sheet as the active device component. Hence, it is important to passivate the leads in order to avoid their coming into contact with the liquid [7, 11, 21]. This is often done by depositing a polymer such as poly(methylmethacrylate) (PMMA), SU-8, or by a thick oxide layer using vacuum deposition, both methods augmented by lithography. After passivation, the graphene sheet is exclusively in contact with the solution, and the observed changes in FET characteristics can be solely associated with processes occurring at the GLI. In cases where the electrodes are not passivated, it is clear that the electrodes will also participate in interfacial processes, which will have a dominant effect on the FET characteristics. This is because the contacts are typically more conducting than graphene and are prone to several electrochemical processes. For example, when gold is used as a contact and is not passivated, electrochemical oxidation and reduction of the gold surface occur [22] when the gate voltage is swept in certain gate voltage ranges. This will induce additional leakage currents, which will make the interpretation of mechanism of sensing and device operation ambiguous. Although transfer curves typical for graphene are often observed using non-passivated devices, the use of such devices for sensing is rather limited, because of the problems mentioned above. A consequence is that reproducibility and reliability of these devices are rather poor in comparison to passivated devices.

11.5 Measurement Modes

The FET response at graphene under liquid gating can be measured in several ways. The simplest method is to keep the voltage at the reference electrode constant and measure changes in the current flowing through graphene. Figure 11.3a shows an example of such a measurement in buffers of different pH. While it is well suited to take this approach in an application scenario, for understanding sensing mechanisms it is necessary to obtain the complete transfer characteristics in a relevant gate voltage range, such as that shown in Figure 11.3b. The gate voltage range is dictated by the solvent, the choice of the reference electrode, and possible reactions occurring at the graphene surface. Typically in aqueous solutions, we are limited by hydrogen evolution, oxygen reduction, and water oxidation at graphene, which limits the operable gate voltage range in neutral pH typically to 0.6 V (oxygen reduction) to −1.0 V

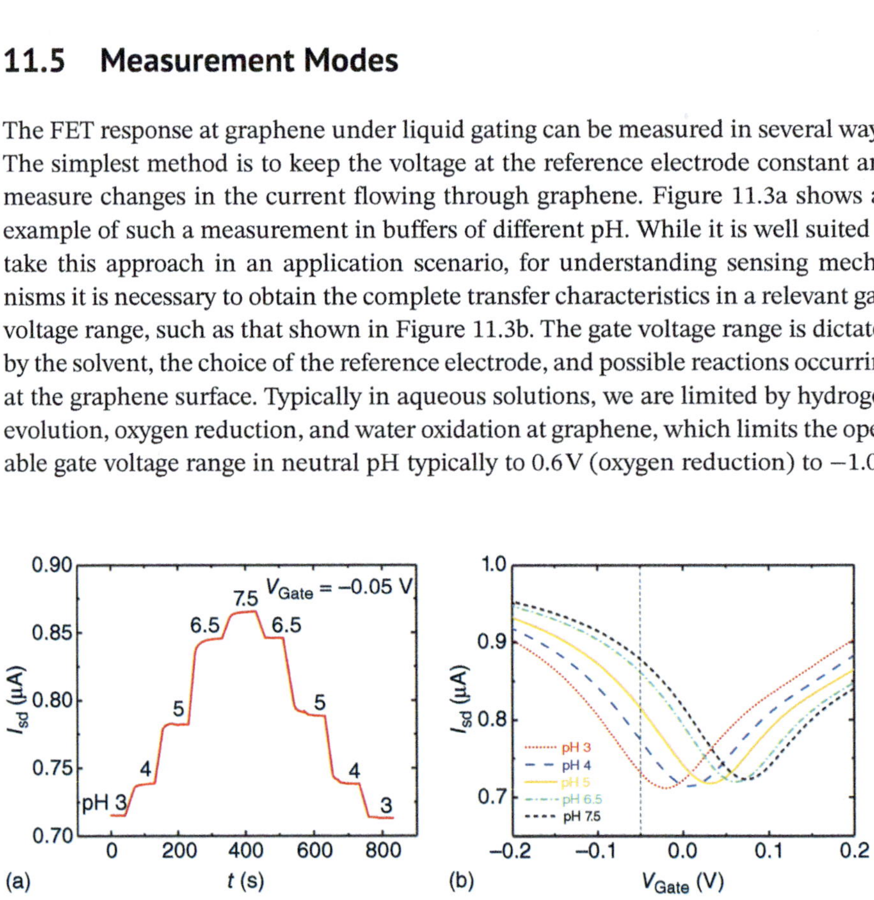

Figure 11.3 Two different measurement modes for sensing pH using a liquid-gated graphene FET: (a) constant gate voltage, (b) complete gate dependence/transfer curve. The dashed vertical line in (b) indicates the gate voltage where the measurement in (a) was carried out. Experimental details: chemical vapor deposition grown graphene on Si/SiO_2 substrate, channel dimensions: w = 20 μm, l = 10 μm, with passivated contacts, Ag/AgCl (3 M KCl) reference used as gate electrode, details of buffers used can be found in ref. [23]. Source: Adapted from Neubert et al. [23].

(water oxidation) at Ag/AgCl. As we know from electrochemistry, the applied potential at any electrode in solution is indicated against a certain reference electrode.

This convention is opposite of what is typically used in liquid-gated FET measurements. In this electrochemical convention, the above-mentioned voltage range will correspond to a range of −0.6 to +1.0 V vs. Ag/AgCl as measured at graphene against an Ag/AgCl reference. Since the reactions discussed above are pH-dependent, the usable voltage range varies from one pH to another [12, 24]. Also, depending on the preparation strategy, the amount of contaminants, the measurement environment, etc., this range may differ from one study to another. Outside this voltage range, the currents due to these electrode reactions have such an influence on the transfer characteristics, that it is difficult to obtain reliable and reproducible data for a sensing application. Especially when the electrode area is large, the electrochemical currents become dominant, which will affect the stability of the electrode potential (this is discussed later in detail). Unlike in FETs operating in air, through appropriate measurement techniques, it is possible to decouple the two current channels to some extent. An AC mode of measurement can be used to measure the impedance of the graphene channel between source and drain as a function of the DC gate voltage. An even more reliable method is to use a proper potentiostatic control using an electrochemical potentiostat, coupled with an impedance analyzer. Here in addition to source, drain, and gate electrodes, a counter electrode is introduced. Figure 11.4b shows the layout for such a measurement setup in comparison to a standard FET measurement setup in Figure 11.4a [23]. The graphene, the counter and the gate (reference) electrodes form an electrochemical cell. Using this setup, it is possible to precisely control the potential at the GLI and the conductance of the graphene flake can be monitored using an AC voltage with minimal disturbance from the electrochemical currents flowing through the electrolyte.

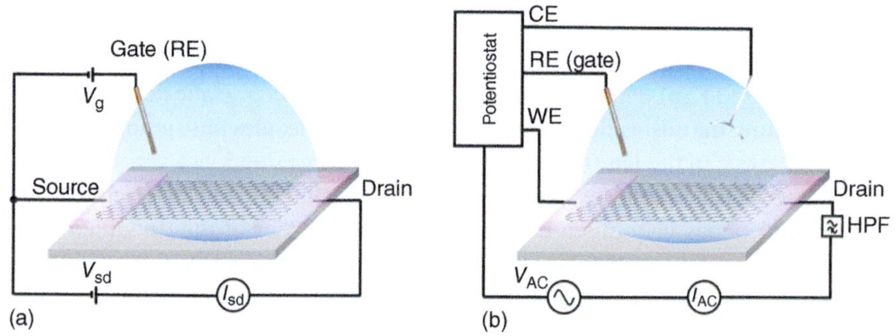

Figure 11.4 Schematic of measurement setups for sensing using liquid-gated graphene FETs: (a) standard FET setup, (b) FET setup with potentiostatic control, where the current measurement through the channel is decoupled from the application of gate voltage. HPF, high pass filter; WE, working electrode; CE, counter electrode; RE, reference electrode. Further details can be found in ref. [23]. Source: Adapted from Neubert et al. [23].

11.6 Using FETs for Sensing in Liquid – Sensing Mechanisms

The central idea behind the sensing mechanism in graphene FETs is the modulation of physicochemical properties of the GLI through the adsorbed or bound analyte molecules, leading to changes in FET's characteristics, which can be used as a sensor signal. The interface is very sensitive to adsorption of molecules, changes in local charge distribution and charge transfer processes at the GLI. Hence, in the absence of any kind of modification on pristine graphene, these processes may induce nonspecific changes in the FET characteristics. In order to be able to detect a specific analyte, the graphene surface needs to be modified with an appropriate chemistry, which will introduce a selective (ideally exclusive) interaction with the analyte molecule of interest [1–3, 25]. In most cases, this is achieved by using a receptor that is chemically complementary to the analyte and changes in FET characteristics are observed when the analyte molecules bind to the immobilized receptors. Depending on the nature and strength of these changes, information can be acquired about the analyte molecules and their concentration in the test solution. In addition, the attachment of receptors can be followed using the FET transfer characteristics itself [26–28]. Achieving selectivity is the main challenge when designing sensors for new analytes based on graphene FETs. The receptors can be attached either covalently or noncovalently to the graphene surface. Covalent attachment directly attacks the sp^2-carbon framework, which leads to a deterioration of graphene conductance and decrease in charge carrier mobility. With noncovalent attachment, on the other hand, the favorable transport properties remain unaffected. Detailed reviews on chemical functionalization methods for graphene can be found in the literature [29].

When analyte or receptor molecules attach to the graphene surface without chemically affecting the sp^2-carbon framework (i.e. noncovalent interaction with graphene), then two possible changes can occur at the GLI. The most prominent case is a change in the charge distribution or potential at the GLI due to the adsorbed species. This potential change acts as an offset gate voltage causing a shift in the transfer characteristics (see Figure 11.5a). This effect is often referred to as electrostatic doping [21]. In a simple picture, the adsorption of positively charged molecules onto graphene shifts the transfer curve to the left (toward negative gate voltage), and vice versa for negatively charged species [25]. However, it is to be kept in mind that the interfacial charge distribution depends very much on the acid–base properties of the (modified) graphene surface as well as the net charge distribution of the analyte. The latter is dependent on the pH and ionic strength of the solution as dictated by the isoelectric point/point of zero charge (pI/pzc) and acid/base dissociation constants (K_a or K_b) of the analyte species [1, 30]. Hence, the shift in transfer curve will depend on the net change in surface pK_a or pI after adsorption of the analyte molecules. This is in marked contrast to charged analytes adsorbing on the graphene surface in air, where mainly the charge of the adsorbed species needs to be considered in the dry state.

The second possibility is a change in the dielectric constant of the EDL capacitor due to the adsorbed analyte species, which would result in a change in the gate

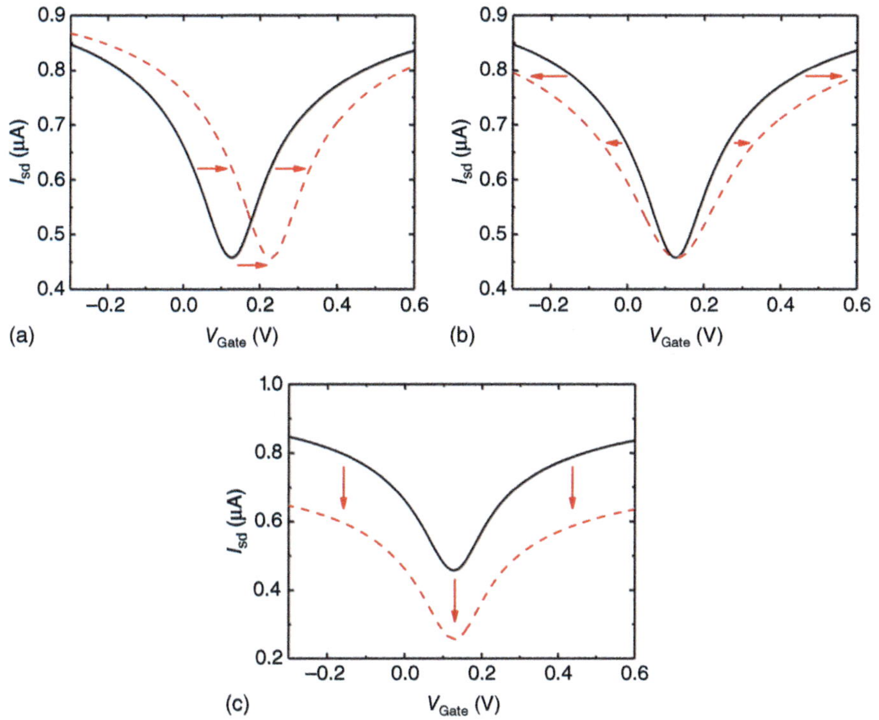

Figure 11.5 Typical sensor responses expected in liquid-gated graphene FETs depending on the underlying sensor mechanism: (a) electrostatic doping through analyte-induced change in interfacial charge density or potential shifting the transfer curve to higher gate voltage. (b) Effect of change in gate coupling efficiency induced by analyte adsorption. (c) Effect of analyte-induced change in charge carrier mobility shifting the transfer curve to lower currents.

coupling efficiency [21]. The ensuing effect is a change in transconductance, or in other words a change in the slope of the gate dependence characteristics as shown in Figure 11.5b. This effect, is however, typically weak, especially in liquid-gated graphene FETs, because the ON/OFF ratio is very low. Drastic changes in the double layer and significant modification of the graphene surface are necessary in order to observe significant changes in gate coupling, e.g. heavy accumulation of bulky proteins or the formation of an intact molecular layer throughout the entire graphene surface impermeable to the electrolyte [21]. As mentioned earlier, here we have assumed that the contacts are passivated, and hence only interactions occurring at the graphene surface can be observed in the transfer characteristics. Hence, the influence of contacts and Schottky barrier effects [21, 31–33] is neglected here in obtaining an understanding of the sensing mechanisms. When contacts are not passivated, often contradictory changes in FET characteristics are observed for the same analyte [34]. Moreover, it is difficult to pinpoint a clear mechanism behind the sensor responses.

In back-gated FETs operating in air, a third sensing mechanism is often widely observed, namely charge transfer doping [35–37]. Adsorption of analyte molecules with the capability to undergo charge transfer will shift the transfer curve according to the type of doping induced. Electron donors shift the transfer curve to negative voltages, while electron acceptors shift the FET curve to positive voltages. The underlying mechanism is a difference between the Fermi level of graphene and the HOMO/LUMO of the adsorbed molecule [38, 39]. The shifts observable in the transfer curves are similar to the case of changes in interfacial charge distribution (Figure 11.5a). Hence, only by observing this kind of shift in the back-gated FET response to analytes is it possible to distinguish between electrostatic doping and charge transfer doping unambiguously. In liquids, however, the situation with this kind of interaction is quite different. Electron transfer processes at electrodes have been studied for ages in the area of electrochemistry [12] and several works have focused on a single graphene sheet as an electrode in electrochemistry [40]. For an electron transfer to occur at the electrode–liquid interface, the molecule has to be redox-active. Moreover, it is not required that the molecules be adsorbed on the surface. Redox-active species diffusing toward the electrode (mass transport) can also induce charge transfer. Moreover, the efficiency of electron transfer can be tuned in several ways, e.g. by varying the applied potential, using a catalytically active surface, etc. For a full understanding of this interaction, we need to take a systematic electrochemical perspective on the operation of graphene FETs in liquid, which will be topic of Section 11.7.

A fourth mechanism, relevant also in liquids, concerns changes in mobility due to a chemical interaction of molecular species modifying the sp^2-carbon framework of graphene. This can be best understood by looking at the case of covalent functionalization. A good example is the coupling of substituted phenyl radicals using diazonium salts in solution [26, 27]. The covalent attachment of the substituted phenyl moieties leads to a decrease in conductance (see Figure 11.5c), caused by a reduction in charge carrier mobility because of the functionalities introduced that function as scattering centers. Typically, such a modification of graphene is irreversible. Hence, when engineering graphene-based sensors, it is advisable to avoid the occurrence of such a mechanism with the analyte molecules. However, on sensors based on chemically derived graphene (CDG), such as reduced graphene oxide, it is possible that such a mechanism can take place when analytes interact with the sensor surface [41, 42]. This is because, in CDG, we do not have an intact sp^2-carbon framework. Such sheets show high defect density with several chemical functionalities. These defect sites act as scattering centers for electronic transport. When analyte molecules interact with such defect sites, the scattering strength is modulated, resulting in a change in charge carrier mobility and correspondingly the resistance.

In order to identify which of these mechanisms is active in a given graphene FET sensor, it is necessary to perform several control experiments. It is likely that more than one mechanism is responsible for the observed sensor response. Typically, information about charge carrier density and mobility is estimated using two-probe (source–drain) measurements in liquids. A more reliable estimation is possible using four-probe and magnetoresistance measurements, such as in Hall-bar or van der Pauw configurations [43]. Furthermore, it has to be ensured that the ionic strength and the pH of the solution are well-defined, when aiming to understand the mechanism of sensor response [18].

11.7 The Electrochemical Perspective

The mechanisms discussed above were mostly derived from ideas known from air-operated graphene FET sensors [21] and applied to the liquid environment. However, as indicated earlier, we can also take a different approach from an electrochemical point of view to understand the sensing mechanisms. Indeed, this is actually inevitable because, whenever an electrode is immersed in liquid an electrode potential develops and a proper discussion of several electrode processes needs to be considered when deciphering the mechanism of interaction. We will start here by giving a short introduction to two types of electroanalytical techniques and discussing how they are relevant for the operation of graphene FETs in liquid.

One of the basic techniques relevant for sensors is potentiometry [44], where the potential measured between an indicator electrode and a reference electrode gives an idea of the activity of the analyte in solution. Two kinds of potentiometric sensors can be distinguished. The first case involves the detection of redox-active analytes in solution. Here, the potential of the indicator electrode (against the reference electrode) is directly related to the relative fraction of the activities of reduced and oxidized species as given by the Nernst equation [12, 44]. Hence, by measuring the voltage, the relative fraction of the amount of oxidized or reduced species can be estimated. The second case concerns the detection of analyte molecules that do not necessarily have to be redox active. In so-called ion-selective-electrodes (ISEs), the indicator electrode and the reference electrode are separated by a membrane, which is in equilibrium with the analyte of interest. In this case, the indicator electrode is also a reference electrode. The membrane is in contact with a standard solution and the indicator electrode on one side, while the test solution along with the reference electrode forms the other side. A classic example of an ISE is the glass electrode used routinely for measuring pH (see Figure 11.6). The heart of a pH-ISE comprises a glass membrane (which is proton selective) in contact with a standard solution with a constant activity of H^+-ions (e.g. 0.1 M HCl). The potential of the glass membrane varies according to the activity of H^+-ions in the test solution, which is detected by the potential difference between the two reference electrodes. In order to detect an analyte of interest, the glass membrane is replaced by an ion-selective membrane (ISM), which selectively binds the analyte of interest.

An extension of the ISE is the ion-selective field effect transistor (ISFET), where the ISM is directly immobilized on a semiconductor surface, typically silicon [45]. The ISFET is actually a liquid-gated silicon FET, with an additional ISM on the silicon surface (see Figure 11.7). In other words, it can be seen as a combination of an ISE and an FET. The potential at the ISM is transferred to a change in conductance of the silicon transistor channel. Since a silicon FET has a high transconductance, a small change in potential at the modified silicon surface (due to the binding of analytes in the ISM) is transferred to a large change in conductance of the silicon channel. The analyte molecules in equilibrium with the ISM set the potential (or equivalently, the interfacial charge distribution) at the interface, which is then seen as a threshold voltage shift in the transfer characteristics. A liquid-gated graphene FET with a well-modified/

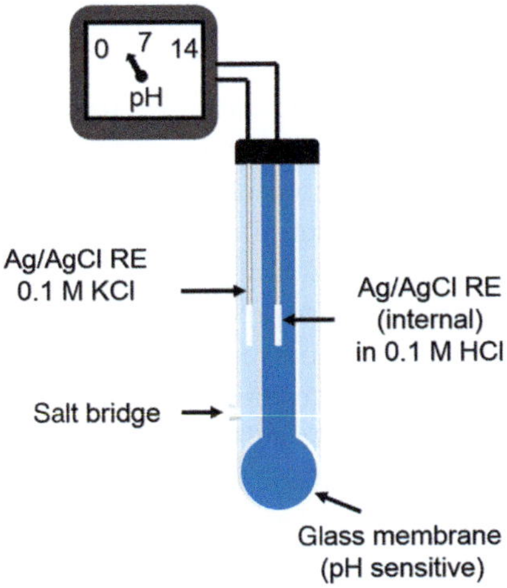

Figure 11.6 Schematic of pH-ISE (ion-selective-electrode), which is used for measuring pH using the principle of potentiometry. RE, reference electrode.

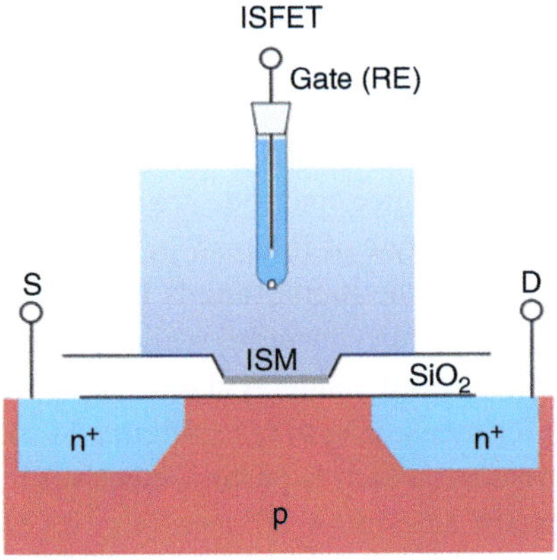

Figure 11.7 Schematic showing the construction of an ion-selective field-effect transistor (ISFET) comprised of an ion-selective membrane (ISM) deposited on the semiconductor surface. S, source; D, drain; p: p-type Si (FET semiconductor channel) n^+: highly doped semiconductor region acting as source/drain contacts.

completely covered receptor layer is a direct parallel to an ISFET. The response to analytes in an ISFET is always akin to the electrostatic doping mechanism discussed earlier (Figure 11.5a). In this context, the receptor layer takes on the role of the ISM [46]. The difference with a silicon FET is that in GFETs the ON/OFF ratio is lower and hence the change in graphene conductance for a given change in interfacial potential is relatively low. On the other hand, the noise level in GFETs has been shown to be quite low [47].

For a silicon ISFET, a basic assumption is that the underlying silicon does not come in direct contact with the solution and is not involved in any electrode reaction. Hence, the measured potential changes can be directly associated with the activity of analytes in solution. In this case, the passage of charge is not allowed, and the interface behaves like a capacitor. Such an electrode is referred to in electrochemistry as a polarized interface [11, 12]. The capacitive nature of this interface is reminiscent of the gate insulator in top-gated FETs. There is also an EDL capacitance in series with capacitance of the ISM. Although the EDL capacitance contributes to the net FET transfer characteristics [48] the sensing response is mainly obtained through analyte interaction in the ISM/receptor layer [46]. Several demonstrations have shown that graphene FET-based sensors indeed function as ISFETs [47, 49, 50]. Typically, an intact receptor layer is immobilized on the graphene surface. The receptors are chosen in such a way that they selectively bind analyte molecules and lead to changes in local electrostatic potential or charge distribution. This change is registered as a change in current when measured at a fixed gate voltage or a shift in the gate voltage characteristics when the entire transfer curve is measured.

When the receptor layer is sparse or permeable to the electrolyte or when there is no such layer, there is a possibility that charge can indeed pass through the electrode–electrolyte interface. In this case, the interface is said to be non-polarized and electron transfer reactions with redox active molecules can directly take place with the graphene sheet [37]. Such a scenario is different from the potentiometric approach discussed above. In electrochemistry, we differentiate between measurements at equilibrium (potentiometry, without passage of a current) and measurements carried out at an applied potential (deviation from equilibrium, non-zero net current) [12]. The latter case includes potentiodynamic methods such as voltammetry, chronoamperometry, and associated methods. In a liquid-gated FET setting, we always apply a certain gate voltage (either set constant or swept over a certain voltage range), which corresponds to the potentiodynamic situation. If the electrode (graphene channel) has the capability to undergo electron transfer, and if there are redox active species present in the solution or on the surface, an electrochemical current may flow depending on the formal potential of the redox species and the (gate) voltage range where the FET is operated (Figure 11.8a). While analytes themselves (if redox active) may contribute to such a current at a non-polarized GLI, there are also other candidates, e.g. water, dissolved oxygen, and hydronium ions, which are often not analytes of interest (other than in pH sensing). Using detailed investigations with deliberately added redox species (ferrocyanide or hexachloroiridate), it could be clearly shown that such an electrochemical current at graphene will also cause a threshold voltage shift in the FET transfer characteristics (see Figure 11.8b), especially when the size of the graphene sheet is large (i.e. not microstructured) and behaves as a macro-electrode [23]. This effect was termed Faradaic effect, since the shift is due to a Faradaic current flowing between a graphene channel and the reference electrode. The transfer curve shift occurs, since graphene behaves as a non-ideal non-polarizable electrode. This effect is different from the charge transfer doping effect in air-operated GFETs, since in liquids the charge transfer at graphene is balanced by the opposing half-cell reaction at the gate (reference electrode). In principle, the presence of such a current can be verified by measuring the leakage current simultaneously with the measurement of gate transfer characteristics. However, it should be kept in mind that the Faradaic current itself is different from the typical leakage current often discussed in the literature, the latter arising mainly due to the capacitive nature of the GLI.

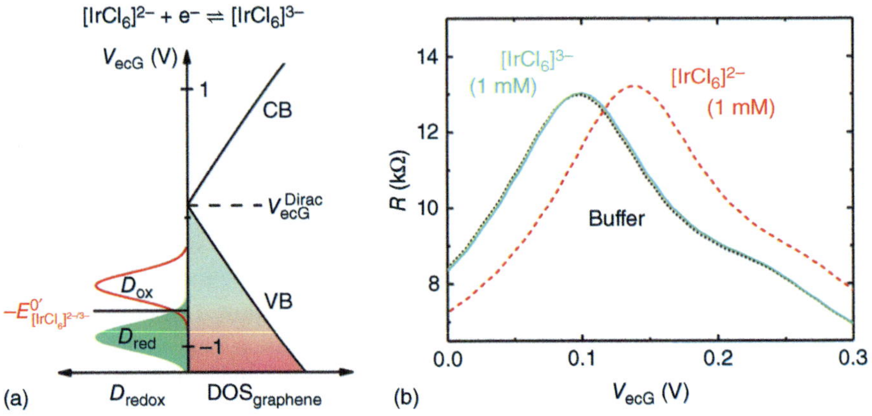

Figure 11.8 Faradaic effect. (a) Schematic showing the alignment of energy levels of graphene with the density of states of a redox-probe $[IrCl_6]^{2-/3-}$ according to Marcus theory. (b) Transfer curves of a liquid-gated graphene FET in buffer (dotted line), and in the presence of 1 mM of oxidized ($[IrCl_6]^{2-}$, dashed line) and reduced ($[IrCl_6]^{3-}$ form, solid line) form of the redox probe. A shift is seen only in the presence of the oxidized form, because this species undergoes reduction in the scanned gate voltage range. V_{ecG}: electrochemical gate voltage. Further details in ref. [23]. Source: Adapted from Neubert et al. [23].

The central goal in graphene-based FETs is, however, to detect mainly changes in interfacial charge or potential occurring at the GLI. Hence, ideally, we would like to minimize the Faradaic effect. There are several ways to achieve this. The Faradaic effect occurs due to an electrochemical reduction or oxidation current flowing between the graphene electrode and redox active species in solution. As we know from basic electrochemistry, the current is proportional to electrode area. Using a small electrode minimizes this current significantly and the threshold voltage shifts due to Faradaic effect become negligible. Hence, it is important to use graphene sheets of small channel dimensions, i.e. micro- or nanostructured, when designing sensors based on graphene FETs. The presence or absence of this effect can be verified by measuring the leakage currents with or without the redox-active species in solution. In addition, by using different scan rates and ranges of the gate voltage sweep, one can get an idea about the occurrence of this effect. Another approach is to ensure that the exchange current density for a given redox reaction at the graphene electrode is low. For example, the reduction of protons to form hydrogen (hydrogen evolution reaction, HER) is an extremely unfavorable reaction at the graphene surface and does not occur in a potential range of up to –0.7 V vs. Ag/AgCl at low pH [51]. Hence, we do not expect to see a Faradaic effect due to protons in solution when the gate voltage is limited to a range above this value. However, when electrodes are modified, for e.g. with metal nanoparticles, hydrogen evolution can be enhanced and the exchange current density becomes sizeable. Often, nanoparticles are used for increasing sensitivity. However, from a mechanistic point of view, for non-polarized graphene FETs operating in an aqueous environment, the addition of nanoparticles may be detrimental to a reliable measurement of interface potential. Oxygen reduction on the other hand occurs already at low potentials

(−0.4 V vs. Ag/AgCl). Hence, care should be taken to keep this current to a minimum, e.g. by using a small electrode. Alternatively, it should be ensured that the concentration of redox-active species remains constant during the sensor trials. Water oxidation is typically not a problem since the FET measurements are usually performed in a lower gate voltage range. Apart from these three cases, care should be taken when working with biomolecules. Several enzymes, proteins, and even nucleic acids are redox active, and it should be ensured that the potential changes due to the analytes are not due to Faradaic effects with such redox active species.

11.8 The GLI and pH Sensing

We will now look at some application examples where the occurrence of one or more of these mechanisms has been documented. pH sensors with a direct analogy to silicon ISFETs have been demonstrated using graphene FETs modified with an intact ISM such as Al_2O_3 or Ta_2O_5 [47, 49]. In sensors where graphene is directly used as a pH sensor or when the ISM is not completely blocking ion transport, we need to take a closer look at the possible mechanisms by which such a sensor can detect pH. At the outset, we can already exclude the possibility of several mechanisms, such as contact or Schottky barrier effects and changes in gate coupling efficiency. Also by using small graphene flakes, charge transfer and Faradaic effects can be excluded. This leaves us with only one possibility, namely the electrostatic doping mechanism, where the interfacial charge-potential landscape is modulated by changes in pH. An ideal unmodified graphene surface is supposed to contain a hydrophobic sheet of sp^2-carbon, without any ionizable groups [49]. However, the graphene sheets prepared in practice usually do indeed have some stray functionalities, which are introduced unintentionally during the fabrication procedure. The charge on these functional groups can be modulated by the solution pH, e.g. via protonation/deprotonation, which is then reflected as a threshold shift in the gate voltage characteristics. In this context, we can understand the GLI as a protein-like surface, where the charge at the interface for a given pH is determined by the isoelectric point (or point of zero charge) of the sensor surface. The pI/pzc of as-prepared graphene is reported to be around 3, rendering the surface negative at pH above 3 [30]. It is not only the functional groups, which bring about this negative charge, but also nonspecific ion adsorption, which was found to play an important role in determining the interfacial charge distribution.

An elaborate theoretical model has also been proposed to explain the pH-dependence of graphene FET characteristics [30]. When one takes a closer look, it is apparent that the sensitivity of the Dirac point to pH in the simplest case, with only one pI will not be linear, since the charge distribution is a sigmoidal function of pH for a given pI/pzc. However, if the surface contains moieties with a range of different pI, then we would obtain a more complex relationship between the measured Dirac point and pH. Often, it has been observed that in a certain pH range, the Dirac point shows a linear dependence on pH [18, 47, 52]. From this viewpoint, it is also clear why graphene devices from different research groups show a broad range of sensitivities. Since every methodology for sensor fabrication uses a different procedure and different reagents for preparing the sensors, it is understandable that the density of the surface chemical moieties and its pI/pzc properties are different from one another, resulting in a broad variation in pH sensitivity of the Dirac point of graphene. In principle, the pH sensitivity can be modulated by deliberately introducing functional groups, whose pK_a is well known [30]. It has to be kept in mind, however, that the net interfacial charge will be a balance between functional groups present on the graphene surface before the functionalization and the additional groups introduced after the functionalization.

Based on this model, it can be seen that the shift in Dirac point with increasing pH should always occur in one direction, since the surface can only become relatively more negative or less positive when one moves toward higher pH. It is also to be kept in mind that the modulation in FET characteristics that we observe can only be sensitive to relative changes in net surface charge density. It will be difficult to quantify the surface charge density directly. However, we can clearly identify the sign of charge change induced by the analyte [30]. In most pH sensors based on graphene demonstrated until now, this explanation can clarify all the observations [18, 47, 52]. There are however a few cases where an opposite shift in Dirac point as a function of pH has been observed. In most of them, this effect can be traced back to an improper setup regarding the choice of gate electrode, passivation, and placement of source and drain electrodes or the choice of buffer compositions containing specifically interacting ions [53, 54].

There is one caveat to using graphene FETs for sensing pH as well as biomolecules. This concerns the ionic strength and buffer capacity of the solution. Since the EDL capacitance (the value determined by the ionic strength) occurs in series with the capacitance due to the functional layer, changes in EDL capacitance caused by ionic strength variations have a strong influence on the FET characteristics. Often, one sees a higher sensitivity for the analyte at a lower ionic strength [16, 18, 30]. Both buffer capacity and ionic strength have an effect on the interfacial charge density. In order to be able to interpret pH sensing effects reliably, it is necessary, at least in the calibration stage, to use buffers with well-defined buffer capacity and ionic strength. The dependence of FET characteristics on the ionic strength is a fundamental limitation for pH sensing using non-polarized graphene FETs. For clarification, reports of graphene FETs as pH sensors should always include sensor responses in solutions of several defined ionic strength, as well as over a sufficient range of pH values. Furthermore, specific interactions of graphene with buffer components other than protons need to be excluded [54].

11.9 Detection of Nucleic Acids

From the foregoing discussions, it is clear that properly designed graphene FET sensors are primarily sensitive to pzc/pI of the GLI. Hence, a general strategy for detecting biomolecules is to use a receptor layer whose pI or pK_a changes upon binding of an analyte. Indeed such a mechanism appears to be plausible for most of the graphene FET sensors for biomolecules and cells reported. The receptor layer is prepared often by using an intermediate anchor layer. In a majority of the studies, the intermediate anchor layer comprises of pyrene butyric acid that sticks to graphene due to π–π interactions. The carboxyl group is activated to obtain an *N*-hydroxysuccinimide ester end group, which is coupled to amine-modified receptors or directly to proteins via formation of an amide bond [16, 55–57]. Other strategies include the use of polyelectrolytes [58], polymer brushes [59], porphyrins [56], aniline derivatives [28, 30], nanoparticles [60], or glutaraldehyde [61].

For the selective detection of single-stranded nucleic acids, it is straightforward to use the complementary strand as the receptor, which can be immobilized on graphene using various functionalization strategies [15, 16, 41, 62]. Often the layer is, however, not completely intact and is penetrable to liquid and electrolytes, which is in contrast to what is required for an ideal ISM. Hence, the exact mechanism is difficult to interpret, due to the complex interplay between several different interfacial interactions and the possibility of more than one mechanism acting simultaneously in this scenario [34, 43]. When realizing sensors for DNA, it is important to demonstrate that the sensor response is recovered upon melting of the hybridized double strand. Typically, the sensor response is measured by moving from low to higher concentrations, where a continuous change in the FET characteristics is often presented. In this situation, it is difficult to distinguish between nonspecific adsorption and specifically bound analytes, and the search for a mechanistic understanding is futile. If one assumes that mainly electrostatic doping mechanisms are involved in the sensor response, it would be expected that the hybridization of DNA leads to a shift in the transfer characteristics along the gate voltage scale in one direction. However, in the literature, one can find shifts to lower [15, 41, 43, 62] as well as higher [16, 34, 63] gate voltages. One possibility is the difference in the chemistry of the graphene surface, due to the varying fabrication methodologies and the exact chemical strategy adopted for preparing the receptor layer, which may yield differing starting pK_a or pI for the devices from one study to another. In addition, what is often overlooked is the possibility of adsorption of analyte molecules on the reference electrode or the occurrence of small leakage currents with polarizable gate electrodes (such as platinum), which may also result in an opposite shift of the transfer curves than what is expected.

11.10 Other Examples

Graphene FETs have also been used for the detection of proteins [55], peptides [57], enzymes [28], etc. Proteins are charged in aqueous solutions based on their isoelectric point (pI). Therefore, proteins with different pI are expected to shift the transfer curve of graphene through the electrostatic doping mechanism depending on the change in effective charge density upon binding. Detection of more complex systems like bacteria [64, 65] and viruses [66] has also been possible. On the other hand, there are graphene sensors, which do not monitor a specific molecule but rather signals from cells, e.g. cell activities and potentials [67, 68], or mimic biological activities [60, 61]. In these and several other examples, graphene has been used for detecting changes in a nonspecific manner. The mechanisms behind such sensing responses are difficult to decipher because of the high complexity of interactions involved.

11.11 Concluding Remarks

Ideally, it would be good to engineer graphene sensors such that only one mechanism is dominant or active. Detection of analytes through changes in interfacial surface charge (electrostatic doping) appears to be the ideal candidate for this purpose. In this situation, it should be ensured that charge transfer is minimized, which may otherwise induce unwanted drifts in sensor operation. Even ubiquitous species such as dissolved oxygen may be involved in modulating the graphene electrode potential. Care must be taken to avoid contributions from such reactions to the FET characteristics, e.g. by using micro-/nanostructured graphene of low surface area. Selectivity can be achieved by choosing a receptor, which not only binds the analyte specifically, but also induces the largest change in the interfacial acid–base properties. Furthermore, the receptor layer can be designed in such a way that charge transfer to the underlying graphene sheet is prohibited, rendering the graphene sheet as polarizable as possible. In any case, measurements of channel-gate (leakage) currents should always be performed to be certain about their contributions to the measured graphene FET response. Finally, from a technical perspective, it is optimal to have the contacts passivated, the graphene channel micro- or nanostructured, and a proper reference electrode used as the gate in order to obtain a clear mechanistic understanding. If the analyte has the capability to undergo charge transfer with graphene, then a more suitable method is to use electrochemical analysis (chronoamperometry or pulsed methods) to detect such species. Such heterogeneous electron transfer can be tuned in several ways, e.g. increasing surface area, introducing edges, modification with catalytically active metal particles, or organic functionalization, and hence it is possible to obtain a very high sensitivity to an analyte of interest. In this case, however, achieving selectivity is not that straightforward. Either using the FET configuration, or using electroanalysis, it can be expected that with a good mechanistic understanding, the dream of using graphene sensors for realistic applications will become a reality.

References

1 Béraud, A., Sauvage, M., Bazán, C.M. et al. (2021). *Analyst 146* (2): 403–428.

2 Fu, W., Jiang, L., van Geest, E.P. et al. (2017). *Adv. Mater. 29* (6): 1603610.

3 Balasubramanian, K. and Kern, K. (2014). *Adv. Mater. 26* (8): 1154–1175.

4 Hwang, M.T., Park, I., Heiranian, M. et al. (2021). *Adv. Mater. Technol. 6* (11): 2100712.

5 Schuck, A., Kim, H.E., Moreira, J.K. et al. (2021). *Sensors 21* (5): 1852-1:12.
6 Fenoy, G.E., Marmisollé, W.A., Knoll, W., and Azzaroni, O. (2022). *Sens. Diagn. 1* (1): 139–148.
7 Balasubramanian, K. (2010). *Biosens. Bioelectron. 26* (4): 1195–1204.
8 Sze, S.M. and Lee, M.K. (2013). *Semiconductor Devices: Physics and Technology*, 3e. Singapore: John Wiley & Sons Singapore Pte. Ltd.
9 Heller, I., Chatoor, S., Männik, J. et al. (2009). *Phys. Status Solidi RRL 3* (6): 190–192.
10 Balasubramanian, K., Lee, E.J.H., Weitz, R.T. et al. (2008). *Phys. Status Solidi A 205* (3): 633–646.
11 Janata, J. (2012). *ECS Solid State Lett. 1* (6): M29–M31.
12 Bard, A.J. and Faulkner, L.R. (2000). *Electrochemical Methods: Fundamentals and Applications*, 2e. New York: Wiley.
13 Islam, S., Shukla, S., Bajpai, V.K. et al. (2019). *Sci. Rep. 9* (1): 276. 1–7.
14 Kwong Hong Tsang, D., Lieberthal, T.J., Watts, C. et al. (2019). *Sci. Rep. 9* (1): 13946-1. 10.
15 Hwang, M.T., Landon, P.B., Lee, J. et al. (2016). *Proc. Natl. Acad. Sci. U.S.A. 113* (26): 7088–7093.
16 Xu, S., Zhan, J., Man, B. et al. (2017). *Nat. Commun. 8*: 14902.
17 Geim, A.K. and Novoselov, K.S. (2007). *Nat. Mater. 6* (3): 183–191.
18 Heller, I., Chatoor, S., Männik, J. et al. (2010). *J. Am. Chem. Soc. 132* (48): 17149–17156.
19 Xia, J., Chen, F., Li, J., and Tao, N. (2009). *Nat. Nanotechnol. 4* (8): 505–509.
20 Rosenblatt, S., Yaish, Y., Park, J. et al. (2002). *Nano Lett. 2* (8): 869–872.
21 Heller, I., Janssens, A.M., Männik, J. et al. (2008). *Nano Lett. 8* (2): 591–595.
22 Burke, L.D., Buckley, D.T., and Morrissey, J.A. (1994). *Analyst 119* (5): 841–845.
23 Neubert, T.J., Wehrhold, M., Kaya, N.S., and Balasubramanian, K. (2020). *Nanotechnology 31* (40): 405201.
24 Iost, R.M., Crespilho, F.N., Zuccaro, L. et al. (2014). *ChemElectroChem 1* (12): 2070–2074.
25 Zhang, X., Jing, Q., Ao, S. et al. (2020). *Small 16* (15): e1902820.
26 Shih, C.-J., Wang, Q.H., Jin, Z. et al. (2013). *Nano Lett. 13* (2): 809–817.
27 Fan, X.-Y., Nouchi, R., Yin, L.-C., and Tanigaki, K. (2010). *Nanotechnology 21* (47): 475208.
28 Zuccaro, L., Tesauro, C., Kurkina, T. et al. (2015). *ACS Nano 9* (11): 11166–11176.
29 Georgakilas, V., Otyepka, M., Bourlinos, A.B. et al. (2012). *Chem. Rev. 112* (11): 6156–6214.
30 Zuccaro, L., Krieg, J., Desideri, A. et al. (2015). *Sci. Rep. 5*: srep11794.
31 Heinze, S., Tersoff, J., Martel, R. et al. (2002). *Phys. Rev. Lett. 89* (10): 106801.
32 Tang, X., Bansaruntip, S., Nakayama, N. et al. (2006). *Nano Lett. 6* (8): 1632–1636.
33 Giubileo, F. and Di Bartolomeo, A. (2017). *Prog. Surf. Sci. 92* (3): 143–175.
34 Campos, R., Borme, J., Guerreiro, J.R. et al. (2019). *ACS Sensors 4* (2): 286–293.
35 Schedin, F., Geim, A.K., Morozov, S.V. et al. (2007). *Nat. Mater. 6* (9): 652–655.
36 Kaverzin, A.A., Strawbridge, S.M., Price, A.S. et al. (2011). *Carbon 49* (12): 3829–3834.

37 Levesque, P.L., Sabri, S.S., Aguirre, C.M. et al. (2011). *Nano Lett. 11* (1): 132–137.

38 Kim, S.J., Park, S.J., Kim, H.Y. et al. (2016). *Appl. Phys. Lett. 108* (20): 203111-1:5.

39 Pinto, H. and Markevich, A. (2014). *Beilstein J. Nanotechnol. 5*: 1842–1848.

40 Macedo, L.J.A., Iost, R.M., Hassan, A. et al. (2019). *ChemElectroChem 6* (1): 31–59.

41 Cai, B., Wang, S., Huang, L. et al. (2014). *ACS Nano 8* (3): 2632–2638.

42 Kurkina, T., Sundaram, S., Sundaram, R.S. et al. (2012). *ACS Nano 6* (6): 5514–5520.

43 Lin, C.-T., Loan, P.T.K., Chen, T.-Y. et al. (2013). *Adv. Funct. Mater. 23* (18): 2301–2307.

44 Wang, J. (2006). Potentiometry. In: *Analytical Electrochemistry*, 165–200. Wiley-VCH.

45 Bergveld, P. (2003). *Sens. Actuators, B 88* (1): 1–20.

46 Palazzo, G., De Tullio, D., Magliulo, M. et al. (2015). *Adv. Mater. 27* (5): 911–916.

47 Fakih, I., Mahvash, F., Siaj, M., and Szkopek, T. (2017). *Phys. Rev. Appl. 8* (4): 044022-1:7.

48 Li, J., Pham, P.H.Q., Zhou, W. et al. (2018). *ACS Nano 12* (10): 9763–9774.

49 Fu, W., Nef, C., Knopfmacher, O. et al. (2011). *Nano Lett. 11* (9): 3597–3600.

50 Fakih, I., Durnan, O., Mahvash, F. et al. (2020). *Nat. Commun. 11* (1): 3226.

51 Wehrhold, M., Neubert, T.J., Grosser, T. et al. (2022). *Chem. Commun. 58*: 3823.

52 Tan, X., Chuang, H.-J., Lin, M.-W. et al. (2013). *J. Phys. Chem. C 117* (51): 27155–27160.

53 Lee, M.H., Kim, B.J., Lee, K.H. et al. (2015). *Nanoscale 7* (17): 7540–7544.

54 Mitsuno, T., Taniguchi, Y., Ohno, Y., and Nagase, M. (2017). *Appl. Phys. Lett. 111* (21): 213103.

55 Wang, X., Hao, Z., Olsen, T.R. et al. (2019). *Nanoscale 11* (26): 12573–12581.

56 Kawata, T., Ono, T., Kanai, Y. et al. (2018). *Jpn. J. Appl. Phys. 57* (6): 065103-1:4.

57 Hao, Z., Zhu, Y., Wang, X. et al. (2017). *ACS Appl. Mater. Interfaces 9* (33): 27504–27511.

58 Piccinini, E., Bliem, C., Reiner-Rozman, C. et al. (2017). *Biosens. Bioelectron. 92*: 661–667.

59 Piccinini, E., Bliem, C., Giussi, J.M. et al. (2019). *Langmuir 35* (24): 8038–8044.

60 Wu, D., Yu, Y., Jin, D. et al. (2020). *Anal. Chem. 92* (5): 4006–4015.

61 Park, S.J., Kwon, O.S., Lee, S.H. et al. (2012). *Nano Lett. 12* (10): 5082–5090.

62 Zheng, C., Huang, L., Zhang, H. et al. (2015). *ACS Appl. Mater. Interfaces 7* (31): 16953–16959.

63 Kim, H.E., Schuck, A., Lee, J.H., and Kim, Y.S. (2019). *Sens. Actuators, B 291*: 96–101.

64 Wu, G., Dai, Z., Tang, X. et al. (2017). *Adv. Healthcare Mater. 6* (19): 1700736-1:9.

65 Kumar, N., Wang, W., Ortiz-Marquez, J.C. et al. (2020). *Biosens. Bioelectron. 156*: 112123-1:8.

66 Kim, J.W., Kim, S., Jang, Y.H. et al. (2019). *Nanotechnology 30* (34): 345502-1:6.

67 Hess, L.H., Jansen, M., Maybeck, V. et al. (2011). *Adv. Mater. 23* (43): 5045–5049.

68 Kireev, D., Brambach, M., Seyock, S. et al. (2017). *Sci. Rep. 7* (1): 6658.

12

Surface Modification Strategies to Increase the Sensing Length in Electrolyte-Gated Graphene Field-Effect Transistors

Juliana Scotto[1], *Wolfgang Knoll*[2,3], *Waldemar A. Marmisollé*[1], *and Omar Azzaroni*[1]

[1]*Universidad Nacional de La Plata (UNLP) – CONICET, Instituto de Investigaciones Fisicoquímicas Teóricas y Aplicadas (INIFTA), Departamento de Química, Facultad de Ciencias Exactas, Diagonal 113 y 64, 1900, La Plata, Argentina*
[2]*Austrian Institute of Technology, Konrad-Lorenz-Straße 24, Tulln an der Donau, 3430, Austria*
[3]*Danube Private University, Steiner Landstraße 124, Krems, 3500, Austria*

12.1 Introduction

In recent years, the use of graphene-based field-effect transistors (GFETs) as bioanalytical tools has attracted interest in the biotechnology research field as well as in the development of diagnostic devices for medical applications since they provide the possibility of real-time and label-free detection with high sensitivity while allowing miniaturization and low-cost mass production [1–9]. In particular, electrolyte-gated graphene-based field-effect transistors (EG-GFETs) are especially well suited for operating in solution, making them the natural choice for biosensing applications.

The general mechanism for biomolecule sensing with EG-GFETs relies on the modulation of the electrical conductivity of the graphene channel caused by the doping effect of analytes binding on the surface. The doping effect in the graphene channel can be achieved by direct interaction of charged biomolecules that are adsorbed close to the surface [3] or by other mechanisms such as conformational changes of the charged recognition elements upon analyte binding [10], changes in the interface impedance caused by the adsorption of dense biomolecules [11], or local pH variations induced by chemical reactions arising from the analyte interaction with the surface [12]. In all cases, the changes in the charge distribution of the surface layer induce a modulation of the effective gate potential, causing a shift in the Dirac point and thus changing the measured drain–source current signal (I_{DS}).

Unfortunately, this interaction is screened by the presence of mobile ions forming the electrical double layer (EDL). Thus, the magnitude of the potential change on the graphene surface strongly depends on the ion concentration of the solution, and

Graphene Field-Effect Transistors: Advanced Bioelectronic Devices for Sensing Applications, First Edition. Edited by Omar Azzaroni and Wolfgang Knoll.

the distance from the surface at which target molecules can be detected is limited by the so-called Debye length,

$$\lambda_D = \kappa^{-1} = \left(\frac{\varepsilon_r \varepsilon_0 k_B T}{2 N_{av} e^2 I} \right)^{1/2} \tag{12.1}$$

where ε_0 and ε_r are the vacuum permittivity and the relative permittivity in solution, respectively, k_B is the Boltzmann constant, T is the temperature, N_{av} is the Avogadro constant, e is the elementary charge, and I is the ionic strength [13]. The decay of the interface potential across the solution can be accounted for by the Debye–Hückel model as

$$\psi(x) = \psi_0 e^{-x/\lambda_D} \tag{12.2}$$

where x is the distance from the charged plane and ψ_0 is the interface potential that can be expressed according to Grahame equation as

$$\psi_0 = \frac{2k_B T}{ze} \sinh^{-1} \left(\frac{\sigma}{\sqrt{8 I \varepsilon_r \varepsilon_0 k_B T}} \right) \tag{12.3}$$

where σ is the surface charge density and z is the charge of the counterions. In physiological conditions (~140 mM ionic strength), this distance is less than 1 nm, hindering the detection of most analytes in biological samples. Moreover, building sensing architectures often requires the immobilization of large recognition bioelements, which significantly increases the distance between the target species and the surface.

As a first approach to surpass this issue, dilution and desalting of biological samples have been performed to enable the detection of a variety of analytes at low ionic strength [14, 15]. However, methods that require samples pretreatment or complex procedures are not suitable for the development of diagnostic devices for real-time detection. Another approach employed to overcome Debye screening limitations consists of decreasing the length of the recognition layer by employing small receptors. For instance, using only the binding fragment of an antibody instead of the whole antibody attached to the surface can considerably reduce the sensing distance to a few nanometers [16, 17]. In addition, aptamers have also been employed as small-size receptors to mitigate the screening issue [18–20]. However, for many sensing systems, the decrease in the recognition layer length achieved through this method is still insufficient to detect analytes in physiological conditions where the thickness of the double layer is below 1 nm. In this context, it is necessary to develop strategies to extend the sensing length facing GFET technology for real-time determinations.

The most common strategy to extend sensing length consists of modifying the graphene surface with soft architectures to reduce the screening inside the region comprised between the target analyte and the sensing electrode surface. This approach has been investigated and applied for the detection of several biomolecules such as proteins, DNA, and small molecules in physiological-like conditions [17, 21–24]. In this chapter, different surface modification strategies will be addressed, as well as the mechanisms enabling the extension of the sensing length in each case.

12.2 Ion-Exclusion and Donnan Potential

A widely explored approach to overcome Debye screening limitations in GFET sensing is to build ion-permeable structures atop the graphene channel to decrease the ion concentration in the binding region. In some architectures, ion-permeable membranes are generated by the immobilization of dense bioreceptors together with linkers employed to attach the receptor to the surface or molecules employed as blocking agents [22, 25]. In other structures, to create sufficiently dense semipermeable membranes the co-immobilization of long polymer chains together with the recognition elements is necessary [10, 17, 24]. The density of the surface layer where the binding process occurs has a very important role in decreasing the screening effect. This concept was demonstrated by Piccinini et al., who studied the screening effect inside a polyelectrolyte multilayer (PEM) as a function of the polymer fraction [26]. In that work, it was shown that changes in the Dirac point due to the adsorption of charged macromolecules could be observed even when the polyelectrolyte deposition was located far beyond the solution Debye length. It was stated that inside the low-hydration film there was a low ion concentration that led to a lower decay of the interface potential through the film compared with bulk solution (Figure 12.1a), enhancing the potential surface change. The relationship between the increase in the Dirac point shift upon charge adsorption and the polymer fraction was determined from a thermodynamic model based on the entropic cost of confining ions inside the polymer film (Figure 12.1b). According to this model, a 1 order of magnitude increase in the effective Debye length is expected inside the PEM (Figure 12.1c). However, these calculations suggested that a very high density of the polymeric film is necessary to increase the sensing range into dimensions that are relevant in biological media, based only on the effect of ion partition due to entropic contributions.

On the other hand, charge immobilization inside an ion-permeable membrane generates the accumulation of counterions to maintain electroneutrality. As a result, a difference in the ion concentration between the film and the solution is established, generating a Donnan potential difference. The Donnan equilibrium can be understood as a series capacitance with the EDL capacitance that does not depend on the distance from the surface at which the membrane equilibrium is set, and therefore its contribution to the surface potential change is not constrained to the Debye length [27]. The Donnan effect has been employed to extend the sensing region, allowing the detection of many biomolecules in physiological conditions [3, 22, 23].

For instance, Hajian et al. functionalized the graphene surface with a clustered regularly interspaced short palindromic repeats (CRISPR) - CRISPR -associated protein 9 (CRISPR-Cas9) complex to specifically bind and quantify target DNA sequences [22]. The detection of DNA beyond the Debye length was accomplished by the construction of an ion-permeable structure atop graphene surface composed by the recognition complex, formed by the catalytically deactivated Cas9 protein and the single-guide RNA, together with the 1-pyrenebutanoic acid (PBA) linker employed to attach the protein to the surface and the amino-polyethylene glycol (NH_2-PEG) and ethanolamine (ETA) used to prevent non-specific adsorption. Then,

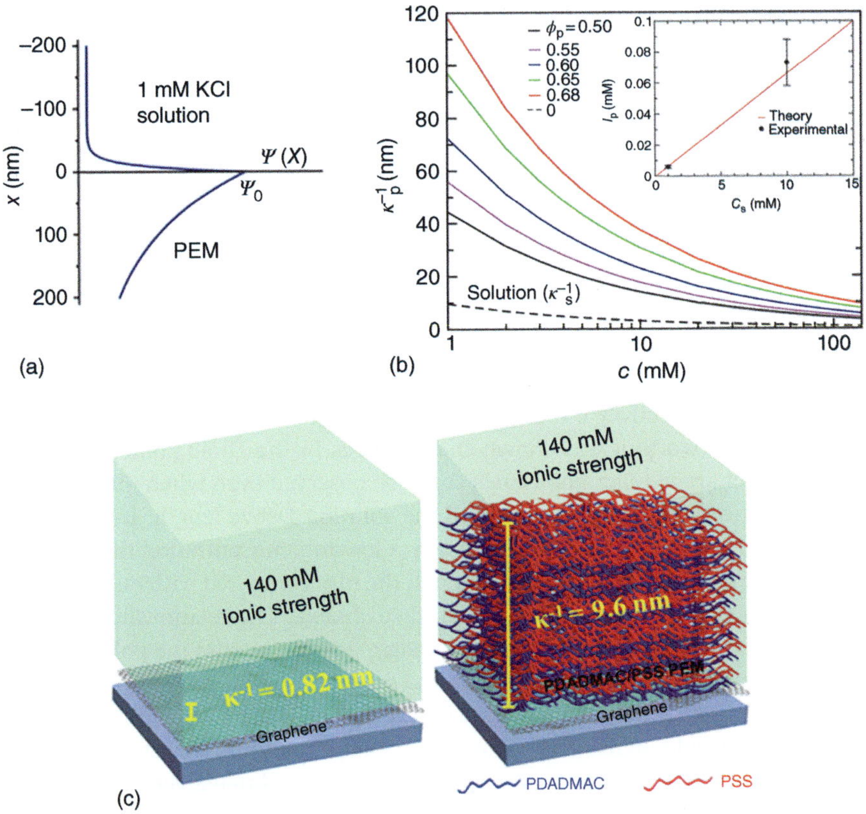

Figure 12.1 (a) Potential profile, $\psi(x)$, as a function of the distance, x, in 1 mM KCl bulk solution and inside the PEM. (b) Theoretical predictions from the thermodynamic model of the Debye screening length as a function of the KCl concentration in the bulk solution, c, for different polymer (volume) fractions, ϕ_p. (c) Scheme of the effective Debye length at a graphene interface in a solution of 140 mM ionic strength with (right) and without (left) the PEM. Source: Reproduced with permission from Piccinini et al. [26]/American Chemical Society.

the selective hybridization of charged DNA targets to the CRISPR-terminated surface inside the membrane generated changes in the Donnan equilibrium, modulating the effective gate potential and, therefore, the drain–source current signal.

A similar strategy was employed by Goldsmith et al. in the detection of human interleukin-6 (IL6) [23]. In that work, the immobilization of the monoclonal antibody (anti-IL6) inside an ion-permeable membrane atop the GFET substrate was performed (Figure 12.2a), allowing the detection of IL6 in phosphate-buffer saline (PBS) ×1 medium in a wider concentration range than the one obtained with standard enzyme-linked immunosorbent assay (ELISA) technique, with a data acquisition time lower than five minutes (Figure 12.2b). The IL6 detection beyond Debye length through a Donnan mechanism was enabled by the use of PEG chains to create a dense recognition surface layer. The specific role of PEG in the extension of the sensing length will be further discussed in Section 12.3.

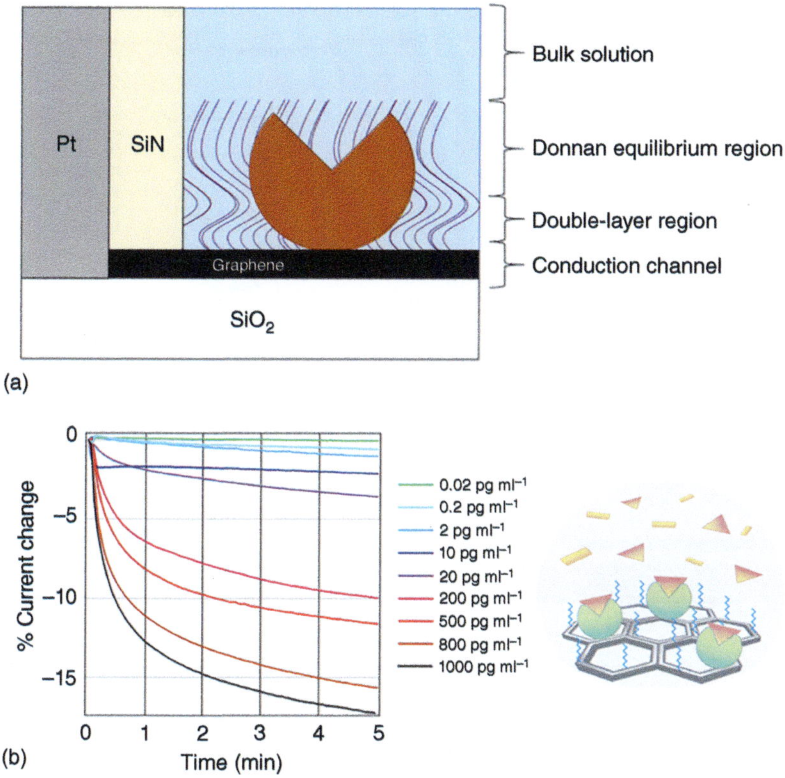

Figure 12.2 (a) Diagram of the IL6 sensor regions. (b) Change in the drain–source current of sensors functionalized with antibodies against IL6 in the presence of different concentrations of IL6 and diagram of the binding process on the modified GFET. Source: Reproduced from Goldsmith et al. [23]/Springer Nature/CC BY 4.0.

12.3 Surface Modification with Polymer Films

Co-immobilization of polymer films with the recognition elements appears to be one of the preferred approaches to increasing Debye length. In particular, graphene modification with PEG is the most employed strategy to enable GFET detection of charged biomolecules in physiological ionic strength conditions. The modification of electrode surfaces with PEG is well known since it has been repeatedly performed in the development of biosensors to attach recognition elements to the surface and to avoid non-specific interactions [3, 7, 23, 28, 29]. However, Gao et al. demonstrated that the use of long chains of PEG could also decrease the electrolyte screening by extending the sensing length. They first showed this concept by studying the enhancement of a silicon nanowire FET signal in the detection of prostate-specific antigen (PSA) [29]. Next, they extended this approach to a GFET by monitoring the specific and non-specific detection of PSA on a PEG-modified graphene surface

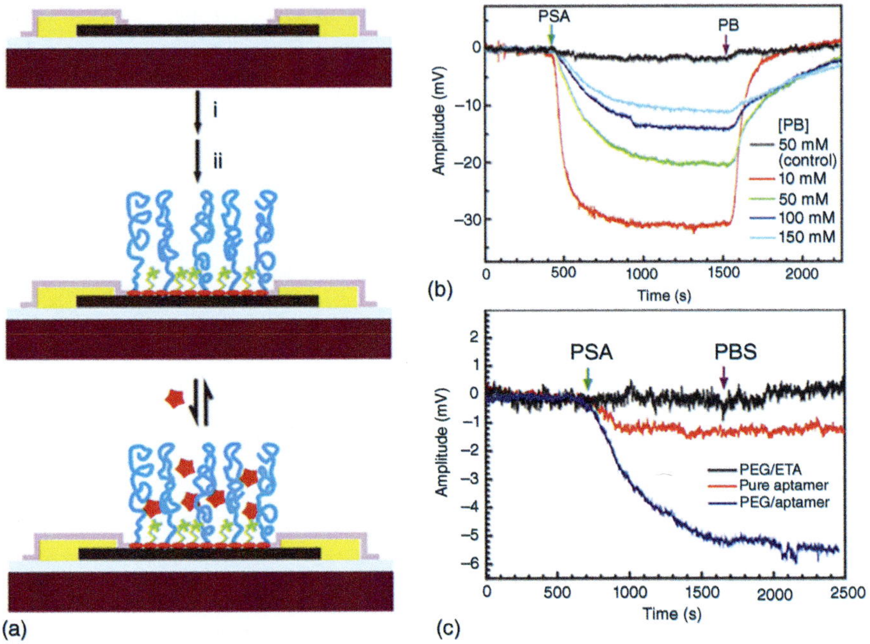

Figure 12.3 (a) Illustration of a graphene FET device modified with PEG (blue) and a small-molecule spacer or a receptor (green) for nonspecific and specific detection of the analyte (red star). (b) Nonspecific PSA detection at different phosphate buffer (PB) concentrations for a PEG-modified GFET and a GFET without PEG (control). (c) Specific detection for 300 nM PSA in high-ionic-strength solution from 1 : 2 PEG:ETA-, pure aptamer-, and 1 : 2 PEG:aptamer-modified graphene FET devices. Source: Reproduced with permission from Gao et al. [10]/National Academy of Sciences.

(Figure 12.3a) [10]. For this purpose, they covalently attached amino-PEG and ETA to the PBA previously adsorbed on the graphene via π–π stacking. Then, they studied the non-specific detection of PSA, obtaining a reversible signal that slowly decreased with the electrolyte concentration (compared with unmodified graphene), holding a well-defined response even at high ionic strengths (150 mM phosphate buffer), while no detectable signal was observed for the same electrolyte concentration in the absence of PEG (Figure 12.3b). In addition, they studied the specific binding of PSA by co-reaction of PBA with PEG and a DNA aptamer. An irreversible signal upon protein binding was obtained and attributed to modulations in the electric field induced by conformational changes of the highly charged aptamer near the sensor surface. Interestingly, the amplitude of the signal obtained with this architecture was five times higher than that obtained in the absence of PEG (Figure 12.3c).

The effect of PEG layers on the FET sensing length was also observed by Gutiérrez-Sanz et al. in the detection of thyroid-stimulating hormone (TSH) [30]. In that work, they covalently attached the antibody fragment (anti-TSH F(ab′)2) via short PEG-based linker to a gold surface and studied the effect of the co-immobilization of long PEG chains. A threefold signal enhancement was observed when the large PEG chains were deposited on the electrode surface for a given electrolyte concentration, enabling sensing in high ionic strength conditions. This strategy was later extended to GFETs technology for the development of TSH immunosensors with femtomolar detection limit in serum sample [17]. Based on those results, Haustein et al. developed an analytical model to explain the effect of PEG on the FET's current response [31]. In this model, it was considered that only a fraction of the charge of the binding analyte is effectively detected by the FET, and a prefactor was introduced to take into account the detectable charge, q_{det}. Then, they modeled the changes in the Dirac potential due to the immobilization of charged analytes inside a surface layer composed of PEG and a recognition molecule, assuming a uniform potential inside the surface layer (the Donnan potential). The obtained results suggested that the enhancement of the GFET signal arises from the increase in the fraction of the analyte charge that is effectively detected as a consequence of the Donnan equilibrium established in the presence of PEG (Figure 12.4a,b). In addition, it was stated that the q_{det} value at a determined ionic strength not only depends on the distance from the surface at which the biding event takes place and the analyte size but also on the length of the polymer chains. Based on their calculations, the PEG length should be at least as long as the combined linker-recognition molecule length to ensure the detection of half of the analyte charge. For the TSH studied system, the theoretical model predicted a threefold increase in the potential shift when PEG was co-immobilized with the recognition element, which was in very good agreement with the experimental results obtained by Gutierrez et al. (Figure 12.4c). PEG-modification approach to enhance the sensing signal has also been demonstrated for other field-effect based biosensors such as silicon nanowires [32], carbon nanotubes [33], and organic [34] FETs.

Following a similar method, Gao et al. functionalized a GFET with poly-L-lysine (PLL) to enhance the sensitivity of a biosensor for breast cancer miRNA detection [24]. The improvement in the performance of the GFET modified with PLL was first seen as an increase in the stability of the electrode and a larger potential shift due to the immobilization of DNA probes. The researchers then evaluated the effect

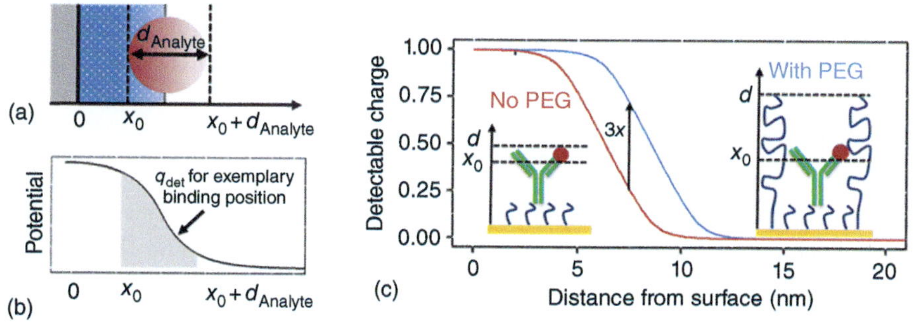

Figure 12.4 (a) Schematic representation of the binding of a charged analyte at a determined distance of the surface, x_0. (b) Potential distribution as a function of the distance from the surface. (c) Detectable charge, q_{det}, for an analyte bound at a distance x_0 from the surface, for a sensor with (blue) and without (red) PEG. Source: Reproduced with permission from Haustein et al. [31]/American Chemical Society.

of PLL modification on breast cancer miRNA hybridization with DNA probes. It was found that the interaction between the miRNA and the modified GFETs generated a n-doping effect, leading to a shift in the Dirac point that was threefold higher than the one obtained in the unmodified electrodes for measurements performed in buffer solutions with concentrations between 0.1× and 1× PBS. This enhancement of the signal due to the presence of PLL is consistent with the one observed for the addition of PEG, as discussed earlier, reinforcing the concept of polymer films capability to extend the sensing length.

12.4 Surface Modification with Lipid Layers

Graphene modification with lipid membranes is another strategy that allows sensing at large distances from the surface. The interest in modifying GFET sensing electrodes with lipid layers arises from the possibility of studying membrane-associated proteins in their natural environment [35, 36]. For instance, supported lipid bilayers (SLBs) have been used to maintain the structure and orientation of membrane-associated receptors for the study of protein–ligand interactions under nondenaturing conditions [36–38].

At first glance, anchoring the recognition element through a lipid layer could hinder the field-effect detection since it separates the binding event from the surface far beyond Debye length. However, the functionalization of FET electrodes with lipids has been successfully employed in the detection of small molecules [39, 40], ions [41, 42], and proteins [37, 43]. To explain the capability of GFETs for detecting charges located above lipid layers, a model based on the effect of the low dielectric constant inside the bilayer was proposed by Hu et al. [37]. In that study, a supported lipid membrane with biotinylated lipids was built on top of a GFET surface as model receptor, and the Dirac point shifts during the streptavidin (SA) binding were monitored (Figure 12.5a,b). Then, the distance between the target analytes and the graphene surface was calculated by considering that the Dirac point shifts were due to the local potential changes upon SA binding (i.e. $\Delta V_{\mathrm{D.P}} = \Delta\psi$) and fitting the obtained values to the theoretical equation for the dependence of the interface potential with the distance, derived from combining Eqs. (12.2) and (12.3):

$$\Delta V_{\mathrm{D.P}} = \frac{2k_{\mathrm{B}}T}{e}\sinh^{-1}\left(\frac{\sigma}{\sqrt{8I\varepsilon_{\mathrm{r}}\varepsilon_0 k_{\mathrm{B}}T}}\right)\mathrm{e}^{-x/\lambda} \tag{12.4}$$

The effective distance obtained by fitting the GFET signal to Eq. (12.4) was smaller than the geometrical distance between the center of the analyte and the surface (that results from the combination of the lipid bilayer length and half of the SA molecule height) in a value that matched the length of the acetyl chain region inside the phospholipid membrane where the screening effect can be neglected. Thus, they attributed the capability of lipid-modified GFETs to sense beyond Debye length to the low decay of the potential inside the lipid membrane as a consequence of the very low dielectric constant in the acyl chain region (Figure 12.5c).

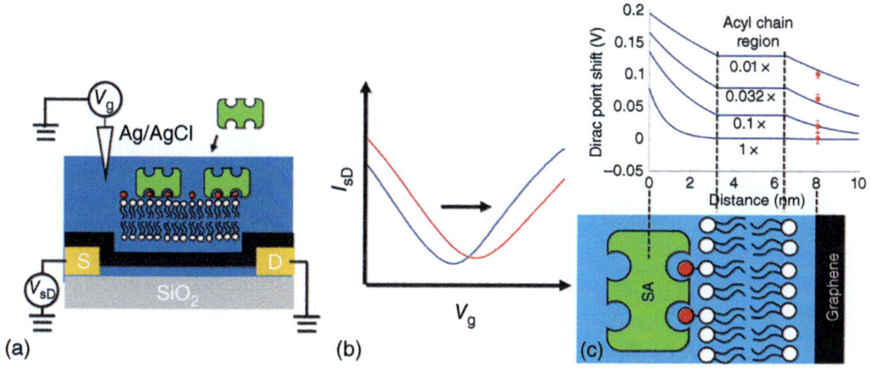

Figure 12.5 (a) Scheme of the detection of membrane-associated biomolecules using a supported lipid bilayer (SLB)-coated GFET platform. (b) Graphene characteristic transfer curves before (blue) and after (red) SA binding. (c) Theoretical predictions of the Dirac point shifts of the SLB-coated GFETs as a function of the distance between the analyte and the graphene surface at different buffer concentrations. Source: Reproduced with permission from Hu et al. [37]/American Chemical Society.

A different strategy to enable GFET biosensing that is also based on the changes in ion screening associated with the presence of lipid bilayers was shown by Ang et al. [43]. In that work, gram-negative bacteria biomimetic membranes were built on a graphene sensor, and the changes in the membrane properties induced by an antimicrobial peptide (Magainin 2) were measured employing field-effect technique. The assembly of biomimetic membranes of different surface charges generated detectable Dirac point shifts, enhanced by the decrease in the electrolyte screening consequence of the low ion permeability of the lipid bilayer. Then, when the surface layer was exposed to the presence of the Magainin 2 peptide, a thinning effect of the membrane was observed, causing the ion permeation to occur closer to the surface and therefore increasing the screening (Figure 12.6). This effect changed the GFET electrical signal, decreasing the minimum conductivity, and gradually shifting the Dirac point toward the bare graphene initial value.

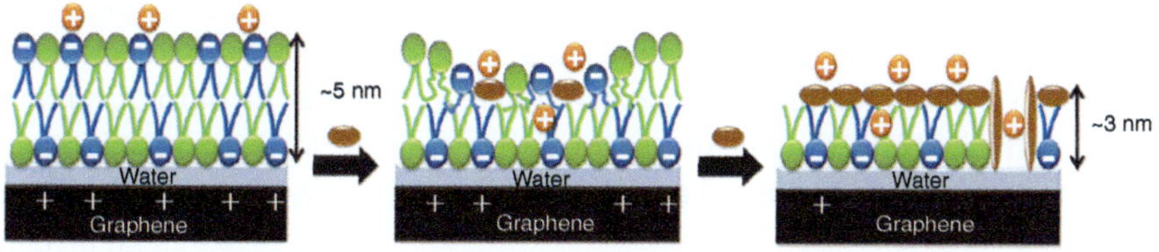

Figure 12.6 Illustration of the membrane thinning effect induced by Magainin 2 peptides (brown ovals) on gram-negative bacteria biomimetic membranes. Source: Reproduced with permission from Ang et al. [43]/American Chemical Society.

12.5 Surface Modification with Mesoporous Materials

The modification of graphene FETs with mesoporous materials has been recently presented as a novel strategy to decrease ion screening, enabling sensing at distances beyond Debye length [44, 45]. These materials have the advantage that they can be synthetized and functionalized to obtain different biocompatible structures with large surface areas and a wide variety of pore sizes that can be accurately controlled [46]. This aspect is relevant since the solution properties inside the mesoporous structure markedly differ from the bulk properties, and they can be tuned by functionalizing the pores or changing their morphology [47]. In particular, the dielectric constant, the ion concentration, and the potential distribution inside the material depend on the size and shape of the pores, which make them promising materials to modulate the potential decay across the interface [48]. Furthermore, the confinement of charged molecules inside the pores leads to ion exclusion through the Donnan effect, providing an additional mechanism to extend the sensing length [46, 49].

As an example, recently Alberti et al. reported the construction of graphene-based transistors modified with mesoporous silica thin films (MTF) [44]. They studied the sensitivity of the modified-GFET to the pH by measuring the Dirac point shifts at different pH values and found an improvement in the linearity of the dependence and an increase in the amplitude of the signal when the MTF was present compared with results obtained on bare graphene. In addition, they sensed the electro-oxidation of dopamine on a MTF-modified Au electrode employed as gate to prove the capability of small molecules to filtrate inside the pores.

Furthermore, to investigate the effect of the mesoporous material in the ion screening, they studied a layer-by-layer assembly generated by the alternate incubation of the MTF-modified graphene in solutions of negatively and positively charged polyelectrolytes of low molecular weight that could penetrate inside the mesoporous film. Interestingly, they observed an enhancement in the Dirac point shift for the first bilayers absorbed inside the pores compared with the signal obtained in bare graphene (Figure 12.7a). Moreover, when pores are completely filled, the field effect generated by polyelectrolyte adsorption outside the MTF can still be observed until the distance is 35 times larger than the Debye length in solution. This outstanding extension of the sensing length was explained by the authors as a consequence of the decrease in the local ion concentration due to the entropic cost of confining ions inside the pore. Finally, they measured the Dirac point changes due to the infiltration of a positively charged protein (lysozyme) inside the pores in 140 mM buffer, showing an enhanced GFET signal due to the presence of the MTF, broadening the use of this strategy toward biological applications (Figure 12.7b,c).

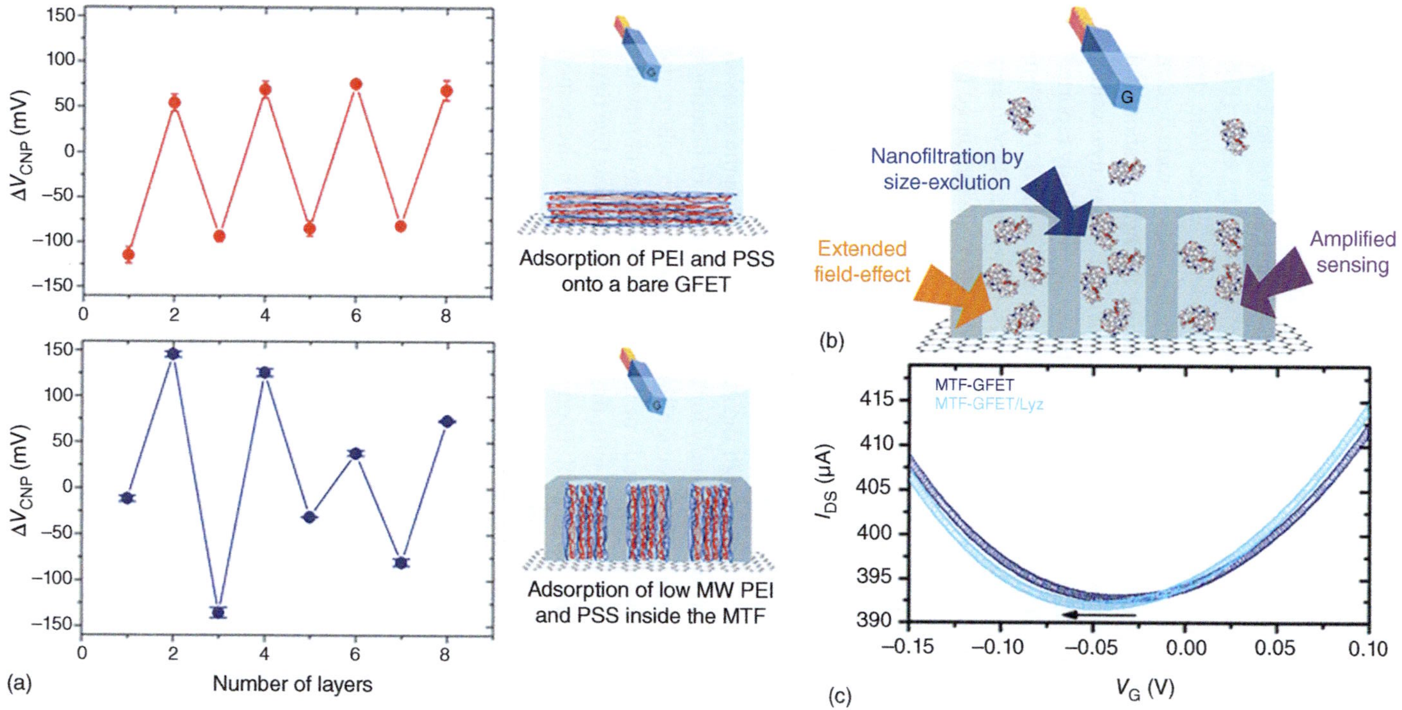

Figure 12.7 (a) Dirac point shifts (ΔV_{CVP}) as a function of the number of polyelectrolyte layers (top) deposited on a bare GFET and (bottom) adsorbed inside the mesoporous MTF-GFET structure, measured in 2 mM KCl. (b) Illustration of lysozyme filtration and adsorption inside the MTF structure. (c) MTF-GFET transfer curves measured before (blue) and after (light blue) the adsorption of lysozyme in a buffer of pH = 5 and ionic strength of 140 mM. Source: Reproduced with permission from Alberti et al. [44]/The Royal Society of Chemistry.

12.6 Kinetic Cost of Extending the Sensing Length

GFETs have been recently shown to be promising tools for the study of binding kinetics since they provide reliable and reproducible monitoring of molecular interactions with high sensitivity and low cost [5, 20, 50, 51]. As an example of this, Xu et al. determined the binding kinetics constants of DNA hybridization to DNA probes attached to the surface via 1-pyrenebutanoic acid succinimidyl ester (PBSE) in a highly stable architecture, obtaining reproducible results in real-time determinations [50]. The association and dissociation constants, k_{on} and k_{off}, were determined by measuring the Dirac point shifts of the GFET during the hybridization at different target DNA concentrations and after the replacement of the analyte solution with pure buffer, obtaining values that were within the range previously reported for the same DNA sequence determined by surface plasmon resonance (SPR) measurements [52]. However, in order to obtain these results, it was necessary to employ low ionic strength solutions (0.01× PBS buffer) to yield a Debye length large enough to enable the detection of the whole DNA chain.

As previously stated, for real-time diagnostics, operating in physiological conditions is crucial, and therefore the extension of the sensing length through the modification of the graphene surface with a suitable coating is often required. However, as a possible disadvantage of this strategy, when the recognition event occurs inside the coating, a delay in the GFET response may be observed due to the slow diffusion of biomolecules and ions through the dense layer covering the graphene electrode, hampering the use of this strategy for the determination of binding kinetics.

For instance, in the case of PSA recognition described earlier, where the use of PEG increased five times the amplitude of the signal, it was also observed that the time required to monitor the process with the modified-GFET was five times longer than without PEG (Figure 12.3c) [10]. Even though the delay in the response may not be limiting for their employment in point-of-care devices (detection results were obtained in less than 15 minutes), it does represent a serious limitation for the application of this technology to the study of binding kinetics. Furthermore, in the case of the TSH detection by anti-TSH F(ab′)2, the analyte quantification was performed after 20 minutes of incubation in the TSH solution, although the results could not be used for the determination of the binding constants [30]. In fact, the equilibrium constant, K_D, estimated from the GFET measurements in that work markedly differs from the one obtained by SPR technique.

The current contributions that alter the temporal response of the GFET signal upon molecular deposition on the surface may arise from different physicochemical processes. For instance, double layer formations on the graphene surface or Donnan capacitances generated by charge immobilization inside a dense membrane may be limited by slow ion or biomolecule diffusion inside the surface layer, altering the kinetic profile of the GFET response. Moreover, in electrolyte-gated GFETs, the double-layer capacitances originated at the gate electrode should also be considered. All these contributions to the GFET current are coupled, and their magnitude depends on many factors such as the electrolyte concentration, the density of the surface layer, and the gate electrode features [53, 54]. Therefore, a thorough analysis of the entire physicochemical sensing systems and a rational design of the measurement setup have to be performed in order to avoid, or at least accurately account for, nonbinding current contributions that delay the GFET response to ensure a reliable correlation between the device signal and the binding process.

To inquire into this aspect, a dual monitoring device was developed by Aspermair et al. for simultaneous determination of surface mass and charge density changes by combining SPR and GFET measurements [55]. In that work, the monitoring of a layer-by-layer formation of a PEM employing the Au-SPR substrate as the gate terminal of the GFET was performed. Then, they separated the obtained GFET signal into two contributions with different kinetic profiles: one signal originated in the electrostatic interaction between the polyelectrolytes and the surface that took place in less than one minute and whose kinetic behavior matched the SPR profile; and a second contribution was generated by the formation of an EDL whose kinetic behavior was limited by slow ion diffusion through the PEM. As a consequence of this additional capacitive contribution, the adsorption process was slowly observed with the GFET, requiring 15 minutes to monitor the polyelectrolyte deposition. However, a recent study showed that these capacitive contributions that delay the current response could be avoided by using a proper measurement setup. Thus, employing a coplanar Ag/AgCl gate electrode, real-time monitoring of the PEM formation with high temporal resolution can be achieved [56].

Moreover, different authors have proven the capability of GFETs for monitoring the specific binding of biomolecules on top of graphene surfaces in physiological conditions with high temporal resolution. As a worth-mentioning example, a graphene foam field-effect transistor (GF-FET) sensor for real-time quantification of adenosine triphosphate (ATP) was developed by Xu et al. [21]. In that work, an ATP aptamer was immobilized on top of the graphene foam surface via PBSE linkage, and the drain–source current was monitored during the addition of increasing ATP concentrations in ×1 PBS buffer (Figure 12.8a). A decrease in the drain–source current was observed when operating under an electron regime, as was expected taking into account the positive shift of the Dirac point generated by the negative electrostatic gating effect of ATP binding (Figure 12.8b). The kinetic profiles were analyzed to calculate the k_{on} and k_{off} rate constants, and the binding equilibrium constant obtained from the ratio between them was in very good agreement with other previously reported values by fluorescence methods [57]. In addition, ATP quantification on the modified-GFET was achieved with a detection limit several times lower than the one obtained by traditional methods and a linear response to the ATP in a broad concentration range.

Another example that illustrates the potential of GFETs for the study of adsorption kinetics on a different kind of system was shown by Blaschke et al. by monitoring the formation of charged lipid layers [58]. First, they measured the GFET response upon the deposition of charged lipids from solutions containing unilamellar vesicles. A well-defined and reproducible negative shift in the Dirac point was

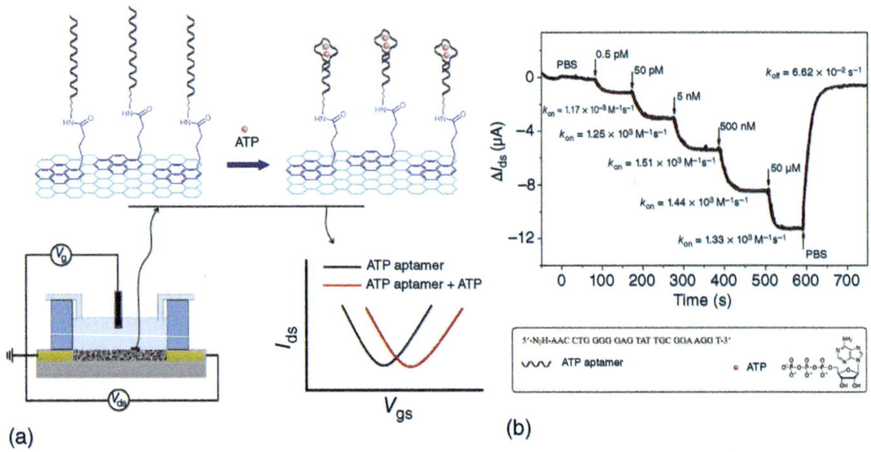

Figure 12.8 (a) Scheme of the GF-FET biosensor for the ATP detection. (b) Real-time monitoring of ATP with the modified GF-FET at different concentrations added progressively from 0.5 pM to 50 μM. Source: Reproduced from Xu et al. [21]/with permission from Elsevier.

observed after modification of the graphene channel with cationic lipids. Then, they *in situ* monitored the formation of the lipid layer by measuring the drain–source current at fixed (positive) gate potential as a function of time, while the transistors were exposed to the cationic lipid solution. In that experiment, a rapid increase of the current was observed due to the instantaneous spreading of the lipids on the surface, demonstrating the capability of the GFET for studying the dynamics of lipid layer formation with high temporal resolution (Figure 12.9). Other examples of successful monitoring of binding kinetics have been recently reported [5, 20, 22, 23, 51, 59, 60].

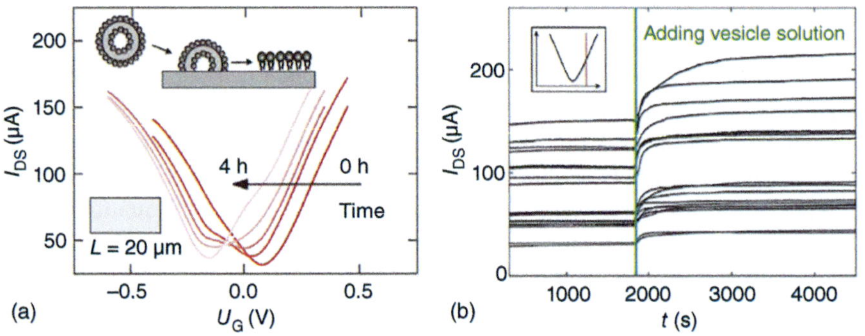

Figure 12.9 (a) Evolution of the GFET transfer curves over time during the formation of a cationic lipid layer by vesicle fusion, as sketched in the inset. (b) I_{DS} at a fixed gate potential recorded during the lipid layer formation for different transistors in parallel. The green vertical line indicates the addition of the lipid solution. Source: Reproduced with permission from Blaschke et al. [58]/American Chemical Society.

12.7 Conclusions

The development of sensing devices based on GFET technology requires overcoming screening limitations in order to enable detection in physiological high-ionic strength conditions. The short Debye length in biological samples often makes GFET sensors insensitive to bioanalytes binding on the recognition surfaces, and thus extending the sensing region becomes mandatory. With this purpose, graphene surfaces have been modified with soft architectures to decrease the screening effect and enhance the interaction of binding analytes with the conductive channel. Different approaches have been employed to achieve this goal, such as modification of the graphene channel with polymer layers, mesoporous materials, or lipids (Scheme 12.1). In addition, the creation of ion-permeable membranes atop the graphene channel generates Donnan potential drops upon charged analyte immobilization, enabling additional capacitive sensing mechanisms beyond the Debye length. In this way, the extension of the sensing limits to allow measuring in biological samples turns GFETs devices into unique tools that can offer real-time, cost-effective, and label-free detection with high specificity and selectivity. Moreover, it has been demonstrated that the extension of the sensing range can be achieved without compromising the temporal resolution of the device signal, favoring the use of GFETs for the determination of binding kinetics and the study of the dynamics of molecular interactions.

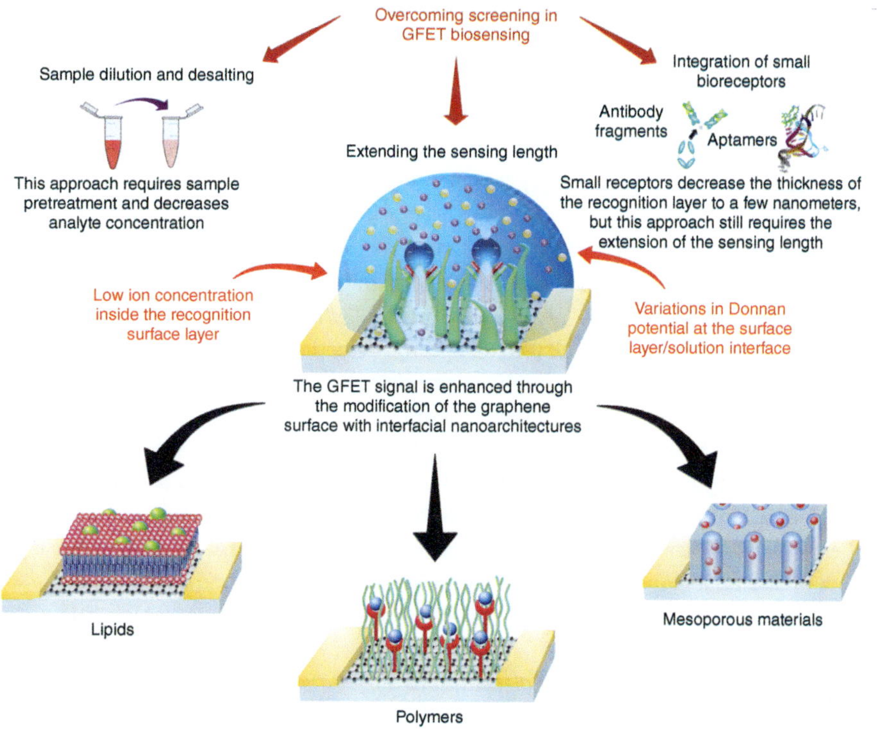

Scheme 12.1 Illustration of different surface modification strategies employed to overcome electrolyte screening in GFET devices toward biosensing applications.

References

1 Lei, Y.M., Xiao, M.M., Li, Y.T. et al. (2017). Detection of heart failure-related biomarker in whole blood with graphene field effect transistor biosensor. *Biosens. Bioelectron. 91*: 1–7. https://doi.org/10.1016/j.bios.2016.12.018.

2 Hao, Z., Zhu, Y., Wang, X. et al. (2017). Real-time monitoring of insulin using a graphene field-effect transistor aptameric nanosensor. *ACS Appl. Mater. Interfaces 9* (33): 27504–27511. https://doi.org/10.1021/acsami.7b07684.

3 Piccinini, E., Fenoy, G.E., Cantillo, A.L. et al. (2022). Biofunctionalization of graphene-based FET sensors through heterobifunctional nanoscaffolds: technology validation toward rapid COVID-19 diagnostics and monitoring. *Adv. Mater. Interfaces 9* (15): 2102526. https://doi.org/10.1002/admi.202102526.

4 Tian, M., Qiao, M., Shen, C. et al. (2020). Highly-sensitive graphene field effect transistor biosensor using PNA and DNA probes for RNA detection. *Appl. Surf. Sci. 527*: https://doi.org/10.1016/j.apsusc.2020.146839.

5 Xu, S., Jiang, S., Zhang, C. et al. (2018). Ultrasensitive label-free detection of DNA hybridization by sapphire-based graphene field-effect transistor biosensor. *Appl. Surf. Sci. 427*: 1114–1119. https://doi.org/10.1016/j.apsusc.2017.09.113.

6 Campos, R., Borme, J., Guerreiro, J.R. et al. (2019). Attomolar label-free detection of DNA hybridization with electrolyte-gated graphene field-effect transistors. *ACS Sensors 4* (2): 286–293. https://doi.org/10.1021/acssensors.8b00344.

7 Afsahi, S., Lerner, M.B., Goldstein, J.M. et al. (2018). Novel graphene-based biosensor for early detection of Zika virus infection. *Biosens. Bioelectron. 100*: 85–88. https://doi.org/10.1016/j.bios.2017.08.051.

8 Kanai, Y., Ohmuro-Matsuyama, Y., Tanioku, M. et al. (2020). Graphene field effect transistor-based immunosensor for ultrasensitive noncompetitive detection of small antigens. *ACS Sensors 5* (1): 24–28. https://doi.org/10.1021/acssensors.9b02137.

9 Kim, D.H., Oh, H.G., Park, W.H. et al. (2018). Detection of alpha-fetoprotein in hepatocellular carcinoma patient plasma with graphene field-effect transistor. *Sensors (Switzerland) 18* (11): https://doi.org/10.3390/s18114032.

10 Gao, N., Gao, T., Yang, X. et al. (2016). Specific detection of biomolecules in physiological solutions using graphene transistor biosensors. *Proc. Natl. Acad. Sci. U.S.A. 113* (51): 14633–14638. https://doi.org/10.1073/pnas.1625010114.

11 Hinnemo, M., Makaraviciute, A., Ahlberg, P. et al. (2018). Protein sensing beyond the Debye length using graphene field-effect transistors. *IEEE Sens. J. 18* (16): 6497–6503. https://doi.org/10.1109/JSEN.2018.2849006.

12 Berninger, T., Bliem, C., Piccinini, E. et al. (2018). Cascading reaction of arginase and urease on a graphene-based FET for ultrasensitive, real-time detection of arginine. *Biosens. Bioelectron. 115*: 104–110. https://doi.org/10.1016/j.bios.2018.05.027.

13 Israelachvili, J.N. (2011). *Intermolecular and Surface Forces*, 3e. New York: Academic Press.

14 Zheng, G., Patolsky, F., Cui, Y. et al. (2005). Multiplexed electrical detection of cancer markers with nanowire sensor arrays. *Nat. Biotechnol. 23* (10): 1294–1301. https://doi.org/10.1038/nbt1138.

15 Stern, E., Vacic, A., Rajan, N.K. et al. (2010). Label-free biomarker detection from whole blood. *Nat. Nanotechnol.* 5: 138–142. https://doi.org/10.1038/nnano.2009.353.

16 Okamoto, S., Ohno, Y., Maehashi, K. et al. (2012). Immunosensors based on graphene field-effect transistors fabricated using antigen-binding fragment. *Jpn. J. Appl. Phys. 51* (6 PART 2): https://doi.org/10.1143/JJAP.51.06FD08.

17 Andoy, N.M., Filipiak, M.S., Vetter, D. et al. (2018). Graphene-based electronic immunosensor with femtomolar detection limit in whole serum. *Adv. Mater. Technol. 3* (12): https://doi.org/10.1002/admt.201800186.

18 Nakatsuka, N., Yang, K.A., Abendroth, J.M. et al. (2018). Aptamer-field-effect transistors overcome Debye length limitations for small-molecule sensing. *Science (80-.) 362* (6412): 319–324. https://doi.org/10.1126/science.aao675.

19 Ohno, Y., Maehashi, K., and Matsumoto, K. (2010). Label-free biosensors based on aptamer-modified graphene field-effect transistors. *J. Am. Chem. Soc. 132* (51): 18012–18013. https://doi.org/10.1021/ja108127r.

20 Wang, X., Hao, Z., Olsen, T.R. et al. (2019). Measurements of aptamer-protein binding kinetics using graphene field-effect transistors. *Nanoscale 11* (26): 12573–12581. https://doi.org/10.1039/c9nr02797a.

21 Xu, S., Zhang, C., Jiang, S. et al. (2019). Graphene foam field-effect transistor for ultra-sensitive label-free detection of ATP. *Sens. Actuators, B* 284: 125–133. https://doi.org/10.1016/j.snb.2018.12.129.

22 Hajian, R., Balderston, S., Tran, T. et al. (2019). Detection of unamplified target genes via CRISPR–Cas9 immobilized on a graphene field-effect transistor. *Nat. Biomed. Eng. 3* (6): 427–437. https://doi.org/10.1038/s41551-019-0371-x.

23 Goldsmith, B.R., Locascio, L., Gao, Y. et al. (2019). Digital biosensing by foundry-fabricated graphene sensors. *Sci. Rep. 9* (1): 1–10. https://doi.org/10.1038/s41598-019-38700-w.

24 Gao, J., Wang, C., Wang, C. et al. (2022). Poly-l-lysine-modified graphene field-effect transistor biosensors for ultrasensitive breast cancer MiRNAs and SARS-CoV-2 RNA detection. *Anal. Chem. 94* (3): 1626–1636. https://doi.org/10.1021/acs.analchem.1c03786.

25 Balderston, S., Taulbee, J.J., Celaya, E. et al. (2021). Discrimination of single-point mutations in unamplified genomic DNA via Cas9 immobilized on a graphene field-effect transistor. *Nat. Biomed. Eng. 5* (7): 713–725. https://doi.org/10.1038/s41551-021-00706-z.

26 Piccinini, E., Alberti, S., Longo, G.S. et al. (2018). Pushing the boundaries of interfacial sensitivity in graphene FET sensors: polyelectrolyte multilayers strongly increase the Debye screening length. *J. Phys. Chem. C 122* (18): 10181–10188. https://doi.org/10.1021/acs.jpcc.7b11128.

27 Palazzo, G., De Tullio, D., Magliulo, M. et al. (2015). Detection beyond Debye's length with an electrolyte-gated organic field-effect transistor. *Adv. Mater. 27* (5): 911–916. https://doi.org/10.1002/adma.201403541.

28 Gol, R.M., Bera, A., Banjo, S. et al. (2014). Effect of amine spacer of PEG on the properties, performance and antifouling behavior of poly(piperazineamide) thin film composite nanofiltration membranes prepared by in situ PEGylation approach. *J. Membr. Sci. 472*: 154–166. https://doi.org/10.1016/j.memsci.2014.08.015.

29 Gao, N., Zhou, W., Jiang, X. et al. (2015). General strategy for biodetection in high ionic strength solutions using transistor-based nanoelectronic sensors. *Nano Lett. 15* (3): 2143–2148. https://doi.org/10.1021/acs.nanolett.5b00133.

30 Gutiérrez-Sanz, Ó., Andoy, N.M., Filipiak, M.S. et al. (2017). Direct, label-free, and rapid transistor-based immunodetection in whole serum. *ACS Sensors 2* (9): 1278–1286. https://doi.org/10.1021/acssensors.7b00187.

31 Haustein, N., Gutiérrez-Sanz, Ó., and Tarasov, A. (2019). Analytical model to describe the effect of polyethylene glycol on ionic screening of analyte charges in transistor-based immunosensing. *ACS Sensors 4* (4): 874–882. https://doi.org/10.1021/acssensors.8b01515.

32 Vu, C.-A., Lai, H.-Y., Chang, C.-Y., Chan, H. W.-H. & Chen, W.-Y. (2022). Optimizing surface modification of silicon nanowire field-effect transistors by polyethylene glycol for MicroRNA detection. *Colloids and Surfaces B: Biointerfaces, 209*, 112142. https://doi.org/10.1016/j.colsurfb.2021.112142.

33 Filipiak, M.S., Rother, M., Andoy, N.M. et al. (2018). Highly sensitive, selective and label-free protein detection in physiological solutions using carbon nanotube transistors with nanobody receptors. *Sens. Actuators, B 255*: 1507–1516. https://doi.org/10.1016/j.snb.2017.08.164.

34 Song, J., Dailey, J., Li, H. et al. (2017). Extended solution gate OFET-based biosensor for label-free glial fibrillary acidic protein detection with polyethylene glycol-containing bioreceptor layer. *Adv. Funct. Mater. 27* (20): https://doi.org/10.1002/adfm.201606506.

35 Glasmästar, K., Larsson, C., Höök, F., and Kasemo, B. (2002). Protein adsorption on supported phospholipid bilayers. *J. Colloid Interface Sci. 246* (1): 40–47. https://doi.org/10.1006/jcis.2001.8060.

36 Glazier, R. and Salaita, K. (2017). Supported lipid bilayer platforms to probe cell mechanobiology. *Biochim. Biophys. Acta, Biomembr. 1859* (9): 1465–1482. https://doi.org/10.1016/j.bbamem.2017.05.005.

37 Hu, S.K., Lo, F.Y., Hsieh, C.C., and Chao, L. (2019). Sensing ability and formation criterion of fluid supported lipid bilayer coated graphene field-effect transistors. *ACS Sensors 4* (4): 892–899. https://doi.org/10.1021/acssensors.8b01623.

38 Sackmann, E. (1996). Supported membranes: scientific and practical applications. *Science (80-.) 271* (5245): 43–48. https://doi.org/10.1126/science.271.5245.43.

39 Murugathas, T., Zheng, H.Y., Colbert, D. et al. (2019). Biosensing with insect odorant receptor nanodiscs and carbon nanotube field-effect transistors. *ACS Appl. Mater. Interfaces 11* (9): 9530–9538. https://doi.org/10.1021/acsami.8b19433.

40 Momoi, M. and Kagawa, Y. (1987). Acetylcholine sensor based on ion sensitive field effect transistor and acetylcholine receptor. *Anal. Lett. 20* (6): 857–870. https://doi.org/10.1080/00032718708062935.

41 Nguyen, T.D., Labed, A., El Zein, R. et al. (2014). A field effect transistor biosensor with a γ-pyrone derivative engineered lipid-sensing layer for ultrasensitive Fe^{3+} ion detection with low pH interference. *Biosens. Bioelectron. 54*: 571–577. https://doi.org/10.1016/j.bios.2013.11.051.

42 Nguy, T.P., Hayakawa, R., Kilinc, V. et al. (2020). Electrolyte-gated-organic field effect transistors functionalized by lipid monolayers with tunable pH sensitivity for

sensor applications. *Appl. Phys. Express 13* (1): https://doi.org/10.7567/1882-0786/ab5322.

43 Ang, P.K., Jaiswal, M., Lim, C.H.Y.X. et al. (2010). A bioelectronic platform using a graphene-lipid bilayer interface. *ACS Nano 4* (12): 7387–7394. https://doi.org/10.1021/nn1022582.

44 Alberti, S., Piccinini, E., Ramirez, P.G. et al. (2021). Mesoporous thin films on graphene FETs: nanofiltered, amplified and extended field-effect sensing. *Nanoscale 13* (45): 19098–19108. https://doi.org/10.1039/d1nr03704h.

45 Yang, L., Jin, Y., Wang, X. et al. (2020). Antifouling field-effect transistor sensing interface based on covalent organic frameworks. *Adv. Electron. Mater. 6* (5): 1–7. https://doi.org/10.1002/aelm.201901169.

46 Walcarius, A. (2015). Mesoporous materials-based electrochemical sensors. *Electroanalysis 27* (6): 1303–1340. https://doi.org/10.1002/elan.201400628.

47 Alothman, Z.A. (2012). A review: fundamental aspects of silicate mesoporous materials. *Materials (Basel) 5* (12): 2874–2902. https://doi.org/10.3390/ma5122874.

48 Wang, H., Varghese, J., and Pilon, L. (2011). Simulation of electric double layer capacitors with mesoporous electrodes: effects of morphology and electrolyte permittivity. *Electrochim. Acta 56* (17): 6189–6197. https://doi.org/10.1016/j.electacta.2011.03.140.

49 Brunsen, A., Díaz, C., Pietrasanta, L.I. et al. (2012). Proton and calcium-gated ionic mesochannels: phosphate-bearing polymer brushes hosted in mesoporous thin films as biomimetic interfacial architectures. *Langmuir 28* (7): 3583–3592. https://doi.org/10.1021/la204854r.

50 Xu, S., Zhan, J., Man, B. et al. (2017). Real-time reliable determination of binding kinetics of DNA hybridization using a multi-channel graphene biosensor. *Nat. Commun. 8*: 1–10. https://doi.org/10.1038/ncomms14902.

51 Xu, S., Wang, T., Liu, G. et al. (2021). Analysis of interactions between proteins and small-molecule drugs by a biosensor based on a graphene field-effect transistor. *Sens. Actuators, B* 326: 128991. https://doi.org/10.1016/j.snb.2020.128991.

52 Gotoh, M., Hasegawa, Y., Shinohara, Y. et al. (1995). A new approach to determine the effect of mismatches on kinetic parameters in DNA hybridization using an optical biosensor. *DNA Res. 2*: 285–293.

53 Chu, C.H., Sarangadharan, I., Regmi, A. et al. (2017). Beyond the Debye length in high ionic strength solution: direct protein detection with field-effect transistors (FETs) in human serum. *Sci. Rep. 7* (1): 1–15. https://doi.org/10.1038/s41598-017-05426-6.

54 Cicoira, F., Sessolo, M., Yaghmazadeh, O. et al. (2010). Influence of device geometry on sensor characteristics of planar organic electrochemical transistors. *Adv. Mater. 22* (9): 1012–1016. https://doi.org/10.1002/adma.200902329.

55 Aspermair, P., Ramach, U., Reiner-Rozman, C. et al. (2020). Dual monitoring of surface reactions in real time by combined surface-plasmon resonance and field-effect transistor interrogation. *J. Am. Chem. Soc. 142* (27): 11709–11716. https://doi.org/10.1021/jacs.9b11835.

56 Scotto, J., Cantillo, A.L., Piccinini, E. et al. (2022). Using graphene field-effect transistors for real-time monitoring of dynamic processes at sensing interfaces. Benchmarking performance against surface plasmon resonance. *ACS Appl. Electron. Mater. 4* (8): 3988–3996. https://doi.org/10.1021/acsaelm.2c00624.

57 Jhaveri, S.D., Kirby, R., Conrad, R. et al. (2000). Designed signaling aptamers that transduce molecular recognition to changes in fluorescence intensity. *J. Am. Chem. Soc. 122* (11): 2469–2473. https://doi.org/10.1021/ja992393b.

58 Blaschke, B.M., Böhm, P., Drieschner, S. et al. (2018). Lipid monolayer formation and lipid exchange monitored by a graphene field-effect transistor. *Langmuir 34* (14): 4224–4233. https://doi.org/10.1021/acs.langmuir.8b00162.

59 Ono, T., Kanai, Y., Inoue, K. et al. (2019). Electrical biosensing at physiological ionic strength using graphene field-effect transistor in femtoliter microdroplet. *Nano Lett. 19* (6): 4004–4009. https://doi.org/10.1021/acs.nanolett.9b01335.

60 Lerner, M.B., Pan, D., Gao, Y. et al. (2017). Large scale commercial fabrication of high quality graphene-based assays for biomolecule detection. *Sens. Actuators, B 239*: 1261–1267. https://doi.org/10.1016/j.snb.2016.09.137.

13

Hybridized Graphene Field-Effect Transistors for Gas Sensing Applications

Radha Bhardwaj and Arnab Hazra

Birla Institute of Technology and Science (BITS)-Pilani, Department of Electrical & Electronics Engineering, Vidya Vihar, Pilani, Rajasthan, 333031, India

13.1 Introduction

Gas sensing technology is emerging into new potential markets and industrial applications with the development of more intelligent and accurate sensors. An excellent solid-state sensor should have good selectivity, sensitivity, fast response, recovery, and long-term stability toward the target gas [1–3]. Gas and volatile organic compound (VOC) sensing applications include the detection of toxic, hazardous, and flammable gases in pharmaceuticals, water treatment, automotive industry, beverage industry, food processing, power plants, and many other applications [1, 2]. Sensors operate due to their surface interactions with target gases and thus modulate their electrical properties [3]. Also, a variety of materials are used in gas sensing field, like semiconducting metal oxides [4, 5], polymers [6, 7], thin metal films [8], solid electrolytes, two-dimensional (2D) nanostructured materials including carbon-nanoforms and transition metal dichalcogenides [9, 10].

In recent days, advanced 2D materials have attracted much attention in gas sensing due to their properties like high electrical conductivity, fast electron transfer kinetics, large surface area, easy functionalization possibilities, room temperature stability, ballistic transport, and so on [9–11]. 2D materials like graphene, transition metal dichalcogenides (e.g. MoS_2 and WS_2), transition metal oxides (e.g. MoO_3 and WO_3), boron nitride, etc., are used as potential gas sensing materials since last two decades [10]. Out of them, graphene and its derivatives, i.e. graphene oxide (GO) and reduced graphene oxide (rGO)-based nanocomposites, have attracted wide attention in the gas sensing field and are effectively used in the detection of a variety of gases/VOCs [12–17]. Besides the above-mentioned advantages, graphene and its derivative-based field-effect transistor (FET) sensors are also trending due to their ideal channel properties, high carrier mobility, large surface area, easy functionalization,

Graphene Field-Effect Transistors: Advanced Bioelectronic Devices for Sensing Applications, First Edition. Edited by Omar Azzaroni and Wolfgang Knoll.

and large-scale integrated device fabrication [10, 11, 18]. In addition of third terminal voltage (gate voltage) in a pristine graphene-based sensor modulates its sensing properties in terms of low-temperature sensing and selectivity toward a particular VOC [19]. Additionally, composite formation of graphene channel with other metals and metal oxides improves the surface properties of the graphene channel in FET structure sensors that is highly beneficial for high sensing throughput [20–22].

This book chapter covers the origin and advanced properties of graphene, the fabrication of graphene field-effect transistor (GFET), the study of their electronic behavior, and the application of GFET in gas/VOC sensing applications. Later on, we discussed the gas/VOC sensing properties of graphene and its derivatives and the improvement of those properties after its composite formation with a variety of other sensing materials like metal oxides, polymers, noble metals, etc. The mechanism behind the graphene composite-based FET-structured gas sensors has also been discussed in this book chapter.

13.2 Graphene

Graphene was first discovered in early 1947 during the investigation of electronic energy bands in crystalline graphite by using tight binding approximation [23]. Graphene atomically thin sheet of carbon atoms packed into a honeycomb crystal plane and sp^2-hybridized, has the ideal 2D structure [23–25]. The high flexibility of graphene and its status as a single atomic plane of carbon allow to it exhibit different morphological variations that includes 0D fullerenes, rolled into 1D carbon nanotubes, and stacked into 3D thin graphene films [24, 25]. Graphene possesses outstanding properties like high crystal quality, high specific surface area, ultrahigh intrinsic strength, ballistic transport, superior thermal conductivity, and extremely high charge carrier mobility [25–27] and offers tremendous possibilities for using graphene in variety of applications. In 2004, Novoselov et al. first demonstrated the field effect in graphene and explained the ambipolar characteristics of graphene which were further explored by various researchers for variety of applications [28]. The uniqueness and abundance of properties of graphene make it desirable for various fields, such as chemical sensors, FETs, transparent conductive electrodes, and photodetectors [23–26, 29].

13.3 Graphene FET

Graphene, with its extremely high intrinsic carrier mobility and ultralong mean free path, excellent intrinsic transport properties, and good electrical and thermal conductivity, occurs to be an ideal channel material for FET applications [28–31]. Since GFET is a capacitance-operated device in which the current density of the graphene channel between two metal contacts (source and drain) is controlled by the third terminal (gate) voltage.

A variety of structures of GFETs have been introduced, namely top-gated GFETs, back-gated GFETs, and back- and top-dual-gated GFETs. In top-gated GFETs, the insulating layer (such as SiO_2 and Al_2O_3) as gate dielectric is deposited on the top surface of graphene, and the device is fabricated after immobilization of metal electrode on the surface of gate dielectric [32, 33]. But the top gate electrode is not a fruitful and flexible approach as it strongly affects the carrier mobility and can destroy the graphene channel layer [28]. The back-gated GFET is most popular and usually grown on Si/SiO_2 substrate by the selective etching of the insulating layer on back side, as shown in Figure 13.1. And then the metal electrodes are deposited by the electron beam evaporation process [19, 34, 35]. Another dual-gate approach (back- and top-gated GFETs) is not frequently used and was mainly established to study the current–voltage characteristics where the back gate is added to a relatively short top gate [28, 36].

A variety of methods have been established for the synthesis of high-quality graphene, and on the basis of that resulting materials have their own specific properties in FET-implemented device. The highly established and most frequently used methods for the synthesis of single or few-layered graphene sheets are micromechanical exfoliation [31, 37, 38] and chemical vapor deposition (CVD) [30, 34–36, 39, 40]. The mechanical cleavage is used to synthesize the pristine single-layered graphene which shows properties like higher carrier mobility and density, room temperature quantum Hall effect, low intrinsic noise, tunable bandgap, and ambipolar field effect [28]. CVD is most commonly reported method for graphene synthesis in FET-based device modulation which can be used to obtain graphene films on some transition metal substrates and also shows advantages like high yield, large area, and easily scalable chemically modified graphene for various applications [28, 36, 39].

A single or few-layered graphene is considered a zero bandgap material which exhibits ambipolar transport behavior in FET structure, where its Dirac fermion in a 2D channel can change from electron to hole (or hole to electron) continuously under the application of an appropriate gate electrostatic [19, 28]. At a certain gate voltage, both the current density and concentration of electric charge carriers are negligible or zero, considered the Dirac point of graphene, and characterized by a "V"-shape transfer characteristic (I_{DS}–V_{GS}) [19]. The relationship between the charge carriers exists away from the Dirac point of graphene, and the applied gate voltage can be written as $n_{e,h} = \alpha V_G$, where n is the concentration of electrons or

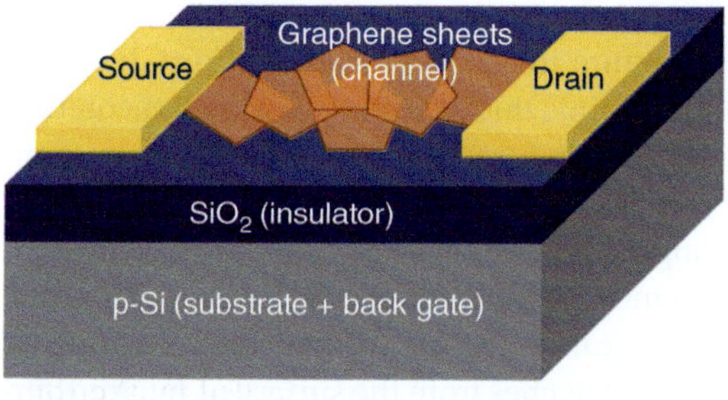

Figure 13.1 Schematic representation of back-gated FET formed on SiO_2/Si substrate.

holes, V_G is the gate voltage, and α is the charge injection rate due to gate voltage [28]. During the gate voltage functioning charge carriers change from holes to electrons under the influence of positive gate voltage, and vice versa, causing a significant change in the current density [19, 28]. The carrier mobility at room temperature is limited by the scattering of the carriers due to the presence of charged impurities like adsorbed gas molecules. Both carrier concentration and carrier mobility influence the drain current under applied gate voltages [19]. So, the scattering of carriers in the graphene can be easily modulated by the effect of applied gate voltage. When the carrier concentration is very small or the gate voltage is biased at the Dirac point, the carrier concentration is smaller than the charge impurity concentration, leading to long-range scattering of the carriers, which results in higher carrier mobility [34].

On the other hand, graphene derivatives (rGO and GO) show p-type conductivity with a good semiconducting bandgap due to the adsorption of oxygen functional groups that eventually increases the hole concentration. To realize the modulation of carrier concentration and current density, we are explaining the rGO-FET model reported in our previous research [19]. The initial hole and electron density in the channel are denoted as p_0 and n_0, respectively, where the applied gate to source voltage is set to zero ($V_{GS} = 0$) and the $p_0 \gg n_0$ in air ambient. When V_{GS} is increased in a positive direction, electron doping $[n(V_{GS})]$ in the channel due to electrostatic potential is taken place, and total carrier concentration in the channel is modulated and can be written as $p_0 - [n_0 + n(V_{GS})]$ in dry air. At Dirac's point the $p_0 = n_0$ and after that, the increase in gate voltage shows $p_0 < n(V_{GS})$ (electron as the majority carrier). This continuous modulation in carrier density under the influence of the third terminal voltage shows a V-shape I_{DS}–V_{GS} characteristic in ambient air.

Graphene, with its unique band structure, has recently attracted significant interest in optoelectronics and photodetection fields. The photonic properties of graphene are equally remarkable for phototransistor [41, 42], thermal transistors [43, 44], gas sensors [29–35], pH sensors [45–48], metal ionic, and other chemical sensors [49–52]. The outstanding properties of GFET make it worthy in the field of biological monitoring, such as DNA, proteins, glucose, certain kinds of bacteria, and so on [28].

13.4 Graphene in Gas Sensing

Since 2004, graphene has a promising future in gas sensor applications due to its outstanding, properties mainly its single-atomic thick 2D structure, high theoretical specific surface area (2630 $m^2 g^{-1}$), remarkable high carrier mobility at room temperature (200 000 $cm^2 V^{-1} s^{-1}$) with a low carrier density of $10^{12} cm^{-2}$, and high chemical stability [28, 31]. The first single-layer graphene sensor was implemented by mechanical exfoliation to detect the individual gas molecule. The breakthrough mechanism involves the change in local carrier concentration (electron transfer theory) when a gas molecule attaches to or detaches from the surface of micrometer-sized graphene sensor. The benefit of single-layer graphene sensor is that every atom is present at the surface and shows sensitivity down to a single molecular level [12]. The achieved sensitivity shows low electronic noise properties, which makes it a promising candidate for chemical sensors [1]. After the primary establishment of graphene sensor, several gases were studied to test the chemical sensing performances of pure graphene. The sensors were capable of detecting a wide variety of gases, such as NO_2 [12, 53–56], NH_3 [54], CO [55], acetone [57], and so on. But poor reversibility of the signal, low sensitivity, high conductivity, and few crystal defects of graphene encouraged the researchers to move toward the graphene derivatives, i.e. GO and rGO for the sensor fabrication. To improve the ambient sensitivity of graphene, the surface was functionalized with different oxygen groups to form GO. And to further increase the defects and disorders, a fractional oxygen functional group was removed from GO to transform GO into rGO [58–60]. These modifications provide reactive sites, resulting in high surface functionality with improved gas sensing properties.

In the later stage, researchers focused on the development of more intelligent and more accurate solid-state sensor devices by improving factors like faster detection, low temperature sensing, high sensitivity, long-term stability, good selectivity, and low ppm detection. Multiple techniques and process steps have been adopted by the researchers for developing more efficient graphene-based solid state gas/vapor sensors viz (i) selection of graphene derivatives, i.e. GO and rGO, along with atomically thin layers or few layers of graphene to achieve high electrical conductivity ($\sim 10^8\,S\,m^{-1}$), high surface to volume ratio (theoretically $2600\,m^2\,g^{-1}$) for better sensitivity and stability [54, 55, 57, 58, 61], (ii) doping of different impurities into graphene derivatives to achieve good sensitivity at relatively low operating temperature [13, 62], (iii) nanocomposites of graphene derivatives with various metal oxides [14, 59, 63–66], (iv) surface modification of graphene derivatives or graphene-metal oxide nanocomposite by using different catalytic metal nanoparticles [67–69].

Graphene and its derivatives (GO and rGO) exhibit p-type semiconducting behavior due to the oxygen adsorption on the active sites in air ambient. Graphene with metal oxides such as WO_3 [13], ZnO [70], In_2O_3 [71], Fe_2O_3 [72], Co_3O_4 [73], and NiO [74] form discrete Schottky junctions due to their work function difference and which directly influence the depletion region and conductivity of the resulting composites in gaseous ambient. When target analytes come into contact with the active layer(s), interact with the chemisorbed oxygen and modulate the depth of the depletion layer and the device current. Functionalization of graphene with metal nanoparticles (Au, Pd, and Pt) exhibits a spillover effect, by means of the chemical sensitization and electronic sensitization effects, where it forms metal-semiconducting Schottky junctions and improves the overall sensing behavior [68, 75].

13.5 Graphene FET for Gas Sensing

The strong field-effect properties of graphene inspired researchers to make graphene-based FET sensors to control carrier concentration and ultimately the sensing properties [12, 19]. A few reports demonstrated the three-terminal FET-structured graphene-based gas sensors due to certain advantages like low-temperature sensing [72], promising selectivity [34], faster detection [35], and excellent gas sensitivity in the range of parts per million to the molecular level [76]. In graphene, tunable charge density and bipolar nature of charge carriers due to its unique atom-thick 2D structure make it advantageous for the fabrication of FETs [19, 28]. The on/off ratio, Dirac point positioning, and type of conductivity during the gaseous interaction play key roles in the improvement of gas sensing [19, 31–35]. The higher on/off ratio in resulting FET leads to higher sensitivity, and due to this, a lot of techniques have been employed to create energy gaps in graphene sheets to achieve a high on/off ratio. It has been reported that GFET is very sensitive to the detection of several types of gases, like CH_3OH [19], SO_2 [30], NH_3 [29, 35, 58], NO_2 [12, 77], and O_2 [76]. In a FET-based gas sensor, the source–drain current (I_{DS}) is dependent on the applied gate voltage (V_{GS}), and it can be effectively modulated under the exposure of the target gas [34]. To improve the gas detection capabilities of GFET, a number of strategies are taken. The significant approaches are (i) modulation of gate potential (zero, positive/negative) [29], (ii) variation in external defects [31], (iii) selection of gate materials [35], (iv) improvement of substrate surface conditions [76], etc. The above mentioned approaches were suitable for the effective improvement in sensing properties in terms of selectivity, low power and low voltage consumption, higher sensitivity and complete recovery of the signals in GFET sensors. The gas-selective nature of GFETs was also demonstrated. Different types of gases have their own specific ratio of unique long-range scattering mobility, and the corresponding Dirac point voltage is called linear factor. A particular type of gas molecule can be detected by using its corresponding linear factor [34].

The GFET sensor is strongly dependent on external defects such as dangling bonds on insulating substrates. GFETs, fabricated on treated SiO_2/Si with H_2 annealing, Al_2O_3 coating, and CF_4 reactive ion etching, showed different transfer characteristics with and without air ambient than GFETs fabricated on bare SiO_2 substrate [76]. The p-type semiconductor (GO and rGO) FET structure exhibits ambipolar behavior, where the majority charge carrier can change from holes to electrons in the channel layer under the influence of appropriate gate electrostatics [19, 28]. In a recent study, graphene-based FET sensor was fabricated and employed for amplified sensitivity toward methanol vapor [19]. In this sensor, single-layered rGO sheets served as the conducting channel by bridging the source and drain electrodes. Modulation of gas sensitivity was studied at four different ambient like synthetic/dry air (reference), argon (inert and oxygen-free), oxygen (oxidizing), and a test gas like methanol to optimize the gate electrostatic (V_{GS}) at different temperature ranges (25–125 °C). In p-type rGO, when the positive gate bias is applied electron doping in rGO due to electrostatic potential is increased and the majority hole carrier concentration decreased to a certain value in air ambient. At a certain V_{GS}, device showed a "V"-shape transfer characteristics (I_{DS}–V_{GS}) with a minimum charge density point (or Dirac point) with the lowest drain current (I_{DS}) in air ambient and with an increase of electron doping at higher V_{GS} than Dirac point, the conductivity of device change from p to n-type. When methanol molecules were adsorbed onto the surfaces of rGO sheets, the local charge carrier concentration decreased or increased depending on the applied V_{GS} and this signal could be monitored by a transistor-like device. Either electrons or holes can serve as the majority charge carriers in rGO FET, depending on the value of gate potential (V_{GS}). The GFET operated near or at the Dirac point, and at high positive V_{GS} in n-mode exhibited rapid sensitivity for the detection of methanol compared with that in p-mode where V_{GS} is lower than the Dirac point. Sensitivity amplification mechanism as a function of modulation in V_{GS} showed the higher sensitivity near the Dirac point and higher V_{GS} mainly due to the fact that positive Dirac point V_{GS} lowers the work function of graphene, decreasing the electron transfer barrier between graphene and methanol, and also at this point the carrier concentration (electron or hole) is negligible in channel, and injection of electrons by adsorbed methanol analytes to the channel leads to a significant change in carrier concentration and the sensitivity [19, 28].

13.6 Hybrid Graphene FET for Gas Sensing

Hybridization of pristine GFET with other sensing materials like metal oxide and functionalization with noble metals (Au, Pd, and Pt) show the synergistic effect of both materials, and the formation of inbuilt heterojunctions results in improved sensing properties. In case of GFETs, gas molecules are adsorbed on graphene sheet and modulate the electrical properties of the channel material (i.e. graphene), but in case of other bulk sensing materials like metal oxides, the gas molecules are adsorbed to the top surface, and thus only the surface region is modulated while the bulk remains unaffected [76]. The hybridization of GFET with other sensing materials can improve the sensing properties, but their bulk properties can hamper the FET characteristics of the resulting graphene composite materials and become a key challenge in the hybridization process. Researchers found in their studies that the controlled addition of any secondary material in the GFET layer is an efficient way to maintain the FET characteristics of the resulting composite material [20–22].

The incorporation of noble metal nanoparticles in GFET modulates the resulting sensing properties in two ways: firstly, by chemical sensitization, which exhibits a spillover (catalytic) effect and secondly, through electrical sensitization, where discrete Schottky junctions are formed and influence the channel properties [20]. In a recent study, noble metals – Au, Pd, and Pt functionalized few-layered GO back gated FET sensors were fabricated as shown in Figure 13.2, and their acetone sensing properties were optimized at 50 °C operating temperature. The transfer characteristics of the noble metal-loaded GO FET sensors showed a similar V-shape behavior as shown in pure GO FET, but a significant shift was observed in the Dirac point positioning due to various surface doping by noble metal nanoparticles in the GO channel, resulting in modulation of majority carrier concentration as displayed in Figure 13.3a. The fabricated sensors showed the significant effect of the noble metal loading and back gate effect on acetone sensing performance in terms of

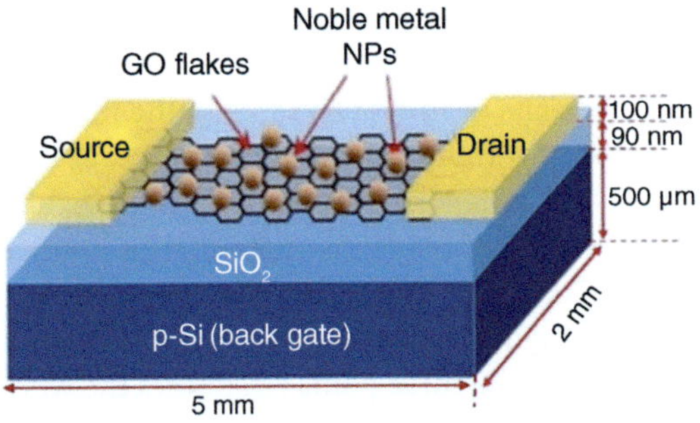

Figure 13.2 Noble metal (Au/Pd/Pt) loaded back-gated GO FET device structure for VOC sensing application. Source: Reproduced with permission from Bhardwaj and Hazra [20]/ Elsevier.

sensitivity and faster response and recovery than the pure GO FET sensor (Figure 13.3b). Among all of the metals, Pd functionalization was found to be most prominent, and the sensor exhibited superior sensitivity (60% at 80 ppm and 15% at 400 ppb) and high selectivity toward acetone when the gate bias was closer to the Dirac point voltage [20]. The sensing mechanism involves the highly oxygen functionalized Pd particles (PdO and PdO_2) among all other metal nanoparticles on the GO surface, Schottky junction formation and different levels of surface doping by noble metal nanoparticles influence the base current of the sensor leading to Dirac point shift. Therefore, a highly sensitive sensing layer promotes more dissociative adsorption of target VOC and contributes a greater number of electrons to the GO channel by lowering their work function. So, the improved response magnitude is the result of an extra injection of electrons and applied gate voltage near Dirac point ($V_{GS} \approx V_{Dirac}$) in noble metal-loaded GO-FET [20].

The binary hybrid of GFET with noble metals and metal oxide was fabricated and found beneficial for gas/VOC sensing applications, but their combined effect in a ternary hybrid can play a pivotal role in sensing different VOC. In this scenario, the six combinations of ternary hybrids were synthesized, which was a combination of triple layers; where GO was acting as a channel or base material to take advantage of FET characteristics, the middle layer was metal oxide (WO_3) for their good semiconducting properties, and the top layer was noble metal nanoparticles (Au, Pd, and Pt), which provide the highly reactive surface for more gaseous interaction named as GO, WO_3/GO, Au/WO_3/GO, Pd/WO_3/GO, Pt/WO_3/GO, and Au,Pd,Pt/WO_3/GO back gated FET sensors [21]. These three participating materials formed various possible Schottky junctions, i.e. WO_3/GO, noble metals/GO, and noble metals/WO_3/GO, which modulate the initial carrier density in the GO channel due to their work function differences and result in different Dirac point voltages for each sensor.

The sensors result in a significantly high response at $V_{GS} \approx V_{Dirac}$ compared to the $V_{GS} = 0\,V$ showing the strong influence of the third terminal on the sensitivity of the sensors as shown in Figure 13.4a. At an optimized gate to source voltage (V_{GS}), which is close to the Dirac point ($V_{GS} \approx V_{Dirac}$) all sensors with various combinations

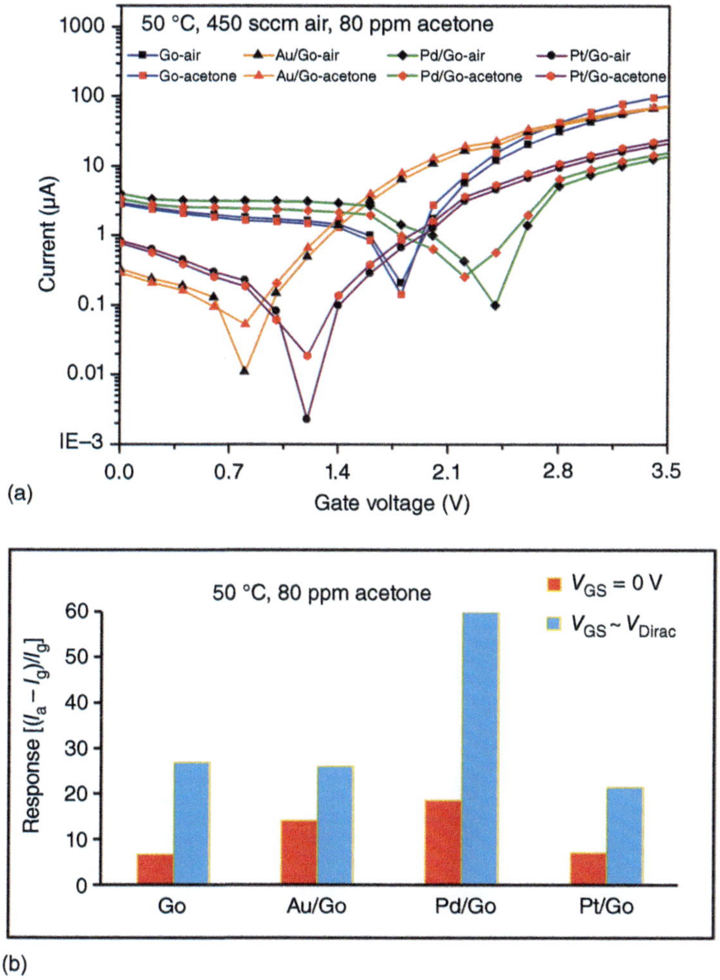

Figure 13.3 (a) The transfer characteristics ($I_{DS}-V_{GS}$) and (b) response comparison at V_{GS} = 0 V and $V_{GS} \approx V_{Dirac}$ of noble metals (Au, Pd, and Pt) loaded FET sensors in air and 80 ppm acetone. Source: Reproduced with permission from Bhardwaj and Hazra [20]/ Elsevier.

were optimized at low concentrations (400 ppb to 80 ppm) of the targeted VOCs even at room temperature as shown in Figure 13.4b. The selectivity of the sensors toward a particular VOC was very diverse but in terms of sensitivity Pt/WO_3/GO FET exhibited the highest response (330%/80 ppm) value toward acetone at applied gate voltage ($V_{GS} \approx V_{Dirac}$) at room temperature (Figure 13.4b). The selective nature of sensors toward different VOCs is directly influenced by the type of ternary heterojunctions, different surface functionality of each noble metal nanoparticle, and the different Dirac point gate voltages [21].

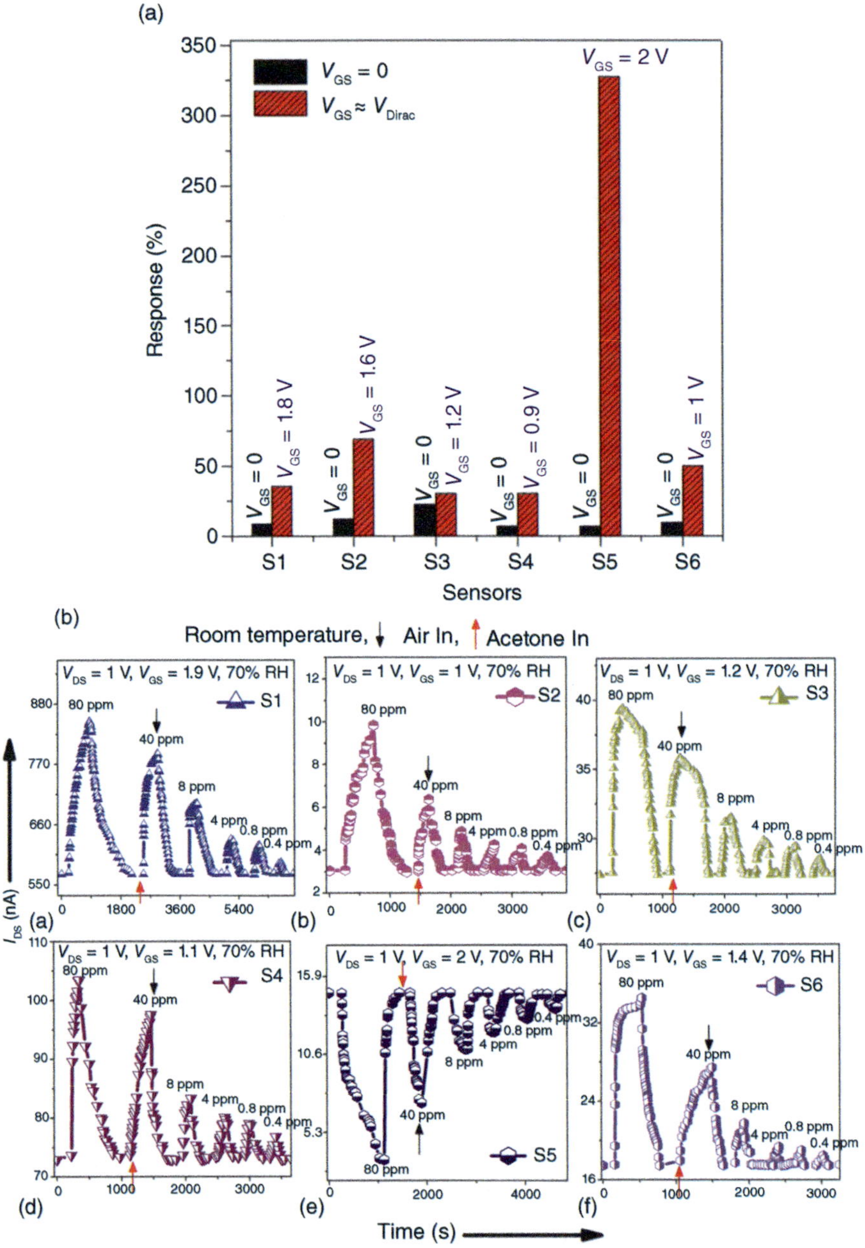

Figure 13.4 (a) Response comparison of all six FET sensors marked as; GO (S1), WO_3/GO (S2), Au/WO_3/GO (S3), Pd/WO_3/GO (S4), Pt/WO_3/GO (S5), and Au,Pd,Pt/WO_3/GO (S6) with and without the influence of third terminal ($V_{GS} \approx V_{Dirac}$) and (b) transient behavior of back-gated FET sensors in the exposure of wide concentration of pure acetone from 80 ppm to 400 ppb at room temperature. Source: Reproduced with permission from Bhardwaj et al. [21]/Wiley.

13.7 Conclusion

In this book chapter, we are focusing on GFET-based devices and their employment in the gas sensing field for better sensing possibilities in real-time applications. After the discovery of graphene, a variety of methods have been employed for the synthesis of graphene from bulk graphite to a single sheet of graphene, which significantly modulates its physical, chemical, and electronic properties. Due to its exceptionally good electrical properties (e.g. high intrinsic carrier mobility and ultralong mean free paths, excellent intrinsic transport, and field effect) of single or few-layered graphene has shown great potential and success in GFETs in various applications. A significant amount of research has been carried out on graphene and its derivative-based gas sensors and their composites with other sensing materials like metal oxides, polymers, etc., based gas sensors in two-terminal configuration. But certain limitations, like high-temperature sensing, lower selectivity, and poor long-term stability, demand more advancement in the gas sensing technology. Graphene in FET configuration is considered the superior gas sensor as compared to its conventional two-terminal structure. So, here we have thoroughly explained the basics of GFET, which involves structure, properties, and transfer characteristics. Later, we discussed the gas sensing behavior of pure GFET and its composite with other materials-based FET sensors for efficiently detecting various gases/VOCs. There is no doubt that GFETs have paid more attention to employing gas/VOC sensors in their pure form and also in their composite form with other materials.

What's more, the various sensing mechanisms of GFET are very critical and are discussed here for a better understanding of GFET-based gas sensing. We believe that GFET-based devices will bring a remarkable change and open new avenues in the real-time gas sensing field.

Acknowledgments

This work was supported in part by Department of Biotechnology grant (Letter No. BT/PR28727/NNT/28/1569/2018), SPARC under Grant (SPARC/2018-2019/P1394/SL), and Core Research Grant of SERB-DST (Letter No. CRG/2021/005550), Govt. of India.

References

1 Yamazoe, N. (2005). *Sens. Actuators, B 108*: 2.
2 Cleaver, K.D. (2001). *Accredit. Qual. Assur. 6*: 8.
3 Fine, G.F., Cavanagh, L.M., Afonja, A., and Binions, R. (2010). *Sensors 10*: 5469.
4 Kanazawa, E., Sakai, G., Shimanoe, K. et al. (2001). *Sens. Actuators, B 77*: 72.
5 Wang, C., Yin, L., Zhang, L. et al. (2010). *Sensors 10*: 2088.

6 Bai, H. and Shi, G. (2007). *Sensors* 7: 267–307.

7 Park, S.J., Park, C.S., and Yoon, H. (2017). *Polymers (Basel) 9*: https://doi.org/10.3390/polym9050155.

8 Pundt, A. (2004). *Adv. Eng. Mater. 6*: 11.

9 Sharma, S. and Madou, M. (2012). *Philos. Trans. R. Soc. London, Ser. A 370*: 2448.

10 Su, S., Sun, Q., Gu, X. et al. (2019). *TrAC, Trends Anal. Chem. 119*: 115610.

11 Wang, T., Huang, D., Yang, Z. et al. (2016). *Nano Micro Lett. 8*: 95.

12 Pearce, R., Iakimov, T., Andersson, M. et al. (2011). *Sens. Actuators, B 155*: 451.

13 Kaur, J., Anand, K., Kaur, A., and Singh, R.C. (2018). *Sensitive and Selective Acetone Sensor Based on Gd Doped WO_3/Reduced Graphene Oxide Nanocomposite.* Elsevier B.V.

14 Liu, F., Chu, X., Dong, Y. et al. (2013). *Sens. Actuators, B 188*: 469.

15 Liu, S., Yu, B., Zhang, H. et al. (2014). *Sens. Actuators, B 202*: 272.

16 Liu, X., Cui, J., Sun, J., and Zhang, X. (2014). *RSC Adv. 4*: 22601.

17 Wang, Z., Zhang, T., Han, T. et al. (2018). *Sens. Actuators, B 266*: 812.

18 Jin, M., Jeong, H.K., Yu, W.J. et al. (2009). *J. Phys. D: Appl. Phys. 42*: https://doi.org/10.1088/0022-3727/42/13/135109.

19 Hazra, A. (2020). *IEEE Trans. Electron Devices 67*: 5111.

20 Bhardwaj, R. and Hazra, A. (2022). *Microelectron. Eng. 256*: 111719.

21 Bhardwaj, R., Thakur, U.N., Ajmera, P. et al. (2022). *ChemNanoMat* 8: e202100448.

22 Gakhar, T. and Hazra, A. (2021). *Measurement 182*: 109721.

23 Yang, G., Li, L., Lee, W.B., and Ng, M.C. (2018). *Sci. Technol. Adv. Mater. 19*: 613.

24 Magne, T.M., de Oliveira Vieira, T., Alencar, L.M.R. et al. (2021). *Graphene and Its Derivatives: Understanding the Main Chemical and Medicinal Chemistry Roles for Biomedical Applications*. Berlin Heidelberg: Springer.

25 Sur, U.K. (2012). *Int. J. Electrochem. 2012*: 1.

26 Santhiran, A., Iyngaran, P., Abiman, P., and Kuganathan, N. (2021). *C 7*: 76.

27 Sang, M., Shin, J., Kim, K., and Yu, K.J. (2019). *Nanomaterials 9*: 1.

28 Zhan, B., Li, C., Yang, J. et al. (2014). *Small 10*: 4042.

29 Lu, G., Yu, K., Ocola, L.E., and Chen, J. (2011). *Chem. Commun. 47*: 7761.

30 Ren, Y., Zhu, C., Cai, W. et al. (2012). *Appl. Phys. Lett. 100*: 1.

31 Kumar, B., Min, K., Bashirzadeh, M. et al. (2013). *Nano Lett. 13*: 1962.

32 Łuszczek, M., Turzyński, M., and Świsulski, D. (2015). *Prz. Elektrotech. 91*: 170.

33 Li, W., Chen, X., Wang, L. et al. (2013). *Sci. Rep. 3*: 1.

34 Liu, Y., Lin, S., and Lin, L. (2015). A versatile gas sensor with selectivity using a single graphene transistor. In: *2015 Transducers – 2015 18th International Conference on Solid-State Sensors, Actuators Microsystems, TRANSDUCERS*, 961. IEEE.

35 Inaba, A., Yoo, K., Takei, Y. et al. (2014). *Sens. Actuators, B 195*: 15.

36 Sharma, P., Bernard, L.S., Bazigos, A. et al. (2015). *ACS Nano 9*: 620.

37 Xu, H., Zhang, Z., Wang, Z. et al. (2011). *ACS Nano 5*: 2340.

38 Meric, I., Han, M.Y., Young, A.F. et al. (2008). *Nat. Nanotechnol. 3*: 654.

39 Lee, J., Chung, H.J., Seo, D.H. et al. (2014). *Nano Res. 7*: 453.

40 Wu, Y., Farmer, D.B., Zhu, W. et al. (2012). *ACS Nano 6*: 2610.

41 Zhang, B.Y., Liu, T., Meng, B. et al. (2013). *Nat. Commun. 4*: 1.

42 Mueller, T., Xia, F., and Avouris, P. (2010). *Nat. Photonics 4*: 297.

43 Trung, T.Q., Tien, N.T., Kim, D. et al. (2012). *Adv. Mater. 24*: 5254.

44 Vicarelli, L., Vitiello, M.S., Coquillat, D. et al. (2012). *Nat. Mater. 11*: 865.

45 Ang, P.K., Chen, W., Wee, A.T.S., and Kian, P.L. (2008). *J. Am. Chem. Soc. 130*: 14392.

46 Cheng, Z., Li, Q., Li, Z. et al. (2010). *Nano Lett. 10*: 1864.

47 Ohno, Y., Maehashi, K., and Matsumoto, K. (2010). *Biosens. Bioelectron. 26*: 1727.

48 Lei, N., Li, P., Xue, W., and Xu, J. (2011). *Meas. Sci. Technol. 22*: https://doi.org/10.1088/0957-0233/22/10/107002.

49 Yang, Q., Yu, A.J., Simonton, J. et al. (2017). *J. Mater. Chem. A 5*: 17841.

50 Zhang, T., Cheng, Z., Wang, Y. et al. (2010). *Nano Lett. 10*: 4738.

51 Maehashi, K., Sofue, Y., Okamoto, S. et al. (2013). *Sens. Actuators, B 187*: 45.

52 Newaz, A.K.M., Markov, D.A., Prasai, D., and Bolotin, K.I. (2012). *Nano Lett. 12*: 2931.

53 Chung, M.G., Kim, D.H., Lee, H.M. et al. (2012). *Sens. Actuators, B 166–167*: 172.

54 Yavari, F., Castillo, E., Gullapalli, H. et al. (2012). *Appl. Phys. Lett. 100*: https://doi.org/10.1063/1.4720074.

55 Joshi, R.K., Gomez, H., Alvi, F., and Kumar, A. (2010). *J. Phys. Chem. C 114*: 6610.

56 Choi, H., Jeong, H.Y., Lee, D.-S. et al. (2013). *Carbon Lett. 14*: 186.

57 Yang, C.M., Chen, T.C., Yang, Y.C. et al. (2017). *Sens. Actuators, B 253*: 77.

58 Kehayias, C.E., MacNaughton, S., Sonkusale, S., and Staii, C. (2013). *Nanotechnology 24*: https://doi.org/10.1088/0957-4484/24/24/245502.

59 Zhang, D., Liu, A., Chang, H., and Xia, B. (2015). *RSC Adv. 5*: 3016.

60 Huang, L., Wang, Z., Zhang, J. et al. (2014). *ACS Appl. Mater. Interfaces 6*: 7426.

61 Arsat, R., Breedon, M., Shafiei, M. et al. (2009). *Chem. Phys. Lett. 467*: 344.

62 Singkammo, S., Wisitsoraat, A., Sriprachuabwong, C. et al. (2015). *ACS Appl. Mater. Interfaces 7*: 3077.

63 Wang, P., Wang, D., Zhang, M. et al. (2016). *Sens. Actuators, B 230*: 477.

64 Wang, C., Zhu, J., Liang, S. et al. (2014). *J. Mater. Chem. A 2*: 18635.

65 Choi, S.J., Jang, B.H., Lee, S.J. et al. (2014). *ACS Appl. Mater. Interfaces 6*: 2588.

66 Zhang, B., Liu, J., Cui, X. et al. (2017). *Sens. Actuators, B 241*: 904.

67 Choi, S.J., Ryu, W.H., Kim, S.J. et al. (2014). *J. Mater. Chem. B 2*: 7160.

68 Ge, S., Zheng, H., Sun, Y. et al. (2016). *J. Alloys Compd. 659*: 127.

69 Zhang, H., Cen, Y., Du, Y., and Ruan, S. (2016). *Sensors (Switzerland) 16*: 1.

70 Liu, S., Wang, Z., Zhang, Y. et al. (2015). *RSC Adv. 5*: 91760.

71 Gu, F., Nie, R., Han, D., and Wang, Z. (2015). *Sens. Actuators, B 219*: 94.

72 Liang, S., Li, J., Wang, F. et al. (2017). *Sens. Actuators, B 238*: 923.

73 Chen, N., Li, X., Wang, X. et al. (2013). *Sens. Actuators, B 188*: 902.

74 Hoa, L.T., Tien, H.N., Luan, V.H. et al. (2013). *Sens. Actuators, B 185*: 701.

75 Shojaee, M., Nasresfahani, S., and Sheikhi, M.H. (2018). *Hydrothermally Synthesized Pd-Loaded SnO_2/Partially Reduced Graphene Oxide Nanocomposite for Effective Detection of Carbon Monoxide at Room Temperature*. Elsevier B.V.

76 Nakamura, M., Kanai, Y., Ohno, Y. et al. (2015). *Jpn. J. Appl. Phys. 54*: 2.

77 Shi, S., Hu, R., Wu, E. et al. (2018). *Nanotechnology* 29: 435502.

14

Polyelectrolyte-Enzyme Assemblies Integrated into Graphene Field-Effect Transistors for Biosensing Applications

Esteban Piccinini[1], Gonzalo E. Fenoy[1], Wolfgang Knoll[2,3], Waldemar A. Marmisollé[1], and Omar Azzaroni[1]

[1]*Instituto de Investigaciones Fisicoquímicas Teóricas y Aplicadas (INIFTA), Universidad Nacional de La Plata – CONICET (Argentina), Departamento de Química, Facultad de Ciencias Exactas, Street 113 and 64, La Plata, 1900 Argentina*
[2]*Austrian Institute of Technology, Konrad-Lorenz-Straße 24, Tulln an der Donau 3430 Austria*
[3]*Danube Private University, Steiner Landstraße 124, Krems 3500 Austria*

14.1 Introduction

Due to their numerous advantages over traditional analytical methods, which include ease of operation, low cost, and compact size, bioelectronic sensors have become a prominent topic of interest in industry and academia. In particular, biosensors that are highly selective represent a powerful tool for real-time measurement of a variety of analytes of considerable interest in different areas, such as food safety, environmental monitoring, drug screening, and diagnosis [1]. Biosensors constitute a subgroup of chemical sensors comprising biological host molecules as recognition elements coupled to a chemical or physical transducer [2]. Nature offers a variety of recognition strategies using, for example, interactions between antibodies and antigens, cell responses, protein interactions, DNA targeting, or enzymatic substrate conversion [3].

In this context, enzymes constitute excellent recognition elements due to their specificity and selectivity toward the analyte [4]. Owing to their availability and biocompatibility, the integration of enzymes on sensor surfaces offers a plausible approach for the detection of specific substrates, whereas the variety of enzymes, or even enzymatic cascade reactions, makes the approach applicable to a wide range of analytes. Depending on the enzymatic reaction product, different transducing strategies can be chosen, ranging from direct electronic communication [5] to electrodes selective to the particular reaction product (e.g. NH_3) [6], or even the detection of pH changes using an interface sensitive to protons [7].

With regard to the sensor platform, graphene-based field-effect transistors (FETs) hold a leading position in signal transduction, offering outstanding capabilities for label-free, highly sensitive, real-time detection [8, 9]. Part of the appeal of

Graphene Field-Effect Transistors: Advanced Bioelectronic Devices for Sensing Applications,
First Edition. Edited by Omar Azzaroni and Wolfgang Knoll.

graphene-based transistors relies on graphene's remarkable electronic properties [10]. Its high carrier mobility and ambipolar field effect [11], together with a great sensitivity toward changes in environmental conditions, make graphene perfectly suitable as a transducing material for use in chemical sensors [12, 13]. Owing to the possibility of being operated in aqueous solutions applying small voltages (which is crucial for biological applications), relatively good biocompatibility [14], and inherent amplification properties [15], graphene-based liquid-gated FETs have gained vast interest in clinical diagnosis [16].

In this context, graphene-based field-effect transistors (GFETs) have been applied for pH sensing [17, 18], DNA [19], and protein detection [20], among other analytes. In particular, owing to its defective structure and the remaining functional groups, reduced graphene oxide (rGO) is extremely advantageous for pH sensing. This pH dependency is ascribed to two factors: (i) the interaction of functional groups at the rGO surface, such as –OH and –COOH groups, with H^+ ions of the electrolyte, giving rise to a change in the surface charge density, and (ii) the change of the Gouy–Chapman diffuse electric double layer, causing electrostatic gating effects. This intrinsic property of rGO can easily be exploited for biosensing purposes, as we will describe in this chapter.

Traditional immobilization techniques for integrating enzymes on solid sensing platforms [21] include entrapment, encapsulation, cross-linking, covalent binding, and adsorption [22]. As the biosensor performance strongly depends on the features of the interfacial architecture [23], the functionalization method has to be chosen with great care in order to retain the enzymatic activity as well as the functionality of the transducing element [24, 25]. Covalent attachment of biomolecules may not only induce damage to the sp^2 structure of graphene [26], but it can also disrupt the folding and thus the activity of the enzyme if essential groups are involved in the immobilization [27].

In this chapter, we will describe the use of enzymes coupled to GFETs via layer-by-layer (LbL) assembly as an approach for designing and constructing biosensors. The approach is based on the intrinsic pH sensitivity of rGO FETs as a strategy to detect the enzymatic conversion of the analyte. Furthermore, we show that when a weak polycation is used for the integration of enzymes onto the graphene surface, the sensor sensitivity is amplified.

14.2 Field-Effect Transistors Based on Reduced Graphene Oxide

In general, a FET consists of two electrodes (source and drain) connected via a semiconducting channel and a gate electrode. The potential applied between gate and source electrodes (V_g) controls the availability of charge carriers within the channel, which thus determines the flow of current between source and drain electrodes. The applied gate voltage (V_g) creates an electric field perpendicular to the channel, attracting either holes or electrons toward the gate, thus creating a conductive channel between source and drain [28].

Due to its ambipolar character, high conductivity, and possibility of functionalization, rGO represents an ideal channel material for the fabrication of FETs. Graphene forms a planar sp^2 hybridized honeycomb structure, with the delocalized electrons being responsible for the outstanding electronic properties of the material [29]. Chemically, rGO differs substantially from pristine graphene, owing to its defective structure and remaining functional groups. In spite of the great homogeneity of graphene grown by chemical vapor deposition (CVD), chemically derived rGO can be highly advantageous as a sensing material. The defective structure and the remaining functional groups in rGO play a key role not only in surface functionalization [30] but also in pH sensing applications. The high sensitivity of rGO-based FETs toward small pH changes near the channel surface is caused by the interaction of carboxyl or hydroxyl functional groups at the rGO surface with H^+ ions in the electrolyte, giving rise to a change in the surface charge density. The field-driven alteration of the surface charge density and the electric double layer leads to electrostatic gating effects [31]. This intrinsic property of rGO can easily be exploited for biosensing purposes.

14.3 Enzyme-Based GFET Sensors Fabricated via Layer-by-Layer Assembly

LbL assembly offers a versatile and simple bottom-up technique for the fabrication of functional thin films on solid substrates through a wide variety of noncovalent interactions: electrostatic [32], molecular recognition [33], and others [34]. Particularly, the LbL assembly through electrostatic interactions [35] has been firmly established as a general method for alternately depositing dense layers of charged molecules onto oppositely charged surfaces [36]. The concept of the approach relies on the attractive forces between oppositely charged molecules, and it is widely accepted that the LbL assembly of polyelectrolytes represents a valuable technique for the fabrication of multicomponent, hybrid, and functional films of composition controlled at the nanometer scale [37].

Furthermore, the incorporation of enzymes into the LbL assemblies led to enormous progress in the field of bioelectronics due to their inherent structural and functional properties [38]. In fact, a large number of electrochemical enzyme-based biosensors using different techniques (e.g. potentiometry, voltammetry, and others) were developed [39]. More recently, it has also been demonstrated that nanofluidic sensors can enhance their ionic signals due to local pH changes induced by the enzymatic reaction in the presence of the target analyte [40, 41]. This notion has offered a particularly suitable framework for exploration in graphene-based transistors [42], provided that the use of interfacial architectures that could alter the magnitude and sign of the surface charge after the enzymatic recognition might confer amplified sensing capabilities to these devices. Taking these concepts into account and being aware of the promising features of enzymatic biosensing in GFETs, recently strategies to design GFET biosensors based on the amplification of an enzymatic reaction using polyelectrolyte multilayers as "reactive signal amplifiers" have been proposed as a step further within this paradigm.

14.3.1 Layer-by-Layer (LbL) Assemblies of Polyethylenimine and Urease onto Reduced Graphene-Oxide-Based Field-Effect Transistors (rGO FETs) for the Detection of Urea

Urea concentration in biological fluids, such as blood or urine, is a relevant indicator of the functional condition of the human organism. Its value is indispensable for the diagnosis and control of a number of kidney and liver diseases [43], and, therefore, the measurement of urea concentration is a routine procedure in clinical practice. The use of solution-gated graphene transistors represents an advantageous approach since this method not only provides a real-time and highly sensitive response but can also be implemented in aqueous solutions at very low voltages, a critical factor in biological sensing.

Within this framework, Piccinini et al. [44] have shown the fabrication of reliable urea sensors by integrating enzyme-containing multilayer assemblies to liquid-gated rGO FETs (Figure 14.1a). These devices were fabricated by using interdigitated gold microelectrodes (Figure 14.1b, top) with 10 μm channels and a total channel width of 52.2 mm. In this configuration, the narrow channel geometry assures the source–drain contact by assembled rGO flakes (Figure 14.1b bottom), resulting in devices with low noise and high transconductance. In order to confer negative charge to the graphene surface, the rGO FETs were incubated overnight in 5 mM sodium 1-pyrenesulfonate (SPS) in dimethylformamide (DMF) and then rinsed three times with DMF and deionized water. LbL assemblies were built up

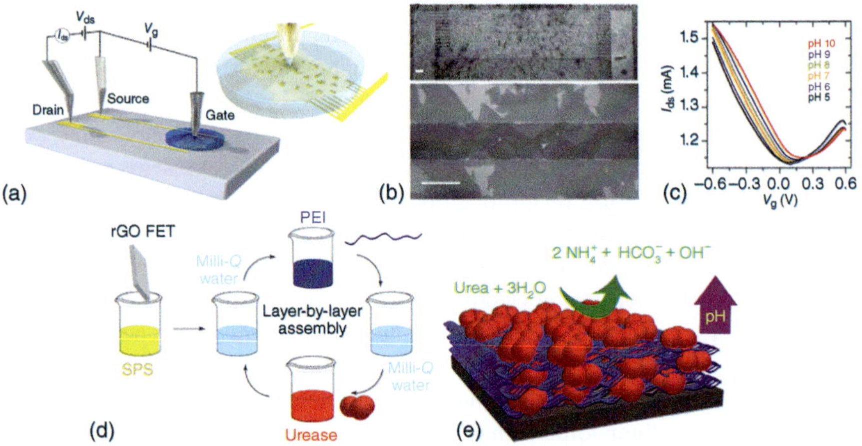

Figure 14.1 (a) Illustration of the solution-gated rGO FET and the interdigitated channel. (b) Scanning electron microscopy (SEM) images of an interdigitated channel modified with rGO: scale bar = 100 μm (top) and scale bar = 10 μm (bottom). (c) Transfer characteristics of an SPS-modified rGO FET obtained at a fixed V_{ds} (0.1 V) in 10 mM KCl and 0.1 mM HEPES solution with pH varying from 5 to 10. An Ag/AgCl gate electrode was used for the measurements. (d) Schematic of the LbL deposition process. (e) Scheme of the urease-PEI multilayer film and the urease-catalyzed hydrolysis of urea. Source: Reproduced with permission from Piccinini et al. [44]/Elsevier.

onto the SPS-modified rGO FETs by alternate adsorption of polyethylenimine (PEI, a polycation) and urease (negatively charged at physiological pH). First, the SPS-modified rGO FETs were incubated in an aqueous solution of 2 mg ml^{-1} polyethylenimine (PEI) at pH 8.5 for 10 minutes and rinsed with deionized water. Then, the GFETs were incubated in 1 mg ml^{-1} urease in buffer (0.1 mM HEPES and 10 mM KCl adjusted to pH 7.4) for 30 minutes, followed by rinsing with deionized water. More than one bilayer was prepared by repeating the PEI and urease adsorption steps as a cycle. The PEI/urease multilayer assembly onto rGO shows a linear growth behavior and a highly stable supramolecular architecture [45]. Moreover, an enzyme surface density value of 515 ng cm^{-2} (per layer) was obtained by surface plasmon resonance (SPR).

A plot of drain–source current (I_{ds}) as a function of the gate–source voltage applied (V_g, using an Ag/AgCl gate electrode) and under a fixed source–drain potential (V_{ds}) for an rGO FET is shown in Figure 14.1c. The application of an external electric potential (e.g. the gate potential) can shift the Fermi energy of the rGO and therefore modulate the charge carrier density, which results in a change in I_{ds}. The minimum I_{ds} (Dirac point, at $V_g = V_i$) at pH 6 (Figure 14.1c) is found at $V_i = 95$ mV. At V_i, there is a transition between a regime with holes as charge carriers (for $V_g < V_i$) and a regime with electrons as charge carriers (for $V_g > V_i$). The charge carrier mobility (μ) of the device was found to be 30 cm^{2}V^{-1}s^{-1} for holes and 15 cm^{2}V^{-1}s^{-1} for electrons.

As discussed earlier, the electrostatic LbL assembly approach was used for the noncovalent immobilization of the enzyme urease onto the channel of the rGO FETs. The assembly process of the urease-PEI multilayer (referred to as (PEI/Urease)$_n$) is depicted in Figure 14.1d. As this type of enzymatic sensor relies on the detection of small changes in pH caused by the hydrolysis of urea (Figure 14.1e), the transfer characteristics of the transistors in a liquid-gated configuration under different pH values represent a critical aspect of the device. Figure 14.1c shows the transfer characteristics of an SPS-modified rGO FET in solutions with constant ionic strength and pH ranging from 5 to 10. Whereas the slope remained almost unchanged for both sides of the V_i, revealing that the charge mobilities of holes and electrons are nearly independent of the pH, a significant shift of V_i to more positive V_g values with increasing pH was observed (Figure 14.2a). Both V_i values and the change in the I_{ds} at constant V_g (ΔI_{ds}) exhibited a linear dependence on pH with a slope of 23 ± 1.8 mV pH^{-1} and 20.3 ± 0.6 μA pH^{-1}, respectively.

Note that after the PEI modification, the V_i shifted to a more negative gate voltage (Figure 14.2b), i.e. −48 mV at pH 6. The change in the transfer characteristics after each step of polyelectrolyte LbL assembly was studied in detail by Wang and Burke [46]. These authors demonstrated that the deposition of a positive (or negative) polyelectrolyte layer on the GFET results in a negative (or positive) shift in the V_i. This behavior was attributed to the electrostatic gating effect of the charged polyelectrolyte layer.

A PEI charge density of 0.011 C m^{-2} was obtained for the first layer, consistent with reported literature values for polyelectrolyte multilayers. Then, the effect of the

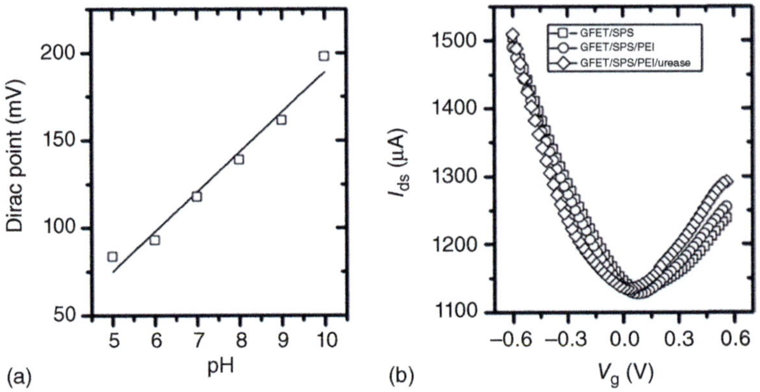

Figure 14.2 (a) Dirac point as a function of the electrolyte pH for an SPS-modified rGO FET and the linear fitting of the data (solid line). (b) I_{ds} as a function of the V_g for GFET/SPS (squares), GFET/SPS/PEI (circles), and GFET/SPS/PEI/Urease (rhombuses) registered at a fixed V_{ds} (0.1 V) and measured in 10 mM KCl, 0.1 mM HEPES at pH 6. Source: Adapted from Piccinini et al. [44].

PEI layer on the pH response was studied. The change in the registered I_{ds} of a rGO/SPS/PEI FET while flowing solutions of increasing pH value is depicted in Figure 14.3a. It is observed that a pH change from 5 to 10 results in remarkable increase in I_{ds}. Furthermore, a detailed analysis reveals a linear relationship between I_{ds} (ΔI_{ds}) and the pH with a sensitivity (slope) of $25.9 \pm 0.6\,\mu A\,pH^{-1}$. Figure 14.3b

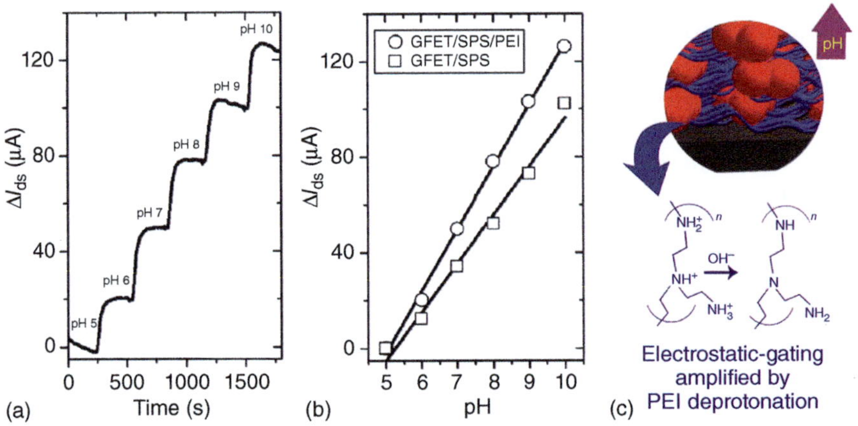

Figure 14.3 (a) Change in the channel current (ΔI_{ds}) of an SPS/PEI-modified rGO FET monitored in real time by varying the pH from 5 to 10. (b) ΔI_{ds} as a function of pH for an SPS-modified rGO FET (squares) and an SPS/PEI-modified rGO FET (circles). Experimental conditions: flow rate = $300\,\mu l\,min^{-1}$, $V_g = -0.2$ V, and $V_{ds} = 0.1$ V. Solid lines are the linear regression fits. (c) Scheme of PEI as a reactive signal amplifier for pH changes. Source: Adapted from Piccinini et al. [44].

shows ΔI_{ds} as a function of pH before and after the PEI assembly. Interestingly, the transistors modified with PEI resulted in a 30% increase in pH sensitivity from $20.3\,\mu A\,pH^{-1}$ (before modification) to $25.9\,\mu A\,pH^{-1}$ (after modification). These results can be interpreted in terms of the nature of PEI. PEI is a weak polycation (pK_a 8–9), and consequently, the degree of protonation of the polymer decreases with increasing pH. When the degree of protonation of PEI decreases (Figure 14.3c), the charge density changes, resulting in p-doping of the graphene layer by the so-called "electrostatic gating effect." A similar behavior was also reported for polymer brushes of *N*,*N*-dimethylaminoethyl methacrylate groups on graphene-based FETs [47]. Therefore, the weak polycation plays a dual role within the interfacial architecture: (i) as a structural building block for the LbL assembly and (ii) as a transducer element to amplify the detection of local pH changes, i.e. "reactive signal amplifiers."

The construction of urea sensors involved the integration of urease onto the GFETs through a polyelectrolyte-enzyme LbL assembly approach. One of the key advantages of this approach is that urease can be properly entrapped in PEI/Urease multilayer assemblies with an efficient preservation of the enzyme activity. The rGO/SPS/PEI transistors were incubated in an aqueous solution of urease, followed by rinsing with deionized water. Noteworthy, the incorporation of urease onto the rGO/SPS/PEI FETs does not affect the pH sensitivity of the device (Figure 14.4a).

Figure 14.4a (top) shows the transfer characteristics in the absence (blue line) and in the presence (red line) of $100\,\mu M$ urea. A clear shift of the V_i to more positive values was observed in the presence of urea, similar to the shift obtained upon increasing the pH of the solution. This pH change results from the enzymatic hydrolysis of urea, which produces NH_4^+, HCO_2^-, and OH^- (Figure 14.1e).

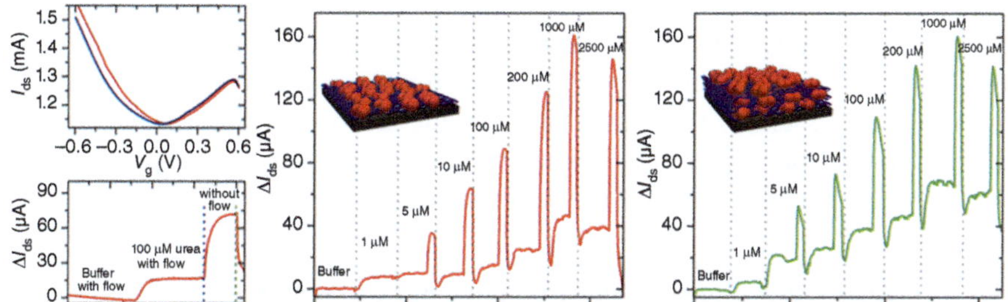

Figure 14.4 (a) Transfer characteristics of a (PEI/Urease)1-rGO FET in the absence (blue line) and in the presence (red line) of $100\,\mu M$ urea (top), and change in the channel current with and without flow (bottom). Channel current response (ΔI_{ds}) monitored in real time for $(PEI/Urease)_1$ (b) and $(PEI/Urease)_3$ (c) LbL assemblies onto rGO FETs at different concentrations of urea. The marked line indicates the elution of the next concentration of urea. Experimental conditions: 10 mM KCl and 0.1 mM HEPES at pH 6, V_g = −0.2 V and V_{ds} = 0.1 V. Source: Reproduced with permission from Piccinini et al. [44].

This change of the local pH near the rGO, coupled to the variation of the charge density of the PEI, results in a significant shift of the V_i to more positive gate voltage.

Figure 14.4a (bottom) depicts the real-time bioelectronic response of the transistors in the presence of 100 μM urea solution. The solution containing the analyte was eluted for four minutes, thus allowing I_{ds} to reach a plateau (flow response). It is observed that I_{ds} increases after the flow is stopped (static response). Noteworthy, once the flow is resumed, the I_{ds} current value returns to the original value observed under flow conditions. This I_{ds} response can be explained in terms of mass transport phenomena taking place in the surroundings of the sensor surface. Under flow conditions, hydroxyl ions resulting from the hydrolysis of urea are rapidly washed away from the graphene surface, mainly because of the forced convection. When the flow is stopped, the transport of hydroxyl ions from the film is only governed by diffusion, thus resulting in an increase in the local concentration of hydroxyl ions near the graphene surface. As a consequence, the enzymatic process promotes significant electrostatic changes in the surroundings of the graphene, which are ultimately transduced as a peak in the recorded I_{ds}. In both conditions (flow and static), an increase in urea concentration leads to an increase in I_{ds} (Figure 14.4b). For urea concentration values above 1 mM, the enzyme could be inhibited by the high substrate concentration, and consequently, the enzymatic transduction mechanism ceases to operate.

It is observed that, by increasing the number of enzyme-polyelectrolyte multilayers, it is possible to magnify the local effects arising from the enzymatic process with a concomitant effect on the transduced bioelectronics signal. Figure 14.4c shows the response corresponding to a graphene transistor modified with $(PEI/Urease)_3$ assemblies.

The effect of the number of multilayers on the device sensitivity can be noted by comparing the ΔI_{ds} response as a function of the logarithmic urea concentration for transistors modified with increasing number of enzyme-containing multilayers (Figure 14.5a). Experimental values reveal that the sensitivity increases by 20% for two bilayers and by 68% for three bilayers, in comparison with one PEI/Urease bilayer. This indicates that, upon increasing the number of bilayers, larger local pH changes take place within the film. Therefore, the use of the LbL assembly technique offers a valuable strategy to create enzyme-containing films with precise control over the enzyme coverage on the graphene surface.

14.3.2 Ultrasensitive Sensing Through Enzymatic Cascade Reactions on Graphene-Based FETs

As discussed in Section 14.3, the strategy of enzymatically converting an analyte and detecting it by the resulting small pH changes near the rGO surface proved very valuable in the sensitive detection of urea. Additionally, it has been shown that electrostatic LbL assembly offers an important avenue for the immobilization of recognition elements on graphene-modified surfaces. Interestingly, these notions can be further extended to coupled enzymatic reactions [48], thereby broadening the

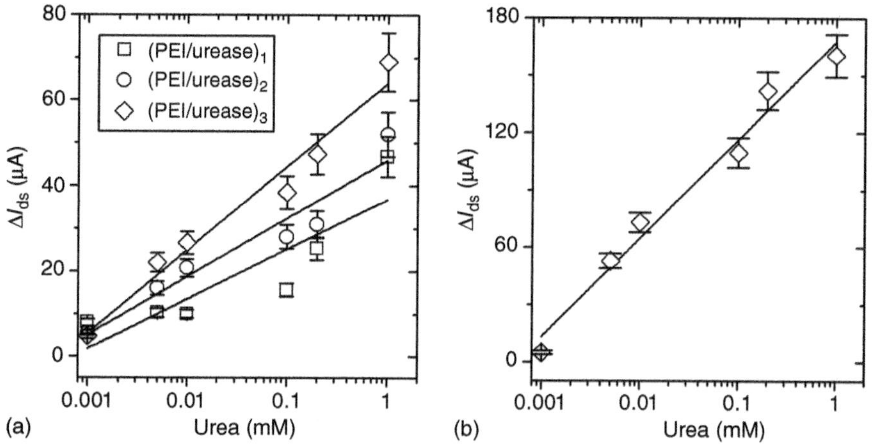

Figure 14.5 (a) ΔI_{ds} for the flow response (flow rate = 300 μl min^{-1}) as a function of the urea concentration for (PEI/Urease)$_1$ (squares), (PEI/Urease)$_2$ (circles), and (PEI/Urease)$_3$ (rhombuses) LbL assemblies on rGO-FETs. (b) ΔI_{ds} for the static response as a function of the urea concentration for a (PEI/Urease)$_3$ LbL assembly on rGO-FETs. Experimental conditions: 10 mM KCl and 0.1 mM HEPES at pH 6, V_g = −0.2 V and V_{ds} = 0.1 V. Error bars represent the standard deviation from measurements performed using three different devices. Solid lines are the linear regression fits. Source: Reproduced with permission from Piccinini et al. [44].

spectrum of detectable analytes. Coupled enzymatic reactions or enzymatic cascade reactions refer to biomolecular systems in which a series of enzymes are coupled in such a way that the product of one enzyme is the substrate for the next.

In this section, we will discuss the formation of enzymatic cascades combining arginase and urease with the aim of sensing L-arginine using GFETs as transducers. L-Arginine is an amino acid found in living organisms that is used in the biosynthesis of proteins. In the human body, it has significant biological functions in the cardiovascular, immune, and endocrine systems [49]. The monitoring of L-arginine in physiological fluids is therefore of great interest in clinical diagnostics, research, and therapy.

The development of the L-arginine sensor involved the immobilization of arginase and urease on the rGO surface through an LbL assembly process (Figure 14.6) [50]. According to this enzymatic cascade, in the first reaction step, arginase hydrolyzes L-arginine to yield ornithine and urea, neither of which provoke a sensor response. However, urea serves as a substrate for the second reaction step, in which it is converted by urease into NH_3 and CO_2, which in turn leads to a pH increase in the surroundings of the rGO-FET surface (Figure 14.6).

Enzyme immobilization through electrostatic LbL assembly using PEI as polycation represents a critical step in the sensor construction. As already discussed, weak polycations play a key role as "reactive signal amplifiers." However, in the

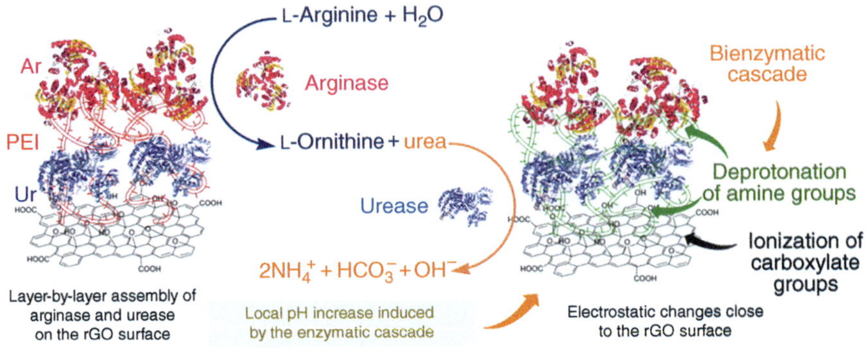

Figure 14.6 Schematic of the LbL-assembled structure using PEI as the cationic building block, and urease and arginase as both recognition elements and anionic building blocks. The figure describes the reaction of the conversion of L-arginine into L-ornithine and the intermediate substrate urea by the enzyme arginase (EC 3.5.3.1) and the following urease (EC 3.5.1.5)-catalyzed hydrolysis of urea.

case of enzymatic cascades, PEI plays a crucial role as it allows for precise tuning of the enzyme spacing and loading within the assembly as well as the optimization of the enzyme ratio. This structural role of PEI is of great importance in the transduction of the multistep cascade reaction taking place on the graphene surface.

According to the sensing strategy based on the enzymatic cascade, the coupling of both enzymes leads to a pH increase in the presence of L-arginine. At negative V_g, the local accumulation of OH^- induces an increase in the hole carrier concentration, which is recorded as an increase in the source–drain current (I_{ds}) (Figure 14.7a,b).

Noteworthy, the multilayer architecture containing both enzymes has an important effect on the sensing performance since it is known that the close proximity of both enzymes participating in a cascade reaction is crucial for optimal catalytic conversion of the analyte [51]. Figure 14.7c describes the response of graphene transistors modified with different configurations of urease and arginase assemblies in the presence of different concentrations of L-arginine. In these structures, urease and arginase were co-immobilized in the outermost layer of the assembly, whereas additional urease was incorporated in the assembly bottom layer to maximize the pH-shifting enzymatic conversion close to the graphene surface.

The experimental data reveal that PEI-Ur-PEI-UrAr-PEI (Ur:Ar 1 : 2) assemblies exhibit the best performance in comparison with PEI-Ur-PEI-UrAr-PEI (Ur:Ar 1 : 1) and PEI-UrAr-PEI. Therefore, the ratio between both enzymes in the bi-enzymatic cascade reaction has a relevant effect on the sensor performance. Assemblies exhibiting a Ur:Ar 1 : 2 ratio in the outermost enzyme layer of the assembly undergo a more extensive analyte conversion, leading to higher intermediate substrate (urea) concentrations. Subsequently, the use of urease in excess ensures a prompt conversion of all intermediate species and renders the observed rate of ammonia formation directly proportional to the concentration of arginine. The optimization of the interfacial architecture involving an increase in the Ur:Ar

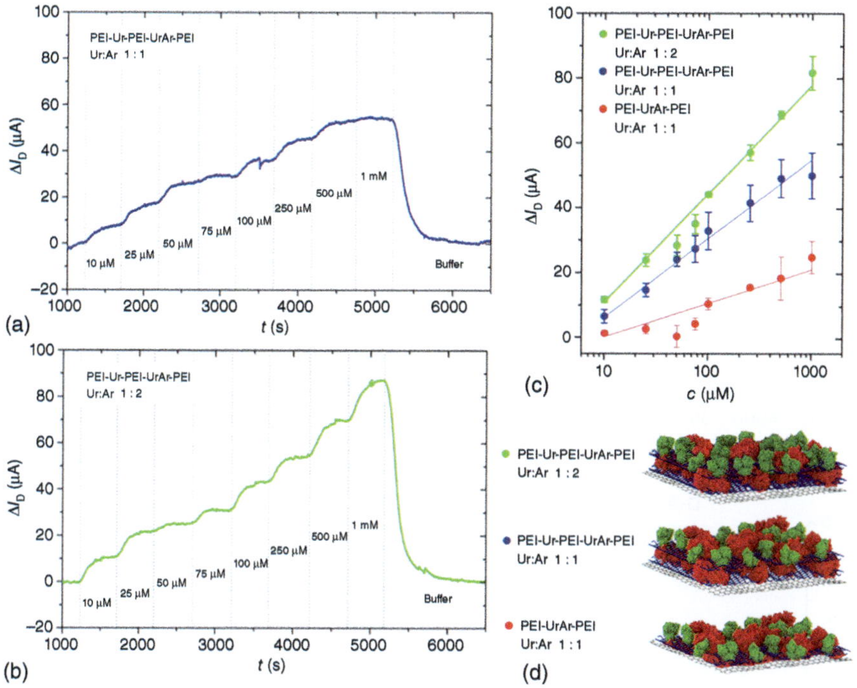

Figure 14.7 (a) Real-time channel current response (ΔI_{ds}) for the PEI-Ur-PEI-UrAr-PEI assembly with a ratio of Ur:Ar of 1 : 1 (b) and Ur:Ar 1 : 2 in the outermost enzyme layer. The dashed lines represent the change in concentration. (c) ΔI_{ds} response of the assemblies PEI-UrAr-PEI Ur:Ar 1 : 1 (R = 0.90) (circles), PEI-Ur-PEI-UrAr-PEI Ur:Ar 1 : 1 (R = 0.99) (rhombuses), and PEI-Ur-PEI-UrAr-PEI Ur:Ar 1 : 2 (R = 0.99) (squares). Error bars represent the standard deviation of three measurements. Experimental conditions: 10 mM KCl at pH 6, V_g: −0.2 V, V_{ds}: 0.1 V, flow rate: 100 μl min^{-1}. Solid lines are the linear regression fits. (d) Illustrative schemes of the different nanoarchitectured systems used for biosensing by enzymatic cascade reactions. Source: Reproduced with permission from Berninger et al. [50].

ratio from 1 : 1 to 1 : 2 resulted in a sensitivity increase from 24 to 33.5 μA per decade (L-arginine) in the 10–1000 μM concentration range.

These rGO-FET-based sensors are compatible with continuous flow systems, endowing the real-time monitoring and detection of the analyte. Another relevant aspect of these GFET sensors is their remarkable selectivity for the analyte enabled by the enzymatic system. Figure 14.8 shows the response of GFETs modified with PEI/Urease/PEI/Urease + Arginase/PEI assemblies in the presence of nontarget amino acids such as L-valine, L-alanine, creatinine, and L-phenylalanine. As it can be seen, the sensor showed no or very little response (maximum 3 μA) to non-target amino acids, even at high concentrations. Moreover, control experiments confirmed that the sensors exhibit excellent stereospecificity in the presence of D-arginine solutions, while the L-arginine antagonist canavanine was also successfully detected (Figure 14.8b).

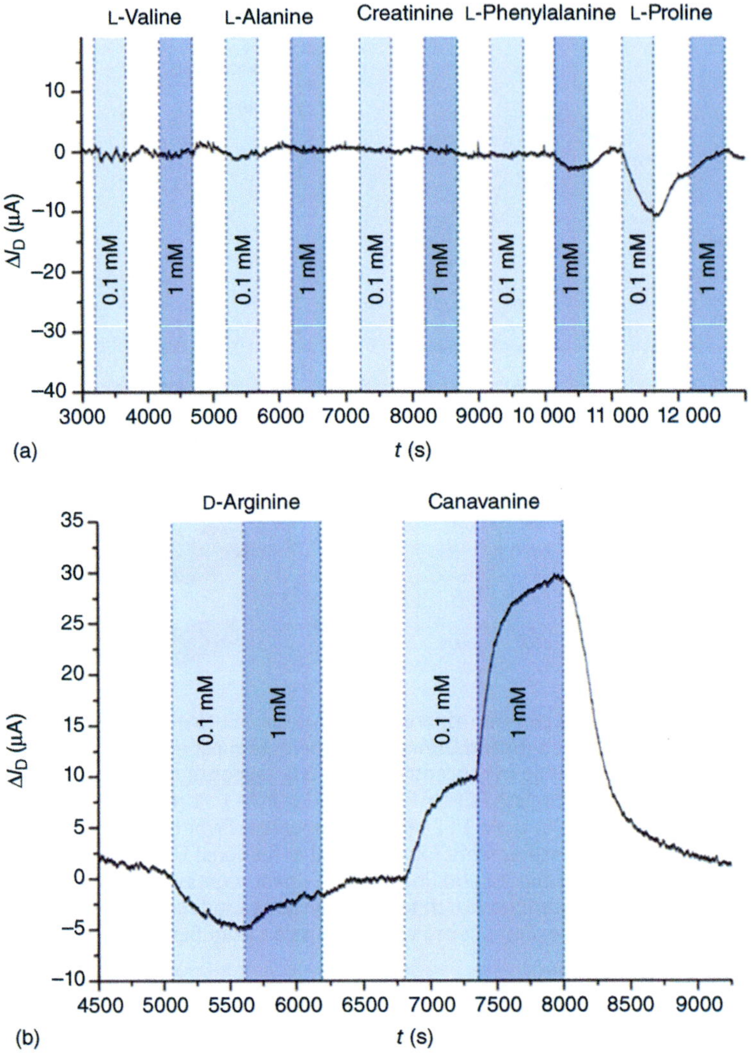

Figure 14.8 (a) Real-time channel current response (ΔI_{ds}) for the LbL assembly PEI-Ur-PEI-UrAr-PEI Ur:Ar 1 : 2 titrating different interfering species: L-valine, L-alanine, creatinine, L-phenylalanine, and L-proline as well as (b) D-arginine and canavanine at concentrations of 100 μM (light gray) and 1 mM (gray). The dashed lines represent the change in concentration. Experimental conditions: 10 mM KCl at pH 6, V_g: −200 mV, V_{ds}: 100 mV, flow rate: 100 μl min^{-1}. Source: Reproduced with permission from Berninger et al. [50].

14.4 Conclusions

In this chapter, we introduced the LbL approach for the construction of polyelectrolyte-enzyme multilayer assemblies onto rGO FETs for biosensing applications. The modification of rGO FETs with a weak polyelectrolyte improves the pH response of the device owing to its transducing properties by electrostatic gating effects. Later, when exposed to urea solutions, graphene transistors modified with (PEI/Urease)$_n$ multilayer assemblies exhibit a shift in the Dirac point due to the change in the local pH close to the graphene surface, produced by the urea catalyzed hydrolysis. This type of urea sensor was able to monitor the analyte in the 1–1000 μM range with a limit of detection (LOD) of 1 μM. In addition, these devices display fast response and good long-term stability. One of the key advantages of the LbL approach is the ability to manipulate enzyme surface coverage onto the graphene channel by controlling the number of assembled bilayers. In this way, an increase in the number of bilayers yields an improvement in the urea response of the transistors.

In addition, the use of the LbL technique for enzyme immobilization offers the opportunity of precisely controlling the nanoarchitecture of the assembly, as opposed to other methods involving covalent chemistries. This not only is relevant in the case of graphene transistors coupled to enzymatic cascades but may also serve as a model approach for the study of other multi-enzyme systems and their micro-environments. Therefore, by optimizing the composition of the LbL assembly, the working stability of the sensor could be improved, resulting in a low standard deviation within measurements. In fact, results suggest that enhancement of catalytic activity can result from enzyme immobilization using polyelectrolytes in a multilayer architecture. In the case of the arginine sensor described in this chapter, experimental evidence shows that enzyme ratio and the proximity between arginase and urease represent key factors that define the sensing performance.

Taken all together, experiments discussed in this chapter introduce a very appealing and useful approach to designing biosensors based on the exquisite synergy arising from the combination of the remarkable physical/electronic features of GFETs, the versatility of electrostatic LbL assembly, and the unsurpassed specificity of enzymatic recognition. We envision that further extrapolation of these notions and concepts relying on the use of "reactive amplification" to other enzymatic systems will open up new opportunities to devise new, highly sensitive GFET biosensors.

References

1 Mello, L.D. and Kubota, L.T. (2002). *Food Chem. 77*: 237–256.
2 Hulanicki, A., Glab, S., and Ingman, F. (1991). *Pure Appl. Chem. 63*: 1247–1250.
3 Wang, Y., Li, Z., Wang, J. et al. (2011). *Trends Biotechnol. 29*: 205–212.
4 Ariga, K., Ji, Q., Mori, T. et al. (2013). *Chem. Soc. Rev. 42*: 6322.
5 (a) Cortez, M.L., Ceolín, M., Azzaroni, O., and Battaglini, F. (2015). *Bioelectrochemistry 105*: 117–122.(b) Cortez, M.L., Ceolín, M., Camacho, L.C. et al. (2017). *ACS Appl. Mater. Interfaces 9*: 1119–1128.(c) Cortez, M.L., Marmisollé, W., Pallarola, D. et al. (2014). *Chem. Eur. J. 20*: 13366–13374.
6 Disawal, S., Qiu, J., Elmore, B.B., and Lvov, Y.M. (2003). *Colloids Surf., B 32*: 145–156.
7 Karacaoğlu, S., Timur, S., and Telefoncu, A. (2003). *Artif. Cells, Blood Substitues, Immobilization Biotechnol. 31*: 357–363.
8 Mao, S., Chang, J., Pu, H. et al. (2017). *Chem. Soc. Rev. 46*: 6872–6904.

9 (a) Scotto, J., Cantillo, A.L., Piccinini, E. et al. (2022). *ACS Appl. Electron. Mater. 4*: 3988–3996.(b) Fenoy, G.E., Piccinini, E., Knoll, W. et al. (2022). *Anal. Chem. 94*: 13820–13828.

10 Choi, W. and Lee, J. (2002). Graphene synthesis and applications. In: *Nanomaterials and Their Applications* (ed. W. Choi and J.-w. Lee), vii–x. Boca Raton, FL: CRC Press.

11 Cai, B., Wang, S., Huang, L. et al. (2014). *ACS Nano 8*: 2632–2638.

12 (a) Fenoy, G.E., Marmisollé, W.A., Azzaroni, O., and Knoll, W. (2020). *Biosens. Bioelectron. 148*: 111796.(b) Fenoy, G.E., Marmisollé, W.A., Knoll, W., and Azzaroni, O. (2022). *Sens. Diagn. 1*: 139–148.(c) Aspermair, P., Ramach, U., Reiner-Rozman, C. et al. (2020). *J. Am. Chem. Soc. 142* (27): 11709–11716.

13 Alberti, S., Piccinini, E., Ramirez, P.G. et al. (2021). *Nanoscale 13*: 19098–19108.

14 Zhu, Z., Garcia-Gancedo, L., Flewitt, A.J. et al. (2012). *Sensors 12*: 5996–6022.

15 Zhang, M., Liao, C., Mak, C.H. et al. (2015). *Sci. Rep. 5*: 8311.

16 Yan, F., Zhang, M., and Li, J. (2014). *Adv. Healthcare Mater. 3*: 313–331.

17 Ohno, Y., Maehashi, K., Yamashiro, Y., and Matsumoto, K. (2009). *Nano Lett. 9*: 3318–3322.

18 Sohn, I.Y., Kim, D.J., Jung, J.H. et al. (2013). *Biosens. Bioelectron. 45*: 70–76.

19 Dong, X., Shi, Y., Huang, W. et al. (2010). *Adv. Mater. 22*: 1–5.

20 Reiner-Rozman, C., Larisika, M., Nowak, C., and Knoll, W. (2015). *Biosens. Bioelectron. 70*: 21–27.

21 (a) Sappia, L.D., Piccinini, E., Marmisollé, W. et al. (2017). *Adv. Mater. Interfaces 4*: 1700502.(b) Sappia, L.D., Piccinini, E., von Binderling, C. et al. (2020). *Mater. Sci. Eng., C 109*: 110575.

22 (a) Amine, A., Mohammadi, H., Bourais, I., and Palleschi, G. (2006). *Biosens. Bioelectron. 21*: 1405–1423. (b) Pallarola, D., Queralto, N., Battaglini, F., and Azzaroni, O. (2010). *Phys. Chem. Chem. Phys. 12*: 8072–8084.(c) Pallarola, D., Queralto, N., Knoll, W. et al. (2010). *Langmuir 26* (16): 13684–13696.

23 (a) Pallarola, D., Queralto, N., Knoll, W. et al. (2010). *Chem. Eur. J. 16*: 13970–13975. (b) Cortez, M.L., Pallarola, D., Ceolín, M. et al. (2012). *Chem. Commun. 48*: 10868–10870. (c) Cortez, M.L., Pallarola, D., Ceolín, M. et al. (2013). *Anal. Chem. 85* (4): 2414–2422.

24 Piccinini, E., Allegretto, J.A., Scotto, J. et al. (2021). *ACS Appl. Mater. Interfaces 13*: 43696–43707.

25 Piccinini, E., Fenoy, G.E., Cantillo, A.L. et al. (2022). *Adv. Mater. Interfaces* 9: 2102526.

26 Niyogi, S., Bekyarova, E., Itkis, M.E. et al. (2010). *Nano Lett. 10*: 4061–4066.

27 Sheldon, R.A. and Van Pelt, S. (2013). *Chem. Soc. Rev. 42*: 6223–6235.

28 Torsi, L., Magliulo, M., Manoli, K., and Palazzo, G. (2013). *Chem. Soc. Rev. 42*: 8612.

29 Wong, H.S. and Akinwande, D. (2011). *Carbon Nanotube and Graphene Device Physics*. Cambridge: Cambridge University Press.

30 Pumera, M. (2011). *Mater. Today 14*: 308–315.

31 Heller, I., Chatoor, S., Männik, J. et al. (2010). *J. Am. Chem. Soc. 132*: 17149–17156.

32 Cortez, M.L., De Matteis, N., Ceolín, M. et al. (2014). *Phys. Chem. Chem. Phys. 16*: 20844–20855.

33 Pallarola, D., Queralto, N., Battaglini, F., and Azzaroni, O. (2010). *Phys. Chem. Chem. Phys. 12*: 8071–8083.

34 Ariga, K., Hill, J.P., and Ji, Q. (2007). *Phys. Chem. Chem. Phys. 9*: 2319–2340.

35 Irigoyen, J., Moya, S.E., Iturri, J.J. et al. (2009). *Langmuir 25*: 3374–3380.

36 Decher, G. and Schlenoff, J.B. (2012). *Multilayer Thin Films: Sequential Assembly of Nanocomposite Materials*, 2e. Weinheim: Wiley-VCH Verlag GmbH & Co.

37 (a) Fenoy, G.E., Rafti, M., Marmisolle, W., and Azzaroni, O. (2021). *Mater. Adv. 2*: 7731–7740. (b) Piccinini, E., González, G.A., Azzaroni, O., and Battaglini, F. (2021). *J. Colloid Interface Sci. 581*: 595–607. (c) Mártire, A.P., Segovia, G., Azzaroni, O. et al. (2019). *Mol. Syst. Des. Eng. 4*: 893–900. (d) Fenoy, G.E., Van der Schueren, B., Scotto, J. et al. (2018). *Electrochim. Acta 283*: 1178–1187. (e) Maza, E., von Bilderling, C., Cortez, M.L. et al. (2018). *Langmuir 34*: 3711–3719. (f) Cortez, M.L., Lorenzo, A., Marmisollé, W.A. et al. (2018). *Soft Matter 14*: 1939–1952.

38 Rydzek, G., Ji, Q., Li, M. et al. (2015). *Nano Today 10*: 138–167.

39 Iost, R.M. and Crespilho, F.N. (2012). *Biosens. Bioelectron. 31*: 1–10.

40 Pérez-Mitta, G., Peinetti, A.S., Cortez, M.L. et al. (2018). *Nano Lett. 18*: 3303–3310.

41 Terrones, Y.T., Laucirica, G., Cayón, V.M. et al. (2022). *Chem. Commun. 58*: 10166–10169.

42 Piccinini, E., Alberti, S., Longo, G.S. et al. (2018). *J. Phys. Chem. C* 122: 10181–10188.

43 Zhybak, M., Beni, V., Vagin, M.Y. et al. (2016). *Biosens. Bioelectron. 77*: 505–511.

44 Piccinini, E., Bliem, C., Reiner-Rozman, C. et al. (2017). *Biosens. Bioelectron. 92*: 661–667.

45 Bliem, C., Piccinini, E., Knoll, W., and Azzaroni, O. (2018). Enzyme multilayers on graphene-based FETs for biosensing applications. In: *Enzyme Nanoarchitectures: Enzymes Armored with Graphene* (ed. C.V. Kumar), 23–46. Elsevier. https://doi .org/10.1016/bs.mie.2018.06.001.

46 Wang, Y.Y. and Burke, P.J. (2014). *Nano Res. 7*: 1650–1658.

47 Hess, L.H., Lyuleeva, A., Blaschke, B.M. et al. (2014). *ACS Appl. Mater. Interfaces 6*: 9705–9710.

48 Pallarola, D., von Bildering, C., Pietrasanta, L.I. et al. (2012). *Phys. Chem. Chem. Phys. 14*: 11027–11103.

49 Popolo, A., Adesso, S., Pinto, A. et al. (2014). *Amino Acids 46*: 2271–2286.

50 Berninger, T., Bliem, C., Piccinini, E. et al. (2018). *Biosens. Bioelectron. 115*: 104–110.

51 Kang, W., Liu, J., Wang, J. et al. (2014). *Bioconjugate Chem. 25*: 1387–1394.

15

Graphene Field-Effect Transistor Biosensor for Detection of Heart Failure-Related Biomarker in Whole Blood

Jiahao Li[1], Yongmin Lei[1], Zhi-Yong Zhang[2], and Guo-Jun Zhang[1]

[1]*Hubei University of Chinese Medicine, School of Laboratory Medicine, 16 Huangjia Lake West Road, Wuhan 430065, China*
[2]*Peking University, Key Laboratory for the Physics and Chemistry of Nanodevices and Center for Carbon-based Electronics, School of Electronics, Department of Electronics, 5 Yiheyuan Road, Beijing 100871, China*

15.1 Introduction

Heart failure (HF), suffering from worldwide diffusion, is defined as a progressive clinical syndrome caused by structural or functional cardiac impairment. As the final battlefield for cardiovascular diseases, HF continues to inflict excruciating disease burdens on global public health, chiefly responsible for the high morbidity, high mortality, dreadful quality of life, and frequent hospitalizations [1–3]. With the incidence, in recent years, showing a younger trend globally [4, 5], the significance of prevention and early diagnosis for HF is underscored throughout the adult life stage. More seriously, in the throes of the ongoing COVID-19 crisis, HF provokes deep angst from clinicians and scientists due to the emerging medical fact that HF patients hold especially increased risk for virus infection [6]. On the other hand, the aggressive COVID-19 infection exacerbates the possibility for deterioration of preexisting HF [7]. According to the literature reported, the mortality rate among HF patients with septic shock was 70–90% [8]. Overall, new or preexisting HF in the setting of COVID-19 can pose a spectrum of formidable challenges that can sophisticate clinical manifestation, treatment, and prognosis.

Consistently, cardiac biomarker-guided prognostication, therapy options, and clinical care irrefutably have important implications in the whole management process of HF, notably in the acute settings – where the majority of patients with HF are initially found. As one of the most well-established and valuable biomarkers, brain natriuretic peptide (BNP, also known as B-type natriuretic peptide) is deemed a reliable indicator for HF confirmation, which is mainly synthesized and secreted by myocytes in the left ventricle [9]. Not only can BNP be used to exclude or ascertain HF in patients with suspected signs and symptoms, but it can also help clinicians

Graphene Field-Effect Transistors: Advanced Bioelectronic Devices for Sensing Applications, First Edition. Edited by Omar Azzaroni and Wolfgang Knoll.

distinguish between HF and lung diseases, both of which are subject to dyspnea [10]. Moreover, BNP has a significant negative predictive value for HF, which implies that People with normal BNP levels are very unlikely to suffer heart failure [11]. Importantly, compared with atrial-type natriuretic peptide (ANP), another common HF biomarker which is not stable in plasma due to a short circulating half-life of 3–5 minutes, BNP is a more attractive indicator with a longer half-life of 23 minutes [12]. Moreover, accumulating evidence [13, 14] has demonstrated that there is a high correlation between ascending levels of cardiac biomarkers and the severity of COVID-19. In a retrospective evaluation of patients admitted with COVID-19, Gavin et al. [15] concluded that BNP may serve as an analytical metric for disease trajectory, predicting the risk stratification. Therefore, timely and accurate BNP monitoring is of substantial importance for HF patients, particularly those combining COVID-19.

Since the FDA approved the first BNP immunoassays in 2000 [16], a plethora of BNP immunoassays, such as radioimmunoassays, enzyme-linked immunosorbent assay (ELISA), chemiluminescence analysis, and immunochromatography, have been commercialized and introduced into clinical practice [17]. Currently, BNP testing in clinical laboratory is confined to above-mentioned traditional immunological methods, which encounter lengthy turn-around times, bulky equipment, and high cost of assays [18]. Fortunately, over the past few years, a multitude of biosensor-based new approaches have been developed to provide future options for BNP detection. Mizutani [19] reported an amperometric BNP biosensor by applying a two-dimensional (2D) cross-linked polysiloxane film as a permselective layer, garnering a dynamic range of 13–60 pM. Teramura et al. [20] constructed a surface plasmon resonance (SPR) biosensor assisted with streptavidin-labeled nanobeads for signal amplification, making the picogram levels ($\mathrm{pg\,ml^{-1}}$) of BNP detectable. More recently, another SPR biosensor [21] based on a sandwich strategy, i.e. antibody–BNP–aptamer structure, was designed for rapid measurement of BNP with a wide dynamic range of $100\,\mathrm{fg\,ml^{-1}}$ to $10\,\mathrm{ng\,ml^{-1}}$. In addition, electrochemical transduction, a widely used sensing technology, also exerts a promoting role in BNP inspection. For instance, Grabowska et al. [22] established an aptamer-modified electrochemical sensing platform for BNP testing with a linear response from $1\,\mathrm{pg\,ml^{-1}}$ to $1\,\mathrm{\mu g\,ml^{-1}}$. For this sensor, the polyethyleneimine/reduced graphene oxide (rGO) nanocomposite film was coated on screen-printed electrodes, and the capture aptamer was immobilized via click chemistry, achieving sub-pM detection limit without any amplification strategy. Despite the fact that various biosensors have been successfully applied to BNP detection, delivering promising solutions for HF diagnosis, they still have several downsides: (i) large sample volume; (ii) laborious sample preparation; (iii) insufficient sensitivity; (iv) cumbersome device fabrication or cumbersome labeling; and (v) not amenable to point-of-care testing. What is more, few of them give a comprehensive picture for practical application and position such biosensor-based examinations closer to clinical reality. It is well known that BNP is one of the most frequent testing items in the emergency department, and the result report is required to be sent faster than normal outpatient examinations.

Among the variety of biosensors available today, field-effect transistor (FET) biosensors, especially in combination with advanced functional nanomaterials, have been considered ideal candidates for next-generation medical diagnostic platforms [23, 24]. FET-type biosensors are a class of potentiometric sensors, which possess distinct advantages over some other systems due to their intrinsic mechanism of action. In a typical FET system, the source and drain electrodes are connected by sensing channels, whose surfaces are coated with sensing elements and bio-probes to capture the targets of interests. Briefly, the accumulation of charged analytes at solution interface leads to a modulation of electrons or holes, resulting in a variation of channel conductance [25], which can be recorded and further processed by a semiconductor parameter analyzer. Owing to the dominance of high speed, low wastage, and outstanding compatibility with integrated circuits, FET sensing platform allows for instant ultrasensitive detection while avoiding the dependence on various labelings [26]. A good channel material is integral to an excellent FET biosensor. As one of the most renowned and attractive 2D carbon nanosheets, graphene stands out by a landslide for its ultrathin sensing layer, high carrier mobility, remarkable conductivity, superior biocompatibility, and other unique properties [27–29]. To date, graphene-based field-effect transistor (GFET) biosensors have been extensively reported to contribute to the diagnosis of many diseases, including thrombotic events [30], neurological degenerative diseases [31], tumors [32], and infectious diseases [33, 34], to name a few.

In this chapter, an integrated FET-based detection platform, which combines a microfilter system and platinum nanoparticles (PtNPs)-decorated GFET nanosensor, was demonstrated for BNP testing in whole blood (Figure 15.1). This technology

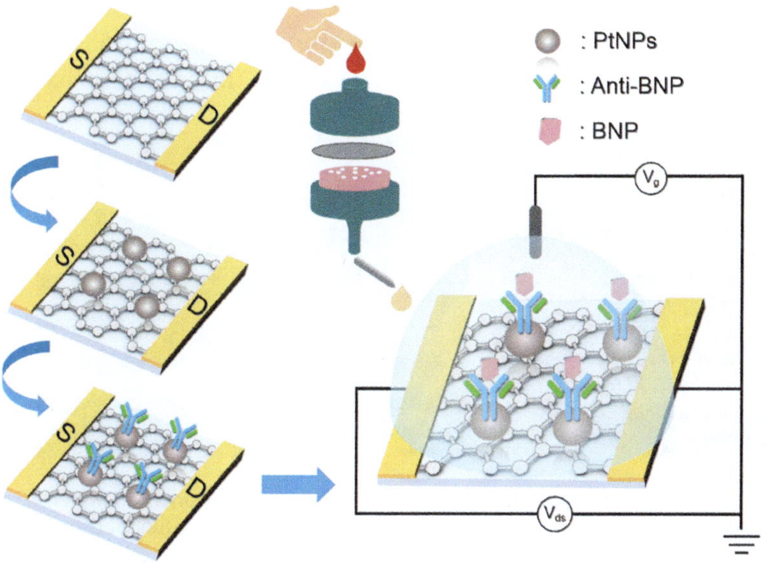

Figure 15.1 Schematic illustration of the PtNPs-decorated GFET biosensor with a custom-made microfilter for BNP detection (unpublished materials).

streamlines BNP analyses by making the sample processing and electrical measurement pipelined, showing its potential for portability and clinical application. It is hoped that this chapter can propose a new solution for BNP detection and HF diagnosis, thus providing feasible strategies and ideas to break through the bottlenecks of biosensor-based analytical tools.

15.2 Experimental Systems and Procedures

15.2.1 Fabrication of GFET Sensor

The FET chip was produced using standard semiconductor technology. Succinctly stated, the source (S) and drain (D) electrodes are formed on SiO_2/Si substrate through photolithography, and are 50 nm thick and 4 μm apart (Figure 15.2). For rGO preparation, a simple chemical reduction method was used with hydrazine as the reductant. Hydrazine can effectively remove the oxygen-containing groups and make the atomic C/O ratio as high as 10.8 for rGO [35]. Compared to electrical reduction, thermal reduction, and catalytic reduction, this approach is not only easy for operation, free of high temperature and high voltage, but also has a high reduction degree. In this way, the resulting stock solution of rGO could be stable for months without aggregation. Then a small volume (0.2 μl) of diluted rGO (0.2 mg ml^{-1}) suspension was drop-cast on the channel region and thermally annealed at 150 °C for two hours. After that, the chip was sonicated with Piranha solution for 10 seconds to obtain few-layer rGO. And the eventually resultant GFET sensor could be used for following functionalization.

15.2.2 Decoration of Platinum Nanoparticles

The GFET device was immersed in 2 ml of 5 mM Na_2PtCl_4 solution, which was in a mixed deionized (DI) water/ethanol (v:v = 1 : 1) solvent. In order to make an

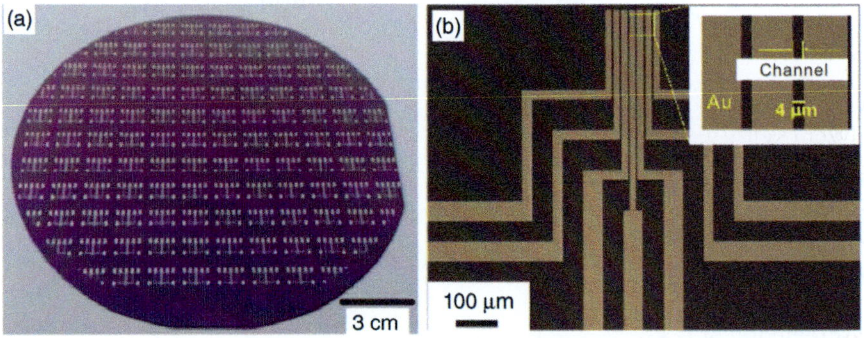

Figure 15.2 (a) Photographic image of the fabricated FET device array on a 4 in. wafer. (b) Optical micrograph showing the core channel regions of the biosensor (unpublished materials).

efficient and complete reaction to deposit a high density of PtNPs on rGO surface, the UV light at 368 nm wavelength was introduced to directly irradiate the sensor chip for 30 minutes. After thoroughly rinsing with ethanol and DI water, followed by drying with N_2 stream, the PtNPs-decorated GFET was acquired.

15.2.3 Surface Functionalization

To immobilize the anti-BNP antibody on the chip, the PtNPs-decorated GFET was treated with a 10 mM solution of thioglycolic acid (TGA) overnight at room temperature to produce carboxylic group on the PtNPs surface through Pt—S bonds. Afterwards, the carboxylic group (—COOH) was activated for 30 minutes by a mixture solution prepared by mixing 0.1 M *N*-hydroxysulfosuccinimide (NHS) and 0.4 M 1-ethyl-3-(3-dimethyllaminopropyl) carbodiimide (EDC) in equal volume. Subsequently, 100 μg ml^{-1} BNP antibody was introduced to the sensor chip and underwent overnight incubation at 4 °C. And then the chip was washed three times with 1 × PBS (phosphate buffer saline) containing 0.2% sodium dodecyl sulfate (SDS), 1 × PBS, and pure water, followed by the blocking step with one hour of exposure to the bovine serum albumin (BSA) solution for the sake of eliminating possible nonspecific binding events. The final nanosensor could be applied to follow-up detection.

15.2.4 Immunodetection in Whole Blood

In this platform, a polycarbonate microfilter with 400-nm aperture was used for preprocessing of samples to remove blood cells. The filtered sample was directly delivered onto the BNP antibody-functionalized GFET biosensor and suffered from a 30 minutes incubation for adequate immunoreaction. After that, the chip was subjected to stringent flushing and drying. The entire process of sensor modification, functionalization, and target capture can be seen in the following diagram (Figure 15.3).

15.2.5 Electrical Measurements

The floating gate GFET devices were measured through a semiconductor parameter analyzer (Keithley 4200-SCS) connected by a probe station. An Ag/AgCl reference electrode was used to apply the liquid gate to stabilize the potential of the electrolyte. For the steady-state measurements, a constant bias V_{DS} of 0.1 V was exerted, and the 0.001 × PBS was used as a liquid gate. Additionally, a negligible gate leakage should be guaranteed, which is at least 3 orders of magnitude lower than the current between S and D electrodes (I_{DS}), to ensure the validity of the I_D–V_G transfer curve measurement. For the real-time measurement, the I_{DS} changed with time and were recorded under the gate voltage of 0.3 V, and a stable current baseline was obtained with 0.001 × PBS buffer as the basic test environment.

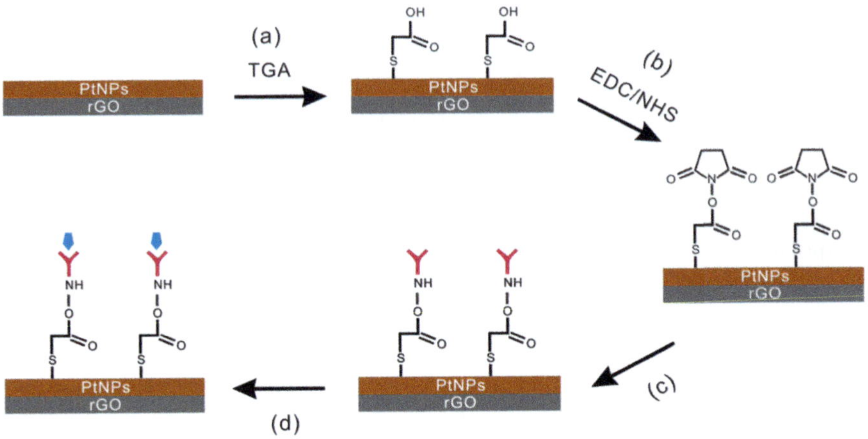

Figure 15.3 Construction scheme of GFET sensor modified with anti-BNP antibody for detection of BNP molecules (unpublished materials).

15.3 Sensing Principle of GFET for BNP Detection

Anti-BNP antibody can be covalently immobilized on the PtNP's surface, via carbodiimide coupling reaction provided by an EDC/NHS system, to capture the BNP protein. Based on its amino acid sequence, the theoretical isoelectric point (pI) of BNP was calculated to be 10.95 [36], making the BNP positively charged in the buffer solution (pH = 7.4). Upon binding to the BNP antibody, which functionalized GFET sensor, BNP gave rise to a decrease in net carrier density, leading to p-doping with a right shift of Dirac point [37]. Therefore, the target level can be reflected by recording the variation of gate voltage at Dirac points, which is denoted as ΔV_{CNP}, the response signal.

15.4 Device Characterization

Aiming to increase the conductivity and electrical response signals, PtNPs were directly synthesized on the rGO film, resulting from the photochemical reduction of $PtCl^{-2}$ to Pt^0 in the presence of alcohols [38–40]. The scanning electron microscopy (SEM) image in Figure 15.4a clearly presented a continuous and few-layer rGO film, of which the surface was coated with a homogeneous and high-density distribution of PtNPs. Moreover, the average particle diameter was about 10 nm, and the UV light's illumination time could be adjusted for this. What is more, the transmission electron microscopy (TEM) and X-ray energy dispersive spectrum (EDS) were also conducted to verify the PtNPs formation (Figure 15.4b,c). As expected, the characteristic peak of Pt (2.2 keV) appeared after PtNP's decoration [42]. And the detected C, O, and Si peaks were mainly from the rGO and the chip substrate, respectively (Figure 15.4c). In order to further prove that the PtNPs on the rGO surface are indeed zero-valent metal nanoparticles rather than existing in intermediate valence

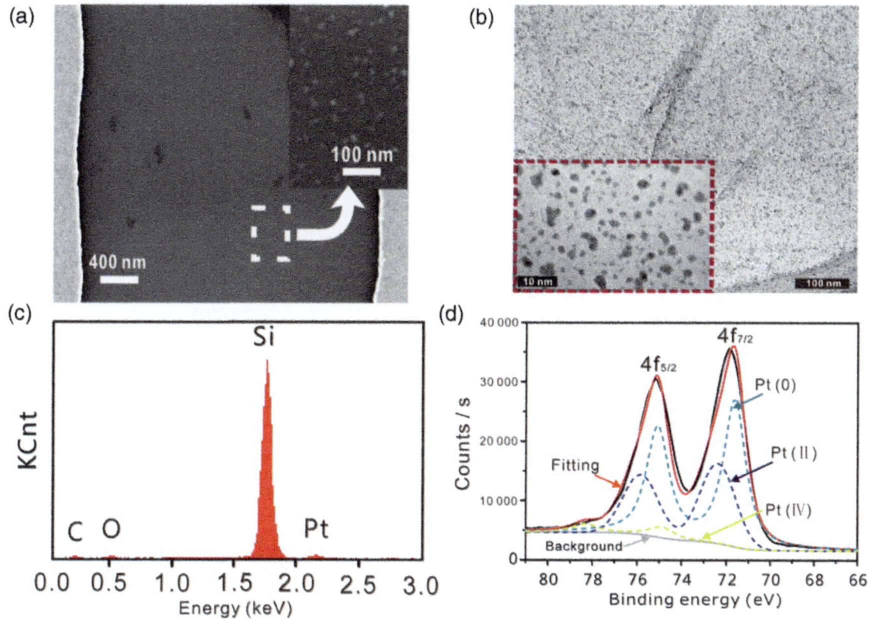

Figure 15.4 (a) SEM image of photochemically synthesized PtNPs on the rGO film spanning across gold electrodes. (b) TEM image of PtNPs decorated on rGO. The diameter of PtNPs is 10 nm. (c) EDS spectrum of the PtNPs-decorated rGO on a Si/SiO_2 substrate. (d) XPS of the PtNPs at Pt4f spectra. Source: Reproduced with permission from Lei e.al. [41]/ Elsevier.

states, X-ray photoelectron spectroscopy (XPS) analysis was further carried out. As shown in Figure 15.4d, the double peaks at 71.7 and 75.2 eV (black line) were attributable to $4f_{7/2}$ and $4f_{5/2}$ of metallic Pt, respectively, which are in good agreement with the previously reported results [43, 44]. It should be noted that Pt 4f could be fitted with three chemical states: Pt (0) at 71.4 eV, Pt (II) at 72.3 eV, and Pt (IV) at 74.4 eV. And apparently, the composition of Pt (0) was larger than that of Pt (II) and Pt (IV), indicating that the dominant valence of the Pt species on the rGO was Pt (0). These results suggest that the PtNPs-decorated GFET chip was successfully constructed, and the UV method is highly efficient for the preparation of dense PtNPs on the rGO surface.

Furthermore, in order to verify that the device is feasible for BNP testing, the transfer characteristic curves of the GFET for stepwise functionalization process and BNP binding were measured, as shown in Figure 15.5a. Invariably, similar ambipolar natures were observed for all steps. Specifically, the presence of modified PtNPs on the rGO channel led to a clear increase in current when compared to unaltered rGO. It may be attributed to the increased active surface area of the well-distributed PtNPs, which facilitates the electron transfer between rGO and electrode [45]. And the Dirac point was found to be almost unchanged at around 0.36 V, which is consistent with Yin's work [40]. After probe immobilization, the negatively charged anti-BNP [46] induced n-doping with a left shift of Dirac point.

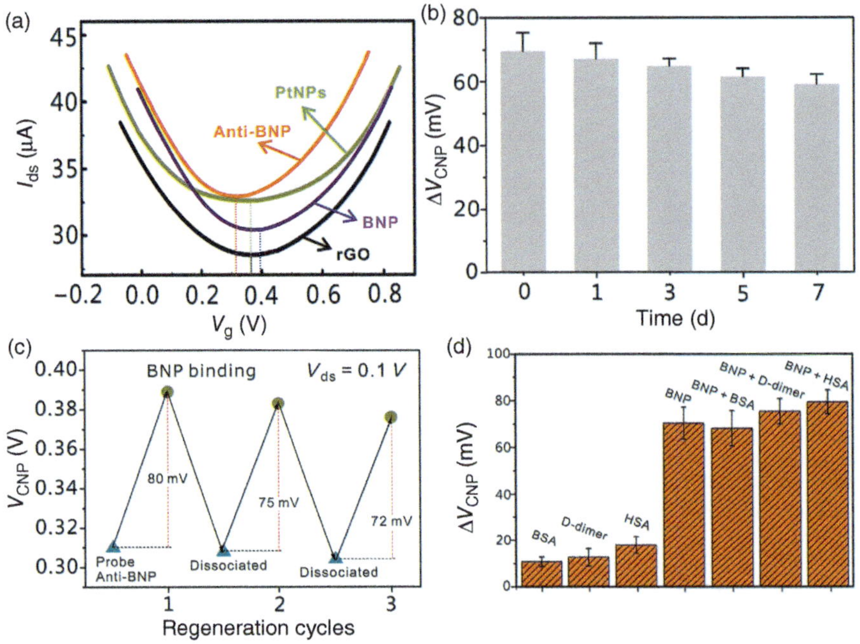

Figure 15.5 (a) Plots of transfer curves for the GFET biosensor in the process of rGO formation, PtNPs decoration, anti-BNP immobilization, and BNP binding. (b) Stability test for the PtNPs-decorated GFET biosensor during one week. (c) Reproducibility study for BNP detection. (d) Variation of V_{CNP} of the anti-BNP immobilized GFET biosensor at the detection of different proteins with the same concentration. Source: Reproduced from Lei et al. [41]/ with permission from Elsevier.

However, the Dirac point travelled to the right from 0.32 to 0.4V upon the incubation with positively charged BNP, on account of p-doping as mentioned in the previous part. These results confirm that anti-BNP was successfully modified on GFET chip and that the target was captured by the functionalized sensor with significant electrical response.

15.5 Sensing Performance

15.5.1 Stability and Reproducibility

To evaluate the stability of the PtNPs-decorated GFET biosensors, the functionalized device was stored in a vacuum oven for a period of time and used for the detection of 200 nM BNP. As described in Figure 15.5b, the variation of gate voltage for the Dirac point retained more than 85% of its original ΔV_{CNP} value with only tiny discrepancy over one week, exhibiting excellent stability. The response signal slightly decreased as a function of the storing time, which is probably due to the nonspecific adsorptions of air and water in the environment.

To determine whether the biosensor could be reused, which is a considerable dimension for cost savings, the reproducibility of the developed platform was investigated. For this set of experiments, the BNP-bound FET biosensor was treated with highly concentrated salt solutions (NaCl, 1 M) to dissociate the binding between captured BNP and the antibody [47, 48]. Afterwards, 200 nM BNP was introduced to the chip once more. As such, the transfer characteristic curves were repeatedly measured to obtain the gate voltage displacement during three cycles. In each process of dissociation and binding, as observed in Figure 15.5c, the changes in V_{CNP} were found to be 80, 75, and 72 mV, respectively. The V_{CNP} value for the third cycle of measurement could still sustain 90% similarity to that of the first cycle of BNP detection, indicating very close results with marginal difference. Therefore, this sensor possesses good reusability and re-generability.

15.5.2 Selectivity

It is well known that one of the major challenges in developing a credible biosensor lies in selectivity, which is the prerequisite for a high degree of accuracy and the minimization of misdiagnosis. To interrogate the capability of the sensor to distinguish between target and other nonspecific proteins, the same concentration (100 nM) of BSA, D-Dimer, human serum albumin (HSA), BNP, BNP + BSA, BNP + D-Dimer, and BNP + HSA solutions were successively exposed to the GFET sensor chip, which had undergone entire functionalization process, including PtNPs decoration and anti-BNP antibody attachment. And then the transfer characteristic curves were recorded after every incubation was conducted. As illustrated in Figure 15.5d, when the contaminant proteins (BSA, D-Dimer, and HSA) were applied to the devices, the ΔV_{CNP} values (11.5 ± 2.0, 13.2 ± 3.7, 17.3 ± 2.8 mV) were substantially smaller than those obtained from target protein (BNP; 70.4 ± 7.2 mV). Furthermore, in the reaction with BNP + BSA, BNP + D-Dimer, and BNP + HSA mixtures, dramatic variations of V_{CNP} were also obtained, which were calculated to be 68.0 ± 7.7, 75.0 ± 4.9, and 79.5 ± 4.2 mV, respectively. Importantly, these signal values were similar to the response intensity of the BNP standard solution, meaning that BNP could be detected with high fidelity even in a mixture with other interfering proteins. It can thus be concluded that the developed assay enjoys remarkable selectivity for BNP testing.

15.5.3 Sensitivity

To study the sensitivity of the established biosensor for BNP detection, the device was applied to test a range of various concentrations of BNP in real time. The dynamic responses of the sensor during successive injections of blank and target were depicted in Figure 15.6a. It can be clearly seen that a very fast response (10 seconds) was achieved when the sensor was exposed to the diverse content of BNP solution, providing a real-time snapshot of the BNP level. As the target level was ramped higher, the I_{DS} decreased gradually as a result of the binding event for

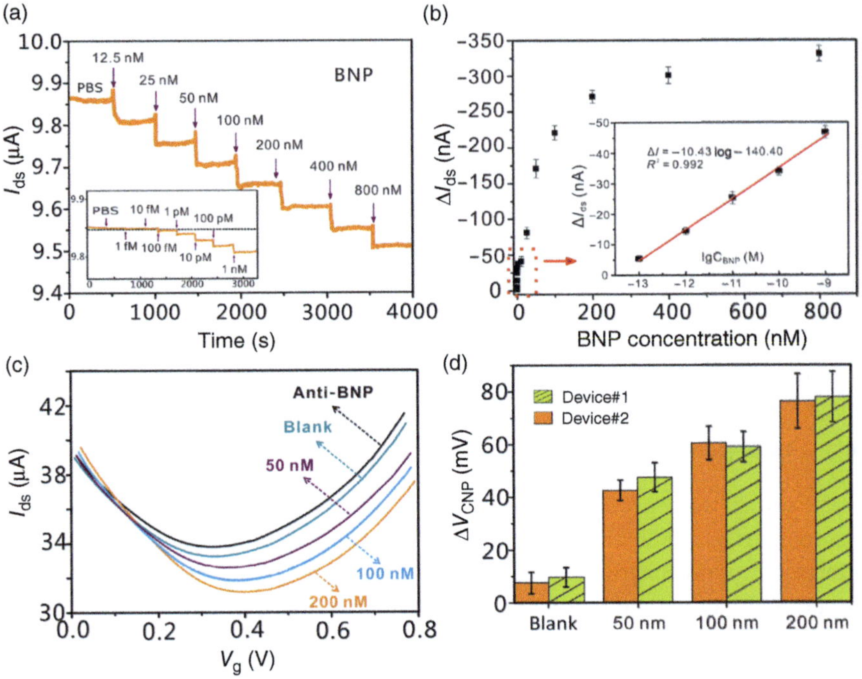

Figure 15.6 (a) Real-time electrical testing for different concentrations of BNP solution in PBS (12.5, 25, 50, 100, 200, 400, and 800 nM). Inset: Real-time electrical testing of lower concentrations of BNP. (b) The ΔI values of the PtNPs-decorated GFET biosensors to a series of BNP concentrations. Inset: The response values of the biosensors to BNP at low concentration range (100 fM, 1 pM, 10 pM, 100 pM, and 1 nM). (c) I_D–V_G transfer curves of the PtNPs-decorated GFET biosensor upon incubation with varying concentrations of BNP in serum-spiked samples. (d) Variations of V_{CNP} for the two representative devices at the detection of varying concentrations BNP in serum-spiked samples (50, 100, and 200 nM), respectively. Error bars show the standard deviations of measurements taken from three independent samples. Source: Reproduced from Lei et al. [41]/with permission from Elsevier.

positively charged BNP. The specific binding caused a structural rearrangement of the charge distribution, finally leading to a decrease in the negatively charged base state at the sensor interface and reducing the hopping rate of charge carriers (holes), which is probably responsible for the decline in current [49]. As shown in the inset, on the basis of threefold signal-to-noise ratio, this sensor could reliably detect BNP down to 100 fM.

Moreover, the logarithmic value of BNP concentration, defined as $\lg C$ shows a good linear response within the range of 100 fM to 1 nM (Figure 15.6b). The regression equation was expressed as ΔI (nA) = $-10.43 \lg C_{BNP}$ (M) $-$ 140.40 with a high correlation coefficient value (R^2) of 0.9917 between target concentration and signal output. In general, of note, the reference range of blood BNP level under normal conditions is 6 pM. And for people at potential risk of HF, the BNP level increases to 600 pM [50–52]. This signified that the sensing platform opened up the exciting possibility of detecting BNP in blood samples at the early stage of HF with extraordinary sensitivity.

15.6 Clinical Application Prospects

Most crucial of all, the practicability and validity of the yielded sensor were verified in real clinical samples. To demonstrate the detection capability in whole blood, BNP was spiked into healthy human blood specimens that were collected from the Wuhan No.1 Hospital (Wuhan, China). Then the acquired, contrived samples containing diverse concentrations of BNP were measured by the PtNPs-decorated GFET sensor. Considering that the whole blood harbors a complex matrix environment that can bring much endogenous interference, especially blood cells, sample pretreatment is entailed prior to BNP detection. Therefore, an integrated system is expected for rapid, scalable, and portable testing. To this end, the custom-made microfilter was employed to readily remove blood cells. After that, the filtered sample was dropped onto the anti-BNP functionalized GFET biosensor. The testing results with three different concentrations (50, 100, and 200 nM) of BNP in blood are shown in Figure 15.6c. As a control, the same volume of blood alone in the absence of BNP was tested, and the V_{CNP} barely shifted, retaining around 0.32 V. Thus, the whole blood sample suffering from pretreatment with the microfilter did not cause nonspecific binding. After the blood samples containing 50, 100, and 200 nM BNP were filtered and applied to the sensor, the transfer curves indicated obvious rightward movement with the ΔV_{CNP} values of 44, 58, and 73 mV, respectively. To assess the repeatability of the measurements, two GFET devices were used for blood BNP testing. The corresponding variation of ΔV_{CNP} was summarized in Figure 15.6d. It can be seen that the two different devices showed similar performance at the same BNP concentration, and the response enhanced along with the ascending BNP level in blood. These results elucidate that the method excels in detection precision and repeatability in practical application, widening the potential of biosensor-based methodology in clinical settings. It is thus conceivable that this GFET sensing platform provides immense value in early and sensitive diagnosis for HF patients, and assists in making clinical decisions faster, especially in an emergency situation.

15.7 Advantages, Limitations, and Outlook of the FET-Based BNP Assay

HF epidemiology reveals that this syndrome is always confronted with high mortality and high morbidity regardless of socioeconomic class, gender, age, and ethnical factors, which is mainly owing to the less-than-optimal availability of diagnostic tools and disease management technologies [53]. As such, the ensuing grievous public health issue brings the detection of cardiac biomarkers, most importantly BNP, into the spotlight. Herein, this chapter introduces a PtNPs-decorated GFET sensing platform for the real-time monitoring of BNP, which enjoys the following attractive merits: (i) an exquisite sensitivity as high as 100 fM; (ii) the fast real-time response (10 seconds) providing snapshots for the BNP levels; (iii) allowing a well-defined distinction between BNP and other nonspecific proteins; (iv) an integration of sample processing and analysis would streamline operations while also reducing laboratory testing time; (v) marvelous performance of BNP detection in whole blood with good repeatability. In summary, this FET-based dedicated analytical platform for BNP is capable of fulfilling its role in the rapid, accurate, sensitive, and confident diagnosis of HF. It is hoped that the developed GFET biosensor represents an auspicious orientation for portable and on-site examination. And in terms of clinical sense, given the paramount diagnostic relevance of BNP for HF, this approach can be used as a rapid screening tool for the ascertainment and exclusion of HF, thus guiding the most timely, appropriate, and effective therapeutic intervention. Undisputedly, the expedited and point-of-care testing of BNP can give a patient a better chance of survival and bring hope to the whole family. And it is the goal toward which the scientists expend their persistent efforts.

However, there remain several formidable challenges to overcome before clinical adoption and practical transformation: (i) Equipment calibration and quality control are indispensable for real clinical use. Only if the calibration and quality control (QC) are satisfactory can the test results of the instrument be sent to associated doctors and patients [54–57]. Whereas, as we can see, a common problem in biosensors is that there are no formal QC rules and standard calibration operating procedures, which makes the detection of the clinical sample less convincing. (ii) To better interpret the test results, reliable reference intervals, medical decision levels, and cutoff values for specific methods should be rigorously validated from robust statistics via mass population screening, abundant prospective cohorts, and large-scale clinical analysis. (iii) Most studies, when evaluating the potential of biosensors for practical applications, simply use the constructed methods to detect clinical samples that have been validated with gold standards. Technically speaking, however, such assessment settings should be implemented with synchronized blind method – that is, all the samples are tested simultaneously by the gold standard and the sensor to be evaluated, and the operator does not know who is diseased or not [58]. (iv) Considering that BNP is a volume-dependent hormone, which is also elevated in other diseases, such as Cushing's syndrome [59, 60], BNP cannot be used as the only indicator for HF diagnosis. Therefore, a high-throughput FET array is glamorous for multiplexed, multimodal, and pleiotropic analysis of cardiac biomarkers, which is another important area for improvement at the device level.

Going forwards, we envision that under the joint efforts of researchers in medicine, physics, analytical chemistry, materials science, and information science, the FET-based BNP assay can be revolutionized for the rapid, sensitive, scalable, systematic, and clinically interpretable detection. In the future study, the scientific applications of the developed FET biosensing technology are potentially more extensive, by combining with the portable semiconductor analyzer, advanced microfluidics, array-type devices, miniaturized detectors, intelligent signal outputs, comprehensive QC system, and a clear correspondence between signal responses and HF risk stratification. To reach all the objectives described here, future research and development in GFET biosensors, especially for diagnosis of fatal diseases such as HF, will need to focus on following key points: (i) integration and miniaturization; (ii) capability of batch processing; (iii) new recognition receptors; (iv) multianalyte profiling; (v) high sensitivity; (vi) fast response and real-time monitoring; and (vii) robust clinical validation.

References

1 Chen, Z., Yang, G., Offer, A. et al. (2012). Body mass index and mortality in China: a 15-year prospective study of 220 000 men. *Int. J. Epidemiol.* 41 (2): 472–481.

2 Finucane, M.M., Stevens, G.A., Cowan, M.J. et al. (2011). National, regional, and global trends in body-mass index since 1980: systematic analysis of health examination surveys and epidemiological studies with 960 country-years and 9.1 million participants. *Lancet* 377 (9765): 557–567.

3 Nguyen, H.N., Fujiyoshi, A., Abbott, R.D., and Miura, K. (2013). Epidemiology of cardiovascular risk factors in Asian countries. *Circ. J.* 77 (12): 2851–2859.

4 Tomasoni, D., Adamo, M., Lombardi, C.M., and Metra, M. (2019). Highlights in heart failure. *ESC Heart Failure* 6 (6): 1105–1127.

5 Arrigo, M., Jessup, M., Mullens, W. et al. (2020). Acute heart failure. *Nat. Rev. Dis. Primers* 6 (1): 16.

6 Shchendrygina, A., Nagel, E., Puntmann, V.O., and Valbuena-Lopez, S. (2021). COVID-19 myocarditis and prospective heart failure burden. *Expert Rev. Cardiovasc. Ther.* 19 (1): 5–14.

7 Walker, C., Deb, S., Ling, H., and Wang, Z. (2020). Assessing the elevation of cardiac biomarkers and the severity of COVID-19 infection: a meta-analysis. *J. Pharm. Pharm. Sci.* 23: 396–405.

8 Bader, F., Manla, Y., Atallah, B., and Starling, R.C. (2021). Heart failure and COVID-19. *Heart Failure Rev.* 26 (1): 1–10.

9 Saito, Y., Nakao, K., Itoh, H. et al. (1989). Brain natriuretic peptide is a novel cardiac hormone. *Biochem. Biophys. Res. Commun.* 158 (2): 360–368.

10 Cao, Z., Jia, Y., and Zhu, B. (2019). BNP and NT-proBNP as diagnostic biomarkers for cardiac dysfunction in both clinical and forensic medicine. *Int. J. Mol. Sci.* 20 (8): 1820.

11 Rezaei, B., Ghani, M., Shoushtari, A.M., and Rabiee, M. (2016). Electrochemical biosensors based on nanofibres for cardiac biomarker detection: a comprehensive review. *Biosens. Bioelectron.* 78: 513–523.

12 Maisel, A.S., Duran, J.M., and Wettersten, N. (2018). Natriuretic peptides in heart failure: atrial and B-type natriuretic peptides. *Heart Failure Clin.* 14 (1): 13–25.

13 Rey, J.R., Caro-Codón, J., Rosillo, S.O. et al. (2020). Heart failure in COVID-19 patients: prevalence, incidence and prognostic implications. *Eur. J. Heart Failure* 22 (12): 2205–2215.

14 Zhou, F., Yu, T., Du, R. et al. (2020). Clinical course and risk factors for mortality of adult inpatients with COVID-19 in Wuhan, China: a retrospective cohort study. *Lancet* 395 (10229): 1054–1062.

15 Gavin, W., Campbell, E., Zaidi, S.A. et al. (2021). Clinical characteristics, outcomes and prognosticators in adult patients hospitalized with COVID-19. *Am. J. Infect. Control* 49 (2): 158–165.

16 Semenov, A.G. and Feygina, E.E. (2018). Standardization of BNP and NT-proBNP immunoassays in light of the diverse and complex nature of circulating BNP-related peptides. *Adv. Clin. Chem.* 85: 1–30.

17 Alawieh, H., Chemaly, T.E., Alam, S., and Khraiche, M. (2019). Towards point-of-care heart failure diagnostic platforms: BNP and NT-proBNP biosensors. *Sensors (Basel)* 19 (22): 5003.

18 Don-Wauchope, A.C. and McKelvie, R.S. (2015). Evidence based application of BNP/NT-proBNP testing in heart failure. *Clin. Biochem.* 48 (4–5): 236–246.

19 Mizutani, F. (2008). Biosensors utilizing monolayers on electrode surfaces. *Sens. Actuators, B* 130 (1): 14–20.

20 Teramura, Y., Arima, Y., and Iwata, H. (2006). Surface plasmon resonance-based highly sensitive immunosensing for brain natriuretic peptide using nanobeads for signal amplification. *Anal. Biochem.* 357 (2): 208–215.

21 Zhao, J., Liang, D., Gao, S. et al. (2019). Analyte-resolved magnetoplasmonic nanocomposite to enhance SPR signals and dual recognition strategy for detection of BNP in serum samples. *Biosens. Bioelectron.* 141: 111440.

22 Grabowska, I., Sharma, N., Vasilescu, A. et al. (2018). Electrochemical aptamer-based biosensors for the detection of cardiac biomarkers. *ACS Omega* 3 (9): 12010–12018.

23 Liang, Y., Xiao, M., Wu, D. et al. (2020). Wafer-scale uniform carbon nanotube transistors for ultrasensitive and label-free detection of disease biomarkers. *ACS Nano* 14 (7): 8866–8874.

24 Syedmoradi, L., Ahmadi, A., Norton, M.L., and Omidfar, K. (2019). A review on nanomaterial-based field effect transistor technology for biomarker detection. *Mikrochim. Acta* 186 (11): 739.

25 Vu, C.A. and Chen, W.Y. (2019). Field-effect transistor biosensors for biomedical applications: recent advances and future prospects. *Sensors (Basel)* 19 (19): 4214.

26 Choi, J., Seong, T.W., Jeun, M., and Lee, K.H. (2017). Field-effect biosensors for on-site detection: recent advances and promising targets. *Adv. Healthcare Mater.* 6 (20): https://doi.org/10.1002/adhm.201700796.

27 Zhang, Y., Wan, Q., and Yang, N. (2019). Recent advances of porous graphene: synthesis, functionalization, and electrochemical applications. *Small* 15 (48): e1903780.

28 Peña-Bahamonde, J., Nguyen, H.N., Fanourakis, S.K., and Rodrigues, D.F. (2018). Recent advances in graphene-based biosensor technology with applications in life sciences. *J. Nanobiotechnol.* 16 (1): 75.

29 Kim, Y.J. and Jeong, B. (2018). Graphene-based nanomaterials and their applications in biosensors. *Adv. Exp. Med. Biol.* 1064: 61–71.

30 Zhang, C., Xu, J.Q., Li, Y.T. et al. (2016). Photocatalysis-induced renewable field-effect transistor for protein detection. *Anal. Chem.* 88 (7): 4048–4054.

31 Li, Y.T., Jin, X., Tang, L. et al. (2019). Receptor-mediated field effect transistor biosensor for real-time monitoring of glutamate release from primary hippocampal neurons. *Anal. Chem.* 91 (13): 8229–8236.

32 Wu, D., Yu, Y., Jin, D. et al. (2020). Dual-aptamer modified graphene field-effect transistor nanosensor for label-free and specific detection of hepatocellular carcinoma-derived microvesicles. *Anal. Chem.* 92 (5): 4006–4015.

33 Jin, X., Zhang, H., Li, Y.T. et al. (2019). A field effect transistor modified with reduced graphene oxide for immunodetection of Ebola virus. *Mikrochim. Acta* 186 (4): 223.

34 Li, J., Wu, D., Yu, Y. et al. (2021). Rapid and unamplified identification of COVID-19 with morpholino-modified graphene field-effect transistor nanosensor. *Biosens. Bioelectron.* 183: 113206.

35 Ambrosi, A., Chua, C.K., Bonanni, A. et al. (2012). Lithium aluminum hydride as reducing agent for chemically reduced graphene oxides. *Chem. Mater.* 24 (12): 2292–2298.

36 Gundry, R.L. and Van Eyk, J.E. (2007). Unraveling the complexity of circulating forms of brain natriuretic peptide. *Clin. Chem.* 53 (7): 1181–1182.

37 Kim, D.J., Sohn, I.Y., Jung, J.H. et al. (2013). Reduced graphene oxide field-effect transistor for label-free femtomolar protein detection. *Biosens. Bioelectron.* 41: 621–626.

38 Cameron, R.E. and Bocarsly, A.B. (1985). Photoactivated oxidation by alcohols by oxygen. *J. Am. Chem. Soc.* 107 (21): 6116–6117.

39 Cameron, R.E. and Bocarsly, A.B. (1986). Multielectron-photoinduced reduction of chloroplatinum complexes: visible light deposition of platinum metal. *Inorg. Chem.* 25 (16): 2910–2913.

40 Yin, Z., He, Q., Huang, X. et al. (2012). Real-time DNA detection using Pt nanoparticle-decorated reduced graphene oxide field-effect transistors. *Nanoscale* 4 (1): 293–297.

41 Lei, Y.-M., Xiao, M.-M., Li, Y.-T. et al. (2017). Detection of heart failure-related biomarker in whole blood with graphene field effect transistor biosensor. *Biosens. Bioelectron.* 91: 1–7.

42 Donev, E.U. and Hastings, J.T. (2009). Electron-beam-induced deposition of platinum from a liquid precursor. *Nano Lett.* 9 (7): 2715–2718.

43 Fang, B., Kim, M.S., Kim, J.H. et al. (2011). High Pt loading on functionalized multiwall carbon nanotubes as a highly efficient cathode electrocatalyst for proton exchange membrane fuel cells. *J. Mater. Chem.* 21 (22): 8066–8073.

44 Pietron, J.J., Pomfret, M.B., Chervin, C.N. et al. (2012). Direct methanol oxidation at low overpotentials using Pt nanoparticles electrodeposited at ultrathin conductive RuO_2 nanoskins. *J. Mater. Chem.* 22: 5197–5204.

45 Ren, F., Wang, H., Zhai, C. et al. (2014). Clean method for the synthesis of reduced graphene oxide-supported PtPd alloys with high electrocatalytic activity for ethanol oxidation in alkaline medium. *ACS Appl. Mater. Interfaces* 6 (5): 3607–3614.

46 Hamilton, R.G., Roebber, M., Reimer, C.B., and Rodkey, L.S. (1987). Isoelectric focusing-affinity immunoblot analysis of mouse monoclonal antibodies to the four human IgG subclasses. *Electrophoresis* 8 (3): 127–134.

47 Duan, X., Li, Y., Rajan, N.K. et al. (2012). Quantification of the affinities and kinetics of protein interactions using silicon nanowire biosensors. *Nat. Nanotechnol.* 7 (6): 401–407.

48 Rahim Ruslinda, A., Tanabe, K., Ibori, S. et al. (2013). Effects of diamond-FET-based RNA aptamer sensing for detection of real sample of HIV-1 Tat protein. *Biosens. Bioelectron.* 40 (1): 277–282.

49 Bishop, J., Chagovetz, A.M., and Blair, S. (2008). Kinetics of multiplex hybridization: mechanisms and implications. *Biophys. J.* 94 (5): 1726–1734.

50 Dao, Q., Krishnaswamy, P., Kazanegra, R. et al. (2001). Utility of B-type natriuretic peptide in the diagnosis of congestive heart failure in an urgent-care setting. *J. Am. Coll. Cardiol.* 37 (2): 379–385.

51 Jang, H.J., Ahn, J., Kim, M.G. et al. (2015). Electrical signaling of enzyme-linked immunosorbent assays with an ion-sensitive field-effect transistor. *Biosens. Bioelectron.* 64: 318–323.

52 Seino, Y., Ogawa, A., Yamashita, T. et al. (2004). Application of NT-proBNP and BNP measurements in cardiac care: a more discerning marker for the detection and evaluation of heart failure. *Eur. J. Heart Failure* 6 (3): 295–300.

53 Ziaeian, B. and Fonarow, G.C. (2016). Epidemiology and aetiology of heart failure. *Nat. Rev. Cardiol.* 13 (6): 368–378.

54 Ohman, S. (1997). Quality control for the clinical chemistry laboratory. *Qual. Assur.* 5 (2): 79–93.

55 Kazmierczak, S.C. (2003). Laboratory quality control: using patient data to assess analytical performance. *Clin. Chem. Lab. Med.* 41 (5): 617–627.

56 Howanitz, P.J. and Howanitz, J.H. (1983). Quality control for the clinical laboratory. *Clin. Lab. Med.* 3 (3): 541–551.

57 Loh, T.P., Bietenbeck, A., Cervinski, M.A. et al. (2020). Recommendation for performance verification of patient-based real-time quality control. *Clin. Chem. Lab. Med.* 58 (8): 1205–1213.

58 Horvath, A.R. and Pewsner, D. (2004). Systematic reviews in laboratory medicine: principles, processes and practical considerations. *Clin. Chim. Acta* 342 (1–2): 23–39.

59 Shibusawa, N., Yamada, M., Hashida, T. et al. (2013). Dilated cardiomyopathy as a presenting feature of Cushing's syndrome. *Intern. Med.* 52 (10): 1067–1071.

60 Peppa, M., Ikonomidis, I., Hadjidakis, D. et al. (2009). Dilated cardiomyopathy as the predominant feature of Cushing's syndrome. *Am. J. Med. Sci.* 338 (3): 252–253.

16

Enzymatic Biosensors Based on the Electrochemical Functionalization of Graphene Field-Effect Transistors with Conducting Polymers

Gonzalo E. Fenoy[1], *Esteban Piccinini*[1], *Wolfgang Knoll*[2,3], *Waldemar A. Marmisollé*[1], *and Omar Azzaroni*[1]

[1]*CONICET, Universidad Nacional de La Plata, Instituto de Investigaciones Fisicoquímicas Teóricas y Aplicadas (INIFTA), Departamento de Química, Facultad de Ciencias Exactas, 64 and 113, La Plata, 1900, Argentina*
[2]*AIT Austrian Institute of Technology GmbH, Biosensor Technologies, Konrad-Lorenz-Strasse 24, 3430, Tulln, Austria*
[3]*Danube Private University, A-3500, Krems, Austria*

16.1 Introduction

Biosensors based on functionalized graphene field-effect transistors (GFETs) are gaining tremendous attention in both academia and industry due to their promising applications in disease diagnostics [1, 2]. In fact, an increasing number of studies shows that GFET-based sensors show distinct advantages regarding the rapid and sensitive detection of different analytes by employing a simple and miniaturized structure [3, 4], which has propelled their popularity among other sensing platforms for biomedical applications [5].

In this context, the rational integration of molecular recognition elements on the graphene channel is the key to achieving biosensing devices with high sensitivity and selectivity [6]. The construction of an interfacial architecture able to functionally combine bioreceptors and antifouling elements while maintaining the surface sensitivity of transistors-based biosensors is not simple, and several strategies addressing these issues have been developed during the last two decades [7].

Particularly, conducting polymer (CP) films offer an interesting alternative for the development of biosensors since their features and chemistry can be easily customized in order to obtain a suitable surface for the immobilization of the desired receptor molecules [8, 9]. Among all the CPs film fabrication techniques, chemical *in situ* polymerization and electropolymerization are the most employed strategies due to their simplicity and reliability [10, 11]. Next, taking into account the high electrical conductivity of graphene while bridging both source and drain electrodes [12], its functionalization through the electrosynthesis of CPs represents an attractive approach for the design of biosensing surfaces.

Graphene Field-Effect Transistors: Advanced Bioelectronic Devices for Sensing Applications, First Edition. Edited by Omar Azzaroni and Wolfgang Knoll.

This approach offers the advantages of a very simple control over the properties of the final polymeric coating as many experimental parameters have considerable influence on the physicochemical and electrochemical properties of the film, including the type of electrochemical method, the potential limits, the nature of the working electrode (WE) substrate, and the electrolyte [13]. Therefore, several relevant features of the resulting CP layer, such as morphology, thickness, dopants type and proportion, and oxidation state, are easily tuned. In addition, the obtained polymer films are frequently more stable than those obtained by other techniques, and the synthesis of copolymers is straightforward [14]. In this regard, the considerable chemical flexibility in the macromolecular structures through synthetic modifications of the monomer can easily provide polymer films with multiple predesigned functional properties.

Due to the above-mentioned reasons, in recent years, there has been an increasing interest in the functionalization of graphene surfaces by the electrosynthesis of CPs for the generation of biosensing devices [15, 16]. For instance, GFETs functionalized by polypyrrole (PPy) have exhibited excellent performance toward glucose sensing, in addition to fulfilling key essential requirements such as low preparation cost, simple deposition process, operation at room temperature, and reproducible response [17].

On the other hand, enzymes constitute excellent bioreceptors to be coupled with GFETs due to their specificity and selectivity toward the target analyte. In contrast to GFET-based immunosensors [18, 19], whose signal transduction mechanism is based on local electric-field changes due to the recognition of the target antigen, enzymatic-GFETs' response roots on local pH changes caused by the enzymatically catalyzed chemical reaction [20, 21].

In this chapter, we will describe the use of electropolymerized poly (3-amino-benzylamine-*co*-aniline) (PABA) nanofilms as platforms to create enzymatic GFET biosensors. PABA is a CP with a considerable proportion of primary amine groups that can be electropolymerized on the graphene channels, thus leading to nanofilms with precise control over the film thickness and the chemical homogeneity. These nanofilms are used as platforms for further electrostatic immobilization of enzymes on the bioelectronic devices. The approach is based on the intrinsic pH sensitivity of PABA films coupled to GFETs as a strategy to detect the enzymatic conversion of the analyte.

16.2 Functionalization of Graphene Transistors with Poly(3-amino-benzylamine-*co*-aniline) Nanofilms

Functionalization of GFETs with PABA films can be simply accomplished by electropolymerization of 3-amino-benzylamine-*co*-aniline copolymers in acidic medium [22]. The electrosynthesized polymer structure bears pendant amino groups that allow for the incorporation of different molecules and functionalities (Figure 16.1a, right). For the electropolymerization, both drain and source electrodes are shorted and used as WE, while the three-electrode cell setup is completed with the addition of a Pt wire counter electrode (CE) and an Ag/AgCl reference

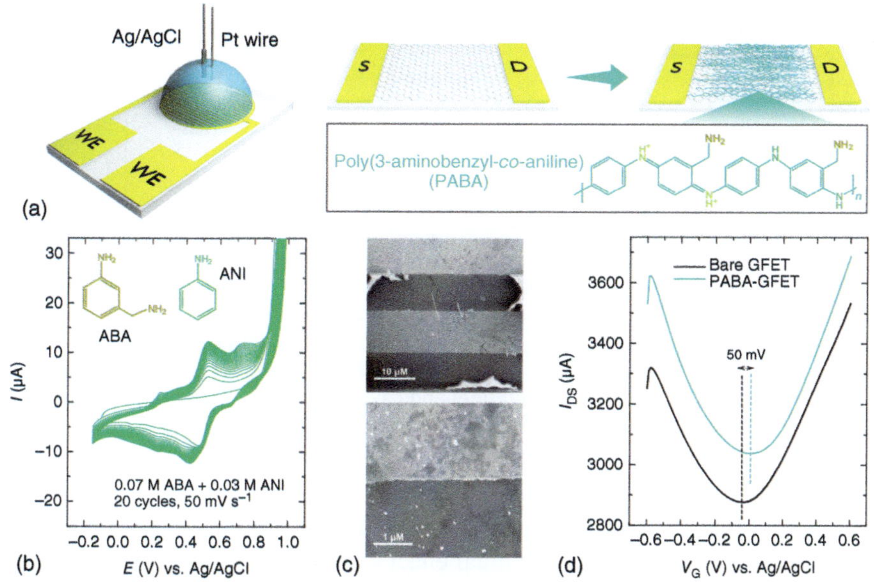

Figure 16.1 (a) Scheme of the setup employed to perform PABA electropolymerization on the GFETs and structure of the polymer. (b) Voltammograms of the electropolymerization of PABA (0.5 M H_2SO_4, 50 mV s^{-1}, 20 cycles) on a GFET. (c) SEM images of a PABA-GFET. (d) Transfer characteristics of the same GFET before and after PABA electrodeposition (10 mM KCl, 0.1 mM HEPES, pH 7). Source: Reproduced with permission from Fenoy et al. [22]/Elsevier.

electrode (Figure 16.1a, left). For the electropolymerization process to be successfully performed, a complete coverage and proper electronic conductivity of the reduced graphene oxide (rGO) located between and on the Au pads is necessary to ensure a suitable surface. In this regard, several works have demonstrated by scanning electrochemical microscopy (SECM) and Kelvin force microscopy (KFM) measurements that the assembly of graphene oxide flakes onto Au pads and their subsequent reduction led to platforms with suitable and reproducible electronic conductivity in the field-effect transistor (FET) channel [23, 24].

The electrosynthesis of the polymer is performed by cyclic voltammetry in an acidic medium in the presence of both aniline (ANI) and 3-amino-benzylamine (ABA) monomers (Figure 16.1b). The good electronic conductivity of the rGO channel yields cyclic voltammetry (CV) features very similar to those reported on gold electrodes [25, 26], even when comparing the current density values during the electropolymerization. A film thickness can be estimated from previously reported electrochemical quartz crystal microbalance (EQCM) results and its correlation with the voltammetric charge measured [27], yielding a 7-nm-thick PABA film deposited on the GFETs after 20 voltammetric cycles. The effective deposition of the copolymer can also be observed by SEM (Figure 16.1c).

As expected, the electrodeposition of PABA modifies the transfer characteristics of the GFET (Figure 16.1d). In this regard, the electropolymerization process promotes a shift of 50 mV in the Dirac point to more negative values. This behavior can

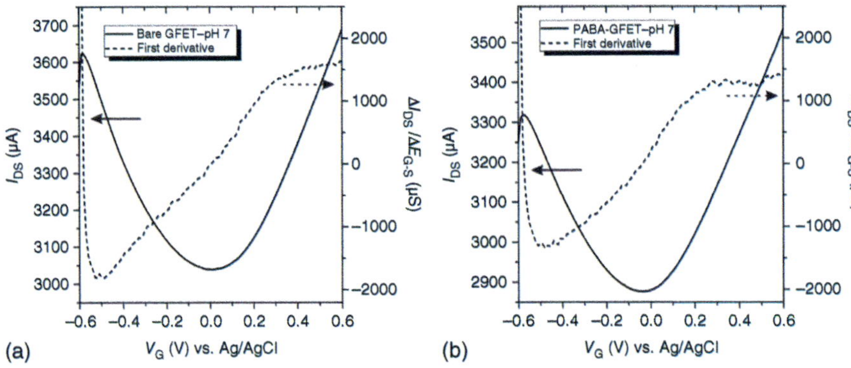

Figure 16.2 Transfer characteristics (full line) and calculated transconductance values (dashed line) for a (a) bare and a (b) PABA-GFET (10 mM KCl, 0.1 mM HEPES, pH 7, V_{DS} = 0.1 V). Source: Adapted from Fenoy et al. [22].

be ascribed to the electrostatic gating effect of the positively charged copolymer onto the graphene channel [28]. A similar effect was observed in the case of GFETs modified with polyelectrolyte multilayers as well as polyelectrolyte brushes [21].

The effect of the PABA electropolymerization on the GFETs was also evaluated by analyzing the change of relevant electrical features of the transistors, such as charge mobility and transconductance, before and after polymer electrodeposition. In this regard, charge carrier mobility was calculated according to:

$$\mu = \frac{L}{WC_G V_{DS}} \Delta I_{DS} / \Delta V_G$$

where L is the channel length (10 μm), W is the channel width (240 mm), C_G is the top gate capacitance (57 nF cm^{-2}), and $\Delta I_{DS}/\Delta V_G$ is the transconductance. The values obtained for holes and electrons mobility are 12 cm^{2}V^{-1}s^{-1} and did not show significant changes upon electropolymerization.

On the other hand, transconductances of bare GFETs and PABA-GFETs were calculated from the transfer characteristics curves, as shown in Figure 16.2. Similar to what was observed for the charge carrier mobility, the transconductance of the GFETs was not affected by the electrodeposition of the polymer. Later, if we consider that both the transconductance and the charge carrier mobility are decisive parameters for the development of highly sensitive sensors [5, 29], the fact that both parameters remain almost unchanged after the electropolymerization process indicates that the polymer deposition has no negative impact on the GFETs performance, allowing for their further use as biosensing platforms.

Since the signal transduction mechanism of enzymatic GFET-based biosensors is based on the detection of pH changes caused by enzyme-catalyzed hydrolysis, the sensitivity to pH variations is of paramount importance for the design of this type of sensing device. Hence, the pH response of the bare and PABA-GFETs was studied in both static and flow conditions (Figure 16.3). When measuring in the static configuration (evaluation of the transfer characteristics), it can be observed that as the pH

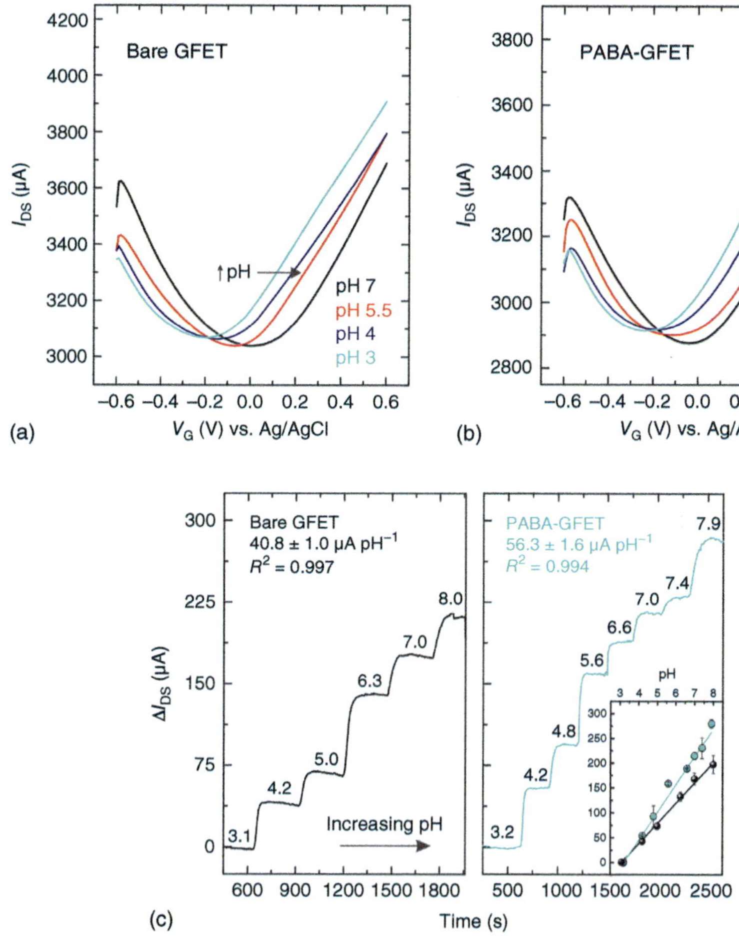

Figure 16.3 (a) Bare GFET and (b) PABA-GFET transfer characteristics while changing the pH of the electrolyte solution from 3 to 7 in 0.1 mM HEPES and 10 mM KCl buffer (V_{DS} = 0.1 V). (c) pH sensitivity curves at a fixed gate potential (V_G = −0.2 V, V_{DS} = 0.1 V) for the same GFET before and after PABA electropolymerization. Source: Reproduced with permission from Fenoy et al. [22]/Elsevier.

decreases, the Dirac point of the GFETs shifts to more negative V_G values due to H_3O^+ adsorption (n-type doping). On the contrary, under alkaline conditions, a shift of the Dirac point shift to more positive voltage values due to OH^- adsorption (p-type doping) is observed (Figure 16.3a) [30].

On the other hand, when monitoring the pH response of the devices in continuous flow mode, a gate potential at a value at which no chemical reaction takes place was employed ($V_G = -0.2$ V). This gate potential is lower than the potential of the Dirac point (V_i) and, as a consequence, the GFET are biased in the hole regime. It is observed that an increasing drain–source current (at a fixed positive V_{DS}) is registered upon decreasing the solution pH. At this operation mode, the obtained pH sensitivity of the bare GFET system was 41 μA pH^{-1} unit. The analogous experiment performed employing PABA-modified GFETs revealed a 37% increase in pH sensitivity (from 41 to 56 μA pH^{-1} unit) due to the presence of the polymer layer (Figure 16.3c). A similar phenomenon was observed in the case of GFETs modified with a layer of polyethyleneimine (PEI) [20], as with cationic polymer brushes [21], and it was ascribed to changes in the degree of protonation of the polymers that, in turn, induce a shift in the graphene Fermi level and affect the pH response of the transistors. In those works, the rGO/graphene channels were modified with polymers with pK_a values around 8 (8–9 for PEI and around 8 for *N*,*N*-dimethylaminoethyl methacrylate brushes). In the case of PABA, the molar composition of the copolymer has been determined to be about 30% in ABA and 70% in ANI, so it is necessary to consider the contribution of both monomer units to the acid–base equilibrium in the whole copolymer. In the case of polyaniline, pK_a values are 2.5 (assigned to $—NH_2^+—$) and 5.5 ($—NH^+=$ species) [31, 32]. On the other hand, in polyaminobenzylamines, the pendant amino groups are still protonated at neutral pH [15, 33]. Therefore, the multiple protonation states of the electrodeposited copolymer could be the main reason behind the improvement of the pH response of PABA-GFETs in comparison with PEI-GFETs (38% vs. 28%). Finally, this result shows that the electrodeposited copolymer does not only act as a priming layer for further enzyme anchoring, but it also improves the pH sensitivity of the biosensing platform.

16.3 Construction of Acetylcholine Biosensors Based on GFET Devices Functionalized with Electropolymerized Poly(3-amino-benzylamine-*co*-aniline) Nanofilms

Acetylcholine (Ach) plays a crucial role in the cholinergic system as a transmitter of impulses on the cholinergic synapse [34]. Moreover, the alteration of central cholinergic transmission is coupled to a number of brain disorders, including Alzheimer's disease, Parkinson's disease, addiction, epilepsy, schizophrenia, and depression [35]. It has been demonstrated that abnormal levels of Ach are related to neurological disorders [36] and, consequently, considerable research efforts have been devoted to develop highly sensitive techniques to detect Ach in low concentrations in real-time conditions. In this context, GFETs have gained increasing relevance as devices fulfilling this requirement [5, 37]. As already mentioned, the sensing principle of GFETs relies on changes in the electrical conductance of the graphene channel upon adsorption of ions or molecules on the sensor surface. Since local pH changes are able to modify the doping of graphene, GFETs work as sensitive interfacial pH sensors [3]. If the pH variation is generated by an enzymatic reaction, immobilizing an enzyme close to the graphene surface could allow detecting changes in the analyte concentration by the conductivity change induced by the local pH variation. This constitutes the signal transduction mechanism of most enzymatic GFET biosensors [6, 38].

In this regard, it is essential to bear in mind that the performance of the biosensor greatly depends on the properties of the interfacial architecture, and, therefore, on the enzyme immobilization strategy. To date, the majority of the research efforts in graphene-based FET biosensors have focused on the immobilization of enzymes by covalent binding to functional groups of previously modified graphene or to linker molecules [5, 38, 39]. However, the covalent attachment of biomolecules presents some drawbacks, such as poor reproducibility and the possibility of disrupting the folding and functionality of the enzyme as crucial groups can be involved in the immobilization, thus affecting the biosensor's performance [40]. In addition, direct covalent attachment to the graphene surface can also provoke damage to the sp^2 structure of the graphene, impairing the device performance [41].

Here is where noncovalent enzyme integration strategies such as electrostatic immobilization appear as an interesting alternative for the fabrication of GFET-based biosensing platforms [42]. These methods offer a simple approach for the rational fabrication of interfacial biosensing architectures, preserving the biological activity of the enzyme and the accessibility to its active sites due to the very mild conditions employed for the immobilization procedure [43].

In this context, it has been recently shown that GFETs previously functionalized with the amino-bearing polymer PABA can be employed for the construction of Ach sensors through the immobilization of acetylcholinesterase (AchE) via electrostatic interactions by employing a solution of higher pH than its isoelectric point (PI) (5.3), at which the protein is negatively charged [22].

The immobilization process resulted in a shift of the Dirac point to more positive potentials (Figure 16.4a), which is coherent with a p-doping effect arising from the adsorption of negatively charged species. Interestingly, the immobilization procedure yielded no significant changes in the features of the transfer curve of the graphene-based transistor, indicating no deterioration of the conducting channel. In contrast to these results, Sohn et al. have reported a decrease in the sensor sensitivity of approximately 50% after the covalent immobilization of AchE on rGO-GFETs [44]. The quantification of the amount of immobilized AchE was performed by surface plasmon resonance (SPR) measurements (Figure 16.4b), yielding a mass surface coverage of $290\,ng\,cm^{-2}$, which is in good agreement with other AchE surface coverages reported in the literature [45].

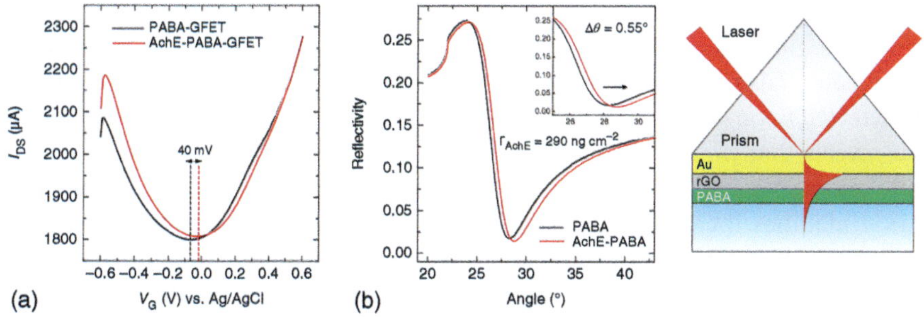

Figure 16.4 (a) Transfer characteristics of a PABA-GFET before and after AchE electrostatic deposition (V_{DS} = 0.1 V, 10 mM KCl, 0.1 mM HEPES, pH 7). (b) SPR curves of a PABA-rGO-Au substrate before and after enzyme immobilization (left) and SPR setup scheme (right). Source: Adapted from Fenoy et al. [22].

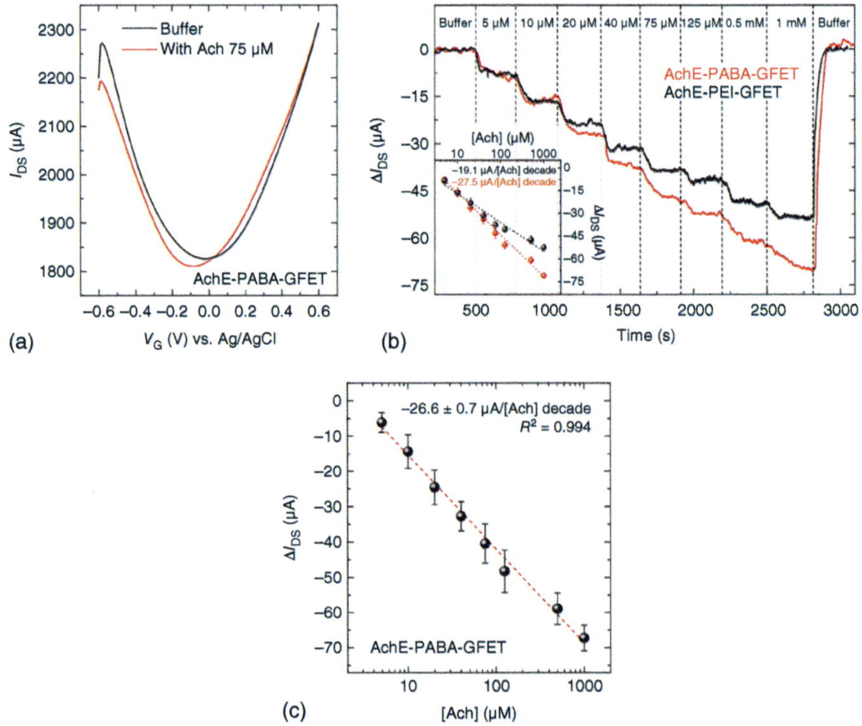

Figure 16.5 (a) Transfer characteristics of an AchE-PABA-GFET in the absence and presence of 75 μM Ach (V_{DS} = 0.1 V, 10 mM KCl, 0.1 mM HEPES, pH 7). (b) Flow response of an AchE-PABA-GFET and an AchE-PEI-GFET (300 μl min^{-1}, V_G = −0.2 V, V_{DS} = 0.1 V) at different acetylcholine concentrations, whereas the inset shows the sensitivity comparison between both biosensors (two measurements performed). (c) Reproducibility assay for the AchE-PABA-GFETs showing the results obtained for three different electrodes. Source: Reproduced with permission from Fenoy et al. [22]/Elsevier.

Next, the sensing features of the enzyme-modified platform were assessed by analyzing the response in presence of the neurotransmitter Ach. Figure 16.5a shows the transfer characteristics of an AchE-PABA-GFET in the absence (solid black line) and presence (dashed line) of 75 μM Ach. A clear shift of the Dirac point to more negative potentials in the presence of the substrate is observed, in agreement with shifts detected upon decreasing the pH of the solution (see Figure 16.3). This observation reveals that the AchE effectively promotes the hydrolysis of Ach, yielding choline and acetic acid that, in turn, generate a local pH decrease. Then, this pH variation modifies the charge density of the PABA nanofilm decorating the graphene sensor, thus triggering a negative shift of the Dirac point.

The response of the biosensor was also studied in a flow setup. Figure 16.5b shows the real-time flow response of an AchE-PABA-modified transistor in the presence of increasing Ach concentration. The decrease in the local pH due to enzymatic hydrolysis of Ach shifts the Dirac point of the GFETs to more negative values, promoting, therefore, a decrease in I_{DS} (as the transistors were biased in the hole regime, where $V_G < V_i$). The continuous flow response of these transistors shows that they are feasible for the real-time sensing of Ach. Moreover, the electrolyte gating setup allows the device to operate at very low-gate voltages ($V_G = -0.2$ V), which is a critical requirement to minimize the disruption of biological samples or any possible electrochemical reaction [46].

The flow response of the transistors exhibits a logarithmic sensitivity with the Ach concentration. By assuming that the proton concentration is proportional to Ach, the logarithmic ratio can be comprehended by evoking the linear response of the GFETs to the pH changes. A decrease of the drain–source current of approximately 70 μA was observed when flowing through the cell at an Ach concentration of 1 mM. Above this concentration, the enzyme could be inhibited by the high substrate concentration (the so-called substrate inhibition). Moreover, the response of these enzymatic GFET sensors is totally reversible, as the injection of buffer solution after the highest Ach concentration yielded an increase in the drain–source current to almost the initial value ($\Delta I_{DS} \sim 0$), implying that the transistors could be reused for further measurements.

The performance of PABA-GFETs was also compared with that of polyethylene-modified GFETs (PEI-GFETs) produced by electrostatic immobilization of the polyelectrolyte for the further anchoring of AchE and Ach sensing (Figure 16.5b). In this regard, the sensitivity of AchE-PABA-GFETs (27.5 μA Ach^{-1} decade) becomes higher than that of AchE-PEI-GFETs (19 μA Ach^{-1} decade), meaning a 44% improvement. Noteworthy, the performance enhancement occurs particularly when higher Ach concentrations are injected, i.e. lower local pH values. This result is related to the higher sensitivity of PABA-GFETs due to the multiple protonation states available in the PABA nanofilm. As it was already mentioned, the different amine moieties from both aniline and ABA monomers are present in the copolymer, displaying a wide pK_a distribution. Particularly, secondary amines present in PABA become deprotonated at more acidic pH values than primary ones, improving the pH response in a more acidic medium.

The reproducibility of the biosensing platform has been also investigated by preparing three different AchE-PABA-GFETs and monitoring the real-time Ach flow response in the range of 5–1000 μM (Figure 16.5c). The transistors sensitivities exhibited a very low relative standard deviation (RSD) of 2.6%, thus indicating a notable device-to-device reproducibility. The biosensors showed a fast response time, with an average value of 130 seconds for 95% of the maximum response. The stability of the devices has been studied by storing the sensors at 4 °C in buffered HEPES solution, showing a 94% retention of the original response after four days of storage. The limit of detection (LOD) of the sensing platform was calculated as three times the standard deviation for the buffer injection (blank), yielding a value of 2.3 μM.

Control experiments in the absence of the enzyme prove the specific response to Ach due to the AchE integration into the GFET (Figure 16.6). In addition, the selectivity of the biosensors relative to a variety of potential interferences known to be present in the extracellular milieu has been tested (Figure 16.7). In this regard, glucose, ascorbic acid (AA), and neurotransmitters dopamine (DA) and serotonin (SA), as well as other endogenous electroactive species such as L-cysteine (l-Cys), were injected after

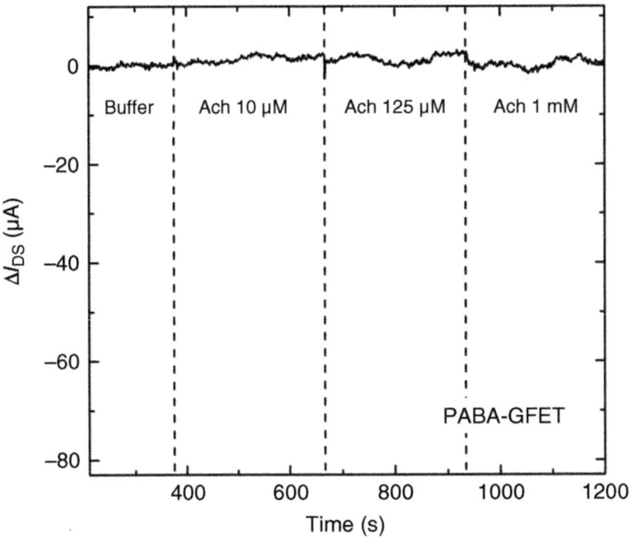

Figure 16.6 Flow response of a PABA-GFET without enzyme (300 μl min^{-1}, V_G = −0.2 V, V_{DS} = 0.1 V) at different acetylcholine concentrations. Source: Adapted from Fenoy et al. [22].

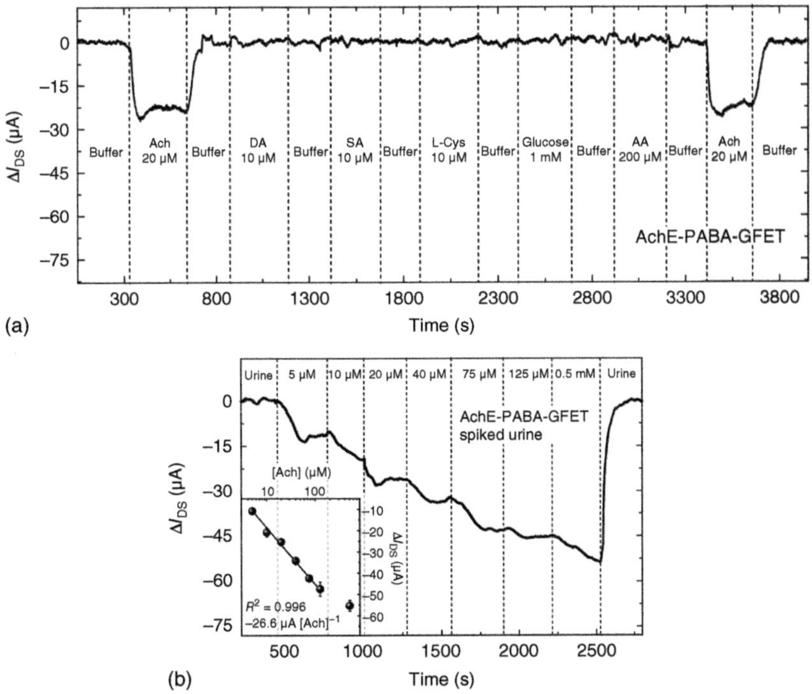

Figure 16.7 (a) Flow response of an AchE-PABA-GFET (300 μl min^{-1}, V_G = −0.2 V, V_{DS} = 0.1 V) while flowing acetylcholine and different interferences. (b) Flow response of an AchE-PABA-GFET (300 μl min^{-1}, V_G = −0.2 V, V_{DS} = 0.1 V) while flowing spiked urine samples. The inset shows the linear fitting of the response in a logarithmic scale (results for two measurements). Source: Reproduced with permission from Fenoy et al. [22]/Elsevier.

20 μM Ach flow detection and buffer washing. The concentrations of potential interferences tested were at levels that occur in brain extracellular fluid [36]. The transistors did not show significant I_{DS} changes in the presence of the interferences, and the response to Ach was recovered by 97.5% after injecting the entire set of interferences.

Next, with the aim of evaluating the response of the biosensor in a real biological fluid, flow measurements of diluted urine samples spiked with Ach have been also performed (Figure 16.7). The devices showed similar sensitivity to that observed in the buffer solution, proving their capability to detect the neurotransmitter in more complex matrices.

16.4 Glucose Detection by Graphene Field-Effect Transistors Functionalized with Electropolymerized Poly(3-amino-benzylamine-*co*-aniline) Nanofilms

Diabetes mellitus is described as a disease of abnormal glucose metabolism that affects a large part of the world's population [47, 48]. Several serious complications can arise when this malfunction is not properly treated, such as kidney failure, blindness, and heart attack [49, 50], making the accurate tracking of glucose levels of critical relevance for the treatment of the affected individuals. In this regard, and while glucose sensors have been used in the management of diabetes for more than 50 years [51], it is still necessary to develop point-of-care (POC) biosensors which endow patients with the ability to perform the precise monitoring of their glucose concentration simply and freely, without the need of visiting specific facilities or consulting specialized staff [52]. In this context, while the POC testing of blood glucose can cause some problems related to blood drawing, the POC monitoring of glucose in urine can be simply performed [53]. Moreover, its relevance is even greater since glycosuria (the existence of glucose in urine) is a serious condition that indicates the aggravation of diabetes [54].

As previously stated, the immobilization of the biorecognition elements presents a crucial step for the development of highly sensitive biosensing devices. In this regard, several strategies have been developed for the immobilization of glucose oxidase (GOx) such as covalent attachment, physical adsorption, and encapsulation [55–57]. Among them, the electrostatic immobilization of GOx appears as an attractive strategy since it usually does not affect the functionality of the enzyme [58]. Moreover, the anchoring process does not affect the transducing properties of the FET channel, as reported for other covalent approaches [55, 59].

On the other hand, and as discussed in Section 16.3, the electrodeposition of PABA films on GFETs offers an effective approach for the generation of platforms capable of immobilizing enzymes by electrostatic interactions, yielding a stable interfacial architecture, and preserving the functionality of the biomolecule. In addition, PABA possesses different classes of amine and imine moieties, which are capable of transducing pH changes over a wide pH range, improving the GFETs' pH response. In this context, it is worth remarking that it is well-known that GOx interacts with surfaces bearing a high number of reactive amino groups, such as PEI [60].

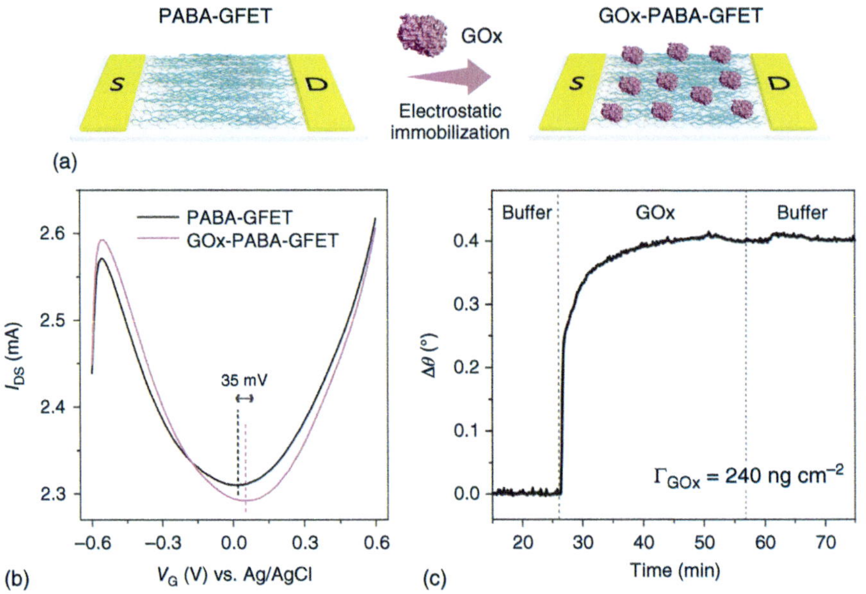

Figure 16.8 (a) Scheme of the GOx immobilization on the PABA-GFETs. (b) Transfer characteristics of a PABA-GFET before and after GOx immobilization (10 mM KCl, 0.1 mM HEPES, pH 7). (c) Change in the SPR minimum reflectivity angle of a PABA–rGO-Au substrate upon GOx adsorption. Source: Reproduced with permission from Fenoy et al. [22]/Elsevier.

Next, in this section, the electrostatic immobilization of GOx on PABA-modified GFETs is reviewed as a simple and effective strategy to fabricate GFETs-based glucose sensors.

The electrostatic assembly of GOx on the PABA-GFETs was accomplished by exposing the PABA-modified GFETs for 30 minutes to 1 mg ml^{-1} solution of the enzyme in buffer (Figure 16.8a). When performed on the GFETs, Figure 16.8b shows that, as in the case of AchE, the assembly of GOx promotes a shift of the Dirac point to more positive potentials, consistent with the adsorption of negatively charged species (i.e. a p-doping effect).

In addition, GOx immobilization on PABA-modified Au–rGO substrates was also characterized by SPR spectroscopy (Figure 16.8c). By monitoring the change in the minimum reflectivity angle (θ) while the different solutions were flowed through the cell, a mass surface coverage of 240 ng cm^{-2} (equivalent to 1.5 pmol cm^{-2}) was obtained. This value is in good agreement with typical GOx surface coverages on amino-functionalized surfaces [61].

As previously explained, the signal generation mechanism of this type of biosensing device relies on the generation of local pH variations caused by the enzyme-catalyzed reaction (in this case, glucose oxidation), which, in turn, causes changes in the surface charge density stemming from the chemical equilibrium of different amine groups in the PABA layer. When employing a static setup, this sequential process shifts the Dirac point of the GOx–PABA transistors to more negative voltages, as shown in Figure 16.9b. This shift is in line with the shifts observed in

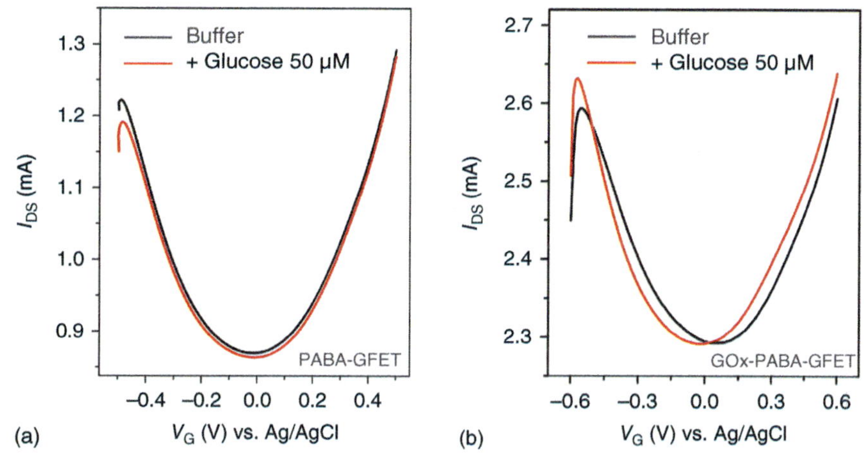

Figure 16.9 Transfer characteristics of a (a) PABA-GFET and a (b) GOx–PABA-GFET in the absence and presence of 50 μM glucose (10 mM KCl, 0.1 mM HEPES, pH 6, V_{DS} = 0.1 V). Source: Reproduced with permission from Fenoy et al. [22]/Elsevier.

solutions with decreasing pH values and could be related to the generation of gluconic acid, whose dissociation releases protons, causing a decrease in the local pH in close proximity to the graphene channel. When the same experiment is carried out in a PABA-modified GFET, no significant variation in the Dirac point is observed (Figure 16.9a), proving the effect of the enzyme catalysis.

However, it has been also reported for glucose GFET sensors that this shift could also be ascribed to the generation of H_2O_2, which is also produced upon glucose-catalyzed oxidation [62]. As such, to elucidate the signal transduction mechanism of the biosensor, the transfer characteristics of a biosensor were examined in the absence and presence of 50 μM H_2O_2, showing no variation in the position of the Dirac point voltage (Figure 16.10a). In other words, the experimental evidence

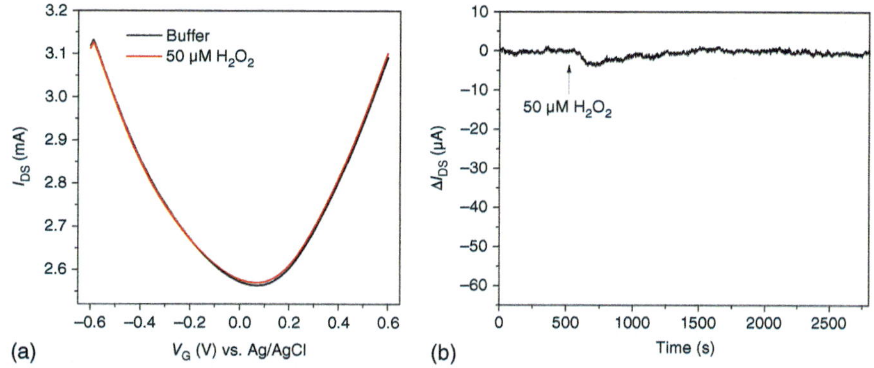

Figure 16.10 (a) Transfer characteristics of a GOx–PABA-GFET upon addition of H_2O_2 (10 mM KCl, 0.1 mM HEPES, pH 6, V_{DS} = 0.1 V). (b) Response of a GOx–PABA-GFET to the flow of H_2O_2 (10 mM KCl, 0.1 mM HEPES, pH 6, 300 μl min^{-1}, V_G = −0.2 V, V_{DS} = 0.1 V). Source: Adapted from Fenoy et al. [42].

shows that the presence of H_2O_2 does not affect the device response. In a similar way, an analogous experiment performed in a flow configuration displayed no significant variation in I_{DS} upon adding H_2O_2 to the system (Figure 16.10b). As such, signal transduction mechanism is unambiguously driven by the change in pH caused by the enzymatic generation of gluconic acid.

One remarkable aspect of this type of glucose sensor is that its transduction mechanism is based on local pH changes and not on interfacial electron transfer process. As such, it is not necessary to incorporate redox-active species that operate as mediators in the film architecture in order to establish an efficient electron transfer between the enzyme and the conducting support, a drawback commonly found in a vast amount of glucose sensing devices [63]. This mechanism avoids usual problems that appear upon mediator integration, such as response to interferences and the toxicity of the mediators, limiting the *in vivo* application of the device [64].

To evaluate the flow response of the biosensors for their possible use as real-time sensing devices, the GOx-modified transistors were exposed to the flow of different glucose solutions, and the bioelectronic response was evaluated (Figure 16.11). As

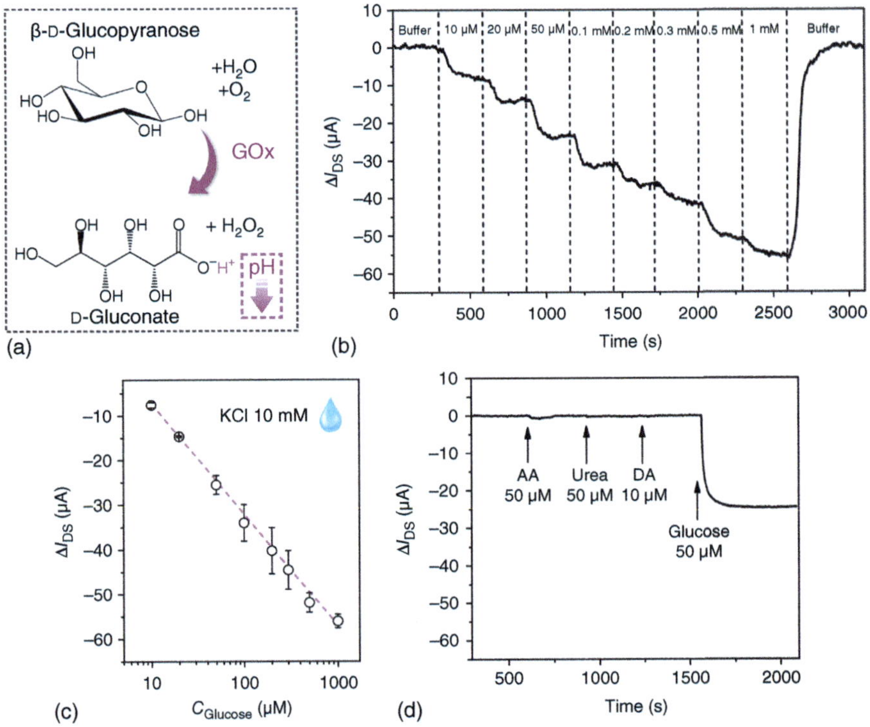

Figure 16.11 (a) Scheme of the glucose oxidation reaction on the transistors. (b) Flow-response of a GOx–PABA-GFET (300 μl min^{-1}, V_G = −0.2 V, V_{DS} = 0.1 V) at different analyte concentrations. (c) Linear fitting for the GOx–PABA-GFETs showing the results obtained for two different electrodes. (d) Response of a GOx–PABA-GFET to different potential interference injections (V_G = −0.2 V, V_{DS} = 0.1 V, 10 mM KCl, 0.1 mM HEPES, pH 6) (n = 2). Source: Reproduced with permission from Fenoy et al. [42].

already discussed, the enzymatic process (Figure 16.11a) causes a decrease in the local pH, shifting the Dirac point of the biosensors to more negative gate potentials. Thus, if the gate voltage is set to a more negative value than the Dirac point (i.e. the hole regime, $V_G < V_i$), a decrease in the recorded drain–source current is observed upon glucose addition (at constant V_G and V_{DS}) (Figure 16.11b).

Experiments show that the flow response of the biosensors exhibits a logarithmic dependence on the glucose concentration (Figure 16.11b,c), which can be understood by recalling the linear response of the pH and assuming that the proton concentration is proportional to the glucose concentration. Moreover, this logarithmic dependence of the glucose response supports the idea of the pH-based signal transduction mechanism of the biosensors. The sensitivity of the GOx–PABA-GFETs in terms of the changes in the drain–source current (ΔI_{DS}) is −25 μA per decade of glucose concentration.

The flow response of the biosensors indicates an average response time of 190 seconds for achieving 95% of the maximum signal. The device-to-device reproducibility showed an average RSD of 7%, while the repeatability of the devices exhibited a RSD of 7.2%. The stability of the biosensors was also investigated, showing that the devices maintained 92% of the original response after three days of storage at 4 °C in buffer HEPES solution. In addition, the flow of buffer solution after running increasing concentrations of glucose solution leads to a recovery of the recorded drain–source current that is close to the original value ($\Delta I_{DS} \sim 0$). This implies that the biosensor response to glucose is totally reversible, and the devices could be reused for further measurements.

The LOD of the GOx–PABA-GFET sensor was determined to be three times the standard deviation for the buffer injection (blank), yielding a value of 4.1 μM. This low LOD would allow the determination of glucose in a variety of body fluids containing the analyte in low concentration regimes, e.g. tears (100–600 μM), saliva (30–80 μM), and sweat (20–600 μM) [51]. The device was also exposed to different potential interferences such as AA, DA, and urea (Figure 16.11d). No appreciable variation in I_{DS} is observed upon injection of different analytes, which are commonly present in biological fluids such as blood, cerebrospinal fluid, and urine.

The response of the device was also studied in diluted urine as a more complex biological fluid (Figure 16.12). The device cell was filled with the diluted real sample, and, in order to determine its glucose concentration, subsequently injections of the same spiked sample were performed. Next, the linear fitting of the response of the biosensor in the biological fluid yielded a sensitivity of −13 ± 2 μA per decade of glucose concentration. By employing this sensitivity, a glucose concentration in the original sample of 40 ± 5 μM was found, consistent with the reported analyte concentration found in healthy individuals [51].

Finally, these results show that the CP-based biosensors can be used for the determination of glucose in complex biological samples without the requirement of arduous sample processing steps, paving the way for the development of POC GFET-based glucose biosensors.

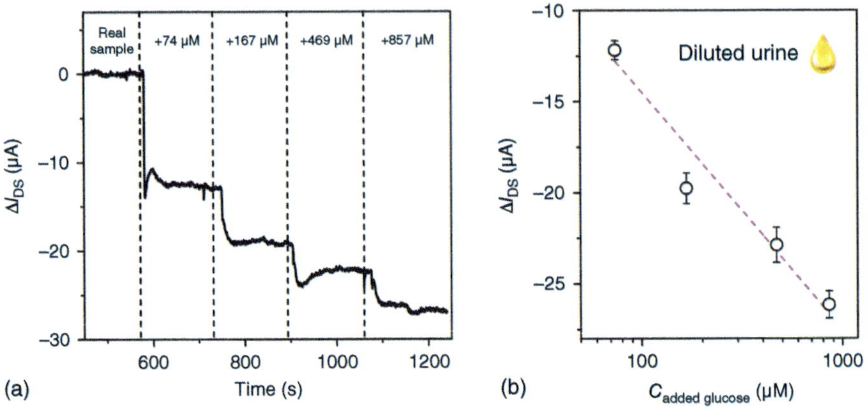

Figure 16.12 (a) Response of a GOx–PABA-GFET to the injection of spiked urine samples (V_G = −0.2 V, V_{DS} = 0.1 V) and (b) linear fitting of the response (dashed line, n = 2). Source: Reproduced with permission from Fenoy et al. [22]/Elsevier.

16.5 Conclusions

In this chapter, we described the use of CP films based on PABA as platforms to create enzymatic GFET biosensors. It was shown that PABA films can be easily electropolymerized on the channels of graphene transistors with precise control over the film thickness and the chemical homogeneity. The chemical richness of the electrodeposited PABA films improves the pH response of the transistors due to their multiple protonation states arising from the different pK_a values of the founding monomers, which also constitute the signal transduction mechanism of the biosensors. Moreover, since PABA bears extra primary amine groups (when compared with polyaniline), the electrostatic immobilization of enzymes such as AchE and GOx proved that the PABA-modified surfaces are suitable for enzyme anchoring without affecting their functionality. The amount of immobilized enzyme was quantified by SPR measurements, yielding values in line with previously reported results regarding AchE and GOx immobilization on amino-containing surfaces.

Next, the enzyme-modified PABA-GFETs were evaluated as Ach and glucose sensors. When analyzing the response of AchE-PABA-modified GFETs in presence of the enzymatic substrate, the biosensors showed a shift in the Dirac point to more negative values due to the enzyme-catalyzed hydrolysis, which lowers the local pH near the graphene channel surface. The evaluation of the response in a flow configuration allowed real-time sensing of the Ach in the range 5–1000 μM, and the AchE-PABA-GFETs showed excellent selectivity to interferences, fast response time, and good reproducibility. The biosensor response in real urine samples showed a similar sensitivity to that observed in buffer solutions. Compared with other FET-based reported Ach sensors, developed devices present better response in flow conditions and a lower LOD.

On the other hand, the immobilization of GOx on PABA-modified GFETs was also proven to be a successful strategy to build biosensing devices. The presence of glucose caused a diminution in the pH due to GOx-catalyzed oxidation, which allowed the real-time sensing of the analyte in the range of 10–1000 μM. The biosensing platforms showed proper reproducibility and a fast response time. When the performance of the devices was compared with that of previously reported GFET sensors, the developed transistors showed a low LOD and wide detection range, while allowing operation at small gate–source and drain–source potentials. In addition, the developed devices were able to effectively detect glucose in a complex biological fluid such as diluted urine.

These two examples show how electropolymerization can be employed to finely tune the features of the GFET surface with the aim of developing highly sensitive biosensing devices. This approach constitutes a promissory technique for the fabrication of biosensing platforms due to the benefits of voltammetric electrosynthesis, such as precise control of thickness, film stability, and the lack of a requirement for chemical primers. In addition, PABA constitutes a scarcely explored electroactive polymer showing numerous advantages for surface functionalization compared to the well-known polyaniline (PANI), particularly when dealing with soft acidic or neutral media, as those required for most enzymes and biorecognition elements to properly perform. Finally, the exhibited approach could be extended to the integration of other enzymes, biorecognition elements, or even nanomaterials, expanding the possibilities for the development of novel biosensing devices.

References

1 Tung, T.T., Nine, M.J., Krebsz, M. et al. (2017). Recent advances in sensing applications of graphene assemblies and their composites. *Adv. Funct. Mater. 27* (46): 1702891.

2 Balderston, S., Taulbee, J.J., Celaya, E. et al. (2021). Discrimination of single-point mutations in unamplified genomic DNA via Cas9 immobilized on a graphene field-effect transistor. *Nat. Biomed. Eng. 5* (7): 713–725.

3 Zhang, X., Jing, Q., Ao, S. et al. (2020). Ultrasensitive field-effect biosensors enabled by the unique electronic properties of graphene. *Small 16* (15): 1–24.

4 Fakih, I., Durnan, O., Mahvash, F. et al. (2020). Selective ion sensing with high resolution large area graphene field effect transistor arrays. *Nat. Commun. 11* (1): 3226.

5 Fu, W., Jiang, L., van Geest, E.P. et al. (2017). Sensing at the surface of graphene field-effect transistors. *Adv. Mater. 29* (6): 1603610.

6 Béraud, A., Sauvage, M., Bazán, C.M. et al. (2021). Graphene field-effect transistors as bioanalytical sensors: design, operation and performance. *Analyst 146* (2): 403–428.

7 Wu, G., Dai, Z., Tang, X. et al. (2017). Graphene field-effect transistors for the sensitive and selective detection of *Escherichia coli* using pyrene-tagged DNA aptamer. *Adv. Healthcare Mater. 6* (19): 1700736.

8 Xia, L., Wei, Z., and Wan, M. (2010). Conducting polymer nanostructures and their application in biosensors. *J. Colloid Interface Sci. 341* (1): 1–11.

9 Scotto, J., Fenoy, G.E., Sappia, L.D., and Marmisolle, W.A. (2018). Conducting polymers-based electrochemical platforms: from biosensing to energy storage. *An. Asoc. Quim. Argent. 105* (2): 135–155.
10 Fenoy, G.E., Azzaroni, O., Knoll, W., and Marmisollé, W.A. (2021). Functionalization strategies of PEDOT and PEDOT:PSS films for organic bioelectronics applications. *Chemosensors 9* (8): 212.
11 Inzelt, G. (2012). Chapter 4: Chemical and electrochemical syntheses of conducting polymers. In: *Conducting Polymers: A New Era in Electrochemsitry*, 2e (ed. F. Scholz), 149–171. Berlin Heidelberg: Springer-Verlag.
12 Novoselov, K.S., Geim, A.K., Morozov, S.V. et al. (2004). Electric field effect in atomically thin carbon films. *Science (80-.) 306* (5696): 666–669.
13 Cosnier, S. and Karyakin, A. (ed.) (2010). *Electropolymerization: Concepts, Materials and Applications*. Weinheim, Germany: Wiley-VCH Verlag GmbH & Co. KGaA.
14 Du, X. and Wang, Z. (2003). Effects of polymerization potential on the properties of electrosynthesized PEDOT films. *Electrochim. Acta 48* (12): 1713–1717.
15 Zuccaro, L., Krieg, J., Desideri, A. et al. (2015). Tuning the isoelectric point of graphene by electrochemical functionalization. *Sci. Rep. 5* (1): 11794–11806.
16 Han, Q., Pang, J., Li, Y. et al. (2021). Graphene biodevices for early disease diagnosis based on biomarker detection. *ACS Sensors 6* (11): 3841–3881.
17 Park, J.W., Lee, C., and Jang, J. (2015). High-performance field-effect transistor-type glucose biosensor based on nanohybrids of carboxylated polypyrrole nanotube wrapped graphene sheet transducer. *Sens. Actuators, B 208*: 532–537.
18 Andoy, N.M., Filipiak, M.S., Vetter, D. et al. (2018). Graphene-based electronic immunosensor with femtomolar detection limit in whole serum. *Adv. Mater. Technol. 3* (12): 1–12.
19 Piccinini, E., Fenoy, G.E., Cantillo, A.L. et al. (2022). Biofunctionalization of graphene-based FET sensors through heterobifunctional nanoscaffolds: technology validation toward rapid COVID-19 diagnostics and monitoring. *Adv. Mater. Interfaces* 9 (15): 2102526.
20 Piccinini, E., Bliem, C., Reiner-Rozman, C. et al. (2017). Enzyme-polyelectrolyte multilayer assemblies on reduced graphene oxide field-effect transistors for biosensing applications. *Biosens. Bioelectron.* 92: 661–667.
21 Hess, L.H., Lyuleeva, A., Blaschke, B.M. et al. (2014). Graphene transistors with multifunctional polymer brushes for biosensing applications. *ACS Appl. Mater. Interfaces 6* (12): 9705–9710.
22 Fenoy, G.E., Marmisollé, W.A., Azzaroni, O., and Knoll, W. (2020). Acetylcholine biosensor based on the electrochemical functionalization of graphene field-effect transistors. *Biosens. Bioelectron. 148*: 111796.
23 Reiner-Rozman, C., Schodl, J., Nowak, C., and Kleber, C. (2015). Scanning electrochemical microscopy as a characterization tool for reduced graphene oxide field effect transistors. *e-J. Surf. Sci. Nanotechnol. 13*: 366–372.
24 Kotlowski, C., Aspermair, P., Khan, H.U. et al. (2018). Electronic biosensing with flexible organic transistor devices. *Flexible Printed Electron. 3* (3): 34003.

25 Fenoy, G.E., Giussi, J.M., von Bilderling, C. et al. (2018). Reversible modulation of the redox activity in conducting polymer nanofilms induced by hydrophobic collapse of a surface-grafted polyelectrolyte. *J. Colloid Interface Sci. 518*: 92–101.

26 Marmisollé, W.A., Gregurec, D., Moya, S., and Azzaroni, O. (2015). Polyanilines with pendant amino groups as electrochemically active copolymers at neutral pH. *ChemElectroChem 2* (12): 2011–2019.

27 Fenoy, G.E., Scotto, J., Azcárate, J. et al. (2018). Powering up the oxygen reduction reaction through the integration of O_2-adsorbing metal–organic frameworks on nanocomposite electrodes. *ACS Appl. Energy Mater. 1* (10): 5428–5436.

28 Wang, Y.Y. and Burke, P.J. (2014). Polyelectrolyte multilayer electrostatic gating of graphene field-effect transistors. *Nano Res. 7* (11): 1650–1658.

29 Ohno, Y., Maehashi, K., Yamashiro, Y., and Matsumoto, K. (2009). Electrolyte-gated graphene field-effect transistors for detecting pH and protein adsorption. *Nano Lett. 9* (9): 3318–3322.

30 Salvo, P., Melai, B., Calisi, N. et al. (2018). Graphene-based devices for measuring pH. *Sens. Actuators, B 256*: 976–991.

31 Marmisollé, W.A., Florit, M.I., and Posadas, D. (2014). Acid-base equilibrium in conducting polymers. The case of reduced polyaniline. *J. Electroanal. Chem. 734*: 10–17.

32 Ray, A., Richter, A.F., MacDiarmid, A.G., and Epstein, A.J. (1989). Polyaniline: protonation/deprotonation of amine and imine sites. *Synth. Met. 29* (1): 151–156.

33 Raffa, D.L., Leung, K.T., and Battaglini, F. (2006). Electrochemical copolymerization of aniline and *ortho*-aminobenzylamine. Studies on its conductivity and chemical derivatization. *J. Electroanal. Chem. 587* (1): 60–66.

34 Sattarahmady, N., Heli, H., and Moosavi-Movahedi, A.A. (2010). An electrochemical acetylcholine biosensor based on nanoshells of hollow nickel microspheres–carbon microparticles–nafion nanocomposite. *Biosens. Bioelectron. 25* (10): 2329–2335.

35 Kong, D., Jin, R., Zhao, X. et al. (2019). Protein-inorganic hybrid nanoflower-rooted agarose hydrogel platform for point-of-care detection of acetylcholine. *ACS Appl. Mater. Interfaces 11* (12): 11857–11864.

36 Mitchell, K.M. (2004). Acetylcholine and choline amperometric enzyme sensors characterized in vitro and in vivo. *Anal. Chem. 76* (4): 1098–1106.

37 Andronescu, C. and Schuhmann, W. (2017). Graphene-based field effect transistors as biosensors. *Curr. Opin. Electrochem. 3* (1): 11–17.

38 Chae, M.S., Yoo, Y.K., Kim, J. et al. (2018). Graphene-based enzyme-modified field-effect transistor biosensor for monitoring drug effects in Alzheimer's disease treatment. *Sens. Actuators, B 272*: 448–458.

39 Viswanathan, S., Narayanan, T.N., Aran, K. et al. (2015). Graphene-protein field effect biosensors: glucose sensing. *Mater. Today 18* (9): 513–522.

40 Vakurov, A., Simpson, C.E., Daly, C.L. et al. (2004). Acetylcholinesterase-based biosensor electrodes for organophosphate pesticide detection: I. Modification of carbon surface for immobilization of acetylcholinesterase. *Biosens. Bioelectron. 20* (6): 1118–1125.

41 Niyogi, S., Bekyarova, E., Itkis, M.E. et al. (2010). Spectroscopy of covalently functionalized graphene. *Nano Lett. 10* (10): 4061–4066.
42 Fenoy, G.E., Marmisollé, W.A., Knoll, W., and Azzaroni, O. (2022). Highly sensitive urine glucose detection with graphene field-effect transistors functionalized with electropolymerized nanofilms. *Sens. Diagn.* 1: 139–148.
43 Fenoy, G.E., von Bilderling, C., Knoll, W. et al. (2021). PEDOT:tosylate-polyamine-based organic electrochemical transistors for high-performance bioelectronics. *Adv. Electron. Mater.* 7: 2100059.
44 Sohn, I.Y., Kim, D.J., Jung, J.H. et al. (2013). pH sensing characteristics and biosensing application of solution-gated reduced graphene oxide field-effect transistors. *Biosens. Bioelectron. 45* (1): 70–76.
45 Milkani, E., Lambert, C.R., and McGimpsey, W.G. (2011). Direct detection of acetylcholinesterase inhibitor binding with an enzyme-based surface plasmon resonance sensor. *Anal. Biochem. 408* (2): 212–219.
46 He, Q., Sudibya, H.G., Yin, Z. et al. (2010). Centimeter-long and large-scale micropatterns of reduced graphene oxide films: fabrication and sensing applications. *ACS Nano 4* (6): 3201–3208.
47 Jean-Marie, E. (2010). Diagnosis and classification of diabetes mellitus. *Diabetes Care 33* (Supplement_1): S62–S69.
48 Sehit, E. and Altintas, Z. (2020). Significance of nanomaterials in electrochemical glucose sensors: an updated review (2016–2020). *Biosens. Bioelectron. 159*: 112165.
49 Gregg, E.W., Sattar, N., and Ali, M.K. (2016). The changing face of diabetes complications. *Lancet Diabetes Endocrinol. 4* (6): 537–547.
50 Nathan, D.M. (1993). Long-term complications of diabetes mellitus. *N. Engl. J. Med. 328* (23): 1676–1685.
51 Nery, E.W., Kundys, M., Jeleń, P.S., and Jönsson-Niedziółka, M. (2016). Electrochemical glucose sensing: is there still room for improvement? *Anal. Chem. 88* (23): 11271–11282.
52 Gubala, V., Harris, L.F., Ricco, A.J. et al. (2012). Point of care diagnostics: status and future. *Anal. Chem. 84* (2): 487–515.
53 Radhakumary, C. and Sreenivasan, K. (2011). Naked eye detection of glucose in urine using glucose oxidase immobilized gold nanoparticles. *Anal. Chem. 83* (7): 2829–2833.
54 Rennke, H.G. and Denker, B.M. (2019). *Renal Pathophysiology: The Essentials.* Lippincott Williams & Wilkins.
55 Chen, C., Xie, Q., Yang, D. et al. (2013). Recent advances in electrochemical glucose biosensors: a review. *RSC Adv. 3* (14): 4473–4491.
56 Cortez, M.L., Marmisollé, W., Pallarola, D. et al. (2014). Effect of gold nanoparticles on the structure and electron-transfer characteristics of glucose oxidase redox polyelectrolyte-surfactant complexes. *Chem. Eur. J. 20* (41): 13366–13374.
57 Pallarola, D., von Bildering, C., Pietrasanta, L.I. et al. (2012). Recognition-driven layer-by-layer construction of multiprotein assemblies on surfaces: a biomolecular toolkit for building up chemoresponsive bioelectrochemical interfaces. *Phys. Chem. Chem. Phys. 14* (31): 11027–11039.

58 Lvov, Y., Ariga, K., Kunitake, T., and Ichinose, I. (1995). Assembly of multicomponent protein films by means of electrostatic layer-by-layer adsorption. *J. Am. Chem. Soc. 117* (22): 6117–6123.

59 Sheldon, R.A. and van Pelt, S. (2013). Enzyme immobilisation in biocatalysis: why, what and how. *Chem. Soc. Rev. 42* (15): 6223–6235.

60 Bankar, S.B., Bule, M.V., Singhal, R.S., and Ananthanarayan, L. (2009). Glucose oxidase – an overview. *Biotechnol. Adv.* Elsevier Inc. 27: 489–501.

61 Hodak, J., Etchenique, R., Calvo, E.J. et al. (1997). Layer-by-layer self-assembly of glucose oxidase with a poly(allylamine)ferrocene redox mediator. *Langmuir 13* (10): 2708–2716.

62 Kwak, Y.H., Choi, D.S., Kim, Y.N. et al. (2012). Flexible glucose sensor using CVD-grown graphene-based field effect transistor. *Biosens. Bioelectron. 37* (1): 82–87.

63 Wang, P., Amarasinghe, S., Leddy, J. et al. (1998). Enzymatically prepared poly(hydroquinone) as a mediator for amperometric glucose sensors. *Polymer (Guildf) 39* (1): 123–127.

64 Kim, J., Campbell, A.S., and Wang, J. (2018). Wearable non-invasive epidermal glucose sensors: a review. *Talanta* 177: 163–170.

17

Graphene Field-Effect Transistors for Sensing Stress and Fatigue Biomarkers

Biddut K. Sarker[1,2], Cheri M. Hampton[1,2], and Lawrence F. Drummy[1]

[1] *Materials and Manufacturing Directorate, Air Force Research Laboratory, WPAFB, 2179 12th Street, OH, 45433, United States*
[2] *UES Inc., 4401 Dayton Xenia Rd, Dayton, OH, 45432, United States*

17.1 Introduction

A biosensor is an analytical device that usually consists of two key elements: a bioreceptor or biological recognition element (BRE), and a transducer (physiochemical detector element) [1, 2]. A bio-recognition element is a biological component (e.g. antibody, DNA, or enzyme) that is used to functionalize the surface of a transducer [3, 4]. The function of a BRE is to provide high selectivity toward the target biomolecules among the various interferences that exist in biological fluids, while the function of a transducer is to generate a measurable analytic signal proportional to the concentration of the analyte/biomolecules. Depending on the type of transducer and its detection method, biosensors are largely classified into different categories such as electrical, optical, photonic, electrochemical, and piezoelectric. Electrical methods, and in particular field-effect transistor (FET)-based biosensing, have attracted significant attention in recent years due to their many advantages including high sensitivity, high selectivity, label-free detection, high speed, simple operation, compactness, and rapid analysis [5–11]. Moreover, FET-based biosensing is highly translational to wearable electronics for rapid, real-time detection of biomarkers that correlate with certain physiological conditions pertinent to human health and performance.

Generally, FET-based biosensing involves the attachment of a target-specific BRE to the surface of a transducer for selectivity toward the target biomolecules of interest (Figure 17.1a). Many nanomaterials (e.g. one-dimensional [1D] nanotubes and nanowires, two-dimensional nanosheets, etc.) have been employed as a transducer for fabricating FET-based biosensors [2]. Graphene has attracted significant attention for the development of FET-based biosensors in the last few years due to its

Graphene Field-Effect Transistors: Advanced Bioelectronic Devices for Sensing Applications,
First Edition. Edited by Omar Azzaroni and Wolfgang Knoll.

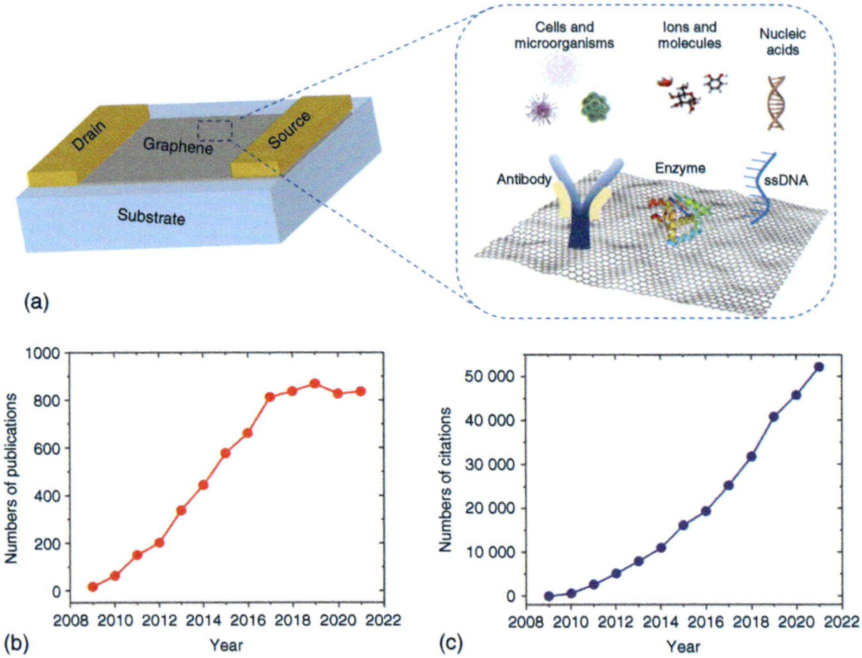

Figure 17.1 (a) Schematic of a graphene field-effect transistor (GFET)-based biosensor. The GFET biosensors consist of mainly two elements: GFET and BRE. The graphene integrated into a FET is used as a transducer which generates a measurable analytic signal proportional to the concentration of the analyte or target molecules (e.g. cells, ions, biomolecules, nucleic acids, etc.). A bioreceptor or BRE is a biological component (e.g. antibody, DNA, peptide, and enzyme) that interacts specifically with the target biomolecules. The graphene surface is chemically functionalized with the bioreceptor. The total number of (b) publications and (c) citations on the graphene-based biosensors show a progressively increasing trend in the last 10 years. Keyword used for this search on the Web of Science website in March 2022 is "graphene biosensor." Source: Part of the figure is adapted from Peña-Bahamonde et al. [7].

extraordinary electronic properties such as high electrical conductivity and high electron mobility [1, 3, 12–17]. In addition to its exceptional electronic properties, graphene has other notable physical properties, including high flexibility, stretchability, and conformability [18–21]. Due to these unique characteristics, graphene has been considered a key potential component for developing stretchable and flexible electronics, including wearable biosensors. Moreover, graphene has outstanding biocompatibility and stability, which makes it an excellent candidate for developing biosensors [3, 4, 22]. Since all the carbon atoms are located on its surface, graphene is extremely sensitive to its surface charge distribution, which enables monolayer graphene to transduce a small detectable change in conductivity due to the interaction of graphene and its associated BREs with the target biomolecules [23]. Transduction occurs within tens of milliseconds, suggesting that graphene field-effect transistor (GFET) biosensors can be used not only for applications in the area of biomedical and healthcare but also they can be integrated into wearable electronics for real-time biomarker monitoring and augmenting human health [24–27]. Another remarkable property of graphene is that it can be easily functionalized to achieve the selectivity of biomolecules by functionalizing the graphene surface with selective biomolecules as a probe while maintaining the intrinsic transport properties of graphene [3, 4]. Since many biomarkers are present in biological fluids, selectively detecting a particular biomarker is one of the main criteria for biosensors. All these properties make GFET-based sensing very attractive and have generated strong interest in the broader scientific community as well as in industry. This is reflected by the recent publications and citations in this area (Figure 17.1b,c).

In this review, we will focus on the recent advancement in GFET biosensing with emphasis on stress and fatigue biomarker detection rather than extending to all recent graphene-based biosensors work comprehensively. We will first cover fundamentals of graphene, GFETs, and sensing mechanisms, as well as various biomarkers related to human performance. Then, we will review graphene synthesis, device fabrication processes, and functionalization methods to achieve selectivity in the sensors. Following that, we will discuss the recent development of the GFET biosensors for monitoring stress- and fatigue-related biomarkers. We will focus on the wearable and flexible GFET biosensor and their application in portable biosensing systems. Then, we will discuss the current challenges for GFET biosensors, including Debye length screening, and a few possible approaches to overcome those challenges. Finally, we will conclude the book chapter by discussing the outlook and future perspective of GFET biosensors.

17.2 Molecular Biomarkers

Biomarkers are, in general, any number of biological (i.e. biochemical, molecular, or cellular) components that can be measured and tracked within the body and correlated with physiological state. Biomarkers can be found in bodily fluids such as blood (plasma), sweat, saliva, and urine (Figure 17.2a). The location and dynamics of the biomarker can be linked to a physiological state or condition (stress, inflammation, and cognitive state) of human performance [28, 30, 31]. Stress is a very common biological state characterized by changes in hormones and neurotransmitters, which impact human physical and mental health. Chronic stress can cause many disorders such as obesity, metabolic syndrome, heart disease, anxiety, depression, fatigue syndrome, burnout, PTSD, memory, and concentration problems [29]. Inflammation is a state of activation of the body's immune system in response to damage from illness or injury and is part of a healthy immune response. Cells secrete small proteins such as cytokines to rapidly signal to other cells by binding to cell-surface receptors. Chronic inflammation, on the other hand, occurs when the body continues to secrete elevated levels of pro-inflammatory proteins in the absence of acute injury and is associated with diseases such as asthma, psoriasis, rheumatoid arthritis, and inflammatory bowel disease.

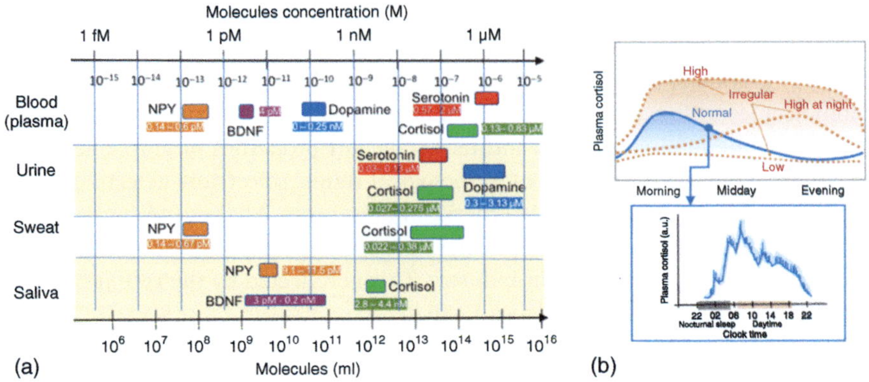

Figure 17.2 (a) The molecular concentration and the equivalent number of molecules per milliliter of key stress biomarkers in different biofluids such as blood, urine, sweat, and saliva. The molar concentration for some biomarkers (e.g. serotonin, dopamine, and cortisol) is high, whereas it is low for some other biomarkers (e.g. NPY and BDNF). The concentration of the same biomarker can slightly vary from biofluid to the biofluid. Source: Reproduced with permission from Steckl and Ray [28]. (b) Regular and irregular circadian pattern cortisol secretion by the adrenal glands in the human body through the day. For a regular circadian pattern, the cortisol level is high in the morning and low at night, whereas for an irregular circadian pattern under stressful situations, the cortisol level is low in the morning and high at night. Source: Adapted from Sheibani et al. [29].

Stress and inflammation are dynamic and vary between individuals [28, 32]. Therefore, ideally, a set of multiple biomarkers needs to be characterized simultaneously to determine the accuracy of the stress level. A recent study by Jenz et al. monitored a panel of biomarkers measured in blood serum samples taken periodically throughout intensive physical training of Battlefield Airmen [33]. Through the use of correlation networks, the researchers were able to establish that differences in individual stress reactivity during training correlated with course performance and graduation status. As the above study used frequent blood draws for biomarker collection, ultimately a noninvasive or minimally invasive method for continuous monitoring of key biomarkers is desired. In this review, we focus our discussion on the detection of stress-related biomarkers (e.g. neuropeptide Y [NPY], cortisol, interleukin-6 [IL-6], tumor necrosis factor-alpha [TNF-α], etc.) using GFET biosensors.

NPY is an important biomarker that is involved in the regulation of important biological and pathophysiological functions such as food uptake, energy homeostasis, circadian rhythm, stress, depression, anxiety, obesity, and cognition [34–36]. NPY is produced as a 97-amino acid long, full-length protein inside brain cells and is cleaved into a 36-amino acid long peptide that circulates in the body. Although NPY can be degraded by proteases, a substantial amount of NPY can be found in blood, sweat, and saliva. The NPY concentration in different bodily fluids is relatively low (0.8–2.9 to 14.2–73.2 pM in normal and stress conditions, respectively) [37]. To detect NPY various types of label-free techniques have been used, including liquid chromatography–mass spectrometry [38], however, these techniques are not real-time and the detection limit of the biosensors based on these techniques could be improved. A major complicating issue for NPY sensing is the strong sensitivity of charge transfer to pH and ion concentration in physiological fluids (e.g. sweat) (see Section 17.5).

Cortisol, a steroid, is commonly known as the "stress hormone" and is produced by the adrenal gland in response to stress. It plays an important role in regulating carbohydrate metabolism, keeps inflammation down, and works with certain parts of the brain to make the human body resist pain, infection, or psychological tension. Cortisol concentration varies throughout the day, with the highest level in the morning (~30 minutes after waking, 0.14–0.69 μM) and the lowest level at night (0.083–0.36 μM) (Figure 17.2b) [29]. Sustained stress can change this trend and can elevate overall biomarker concentration, and therefore significant interest has recently emerged in sensing cortisol biomarkers in biofluid.

IL-6 and TNF-α are both inflammatory cytokine biomarkers. IL-6 can provide valuable information about stress, trauma, and infection of the human body, as well as information on severe diseases (e.g. prostate, breast, and pancreatic cancer). IL-6 also plays an important role in both immune regulation and exercise metabolism. IL-6 is released during exercise and its level is increased as a function of the intensity. It has been reported that plasma IL-6 levels can increase more than 100-fold after a marathon [39]. TNF-α is closely related to stress, trauma, infections, fever, Crohn's disease, and pulmonary tuberculosis [40]. The detection of TNF-α concentration in the human body is highly desirable for screening, early clinical diagnosis, and long-time management of patients.

The detection of biomarkers depends on many factors such as sensing technology, sensor/device configuration, measurement conditions, and biomarker molecular weight, concentration, physical size, and solubility. The concentration of biomarkers can vary by several orders of magnitudes between biofluid types and the collection time of the samples (Figure 17.2b).

17.3 Graphene Field-Effect Transistor Based Biosensors

GFET based biosensors can sense changes in the electrical conductance of the graphene due to modulation of the electric field upon binding of the target charged biomolecules on the graphene sensor surface [3–5, 9]. Fundamentals of graphene, especially the electronic properties and basic configuration of a GFET, will be reviewed before discussing the device, structure, and sensing mechanism of the GFETs.

17.3.1 Graphene

Graphene is an atomically thin 2D crystal with carbon atoms are arranged in a honeycomb lattice with noninteracting π and π^* states [13, 14, 41, 42]. The π bonding states and π^* antibonding states form the valence band and conduction band, respectively. These two bands meet at six points which are called Dirac points or charge neutrality points (CNPs) [15, 41]. Since the valence band and conduction band meet at these Dirac points, graphene is neither a metal nor a semiconductor, and it is therefore considered a zero-bandgap semiconductor. In undoped or intrinsic graphene the valence band is completely filled with electrons, the conduction band is completely empty and the Fermi level and the Dirac point coincide precisely at the intercept of valence band and conduction band. In the case of positively doped graphene (also called p-doped or hole-doped), the electrons are partially drained from the valence band and the Fermi level is displaced to a position below the Dirac point within the valence band, whereas in the negatively doped graphene (n-doped or electron-doped) the extra electrons are forced into the conduction band and the Fermi level resides above the Dirac point (Figure 17.3b).

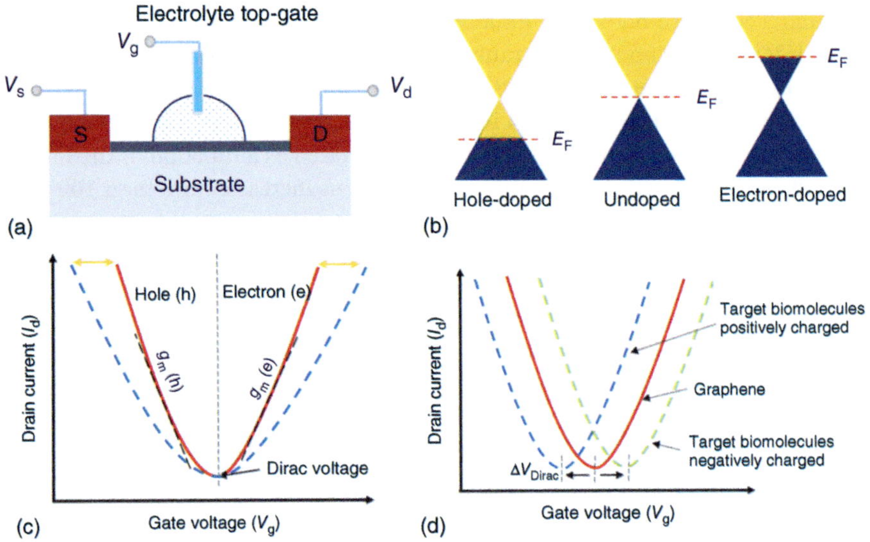

Figure 17.3 (a) The schematic of a typical liquid-gated GFET sensor, showing a gate electrode wire immersed into a liquid solution containing the target analyte or biomolecules. The graphene is connected between the source and drain electrodes. The bias voltage is applied between the source and drain contacts to transport the current through the graphene. The current of the graphene channel is modulated by applying a gate voltage at the gate electrodes. (b) The linear dispersion (Dirac cone) with conduction and valence bands touching at K point (Dirac point). Blue indicates the levels filled with electrodes and yellow indicates the empty level (holes). In an undoped or neutral graphene, the valence band is completely filled with electrons whereas the conduction band is completely empty, and in this case, the Fermi level and Dirac point coincide. When the electrons are partially drained from the valence band and the Fermi level displaced below the Dirac points is called hole-doped graphene. When the extra electrons are forced into the conduction band which moves the Fermi level above the Dirac point it is called electron-doped graphene. (c) Typical transfer characteristics (drain current as a function of gate voltage) of GFET biosensors. The linear slope of the transfer is used to calculate the transconductance and hence the device mobility. The gate voltage corresponding to the minimum current (also known as charge neutrality point) is called Dirac voltage (V_{Dirac}). (d) The change of V_{Dirac} of a GFET due to interaction of graphene with biomolecules. When the positive biomolecules are attached to the graphene surface, the V_{Dirac} is shifted negatively (known as n-doping), whereas when the negative charge biomolecules are attached to the graphene surface, the V_{Dirac} is shifted positively (known as p-doping). Source: Biddut K. Sarker.

As graphene has a unique one-atom-thick carbon lattice, it is a unique electronic material with excellent electronic properties [13–16]. For example, the electrical resistivity of graphene is very low ($10\,k\Omega\,sq^{-1}$) [43], and the room temperature electron mobility is extremely high ($2.5\times10^5\,cm^2V^{-1}s^{-1}$) [13, 17]. The mobility of graphene is the highest among the existing 2D materials and about 2–3 orders of magnitude higher than that for the conventional semiconductor such as silicon [6, 44]. Because of its unique 2D crystal lattice and high charge carrier mobility, graphene shows low electrical noise at room temperature [21]. In addition, the electronic properties of graphene are highly sensitive to its surface charge distribution. When the graphene is configured as a FET, it can transduce its surface charge distribution into a detectable change within tens of milliseconds [24]. Because of these extraordinary electronic properties, graphene is considered an excellent active sensing material for fabricating FET-based biosensors with high sensitivity.

It is important to note that graphene is a semimetal (zero-bandgap), and graphene transistors cannot turn off completely even at the Dirac point. Therefore, the current on–off ratio of GFETs is significantly smaller than that of the semiconductor (fixed bandgap) FETs [42]. Although current on–off ratio has no direct relationship with biosensor performance, the on-current should be high enough compared to the noise current so that a small change in the on-current upon adsorption of biomolecules on the sensor surface can be easily detected. In addition, GFET biosensors with a high current on–off ratio are preferred for their applications in digital electronics where a high on-state current and a low off-state current are desired.

17.3.2 Structure of Graphene Field-Effect Transistors

A GFET typically consists of mono (single)-layer or few-layer graphene in the channel with length (L) and width (W), source (S) and drain (D) electrodes that are in contact with graphene, and a gate electrode which is separated from graphene layer by a thin dielectric layer (Figure 17.3a). Depending on the arrangement of these components, GFET structures can be categorized as either back gated or liquid gated (also known as solution gated or electrolyte gated) configurations. To operate a liquid-gated GFET, a gate voltage (V_G) is applied at the one end of the reference electrode (a thin conducting wire, e.g. Ag/AgCl) and another end of the reference electrode is immersed in the liquid or electrolyte solution, whereas a bias voltage (V_D) is applied to the between drain and source electrodes. The source electrode is usually connected to the ground ($V_S = 0\,V$). When a bias voltage is applied at the drain electrode, the charge carriers from the source contact are injected into the graphene and transported through the conducting channel to the drain electrode. The conductivity of the graphene channel is modulated by a gate voltage applied to the electrolyte on the graphene surface via a reference electrode. The operation and device structure for both liquid-gated and back-gated GFETs are similar. However, to operate back-gated GFETs, a gate voltage is applied to the backside of the substrate of the GFET. In most cases, a highly conductive silicon (Si++) is used as a substrate, and a thin insulating SiO_2 deposited on the conducting silicon substrate is used as a dielectric layer.

The charge transport in graphene is conducted by both the positive (hole) and negative (electron) charge carriers (Figure 17.2b), and the transfer curves (I_D–V_G) of graphene transistors are intrinsically ambipolar (Figure 17.3c). The critical transition point of the transfer curve, where the conductivity of graphene is at its minimum, is called the CNP. The gate voltage at the CNP is called Dirac voltage (V_{Dirac}). Since the Fermi level of graphene is changed with the V_G, the carrier density and carrier type in the graphene channel can be easily modulated by the applied V_G. For $V_G < V_{Dirac}$ the charge transport is conducted by holes, while for the $V_G > V_{Dirac}$, the charge transport is conducted by electrons.

17.3.3 Sensing Mechanism of GFET Biosensors

The graphene surface of the GFET biosensors is typically functionalized with various bioreceptors or BREs to selectively probe the intended target biomolecules [3, 4]. When a target charged biomolecule is bound on the graphene surface, the interfacial electrical field is changed which causes a change in the concentration of charge carriers in the graphene channel of the device, with a corresponding measurable change in the source–drain conductivity or the drain current. The drain current of the GFET can be calculated using the following equation (17.1),

$$I_{DS} = \mu C_i \frac{W}{L} \left| V_G - V_{Dirac} \right| V_{DS}, \quad \text{for } | V_G - V_{Dirac} | \gg V_{DS} \tag{17.1}$$

where C_i is the gate capacitance, V_{Dirac} is the Dirac voltage, μ is the field-effect mobility, W is the channel width, and L is the channel length of the graphene layer. The gate capacitance of a back-gated GFET is formed by the dielectric material between the substrate and graphene. The gate capacitance of an electrolyte device consists primarily of the two electrical double layers (EDLs) formed by the separated charges at the reference electrode/electrolyte interface and electrolyte/graphene interface and quantum capacitance of the graphene. The quantum capacitance of the graphene is given by [6, 45, 46]

$$C_Q = q^2 \frac{2}{\pi} \frac{qV_{ch}}{\left(\hbar v_F\right)^2}, \quad \text{for } qV_{ch} \gg k_B T \tag{17.2}$$

where $q = -1.602 \times 10^{-19}$ C is the electron charge, $\hbar = 1.055 \times 10^{-34}$ J s is reduced Planck constant, $v_F = 1.1 \times 10^6$ m s^{-1} is Fermi velocity of Dirac fermion, $k_B = 1.381 \times 10^{-23}$ J K^{-1} is the Boltzmann constant, $T = 300$ K at room temperature, and V_{ch} is the graphene channel electrostatic potential. Depending on channel electrostatic potential, the quantum capacitance of the graphene can vary from 0 to 10 of μF cm^{-2}. The EDL capacitance strongly depends on the concentration of ions in the electrolyte solution as well as the charge distributions to the channel or gate electrode. The capacitance for an EDL is in the range of 1–40 μF cm^{-2} [6, 47]. The total gate capacitance of an electrolyte-gated GFET is significantly higher than that of a back-gated device (e.g. 345 nF cm^{-2} for 100 nm thick SiO_2 layer) [6].

According to Eq. (17.1), the current of GFET can be changed by four parameters μ, V_{Dirac}, C_i, and V_G for a fixed applied bias voltage (V_{DS}) and fixed device geometry.

When a biomolecule in the electrolyte solution interacts with the graphene surface, one or more of these four parameters can be changed. Possible sensing mechanisms of the GFET biosensors can be summarized below.

The most commonly utilized mechanism for GFET-based biosensing is based on the change in Dirac voltage shift due to changes in the surface potential of graphene [48]. The Dirac voltage for GFETs with an intrinsic graphene layer is expected to be around zero gate voltage. However, it reality the Dirac voltage is either positive if the graphene has large density of holes, (called p-doped graphene) or negative if the graphene has large density of electrons (called n-doped graphene) [9]. Likewise, when biomolecules interact with the graphene, the surface potential of graphene is changed, which results in the Dirac voltage shift in the transfer curve. When positively charged biomolecules are attached to the graphene surface, the Dirac voltage is shifted negatively (direction of the Dirac voltage change on the gate voltage axis), which is called n-doping, whereas when negatively charged biomolecules are attached to the graphene surface, Dirac voltage is shifted positively, which is called p-doping (Figure 17.3d). The sensitivity of this biosensor can be determined by monitoring the magnitude and polarity of the Dirac voltage shift as a function of biomolecule concentrations.

Another sensing mechanism is based on the change in carrier mobility in graphene, and a limited number of studies have employed this sensing mechanism. When the biomolecules interact with the graphene surface, the resulting changes in Fermi level and mobility change the device performance. The biomolecules can bind directly on the graphene surface, which creates additional scattering centers and hence increases disorders, which also effect on the charge carrier mobility. The mobility can be calculated by measuring the transconductance. Transconductance (g_m) is the slope of the linear regime of p-or n- branch of the transfer characteristics curve which is given by [9, 48]

$$g_m = \frac{\partial I_{ds}}{\partial I_{gs}} = \frac{W}{L}\mu C_i V_{DS} \tag{17.3}$$

If L, W, C_i, and V_{ds} are kept constant, the mobility is directly proportional to g_m. Transconductance is an important indicator of transistor performance and it is considered as the direct indicator of the sensitivity of the GFET biosensors. Since the graphene shows ambipolar transport characteristics, there are two maximum transconductance values are observed: one is for electrons, located at right side of transfer curve, denoted by g_m (e), and another is for holes, located at left side of transfer curve, denoted by g_m (h) (Figure 17.3c). A few groups use this sensing mechanism to explain the results of the graphene-based DNA [49] and bacteria sensors [50].

In addition to the sensing mechanisms based on Dirac voltage and mobility change, the change in the amplitude of electrical current of transfer curves either in the p- or n-branch or at the CNP or both can also be used as biosensing matric [9, 51]. As discussed, binding of the biomolecules to the graphene channel can either increase or decrease the current depending the polarity of target biomolecules (Figure 17.3d). In most cases, the current is shifted in both the p-branch and n-branch of the transfer curves due to interaction of biomolecules and sensors. However, sometimes current change is only observed in one branch of transfer curve [52]. The change of the drain current is related to the change of the carrier density (Δn) which is induced by, and proportional to the total number of charge biomolecules (N) absorbing on the graphene surface [4]

$$\Delta I_{\mathrm{DS}} = \frac{W}{L} e\mu V_{\mathrm{DS}}\, \Delta n \propto N \tag{17.4}$$

where e is the electron charge. Equation (17.4) shows the change of drain current is ideally directly proportional to total number of charge absorbed biomolecules but this is not always observed [4]. In addition to the above-mentioned sensing mechanism, gate capacitance change is also used as a sensing metric. The gate capacitance, especially the two EDLs capacitance, is strongly dependent on the ion concentrations and impurities as well as charge distributions near the gate/electrolyte and electrolyte/graphene interfaces [6].

17.4 GFET Biosensor Fabrication

A few key process steps for fabricating graphene field-effect based biosensors are production of high-quality graphene, transfer of the graphene to the required substrates, device fabrication, and functionalization of the graphene surface with BREs to selectively detect the target biomolecules. The processes are discussed in more detail in this section.

17.4.1 Graphene Production

The production of high-quality graphene in a controllable and scalable manner is important for the realization of high-performance graphene-based biosensors. A number of graphene production techniques have been demonstrated including exfoliation (mechanical, liquid, and oxidation/reduction methods), epitaxial growth, and chemical vapor deposition (CVD) [53–57]. Most of the graphene-based biosensors are fabricated using graphene produced by either mechanical exfoliation or the CVD method. Therefore, we will briefly discuss the mechanical exfoliation and CVD methods in this review.

Graphene was first exfoliated mechanically from graphite by a group of researchers at Manchester University in 2004 [13]. They simply peeled a single-layer or multilayer graphene from bulk graphite using adhesive tape, colloquially referred to as the scotch-tape method. Single-layer graphene can be easily exfoliated from bulk graphite because in a single layer of graphene the carbon atoms are bonded by a covalent bonds in the graphene plane, while the adjacent graphene sheets in bulk graphite are bonded by weaker van der Waals bonds. The mechanical exfoliation technique is a simple process, and the quality of the exfoliated graphene is very high. Several biochemical sensors based on exfoliated graphene have been demonstrated to detect various biomolecules [3]. The major limitations of exfoliated graphene are its small size (several microns to tens of microns at best) and its irregular shape. Because the mechanical exfoliation method is not efficient or scalable, it will likely not be used for developing controllable, reproducible biosensors for commercial applications.

In contrast to the mechanical exfoliation method, the production of graphene using the CVD process offers large-area, high-uniformity, high controllability, and relatively high-quality graphene [54–58]. The CVD growth of uniform graphene film on the copper foil was first demonstrated by the Ruoff group in 2009 [55]. Subsequently, a large number of studies have been performed by many research groups to optimize the CVD growth process to achieve high-quality graphene [56, 57, 59]. In a typical CVD graphene process, a copper foil is loaded in the CVD chamber and pumped down to a pressure of 1.375 torrs (Figure 17.4a). The temperature of the chamber is increased to more than 1000 °C. Once the temperature is stabilized, methane and hydrogen gases are flown into the chamber for ~30 minutes to grow single-layer graphene on the copper foil [5].

One of the major challenges for CVD graphene is transferring the graphene from the copper foil to the required substrates for device fabrication [56, 57]. Many different graphene transfer techniques have been demonstrated. The most widely used transfer technique is polymer assisted graphene transfer (Figure 17.4b). In this method, polymer (mainly, polymethyl methacrylate or PMMA) is coated on one side of the graphene-Cu foil to protect the graphene from reactive ion etching (RIE), and the graphene from other side of the Cu foil was removed using RIE. To etch the Cu, polymer/PMMA-coated graphene/Cu was then placed in $Fe(NO_3)_3$ solution for a few hours, and the resulting polymer coated graphene film is transferred to deionized (DI) water. The film is lifted then from surface of the DI water using a substrate such as silicon. The substrate with polymer-coated graphene is annealed in air at ~70 °C for 20 minutes to adhere polymer-coated graphene film to the substrate and the coated substrate is stored at room temperature to completely evaporate any residual solvents. Finally, the PMMA is then removed by placing the substrates with polymer-coated graphene in acetone at ~45 °C for 30 minutes and rinsing the substrate with isopropyl alcohol (IPA) and DI water followed by drying with a dry nitrogen blower. Although CVD graphene offers its potential in commercial high-end graphene-based biosensors applications, there are many challenges that need to be overcome. A few major disadvantages of CVD growth of graphene are the large number of defective grain boundaries, complicated transfer process, and presence of impurities and residual contaminants from the transfer process. These issues strongly influence the intrinsic properties of graphene and limit its further applications, not only in biosensors but also in other fields. We will discuss the graphene quality issue in more detail in Section 17.7.

17.4.2 Device Fabrication

After transferring graphene onto the substrate, graphene is patterned using photolithography and RIE to define the channel length and width for the GFET (Figure 17.4c–e). Then the source and drain electrodes are fabricated using

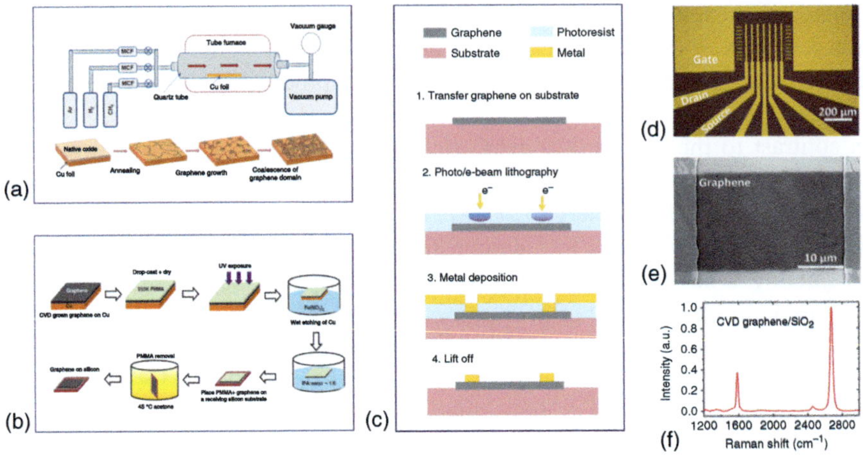

Figure 17.4 (a) Chemical vapor deposition on graphene on a copper foil. A copper foil is loaded in a CVD chamber and the temperature of the chamber rises to more than 1000 °C. Typically methane and hydrogen gas are flown to the chamber for ~30 minutes to grow single-layer graphene on a copper foil. Source: Biddut K. Sarker. Part of the figure (a) is adapted from Kaymak et al. [60]. (b) Process flow of graphene transfer from the copper foil to a device substrate. Source: Adapted from Islam et al. [5]. (c) Typical GFET device fabrication process. Source: Biddut K. Sarker. (d) Optical image of a GFET chip. (e) Scanning electron microscopy (SEM) image of graphene between the source and drain metal contact. Source: Reproduced with permission from Hao et al. [26]/American Chemical Society. (f) Raman spectra of monolayer graphene on a silicon substrate. Source: Biddut K. Sarker.

different nano and microfabrication steps. Typically the electrodes are fabricated using photolithography to define the electrode pattern, and then metal is deposited by e-beam or thermal evaporation. The GFET biosensors are generally characterized by measuring the drain–source current as a function of gate voltage at fixed bias voltage. The gate voltage can be swept either in the forward and/or reverse direction. When the gate voltage is swept in both the forward and reverse direction, the drain–source current for the reverse sweep can be shifted relative to the drain–source current for the forward sweep due to hysteresis. Since Dirac-voltage shifts are mainly used to determine the sensitivity of GFET based biosensors, GFETs with a lower value or almost no hysteresis are required for a high sensitivity device. The hysteresis of the GFETs depends on many factors, including the gate electrode materials [61]. The GFETs with Ag/AgCl gate electrodes can provide a lower hysteresis compared to the GFETs with other gate electrodes such as Au and Pt. In addition, Ag/AgCl gate electrode provides a lower gate leakage current compared than that for the other electrode materials [61].

17.4.3 Graphene Functionalization

The working principle for FET-based biosensors is based on detecting the change of conductance of the transducing material upon exposure to the target biomolecules. The simple and ideal sensing strategy is the direct interaction of transducing material and target. However, this simple sensing strategy is not available for graphene-based sensors because graphene is chemically inert due to its stable planar sp^2 hybridization and thus free of dangling bonds. To achieve the selectivity and sensitivity of graphene-based sensors, the graphene surface needs to be functionalized with the selective BREs such as various inorganic groups, organic molecules, DNA, proteins, peptides, and nanoparticles [3, 4, 62]. These molecules can respond chemically and physically to the target and that response can be transduced into conductance changes of the graphene.

Many methods for the functionalization of graphene have been demonstrated, including covalent functionalization and noncovalent functionalization (Figure 17.5a) [3, 4, 63]. In covalent functionalization the sp^2 bonding of the graphene surface is modified with various functional biochemical molecules. Although covalent functionalization of graphene is very stable, it can reduce the local flatness of the graphene and affect the electron transport properties due to disruptions of the graphene π-conjugation. In contrast to covalent functionalization, nonconvent functionalization has some significant advantages. In the noncovalent method, graphene can be functionalized through π–π interaction without disrupting the graphene lattice, which preserves the graphene's intrinsic properties, including surface flatness and electrical transport properties. Because of these advantages and graphene's extensive π bonding, graphene is commonly functionalized by anchoring selective biochemical molecules onto its surface based on their corresponding noncovalent intermolecular interactions, including π–π, electrostatic, and van der Waals interactions [4].

Graphene can easily be functionalized by drop-casting the solution of the selective biochemical molecule on the graphene surface or incubating the graphene sensors in a solution containing the BRE molecules. For example, our group functionalized the graphene surface of the GFET biosensors by drop casting a small amount (20 μl) of P1N3 peptide solution on the graphene surface, where P1N3 is a 27-amino acid peptide BRE, which has been demonstrated to selectively detect the molecular biomarker NPY [5]. This particular peptide contained a 12 AA graphene-binding domain coupled to a 12 AA NPY-binding domain via a triglycine linker. In the case that a BRE cannot form a strong interaction with the graphene surface, the graphene surface can be noncovalently functionalized with a linker that can π–π bond with the graphene and then couple to the BRE (Figure 17.5b). Due to its simplicity and effectiveness, the noncovalent functionalization method is commonly used in fabricating graphene-based biosensors.

Functionalization of graphene is typically confirmed by characterizing the graphene before and after functionalization using various electrical and surface characterization tools, including XPS, AFM, scanning electron microscopy (SEM), and Raman spectroscopy (Figures 17.4f and 17.6a,b). To confirm that functionalized biomolecules selectively probe the target biomolecules, various control experiments can be performed. For example, Islam et al. functionalized one batch of devices with P1N3 (a BRE selective for NPY) and another batch of devices with P1 (a BRE expected to be non-selective for NPY) as negative control devices (Figure 17.6c,d) [5]. The device functionalized with P1N3 showed a large response (in terms of a large change in the Dirac voltage with any NPY concentration), whereas the device functionalized with P1 (control devices) showed a negligible response. An additional set of control experiments to validate selectivity of the device exposed the sensor to nontarget

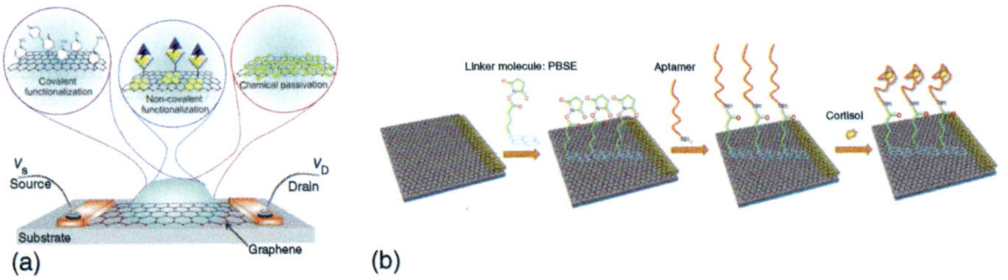

Figure 17.5 (a) Schematic of graphene surface functionalization methods for biochemical sensing. Source: Reproduced with permission from Zhang et al. [3]. (b) Process flow with functionalization steps of graphene surface using PBSE (1-pyrenebutyric acid *N*-hydroxy succinimide ester) molecule and aptamer. The pyrene group, one end of the PBSE, binds with graphene via $\pi-\pi$ interaction whereas the ester group, which is the other side of the PBSE, interacts with aptamer. Source: Adapted from Sheibani et al. [29].

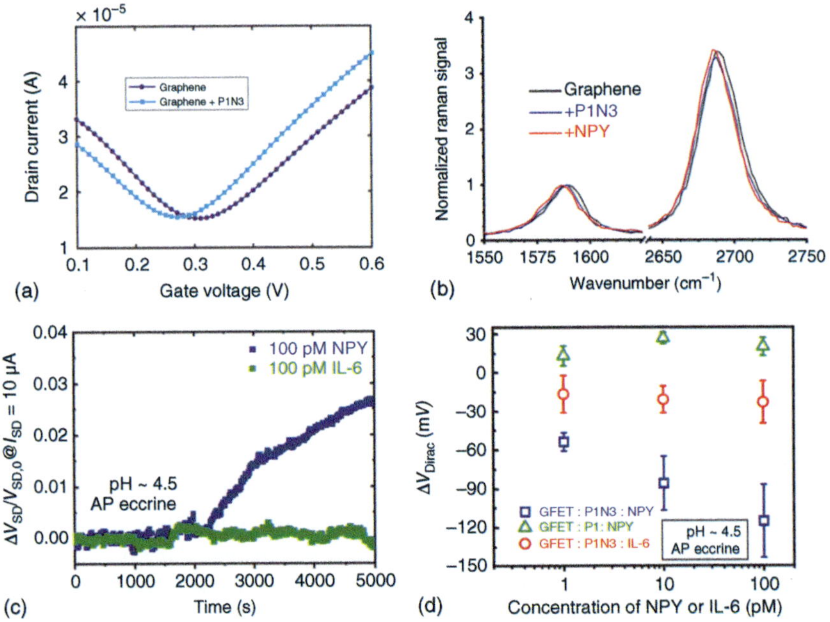

Figure 17.6 (a) Transfer characteristic curves of the GFET before and after functionalization of the graphene using P1N3 peptide. Source: Biddut K. Sarker. (b) Raman spectra of bare graphene (before functionalization) and the graphene after functionalization sequentially with P1N3 and NPY. (c) Response from a P1N3 functionalized GFET sensor tested using artificial eccrine perspiration (AP eccrine, pH ~ 4.5). The P1N3 functionalized GFET sensor shows the response to NPY after the 2000s whereas no response to IL-6 is observed, indicating a good selectivity of the developed sensor is a complex medium like artificial perspiration. (d) Change in V_{Dirac} of the GFET device as a function of the concentration of NPY or IL-6. The GFETs are functionalized with P1N3 or P1 and then sensor response was tested for different concentrations of NPY or IL-6. Here, tests of the P1 functionalized device to NPY response, and P1N3 functionalized device to IL-6 response is the negative control experiments. The response of P1N3 functionalized devices toward NPY is notably steeper than the responses of both negative controls. In these experiments, artificial eccrine perspiration is used as an electrolyte. Source: Adapted from Islam et al. [5].

biomarkers of a similar molecular type. In the case of Islam et al., the response of the BRE specific for NPY, P1N3 functionalized GFET sensors, to the protein IL-6 to showed an insignificant response compared to the sensor's response to NPY. This indicates good selectivity of the P1N3 functionalized GFETs to the target NPY.

17.5 GFET-Based Stress and Fatigue Biosensors

With the goal of real-time biomarker detection and correlation with physiological condition, our research group has recently performed a comprehensive study of GFET-based biosensors for detecting NPY in physiologically relevant environments (Figure 17.7) [5]. The devices were fabricated by transferring CVD graphene onto

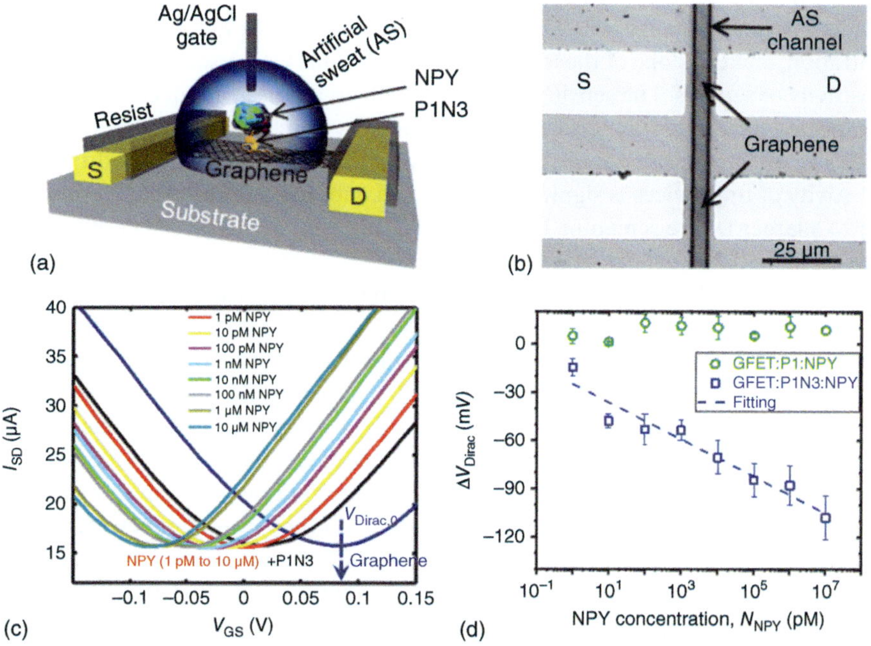

Figure 17.7 (a) Schematic of a liquid electrolyte-gated GFET biosensor to detect Neuropeptide Y (NPY) biomarker. The graphene surface of the GFET is functionalized with a P1N3 (a peptide-based BRE) to selectively detect the NPY biomarkers. A bias voltage is applied between the source (S) and drain (D) electrodes, and a gate voltage is applied to the reference electrode (Ag/AgCl wire). An artificial sweat (AS) is used as the electrolyte. (b) Higher magnification optical image showing AS channel, and two representative GFET sensors. No electrolyte and the gate electrode are not shown in this image. (c) Transfer characteristics (I_{DS} vs. V_G) of a GFET before and after functionalization of graphene with P1N3 and different concentrations of NPYs. Transfer curves and hence the Dirac voltages are shifted negatively with increasing the NPY concentrations. (d) Semi-log plot of Dirac voltage shift as a function of NPY concentrations for two GFETs. One GFET is functionalized using P1N3 peptide and another one is functionalized with P1 peptide for control experiment. The salt (NaCl) concentration in the electrolyte is 50 mM. The slope ΔV_{Dirac} vs. NPY plot provides the sensitivity of the GFETs. Source: Reproduced with permission from Islam et al. [5]/American Chemical Society.

silicon substrates, patterning the graphene channel area using photolithography and RIE, followed by coating the source and drain contacts using photolithographically patterned photoresist. The graphene surface of the devices was functionalized with a peptide-based biorecognition element to selectively detect NPY, and the devices were characterized electrically (measured I_{SD} vs. V_{GS} characteristics) by varying concentrations of NPY in a complex medium containing artificial sweat with varying ionic concentrations (NaCl: 0–100 mM) and pH (4.5–8.0).

The Dirac voltage (V_{Dirac}) was determined for the transfer curves measured after each stage of graphene functionalization with the BRE (denoted P1N3) and NPY. When P1N3 and NPY are introduced on the graphene surface, the Dirac voltage of the device is shifted from the right to the left direction, suggesting that the binding of NPY to P1N3 introduces n-type doping to graphene. No change is observed in the transconductance, which is determined by the slope of the linear part of the transfer curve, indicating that the charge carrier's (electron and hole) mobility does not change. The Dirac voltages are plotted as a function of NPY concentration, and the slope of the log-linear fit to the data (V_o) is used to determine the sensitivity of sensors. The sensitivity reduces with an increase in salt concentration, as expected for a potentiometric sensor. The devices were tested with different artificial sweats having salt concentrations of 0, 50, and 100 mM. It is found that the sensitivity of the devices is significantly reduced with increasing salt concentration due to a larger Debye screening length. We will discuss the Debye screening length issue in more detail in Section 17.7.

In addition to NPY, cortisol is another known biomarker that is correlated with stress. Sheibani et al. developed a GFET-based biosensor for label-free detection of the cortisol stress hormone (Figure 17.8a) [29]. For this sensor, the graphene surface was first functionalized with PBSE (1-pyrenebutyric acid *N*-hydroxy succinimide ester) and then functionalized with a single-stranded nucleic acid aptamer to selectively detect the cortisol. The device was designed as an extended gate FET where the functionalized graphene with aptamer is used as a sensing element, and a commercial metal–oxide–semiconductor field-effect transistor (MOSFET) was used as the transducer. The MOSFET and graphene were physically separated from each other, but the metal underneath the functionalized graphene was electrically connected to the gate terminal of the MOSFET. The device showed a voltage sensitivity of the order of 14 mV/decade, a current sensitivity up to 80% over the four decades of cortisol concentration, and a detection limit of 0.2 nM. Moreover, these sensors were hysteresis-free and showed excellent selectivity toward cortisol.

Hao Z. et al. developed a GFET based nanosensing system for highly sensitive and specific online detection of IL-6 (Figure 17.8b) [24]. In this novel buried-gate geometry where the GFET employs a planar gate electrode (defined on the substrate and buried by a 30 nm HfO_2 dielectric layer) and the graphene conducting channel connects the drain and source electrodes. In contrast to the majority of studies that measure changes in the Dirac voltage, in this study the authors measured I_{DS}–V_G at

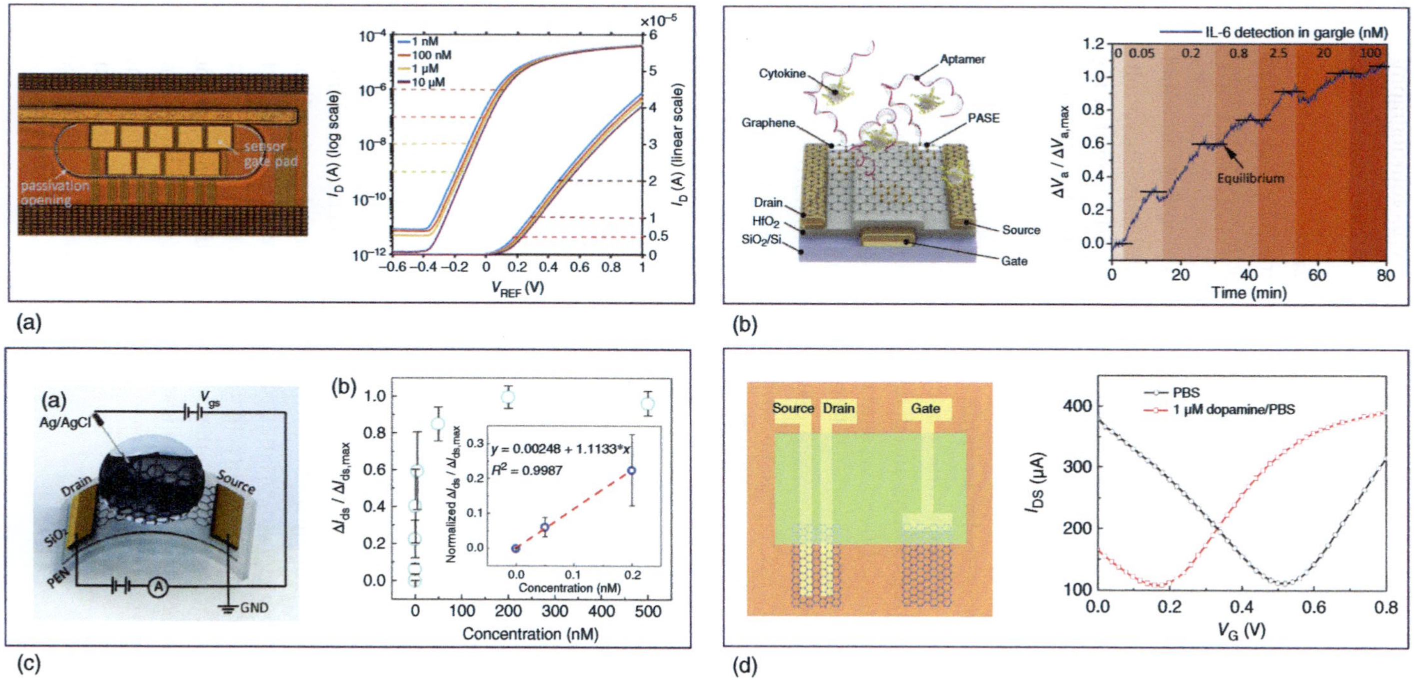

Figure 17.8 (a, left) Optical image of the extended gate CMOS-based cortisol sensor. The CMOS silicon chip is connected to an extended gate graphene electrode. (a, right) Transfer characteristics of this sensor for the different cortisol concentrations. Source: Reproduced from Sheibani et al. [29], Springer Nature/CC BY 4.0. (b, left) Schematic of the GFET based biosensor for cytokine (IL-6) detection. The graphene is functionalized with aptamer and buried-gate geometry is used. (b, right) Time-dependent measurements of changes in the IL-6 concentration using a gargle solution showing the ratio of ΔV_a and $\Delta V_{a,max}$ increases stepwise and sharply with increasing IL-6 concentrations in saliva. Source: Reproduced with permission from Hao et al. [24]/Elsevier. (c, left) Schematic of liquid-gated flexible GFET sensors for detection of tumor necrosis factor-α (TNF-α). (c, right) The ratio of the I_{DS} and $I_{DS,max}$ as a function of the TNF-α concentration. Inset: a linear fit of the I_{DS} vs. TNF-α concentration from 0 to 200 pM. Source: Reproduced with permission from Hao et al. [40]/Royal Society of Chemistry. (d, left) Schematic diagram of a dopamine sensor based on solution-gated GFET with graphene channel as well as a graphene gate electrode. (d, right) Transfer characteristics of a GFET measured before and after the addition of dopamine in the phosphate buffered saline (PBS) solution. Source: Reproduced with permission from Zhang et al. [64]/John Wiley & Sons.

a fixed V_G and the change in the I_{DS} of the original signal is used to determine the concentration of the IL-6. The authors demonstrated a response to the change of IL-6 concentration within 400 seconds in saliva with a detection limit down to 12 pM, which is within the physiological range. Also, by examining the GFET biosensors with a few negative controls such as GH and EGF, the selectivity of biosensors to IL-6 was confirmed. Huang et al. also measured I_{DS}–V_G and then they plotted I_{DS} as a function of the concentration of the IL-6 [39], showing I_{DS} decreased linearly as a function of IL-6 concentration.

Like IL-6, TNF-α is another inflammatory cytokine biomarker and it is closely related to stress, trauma, infections, fever, Crohn's disease, and pulmonary tuberculosis (Figure 17.8c) [40]. The detection of TNF-α concentration in the human body is highly desirable for screening, early clinical diagnosis, and long-time management of patients. Hao et al. demonstrated an aptamer-functionalized GFET sensor for detecting cytokine biomarkers, particularly TNF-α [40]. The GFETs were fabricated on the flexible polymer polyethylene naphthalate (PEN) substrate. The graphene was functionalized with aptamers which upon binding to the target change the charge carrier conduction in graphene and determine the cytokine biomarker concentration. The detection limit of these sensors was as low as 26 pM and could detect the changes in the TNF-α concentration within five minutes.

Dopamine is another important neurotransmitter that plays an important role in the function of the central nervous, hormonal, and cardiovascular systems [64]. Zhang et al. developed high-performance GFET-based dopamine sensors where both the channel and gate electrodes of the GFET were fabricated using graphene (Figure 17.8d). The device is designed as a planar gate electrode configuration. The source, drain, and gate electrodes were deposited on glass substrates, one graphene sheet was transferred between source and drain contacts, and another graphene sheet was transferred onto the gate electrode. The graphene contacted with the gate electrode was functionalized with a thin Nafion film by solution process to achieve selectivity. When the device was immersed into a solution with biomolecules, a gate voltage was applied to the gate electrode which modulated the charge carrier in the graphene between the source and drain electrodes. These GFET devices showed high sensitivity and a low detection limit down to 1 nM for dopamine biomarkers.

We have discussed a few selective GFET-based biosensors for detecting stress biomarkers. In Table 17.1, we have summarized the dynamic range and detection limit of a few selected GFET biosensors for important stress and fatigue biomarkers. It is important to note that it is often difficult to directly compare different GFET biosensor performance because device geometry, measurement conditions, graphene quality, and surface conditions after functionalization, may vary, and all of this information is not always reported.

Table 17.1 Selected examples of the state-of-the-art performance of the GFET-based biosensors for detecting stress biomarkers.

Target	Bioreceptor	Analyte/salt concentration	Target concentration	Dynamic range	Detection limit	Sensitivity	Flexible/ wearable	References
NPY	Peptide	(0–100 mM) × NaCl	1 pM to 10 μM	1×10^7	NR	$17\,mV\,dec^{-1}$	No	[5]
Dopamine	Nafion	PBS	1 nM to 1 μM	1×10^3	1 nM	NR	No	[64]
Dopamine	Aptamer	NR	2.5 aM to 2.5 μM	1×10^{12}	25 aM	NR	No	[65]
Cortisol	PBSE + aptamer	0.05 × PBS	1 nM to 10 μM	1×10^4	0.2 nM	$14\,mV\,dec^{-1}$	Wearable	[29]
IL-6	PTDA	0.01 × PBS	1 pM to 10 nM	1×10^4	100 pM	NR	No	[61]
IL-6	PASE + aptamer	1 × PBS	0–1000 nM	1×10^3	10.5 pM	NR	Portable	[24]
		Gargle solution	0.05–100 nM	1×10^2	11.8 pM	NR	Portable	
		Human saliva	0.05–100 nM	1×10^2	12.2 pM	NR	Portable	
TNF-α	PASE + aptamer + tween 20	NR	0 pM to 100 nM	1×10^5	5 pM	NR	Flexible	[18]
TNF-α	Aptamer	1 × PBS buffer	0–200 pM	2×10^2	26 pM	NR	Flexible	[40]
TNF-α	PASE + aptamer + tween 80	Artificial tears	0 pM to 500 nM	5×10^2	2.75 pM	NR	Wearable	[66]
IFN-γ	PASE + aptamer + tween 80	Artificial tears	0 pM to 500 nM	5×10^2	2.89 pM	NR	Wearable	[66]
Cytokines	Nafion	1 × PBS	0.015–250 nM	3×10^2	740 fM	NR	Wearable	[19]

PBS, phosphate buffered saline; PBSE, 1-pyrenebutyric acid *N*-hydroxysuccinimide ester; PASE, 1-pyrenebutanoic acid succinimidyl ester; PTDA, pyrene-tagged DNA aptamer; NR, not reported.

17.6 Flexible, Wearable GFET Biosensors, and Biosensor Systems

Wearable bioelectronics have received significant attention over the past decade owing to their potential to obtain clinical quality data informing patient conditions and to provide treatment toward personalized medicine in a normal life setting outside of traditional medical facilities. Moreover, the detection of biomarkers using wearable sensors can occur in a real-time, continuous, and rapid manner, which allows collecting large-scale information about the patient's dynamic health status from the physiological to the molecular level. Therefore, wearable bioelectronics are considered an attractive alternative to the typical bulky, expensive, and time-consuming analytical instruments used in medical and healthcare facilities.

Material flexibility is critical for the realization of wearable bioelectronics, and graphene, owing to its inherent properties such as its 2D structure, low density, and conformability, has been considered as a key component for developing stretchable, flexible biosensors (Figures 17.9 and 17.10). Several studies have been conducted on graphene-based flexible biosensors, where graphene is used as a channel or sensing

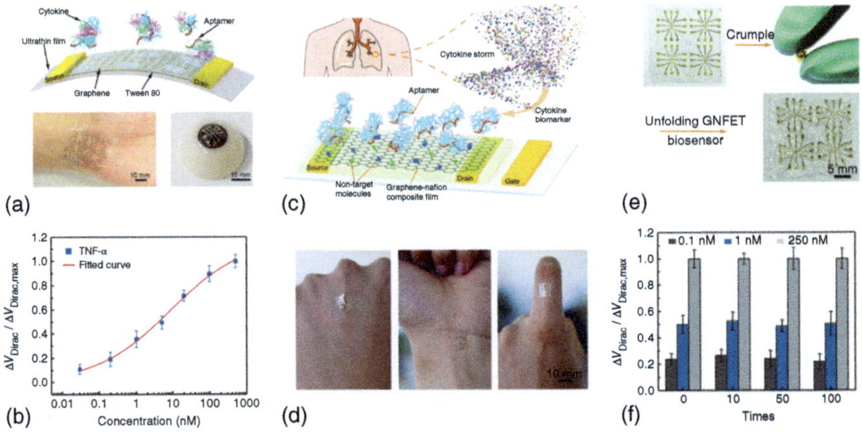

Figure 17.9 (a) Schematic of the aptameric GFET biosensor fabricated on a biocompatible and ultrathin polymer-supported substrate. Optical images show the flexible GFET device conformably attached to the human wrist and artificial eyeball. (b) The normalized Dirac voltage shift as a function of the TNF-α concentration. Source: Reproduced from Wang et al. [66], with permission from MDPI/Public Domain CC BY 4.0. (c) Schematic of the graphene nafion field-effect transistor (GN-FET) sensor for cytokine biomarker detection. (d) Optical image of the flexible biosensors mounted on a human hand. (e) Optical image of the biosensor before and after the crumping process. (f) Normalized Dirac voltage shift of the sensor for various IFN-γ concentrations at different crumpling cycles. Source: Reproduced with permission from Wang et al. [19]/John Wiley & Sons.

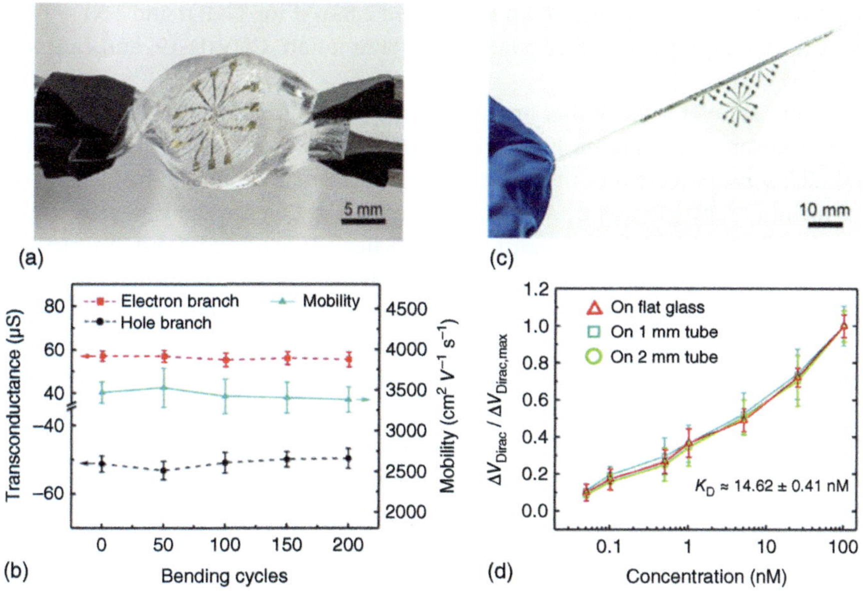

Figure 17.10 (a) Twisting tests of the flexible aptameric GFET nanosensors. (b) Transconductance and carrier mobility of the GFET nanosensors as a function of bending cycles. (c) Bending tests of the flexible aptameric GFET nanosensors. (d) The normalized Dirac voltage shift as a function of the TNF-α concentration for the nanosensors rolled on the tube with radii 1 and 2 mm. Source: Reproduced with permission from Wang et al. [18]/ John Wiley & Sons.

material and a thin polymer film (e.g. PEN) is used as a substrate [40]. In these studies, the thickness of polymer substrates is 100 μm or more which may not provide a large deformation and large curvature. In order to increase the degree of bending deformation towards a larger film curvature, Wang and coworkers employed a biocompatible and ultrathin, polymeric substrate for a flexible and stretchable GFET biosensor (Figure 17.9a,b) [66]. To verify the stretchability, and flexibility, these sensors were tested by rolling on cylindrical surfaces with radii from 2 mm down to 40 μm, by twisting ranging −180° to 180°, and stretching up to 125% [18, 66]. It was found that these sensors were capable of high conformality to nonplanar surfaces including human skin and eyes, have high mechanical flexibility, and can withstand larger deformations including bending and stretching (Figure 17.9a,b). While the device was under large deformation, electrical measurements demonstrated that the transconductance and mobility of the devices did not vary significantly for any of the bending and rolling conditions described above (Figure 17.10) [18]. In addition to the mechanical flexibility and robust electrical properties of the device components, the sensors exhibited excellent sensitivity for cytokine detection in artificial tears, with a limit of detection (LOD) down to 2.75 and 2.89 pM for TNF-α and TNF-β biomarkers (inflammatory cytokines related to inflammation, COVID-19, and cancer), respectively [66].

In another study, the flexibility and conformity of aptameric Graphene–Nafion-based biosensors for cytokine biomarker monitoring were tested (Figure 17.9c–f) [19]. The sensors were mounted on an artificial human hand and the sensing of IFN-γ in human sweat was evaluated. The sensors showed a consistent sensing response both before and after mounting on a flexible artificial hand. To further test the flexibility, the ultrathin biosensor foil was crumpled into a small ball, flattened out, and retested. These biosensors withstood the crumping test and only a small change (less than 3.6%) of the sensor response to the IFN-γ biomarker were observed, exhibiting only a small shift (28 mV) in the Dirac voltage after 100 crumpling cycles.

Although graphene-based flexible biosensors with high sensitivity have been demonstrated by a several research groups, incorporating the GFET biosensor into portable or wearable bioelectronics/devices remains challenging for several key reasons. One of the key issues which hinders the integration of GFETs in wearable bioelectronics is the external reference/gate electrode (typically Ag/AgCl wire) used to operate most liquid gated GFETs. To overcome the limitation of the external reference/gate electrode, a back-gated GFET can be fabricated on a Si/SiO_2 substrate with the native SiO_2 as a dielectric layer and the gate voltage applied to the back the Si substrate. Besides the drawback of the rigidity of the Si wafers, this approach requires a large gate voltage (>30 V) to operate the back-gated GFETs, and this large voltage is not safe for on-body use. In addition to the external gate electrodes, bulky semiconductor measurement instruments are required to supply a small bias voltage (typically less than 100 mV) to the GFET biosensors as well as collect data from the biosensors. These bulky instruments impede the realization of GFET-based wearable or portable biosensor systems.

To address these issues, Zhuang et al. reported a GFET-based fully integrated and portable nanosensing system (Figure 17.11) [24], the size of which was smaller than a smartphone, could be held in one hand, and was capable of on-line detection of IL-6 biomarkers with high sensitivity and specificity. A novel buried-gate geometry with HfO_2 as the dielectric layer was employed for fabricating aptameric GFET-based biosensors. Advantages of the HfO_2 dielectric layer include that no external gate electrode is needed because the HfO_2 dielectric layer also serves as a gate, and the devices can be operated with a lower operating gate voltage (<2 V) than GFETs with a SiO_2 dielectric layer because the dielectric constant of HfO_2 (~16) is higher than the dielectric constant of SiO_2 (~3.9).

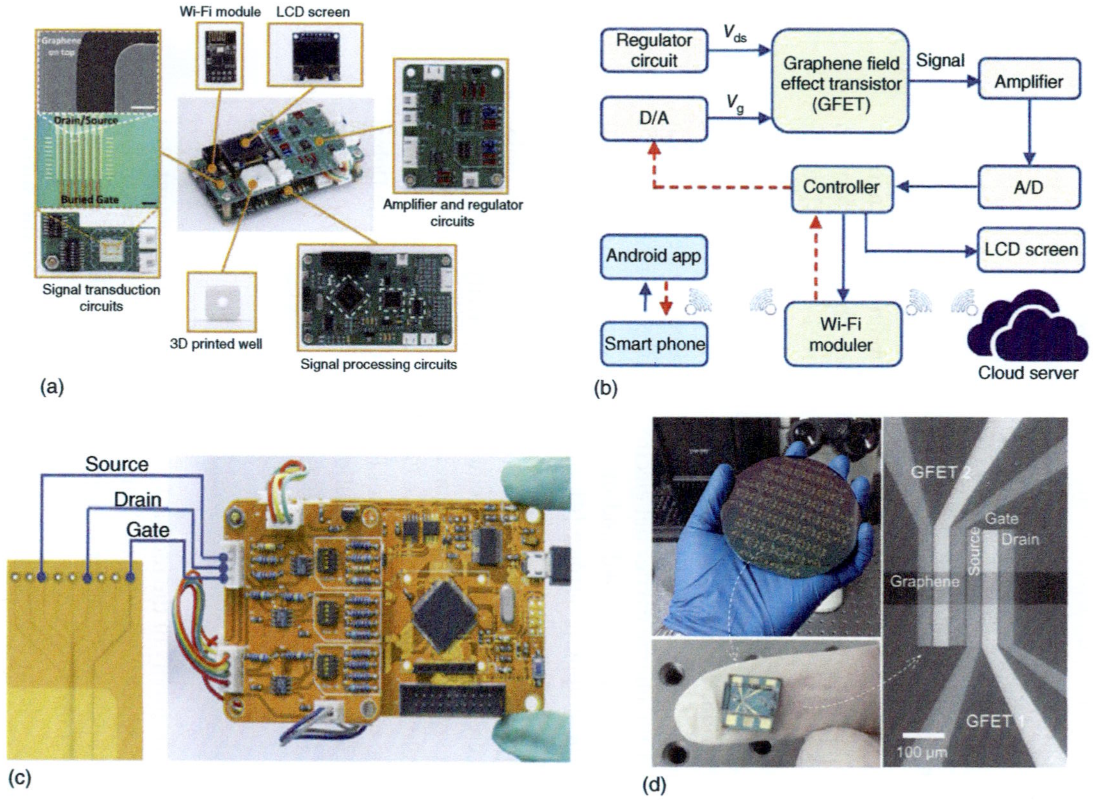

Figure 17.11 (a) Photographs of the flexible GFET-based biosensor with the customized portable and flexible measurement system. The PBA molecules are used to functionalize the graphene surface to selectively detect the biomolecules and GFETs are operated using a solution gate configuration. The source, drain, and gate contacts of GFETs are connected to the electronic measurement system to operate the sensor and collect the signal data. (b) Block diagram of GFET-based portable biosensor system. The gate voltage and source–drain voltage are applied to GFET from the regular circuit and digital to analog (D/A) converter, whereas the output signal of the biosensor system can be transmitted to smartphones and could sever. Source: Reproduced with permission from Hao et al. [24]. (c) Optical images of nanosensing systems showing the difference apart and components of the nanosensing system including GFET, LCD monitors, etc. Source: Reproduced with permission from Huang et al. [20]/Elsevier. (d) Silicon wafer containing a large number of GFET biosensors. Devices were fabricated based on a high-**κ** solid-gate GFET configuration. The conductivity of the graphene channel is modulated by applying gate voltage at the planar gate electrode. PDMS microfluidic channels were used to pack the devices for liquid handling. Source: Reproduced with permission from Li et al. [67]/Elsevier.

17.7 Current Challenges and Future Perspective

Despite the significant progress and advances that have occurred in many key areas related to the design and application of GFET-based biosensors by research groups in laboratory settings, a strong commercial case for their practical application is yet to be made. Before implementing GFET-based biosensors in analytical applications, numerous challenges need to be addressed. In this section, we will discuss some of the major challenges including Debye screening length, device-to-device variability, and device reusability.

17.7.1 Debye Length Screening

Sensing biomarkers in physiological solutions with GFETs remains a challenge due to the presence of mobile ions present in the solution [68–72]. These ions form an EDL capacitance at the graphene-liquid interface (Figure 17.12a–c) [65, 68] which can substantially change the charge carrier concentrations in the graphene channel. The charge from the biomarker in the physiological solution also suffers from ionic screening due to mobile conductive ions in solution, known as the Debye screening effect [68, 70, 73]. The Debye screening effect is characterized by Debye length which measures how far the charge carrier net electrostatic effect persists. Outside of the Debye length charges are screened and only 36.7% of the charges remain effectively apparent at the graphene surface [4]. If the distance between the captured target biomolecules and graphene surface is larger than the Debye length, charge transport of the graphene is not modulated by the charge from the biomolecules due to the Debye screening effect, which reduces the sensitivity of the sensor (Figure 17.12d–f). Therefore, it is very critical that the sensing event needs to have happened within the Debye length of the device channel. The strength of mobile ions in the solution is the main determining factor for the Debye length. The Debye length can be estimated using the following equation [74]

$$\lambda_D = \sqrt{\frac{\varepsilon k_B T}{q^2 c}} \tag{17.5}$$

where ε the dielectric permittivity of the media, k_B is the Boltzmann's constant, T is the temperature, q is the electron charge, and c are the ionic strength of electrolyte solution. Equation (17.5) indicates that the higher ionic strength solution creates a shorter Debye length and vice versa. The typical Debye length for 100 mM buffer solution (physiologically relevant ionic strength) is around 1 nm, and the surface potential is completely screened at a distance of a few nanometers, which is shorter than the size of a typical biomolecule [74]. For example, the size of the antibody is around 10–15 nm and the size of a 30-base aptamer is nearly 10 nm. This intrinsic mismatch of the Debye length and size of biomolecules is a fundamental physical limitation, which prevents many GFETs or other nanomaterials based biosensor platforms from becoming broadly useful [68].

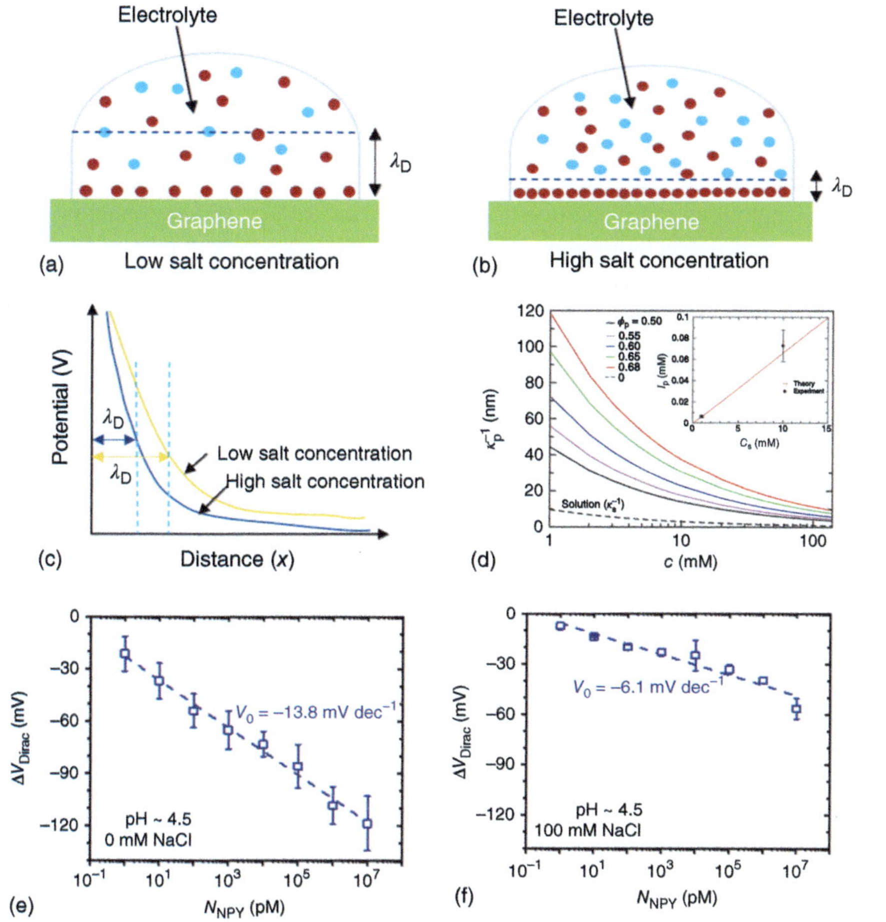

Figure 17.12 The schematic of electrical double-layer formation at the graphene–liquid interface of a liquid gate GFET biosensor for a biomolecule solution with (a) low salt concentration and (b) high-salt concentration. (c) The plot of the electrostatic potential as function distance along the vertical direction to the graphene channel surface shows a decrease of the electrostatic potential is faster for a solution with high salt concentration. Source: Biddut K. Sarker. (d) Theoretically calculated Debye length as a function of the salt (KCl) concentration in the solution for different polymer volume fractions. Source: Reproduced with permission from Piccinini et al. [71]. Semi-log plot of Dirac voltage shift of GFET sensors as a function of NPY concentration with (e) 0 mM NaCl and (f) 100 mM NaCl, showing the decrease of sensor's sensitivity for the NPY solution with high salt concentration.

To overcome the Debye screening effect a number of techniques have been demonstrated, including use of coated graphene with biomolecule permeable polymer, use of deformed graphene, operating of sensors in dry state [75], or in a low ionic strength buffer (<10 mM), modifying the BREs chemically, and functionalizing graphene surface with smaller biorecognition elements (e.g. antibody fragments, aptamers, and peptides). Gao et al. demonstrated that the sensitivity of GFET biosensors for detecting biomolecules in high-ionic strength solution can be improved by coating the graphene surface with a porous and biomolecule-permeable polymer layer (Figure 17.13a,b). The authors coated the surface of the GFET with poly(ethylene glycol) (PEG) and showed real-time reversible detection of a prostate-specific antigen (PSA) from 1 to 1000 nM in 100 mM phosphate buffer [77]. The PEG layer was hypothesized to change the local dielectric properties and increase the effective Debye screening length for a given solution near to the device surface, thereby enabling the sensor effectively detect biomolecules in physiological ionic strength buffer solution. Song et al. studied the effect of the molecular weight of the PEG on the sensitivity of the sensor and found that higher molecular weight PEG increases the drain current of the device and hence provides higher sensitivity [78]. PEG coatings were also applied to other FET-based biosensors to overcome the Debye screening length and increase the sensitivity [2, 69, 77]. In one example, a PEG-coated silicon nanowire FET sensor was used to detect PSA in solution with phosphate buffer (PB) concentrations as high as 150 mM, whereas similar devices without PEG modification only can detect signals for concentrations ≤10 mM (Figure 17.13b).

Hwang et al. demonstrated another approach to increase sensor sensitivity in high ionic strength solution by modulating the Debye screening length for GFET-based biosensors [76]. To modulate the Debye screening length, they fabricated crumpled GFET biosensors using a millimeter scale deformed graphene channel (Figure 17.13c,d). Notionally, the traditional flat GFET biosensors show a constant Debye length, whereas the Debye length for crumpled GFET varies at the peaks and valleys of the deformed graphene. As a result, the Debye screening effect becomes less pronounced for the crumpled GFET biosensor and has the overall benefit of improving sensitivity. The crumpled GFET biosensors shown ultrasensitive detection of DNA/RNA molecules down to a LOD of 600 zM. From molecular dynamics simulations, it was found that nanoscale deformities in the graphene channel can create electrical hot spots, which increase the local Debye length, resulting in a local high potential due to the charge of DNA/RNA. The bandgap may also open up at the hotspots of the deformed graphene, which would allow for an exponential increase of conductivity from a small number of biomolecules and hence increase the device sensitivity [65].

The Debye screening effect can be minimized by functionalizing the graphene surface with a small BREs, such as aptamers [29] or peptides. Binding the aptamer brings target biomolecules closer to the graphene surface, reducing the Debye screening length. Sheibani et al. designed such an extended gate FET for sensing cortisol in biological buffers within the Debye screening length [29]. The device showed a voltage sensitivity on the order of 14 mV/decade, a current sensitivity up to 80% over the four decades of cortisol concentration, and a detection limit of 0.2 nM. Overall, although a few different approaches have been demonstrated to overcome the Debye screening length issue, more investigations are needed to clearly understand this issue and to find a comprehensive solution to this problem.

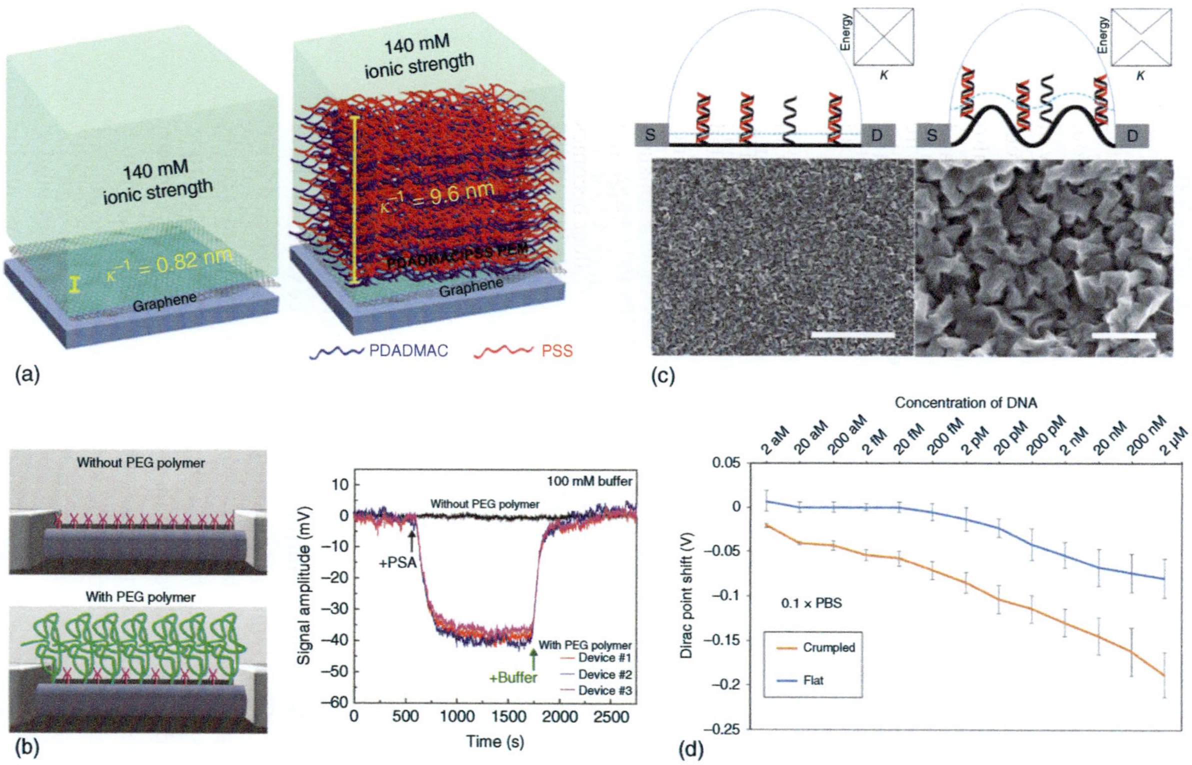

Figure 17.13 (a) Schematic of the Debye screening length at the graphene–liquid interface for a physiological ionic strength without (left) and with (right) a polyelectrolyte multilayer (PEM) film. The Debye screening length is increased by more than 1 order of magnitude due to a PEM film on the graphene surface. Source: Reproduced with permission from Piccinini et al. [71]. (b) Schematic of FET biosensors (top) without and (bottom) with surface modification by a porous and biomolecule permeable polymer (green). The signal amplitude as a function of time of a silicon nanowire FETs without and with surface modification. Source: Reproduced with permission from Gao et al. [69]. (c) The schematic of the GFET biosensors with (top-left) flat graphene and (top-right) crumpled or deformed graphene. The black line represents the BRE, red lines represent the target DNA, and blue dotted lines represent the Debye length. Compared to the Debye length for the flat graphene, the Debye length is increased at the convex region for the crumpled graphene. (Insets) The qualitative energy diagram in K-space shows the bandgap opening for the crumpled graphene. The SEM images of crumpled graphene (bottom). The scale bar is 5 μm (bottom-left) and 500 nm (bottom-right). (d) Dirac voltage shift of the GFET biosensors with a flat and crumpled graphene as a function of DNA concentration. Source: Reproduced from Hwang et al. [76], Springer Nature/CC BY 4.0.

17.7.2 Device-to-Device Variability

Another major challenge for developing high-performance commercial grade GFET biosensors is device-to-device variability. One of the main factors which plays a critical role in the device-to-device variability is the quality of graphene. Graphene obtained by the exfoliation method is considered to be the best quality, almost free of defects and impurities. However, the exfoliation method is a time-consuming, nonscalable, and nonreproducible process. Additionally, exfoliated graphene sheets are typically small and irregularly shaped. Due to these limitations, graphene grown by the CVD method is commonly used for fabricating graphene-based biosensors. In the CVD method, a large area of monolayer graphene is grown on Cu foil, and then transferred to a solid-state substrate. Since the graphene transfer process involves chemical processes, as a result many organic and metallic impurities can be attached to the graphene surface which affect the ultimate performance of the biosensors [54, 56, 57, 79]. Several methods can be used to remove impurities from the graphene surface, including chemical etching.

Additional challenges with CVD graphene include the presence of grain boundaries, ripples, holes, and wrinkles. The CVD growth process may also introduce some variability because slight changes in the growth conditions may change in domain size and defect densities in graphene. In addition, device fabrication methods may also induce many variabilities in the physical and chemical properties of graphene, meaning the quality of graphene is not exactly the same for all the devices even if they are on the same chip, and a slight variability in sensing performance is typically observed from device-to-device variation. Besides the graphene quality, finding the appropriate bioreceptor for the developing GFET biosensors to selectively sense the target biomolecules to enhance the selectivity of GFET biosensors is still a complicated issue due to limited understanding of the interaction of the graphene and bioreceptor. Therefore, more control and optimization are needed in the graphene synthesis, device fabrication, and graphene functionalization processes to develop highly reproducible and commercial-grade GFET-based biosensors.

In addition to the variation in graphene quality and device fabrication variability, biosensor measurement systems can also introduce device-to-device variation. Most GFETs are mostly measured in a liquid-gate configuration where an external reference electrode (e.g. Ag/AgCl thin wire) is immersed in the electrolyte, and it can be difficult to maintain the same distance between graphene and reference electrode. It has been reported that the electric field from the external gate electrodes can vary significantly, and the gate capacitance or capacitance across the EDL may change with the length and diameter of the electrode immersed into the electrolyte [46, 80]. This may result in the performance variation from device-to-device, even with constant GFET device quality.

17.7.3 Short Lifetime and Reusability Issue

Besides the existing challenges with Debye screening length and device-to-device variability, another limiting factor for practical commercial applications of the GFET biosensors is the high cost of device fabrication [81]. Moreover, GFET sensors are not typically reused, as the device sensitivity degrades after sensing biomolecules in physiological solutions with high salt concentration. One promising way to reduce the cost and increase the supply of devices is to regenerate the biosensors using a particular post-sensing treatment or process. When a biosensor is successfully regenerated, it can be reused many times without compromising the sensing performance. Therefore, in addition to the specificity, sensitivity, and LOD, regeneration is another highly desirable property for a biosensor. Since the graphene surface of GFET biosensors is typically functionalized with a BRE to selectively probe the intended target biomolecules, it will be critical that these BREs either are removed and reapplied, or restored to their original function through washing away the target molecules completely and leaving the BREs in place. Wang and coworkers have demonstrated aptameric graphene–nafion biosensors, which can be regenerated up to 80 cycles by immersing the biosensors in ethanol solution after each cycle of the biomarker detection to dissolve the Nafion film and to obtain a brand-new biosensor [19]. Aptamer functionalized GFET-based biosensors can also be regenerated by denaturing the aptamer with guanidinium chloride [77]. One important sensor design consideration is the need to keep in mind that if the sensor has to be regenerated, noncovalent functionalization of the BRE could be done on graphene surface instead of covalent functionalization so that the BREs are easier to remove from the sensor surface [4].

17.8 Conclusion

In this review, we described recent advances in the development of biosensors based on GFET. We discussed the fundamentals of GFET-based biosensors, sensing mechanisms, device fabrication, and functionalization methods. We have reviewed a number of GFET biosensors for monitoring human performance related biomarkers. These GFET biosensors have shown many advantages over other sensing modalities, such as ultrahigh sensitivity, high selectivity, fast response, and large detection limit. We then emphasized the latest development of GFET-based wearable biosensors, fully integrated sensor systems, and highlighted the outstanding advantages including excellent flexibility, stretchability, and conformability. We have also discussed a few major challenges for the practical application of GFET-based biosensors including the Debye screening length, device-to-device variability and device reusability. Finally, in addition to overcoming these challenges, other important factors such as device integration and the biocompatibility of GFETs need to be addressed for bringing GFET-based biosensing platforms closer to reality.

References

1 Pumera, M. (2011). Graphene in biosensing. *Mater. Today* 14 (7–8): 308–315.

2 Zhang, A. and Lieber, C.M. (2016). Nano-bioelectronics. *Chem. Rev.* 116 (1): 215–257.

3 Zhang, X., Jing, Q., Ao, S. et al. (2020). Ultrasensitive field-effect biosensors enabled by the unique electronic properties of graphene. *Small* 16 (15): 1902820.
4 Fu, W., Jiang, L., van Geest, E.P. et al. (2017). Sensing at the surface of graphene field-effect transistors. *Adv. Mater.* 29 (6): 1603610.
5 Islam, A.E., Martineau, R., Crasto, C.M. et al. (2020). Graphene-based electrolyte-gated field-effect transistors for potentiometrically sensing neuropeptide Y in physiologically relevant environments. *ACS Appl. Nano Mater.* 3 (6): 5088–5097.
6 Yan, F., Zhang, M., and Li, J. (2014). Solution-gated graphene transistors for chemical and biological sensors. *Adv. Healthcare Mater.* 3 (3): 313–331.
7 Peña-Bahamonde, J., Nguyen, H.N., Fanourakis, S.K., and Rodrigues, D.F. (2018). Recent advances in graphene-based biosensor technology with applications in life sciences. *J. Nanobiotechnol.* 16 (1): 75.
8 Sadighbayan, D., Hasanzadeh, M., and Ghafar-Zadeh, E. (2020). Biosensing based on field-effect transistors (FET): recent progress and challenges. *TrAC, Trends Anal. Chem.* 133: 116067.
9 Béraud, A., Sauvage, M., Bazán, C.M. et al. (2021). Graphene field-effect transistors as bioanalytical sensors: design, operation and performance. *Analyst* 146 (2): 403–428.
10 Kim, S., Xing, L., Islam, A.E. et al. (2019). In operando observation of neuropeptide capture and release on graphene field-effect transistor biosensors with picomolar sensitivity. *ACS Appl. Mater. Interfaces* 11 (15): 13927–13934.
11 Lowe, B.M., Sun, K., Zeimpekis, I. et al. (2017). Field-effect sensors – from pH sensing to biosensing: sensitivity enhancement using streptavidin–biotin as a model system. *Analyst* 142 (22): 4173–4200.
12 Bonaccorso, F., Sun, Z., Hasan, T., and Ferrari, A.C. (2010). Graphene photonics and optoelectronics. *Nat. Photonics* 4 (9): 611–622.
13 Novoselov, K.S., Geim, A.K., Morozov, S.V. et al. (2004). Electric field effect in atomically thin carbon films. *Science* 306 (5696): 666–669.
14 Geim, A.K. (2009). Graphene: status and prospects. *Science* 324 (5934): 1530–1534.
15 Das Sarma, S., Adam, S., Hwang, E.H., and Rossi, E. (2011). Electronic transport in two-dimensional graphene. *Rev. Mod. Phys.* 83 (2): 407–470.
16 Novoselov, K.S., Jiang, Z., Zhang, Y. et al. (2007). Room-temperature quantum Hall effect in graphene. *Science* 315 (5817): 1379–1379.
17 Novoselov, K.S., Fal'ko, V.I., Colombo, L. et al. (2012). A roadmap for graphene. *Nature* 490 (7419): 192–200.
18 Wang, Z., Hao, Z., Yu, S. et al. (2019). An ultraflexible and stretchable aptameric graphene nanosensor for biomarker detection and monitoring. *Adv. Funct. Mater.* 29 (44): 1905202.
19 Wang, Z., Hao, Z., Wang, X. et al. (2021). A flexible and regenerative aptameric graphene–nafion biosensor for cytokine storm biomarker monitoring in undiluted biofluids toward wearable applications. *Adv. Funct. Mater.* 31 (4): 2005958.
20 Huang, C., Hao, Z., Qi, T. et al. (2020). An integrated flexible and reusable graphene field effect transistor nanosensor for monitoring glucose. *J. Materiomics* 6 (2): 308–314.

21 Yavari, F. and Koratkar, N. (2012). Graphene-based chemical sensors. *J. Phys. Chem. Lett.* 3 (13): 1746–1753.

22 Mao, S. and Chen, J. (2017). Graphene-based electronic biosensors. *J. Mater. Res.* 32 (15): 2954–2965.

23 Hill, E.W., Vijayaragahvan, A., and Novoselov, K. (2011). Graphene sensors. *IEEE Sens. J.* 11 (12): 3161–3170.

24 Hao, Z., Pan, Y., Shao, W. et al. (2019). Graphene-based fully integrated portable nanosensing system for on-line detection of cytokine biomarkers in saliva. *Biosens. Bioelectron.* 134: 16–23.

25 Xu, S., Zhan, J., Man, B. et al. (2017). Real-time reliable determination of binding kinetics of DNA hybridization using a multi-channel graphene biosensor. *Nat. Commun.* 21 (8): 14902.

26 Hao, Z., Zhu, Y., Wang, X. et al. (2017). Real-time monitoring of insulin using a graphene field-effect transistor aptameric nanosensor. *ACS Appl. Mater. Interfaces* 9 (33): 27504–27511.

27 Park, J., Kim, J., Kim, S.-Y. et al. (2018). Soft, smart contact lenses with integrations of wireless circuits, glucose sensors, and displays. *Sci. Adv.* 4 (1): eaap9841.

28 Steckl, A.J. and Ray, P. (2018). Stress biomarkers in biological fluids and their point-of-use detection. *ACS Sens.* 3 (10): 2025–2044.

29 Sheibani, S., Capua, L., Kamaei, S. et al. (2021). Extended gate field-effect-transistor for sensing cortisol stress hormone. *Commun. Mater.* 2 (1): 10.

30 Ray, P. and Steckl, A.J. (2019). Label-free optical detection of multiple biomarkers in sweat, plasma, urine, and saliva. *ACS Sens.* 4 (5): 1346–1357.

31 Heikenfeld, J., Jajack, A., Feldman, B. et al. (2019). Accessing analytes in biofluids for peripheral biochemical monitoring. *Nat. Biotechnol.* 37 (4): 407–419.

32 Schneiderman, N., Ironson, G., and Siegel, S.D. (2005). Stress and health: psychological, behavioral, and biological determinants. *Annu. Rev. Clin. Psychol.* 1 (1): 607–628.

33 Jenz, S.T., Goodyear, C.D., TSgt Graves, P.R. et al. (2021). Blood and affective markers of stress in Elite Airmen during a preparatory training course: a pilot study. *Neurobiol. Stress* 14: 100323.

34 Sajdyk, T.J., Shekhar, A., and Gehlert, D.R. (2004). Interactions between NPY and CRF in the amygdala to regulate emotionality. *Neuropeptides* 38 (4): 225–234.

35 Reichmann, F. and Holzer, P. (2016). Neuropeptide Y: a stressful review. *Neuropeptides* 55: 99–109.

36 Sabban, E.L., Alaluf, L.G., and Serova, L.I. (2016). Potential of neuropeptide Y for preventing or treating post-traumatic stress disorder. *Neuropeptides* 56: 19–24.

37 Cizza, G., Marques, A.H., Eskandari, F. et al. (2008). Elevated neuroimmune biomarkers in sweat patches and plasma of premenopausal women with major depressive disorder in remission: the POWER study. *Biol. Psychiatry* 64 (10): 907–911.

38 Racaityte, K., Lutz, E.S.M., Unger, K.K. et al. (2000). Analysis of neuropeptide Y and its metabolites by high-performance liquid chromatography-electrospray ionization mass spectrometry and integrated sample clean-up with a novel restricted-access sulphonic acid cation exchanger. *J. Chromatogr. A* 890 (1): 135–144.

39 Huang, J., Harvey, J., Fam, W.H.D. et al. (2013). Novel biosensor for interleukin-6 detection. *Procedia Eng.* 60: 195–200.

40 Hao, Z., Wang, Z., Li, Y. et al. (2018). Measurement of cytokine biomarkers using an aptamer-based affinity graphene nanosensor on a flexible substrate toward wearable applications. *Nanoscale* 10 (46): 21681–21688.

41 Avouris, P. (2010). Graphene: electronic and photonic properties and devices. *Nano Lett.* 10 (11): 4285–4294.

42 Avouris, P. and Xia, F. (2012). Graphene applications in electronics and photonics. *MRS Bull.* 37 (12): 1225–1234.

43 Macedo, L.J.A., Iost, R.M., Hassan, A. et al. (2019). Bioelectronics and interfaces using monolayer graphene. *ChemElectroChem* 6 (1): 31–59.

44 Du, X., Skachko, I., Barker, A., and Andrei, E.Y. (2008). Approaching ballistic transport in suspended graphene. *Nat. Nanotechnol.* 3: 491–495.

45 Fang, T., Konar, A., Xing, H., and Jena, D. (2007). Carrier statistics and quantum capacitance of graphene sheets and ribbons. *Appl. Phys. Lett.* 91 (9): 092109.

46 Wang, C., Li, Y., Zhu, Y. et al. (2016). High-κ solid-gate transistor configured graphene biosensor with fully integrated structure and enhanced sensitivity. *Adv. Funct. Mater.* 26 (42): 7668–7678.

47 Mackin, C., McVay, E., and Palacios, T. (2018). Frequency response of graphene electrolyte-gated field-effect transistors. *Sensors* 18 (2): 494.

48 Fan, Q., Wang, L., Xu, D. et al. (2020). Solution-gated transistors of two-dimensional materials for chemical and biological sensors: status and challenges. *Nanoscale* 12 (21): 11364–11394.

49 Lin, C.-T., Loan, P.T.K., Chen, T.-Y. et al. (2013). Label-free electrical detection of DNA hybridization on graphene using Hall effect measurements: revisiting the sensing mechanism. *Adv. Funct. Mater.* 23 (18): 2301–2307.

50 Mohanty, N. and Berry, V. (2008). Graphene-based single-bacterium resolution biodevice and DNA transistor: interfacing graphene derivatives with nanoscale and microscale biocomponents. *Nano Lett.* 8 (12): 4469–4476.

51 Farid, S., Meshik, X., Choi, M. et al. (2015). Detection of interferon gamma using graphene and aptamer based FET-like electrochemical biosensor. *Biosens. Bioelectron.* 71: 294–299.

52 Chen, Y., Michael, Z.P., Kotchey, G.P. et al. (2014). Electronic detection of bacteria using holey reduced graphene oxide. *ACS Appl. Mater. Interfaces* 6 (6): 3805–3810.

53 Sun, H., Chen, D., Wu, Y. et al. (2017). High quality graphene films with a clean surface prepared by an UV/ozone assisted transfer process. *J. Mater. Chem. C* 5 (8): 1880–1884.

54 Mattevi, C., Kim, H., and Chhowalla, M. (2011). A review of chemical vapour deposition of graphene on copper. *J. Mater. Chem.* 21 (10): 3324–3334.

55 Li, X., Cai, W., An, J. et al. (2009). Large-area synthesis of high-quality and uniform graphene films on copper foils. *Science* 324 (5932): 1312–1314.

56 Liang, X., Sperling, B.A., Calizo, I. et al. (2011). Toward clean and crackless transfer of graphene. *ACS Nano* 5 (11): 9144–9153.

57 Lin, L., Deng, B., Sun, J. et al. (2018). Bridging the gap between reality and ideal in chemical vapor deposition growth of graphene. *Chem. Rev.* 118 (18): 9281–9343.

58 Zhou, H., Yu, W.J., Liu, L. et al. (2013). Chemical vapour deposition growth of large single crystals of monolayer and bilayer graphene. *Nat. Commun.* 4 (1): 2096.

59 Pirkle, A., Chan, J., Venugopal, A. et al. (2011). The effect of chemical residues on the physical and electrical properties of chemical vapor deposited graphene transferred to SiO_2. *Appl. Phys. Lett.* 99 (12): 122108.

60 Kaymak, N., Bayram, O., Tataroğlu, A. et al. (2020). Electrical properties of graphene/silicon structure with Al_2O_3 interlayer. *J. Mater. Sci. – Mater. Electron.* 31 (12): 9719–9725.

61 Khan, N.I. and Song, E. (2021). Detection of an IL-6 biomarker using a GFET platform developed with a facile organic solvent-free aptamer immobilization approach. *Sensors* 21 (4): 1335.

62 Mukherjee, S., Meshik, X., Choi, M. et al. (2015). A graphene and aptamer based liquid gated FET-like electrochemical biosensor to detect adenosine triphosphate. *IEEE Trans. Nanobiosci.* 14 (8): 967–972.

63 Saini, D. (2016). Synthesis and functionalization of graphene and application in electrochemical biosensing. *Nanotechnol. Rev.* 5 (4): 393–416.

64 Zhang, M., Liao, C., Yao, Y. et al. (2014). High-performance dopamine sensors based on whole-graphene solution-gated transistors. *Adv. Funct. Mater.* 24 (7): 978–985.

65 Hwang, M.T., Park, I., Heiranian, M. et al. (2021). Ultrasensitive detection of dopamine, IL-6 and SARS-CoV-2 proteins on crumpled graphene FET biosensor. *Adv. Mater. Technol.* 6 (11): 2100712.

66 Wang, Z., Hao, Z., Yu, S. et al. (2020). A wearable and deformable graphene-based affinity nanosensor for monitoring of cytokines in biofluids. *Nanomaterials* 10 (8): 1503.

67 Li, Y., Wang, C., Zhu, Y. et al. (2017). Fully integrated graphene electronic biosensor for label-free detection of lead (II) ion based on G-quadruplex structure-switching. *Biosens. Bioelectron.* 89: 758–763.

68 Kesler, V., Murmann, B., and Soh, H.T. (2020). Going beyond the Debye length: overcoming charge screening limitations in next-generation bioelectronic sensors. *ACS Nano* 14 (12): 16194–16201.

69 Gao, N., Zhou, W., Jiang, X. et al. (2015). General strategy for biodetection in high ionic strength solutions using transistor-based nanoelectronic sensors. *Nano Lett.* 15 (3): 2143–2148.

70 Zheng, Z., Zhang, H., Zhai, T., and Xia, F. (2021). Overcome Debye length limitations for biomolecule sensing based on field effective transistors. *Chin. J. Chem.* 39 (4): 999–1008.

71 Piccinini, E., Alberti, S., Longo, G.S. et al. (2018). Pushing the boundaries of interfacial sensitivity in graphene FET sensors: polyelectrolyte multilayers strongly increase the Debye screening length. *J. Phys. Chem. C* 122 (18): 10181–10188.

72 Zhao, W., Bi, S., Balke, N. et al. (2018). Understanding electric double-layer gating based on ionic liquids: from nanoscale to macroscale. *ACS Appl. Mater. Interfaces* 10 (49): 43211–43218.

73 Stern, E., Wagner, R., Sigworth, F.J. et al. (2007). Importance of the Debye screening length on nanowire field effect transistor sensors. *Nano Lett.* 7 (11): 3405–3409.

74 Kulkarni, G.S. and Zhong, Z. (2012). Detection beyond the Debye screening length in a high-frequency nanoelectronic biosensor. *Nano Lett.* 12 (2): 719–723.

75 Star, A., Tu, E., Niemann, J. et al. (2006). Label-free detection of DNA hybridization using carbon nanotube network field-effect transistors. *Proc. Natl. Acad. Sci. U.S.A.* 103 (4): 921–926.

76 Hwang, M.T., Heiranian, M., Kim, Y. et al. (2020). Ultrasensitive detection of nucleic acids using deformed graphene channel field effect biosensors. *Nat. Commun.* 11 (1): 1543.

77 Gao, N., Gao, T., Yang, X. et al. (2016). Specific detection of biomolecules in physiological solutions using graphene transistor biosensors. *Proc. Natl. Acad. Sci. U.S.A.* 113 (51): 14633–14638.

78 Song, J., Dailey, J., Li, H. et al. (2017). Extended solution gate OFET-based biosensor for label-free glial fibrillary acidic protein detection with polyethylene glycol-containing bioreceptor layer. *Adv. Funct. Mater.* 27 (20): 1606506.

79 Zaretski, A.V. and Lipomi, D.J. (2015). Processes for non-destructive transfer of graphene: widening the bottleneck for industrial scale production. *Nanoscale* 7 (22): 9963–9969.

80 Zhu, Y., Wang, C., Petrone, N. et al. (2015). A solid dielectric gated graphene nanosensor in electrolyte solutions. *Appl. Phys. Lett.* 106 (12): 123503.

81 Goode, J.A., Rushworth, J.V.H., and Millner, P.A. (2015). Biosensor regeneration: a review of common techniques and outcomes. *Langmuir* 31 (23): 6267–6276.

18

Highly Sensitive Pathogen Detection by Graphene Field-Effect Transistor Biosensors Toward Point-of-Care-Testing

Shota Ushiba[1], Takao Ono[2,3], Yasushi Kanai[2], Naruto Miyakawa[1], Shinsuke Tani[1], Hiroshi Ueda[4,5], Masahiko Kimura[1], and Kazuhiko Matsumoto[2]

[1]*Murata Manufacturing Co., Ltd., 1-10-1 Higashikotari, Nagaokakyo, Kyoto 617-8555, Japan*
[2]*SANKEN, Osaka University, Ibaraki, Osaka 567-0047, Japan*
[3]*JST, PRESTO, 4-1-8 Honcho, Kawaguchi, Saitama 332-0012, Japan*
[4]*Laboratory for Chemistry and Life Science, Institute of Innovative Research, Yokohama, Kanagawa 226-8503, Japan*
[5]*World Research Hub Initiative, Institute of Innovative Research, Tokyo Institute of Technology, Yokohama, Kanagawa 226-8503, Japan*

18.1 Introduction

Kazuhiko Matsumoto, the last author of this chapter, has been working on carbon nanotube field-effect transistor (FET) biosensors since 2004 [1] and has applied them for the detection of molecules, including DNA [1] and proteins [2]. In October 2008, Andre Geim (Nobel laureate in physics for the discovery of graphene) visited the author's laboratory and discussed graphene with us. This visit triggered our first demonstration of solution-gated graphene field-effect transistor (GFET) biosensors in 2009 [3], inspiring numerous further studies worldwide. GFETs are an ideal biosensing platform, and thus, we call it "lab-on-a-graphene-FET." There are two reasons for this: (i) the two-dimensional (2D), one-atom-thick nature of graphene exhibits ideally high specific surface area because all of its constituent atoms are present on the surface. Furthermore, being made solely of carbon atoms, graphene has a wide potential window and is stable in water. Therefore, unlike Si-FET biosensors, which require a SiO_2 insulating layer, the target molecules in solution directly contact the graphene surface and induce carriers therein. (ii) In GFETs, the carrier density modulation induced by target molecules can be converted into a large change in the channel current, i.e. high sensitivity, owing to the extremely high electron/hole mobility of graphene [4, 5]. Sensing platforms have been used to detect ions [3], charged molecules [6], biomolecules [3, 7], and photons [8]. In biosensors, specific detection of target molecules is achieved using receptors. A large

Graphene Field-Effect Transistors: Advanced Bioelectronic Devices for Sensing Applications,
First Edition. Edited by Omar Azzaroni and Wolfgang Knoll.

variety of receptors can be immobilized on GFETs, including antibodies [9], fragments of antibodies [10, 11], aptamers [7], and sialic acids [12]; thus, GFETs demonstrate highly sensitive detection of proteins [3, 7], lectins [12], and bacteria [9].

Close collaborations between academia and industry are key practices to favor and accelerate the implementation of cutting-edge technologies in society. Matsumoto's group and Murata Manufacturing Co., Ltd. have established an ongoing collaboration since 2016. We have been running a project, financially supported by the Japan Science and Technology Agency (JST), to develop GFET biosensors for influenza viruses, especially avian influenza. Although avian influenza viruses do not normally infect humans, they have nonetheless been reported to sporadically infect humans with an extraordinarily high mortality rate. The Spanish flu is a prominent example [13–15]. Avian influenza virus infection in humans cannot be diagnosed based on clinical signs and symptoms alone, which indicates the need for laboratory testing. Nonetheless, laboratory testing is time-consuming, and thus, point-of-care testing solutions are in high demand. GFET biosensors are among the best candidates for this owing to their high sensitivity, speed, and low cost.

Subsequent to this progress review in the field of biosensing applications, the chapter is organized as follows: Section 18.2 introduces our GFETs mimicking cell surfaces. We functionalized sialoglycan, which is present on both avian and human cell surfaces and plays an important role in the infection process. Our sialoglycan-functionalized GFETs detected biomolecules specifically bound to sialoglycan. The results demonstrate the proof of concept that GFETs, where the surfaces mimic cells, can detect biochemical reactions and thus should enable us to specifically detect human infectious viruses. Section 18.3 discusses three methods to enhance the sensitivity of GFET biosensors: (i) increasing receptor density, (ii) detection of small molecules by open-sandwich immunoassay (OS-IA), and (iii) detection of enzyme reactions in microdroplets. Finally, in Section 18.4, we focus on two practical issues: (i) baseline drift, which refers to signal variation that occurs in the baseline due to low-frequency variations in the sensor and/or instrument-controlled parameters, and (ii) inspection methods. In the first half of Section 18.4, we introduce two approaches for drift suppression: hardware-based and software-based approaches. Both approaches are useful for accurate estimation of target concentrations in practical applications. In the second half of Section 18.4, we introduce a screening method to facilitate fast and quantitative evaluation of GFET arrays using a deep learning-based algorithm. The screening method will pave the way for large-scale and stable manufacturing of GFET arrays.

18.2 Toward Detection of Pathogens by Mimicking Cell Surfaces

18.2.1 Introduction

Although there are various definitions, life is generally defined by three conditions: self-replication, energy metabolism, and boundaries with the outside world. Therefore, biochemical reactions at the boundary with the environment or outside world (i.e. the interface) are of universal importance for living organisms. A typical example is the reaction associated with an infection, which is an invasion by external organisms of the host interface. Let us consider the influenza virus as an example. Influenza viruses cannot replicate themselves without cells, and thus invade and multiply inside cells, such as tracheal epithelial cells. Sialoglycans, which are one of the receptor molecules, play an important role in this infection process. They exist on the surfaces of avian and human cells alike, with slightly different terminal structures. This difference determines the viral infection response at the cell surface and is associated with human infectivity [13]. A cell-based approach, which includes reactions such as metabolism outside the interface, is one of the traditional approaches to studying virus infectivity in humans. More interface-focused analytical approaches are required. For this purpose, highly sensitive measurement platforms that mimic interfacial structures are useful. In this section, we discuss our work on functionalized sialoglycans on GFETs to selectively measure human infectious viruses. To prove this concept, we used lectins, which specifically bind to sialoglycans, as a model of viruses [12].

18.2.2 Fabrication of Sialoglycan-Functionalized GFETs

Graphene was obtained from mechanically exfoliated kish graphite flakes, and Au/Ni electrodes were subsequently wired onto the obtained graphene flakes. 1-Pyrenebutanoic acid succinimidyl ester (PBASE) was used as a linker to modify sialoglycans on graphene. Two types of glycans (sialylglycopeptide [SGP]) were used in this study: human-type a2.6-SGP and avian-type a2.3-SGP. The lectin from *Sambucus sieboldiana* (SSA) was used as a model for the human infectious influenza virus, and the lectin from *Maackia amurensis* (MAM) was used as a model for the avian infectious influenza virus. A gate voltage (V_G) was applied to the GFET via an aqueous solution using an Ag/AgCl reference electrode (Figure 18.1).

18.2.3 Evaluation of Sialoglycan-Functionalized GFETs

First, we evaluated the sialoglycan-functionalization of graphene (Figure 18.2a). SGP is negatively charged because of terminal sialic acid. Therefore, hole carriers should be induced in the modified GFETs, and the transfer characteristics (drain current–gate voltage [I_D–V_G] characteristic) should shift toward the positive side. In our experiments, the transfer characteristics after modification shifted positively. It should be noted that I_D at the charge neutrality point (CNP) did not decrease. This result indicates that biomolecule modification does not induce additional carrier scattering in graphene. The modification of the SGP was also confirmed by atomic force microscopy (AFM), as shown in Figure 18.2b.

We then monitored the pseudoinfection reactions by introducing lectins. The introduction of SSA lectin into a2.6-SGP-modified GFET increased I_D because of the hole carriers. This indicates that SSA lectin is negatively charged. By monitoring the hole current at a constant gate voltage, SSA lectin was introduced over time with a gradual increase in concentration. The time course showed that I_D increased in a

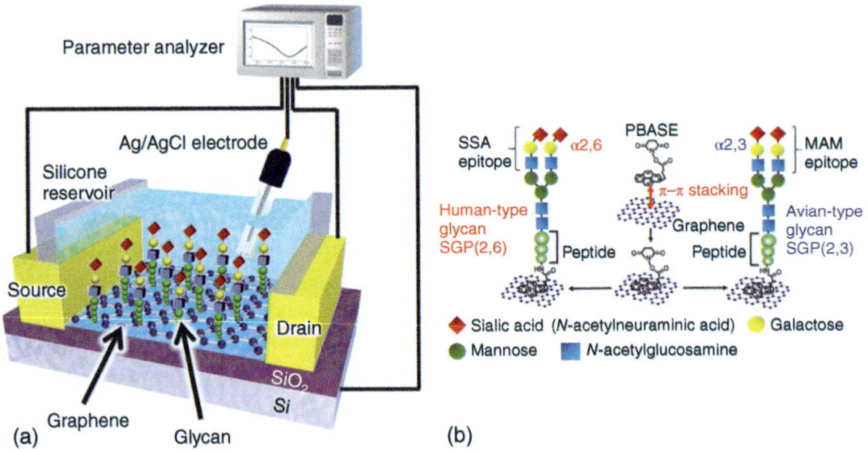

Figure 18.1 Experimental setup and scheme. (a) Schematic image of GFET modified with sialoglycans. (b) Schematic structure of sialoglycans modified to graphene via PBASE linker. Source: Reproduced from Ono et al. [12]/with permission from IOP Publishing.

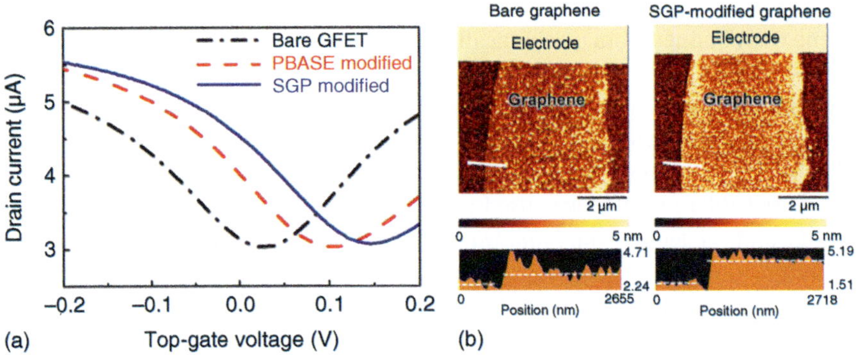

Figure 18.2 (a) I_D–V_G characteristics of a GFET with bare graphene (black dashed-dotted line), PBASE-modified graphene (red dashed line) and SGP-modified graphene (blue solid line). (b) AFM images of graphene channels before and after SGP modification. The height profiles at the edge of the channels (white solid lines) are also shown. Source: Reproduced from Ono et al. [12]/with permission from IOP Publishing.

stepwise fashion. However, no such response occurred with MAM lectin, confirming the specific response of the GFETs. A similar response was also observed in a2.3-SGP-modified GFETs and MAM lectin (Figure 18.3). These results indicate that the glycan-functionalized GFETs reconstructed the model of the viral infection response at the cell surface and achieved infectivity-based detection. The limit of detection (LOD) was estimated to be approximately 130 pM for SSA lectin and 150 pM for MAM lectin. Conversely, bovine serum albumin (BSA), which has no specificity for either glycan, induced a negligible current change, which also supports the conclusion that our GFETs specifically detected target molecules.

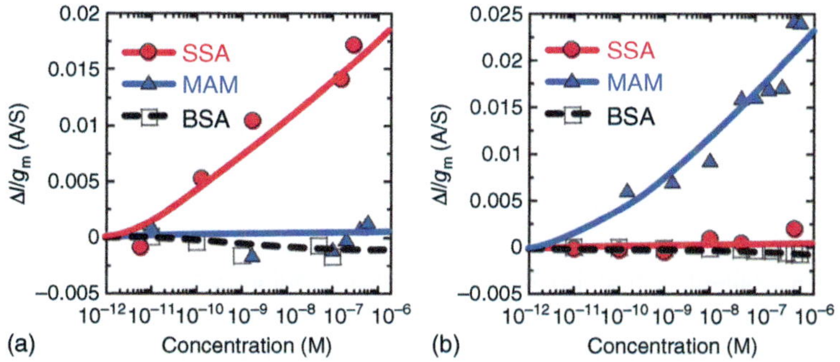

Figure 18.3 Response to pseudo viral reactions on graphene. (a) Response of SSA lectin on 2.6-SGP-modified GFET. (b) Response of MAM lectin on a 2.3-SGP-modified GFET. Source: Reproduced from Ono et al. [12]/with permission from IOP Publishing.

18.3 Signal Enhancement in GFETs

Here, we introduce three methods to enhance the sensitivity of GFET biosensors: (i) by increasing receptor density, (ii) detection of small molecules by OS-IA, and (iii) detection of enzyme reactions in microdroplets.

18.3.1 Sensitivity Enhancement by Increasing Receptor Density

18.3.1.1 Case of Linkers

In our GFETs, the target molecules are specifically adsorbed on the graphene surface according to Langmuir kinetics. Therefore, an increase in receptor density, together with a reduction in the dissociation constant K_D, can contribute to sensitivity enhancement. In this section, we describe a practical example of the proposed approach. We used two types of linkers: PBASE [16, 17] and tetrakis(4-carboxyphenyl)porphyrin (TCPP) [18–21] (Figure 18.4). PBASE has a pyrenyl group at one end, which stacks with graphene, and a succinimidyl group at the other end, which forms covalent bonds with the amino groups of receptors. Graphene can be functionalized

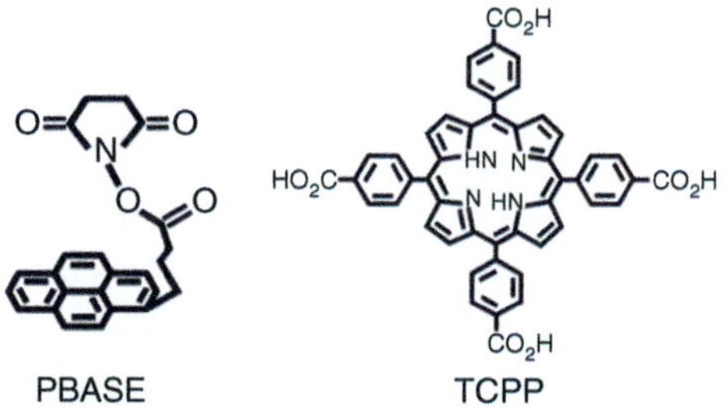

Figure 18.4 Molecular structure of PBASE and TCPP. Source: Reproduced from Kawata et al. [18]/with permission from IOP Publishing.

to capture a specific target without introducing structural defects [22–24]. TCPP, which is a porphyrin-derived molecule, has a larger conjugated system compared to PBASE and has four carboxyl groups that can be converted into succinimide groups. These features are expected to give TCPP the ability to strongly adsorb on graphene, compared to PBASE, and to bind more receptor molecules per unit area, which should lead to enhanced sensitivity.

18.3.1.2 Basis for Evaluation of Linker-Based Performance Enhancement

Human immunoglobulin E (IgE) was used as the model target to validate the performance of the TCPP linker. IgE is normally present in human serum at nanomolar concentrations, which are markedly increased in patients with atopic dermatitis, allergic asthma, and related diseases [25]. To capture IgE, an oligonucleotide IgE aptamer (D17.4ext) was used as a receptor. GFETs were fabricated using graphene obtained by exfoliation from kish graphite and wiring Au/Ni electrodes, as previously mentioned. PBASE and TCPP (50 mM) were dissolved in 2-methoxyethanol and reacted with GFETs for two hours at room temperature. After a moderate rinse, the TCPP-GFETs were soaked in a mixture of 1-ethyl-3-(3-dimethylaminopropyl) carbodiimide hydrochloride (EDC-HCl; 200 mM) and *N*-hydroxysuccinimide (NHS; 50 mM) to form succinimide groups. Next, 1 mM IgE aptamer was reacted with both GFETs at four degrees for 12 hours. Finally, the electrical properties of the GFETs were measured while adding the target IgE at a drain voltage of 100 mV applied via the Au/Ni electrodes and a gate voltage applied via an Ag/AgCl reference electrode. Sodium phosphate buffer (pH 7.4) was used as the buffer solution for the reactions and electrical measurements. The concentration was 100 mM to shield the surface charge of the receptor to the extent possible during reactions so that the electrostatic repulsive force was suppressed, resulting in a denser modification of the receptors. In contrast, for the measurements, the concentration was 10 mM to effectively transmit the electric field-effect signals of the captured targets to the GFETs.

18.3.1.3 Evaluation of Performance Enhancement by Linkers

The numbers of receptors and captured targets were evaluated based on the transfer characteristics of the GFETs. The density was evaluated as the position of the CNP in the bipolar characteristics of the GFETs. The CNP position in the I_D–V_G curve can be treated similarly to the threshold voltage in typical Si-FETs. After aptamer modification, the CNP shifted in the positive voltage direction, corresponding to the negative charge of the aptamer. The shift was larger for the TCPP-GFETs compared to that of the PBASE-GFETs, indicating a denser receptor modification (see Figure 18.5).

When IgE was introduced into the aptamer-modified GFET, the CNP shifted in the negative voltage direction, in contrast to the aptamer modification. This is consistent with the positive charge of IgE at pH 7.4 [17]. I_D at a constant gate voltage (no gate voltage sweep) was monitored over time at varying IgE concentrations. The current change can be converted into a shift in CNP by appropriately tuning the gate voltage; this can be done by dividing the current changes by the GFET's transconductance g_m. Although the absolute values of I_D varied among the devices because of the inevitable size deviation of exfoliated graphene, they can be compared to each other by normalization.

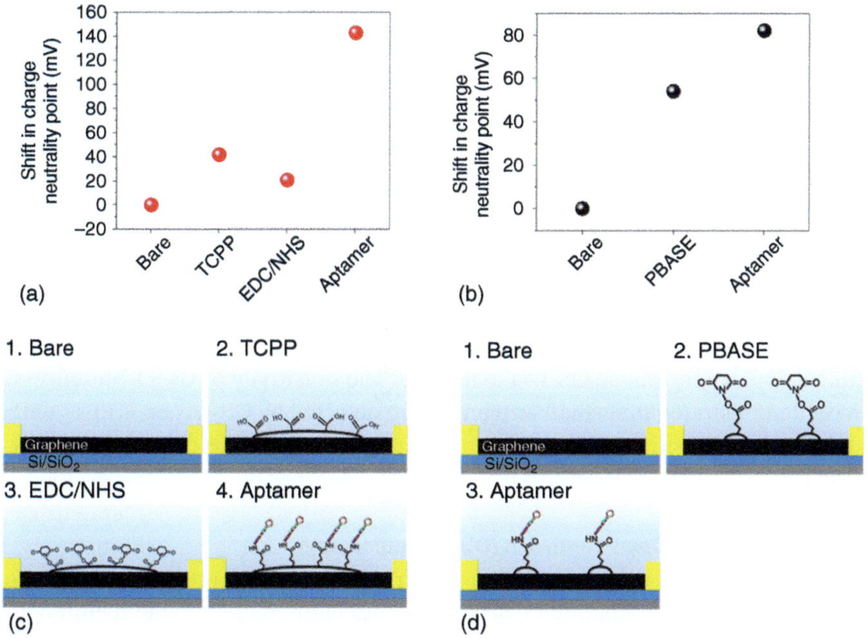

Figure 18.5 Receptor modification overview and modification density. (a, b) CNP shift before and after TCPP (a) or PBASE (b) modification. (c, d) Scheme of the modification process of TCPP (c) or PBASE (d). Source: Reproduced from Kawata et al. [18]/with permission from IOP Publishing.

The results of IgE measurements using TCPP-GFETs and PBASE-GFETs showed three aspects (Figure 18.6): (i) the fitting curves provided similar K_d values: 38 nM for TCPP and 147 nM for PBASE, which are comparable with previously reported values [26–30]. The results indicate that both GFETs worked appropriately as IgE sensors. (ii) At the maximum IgE concentration, the response of TCPP-GFETs was approximately twice that of PBASE-GFETs. The increase in maximum adsorption

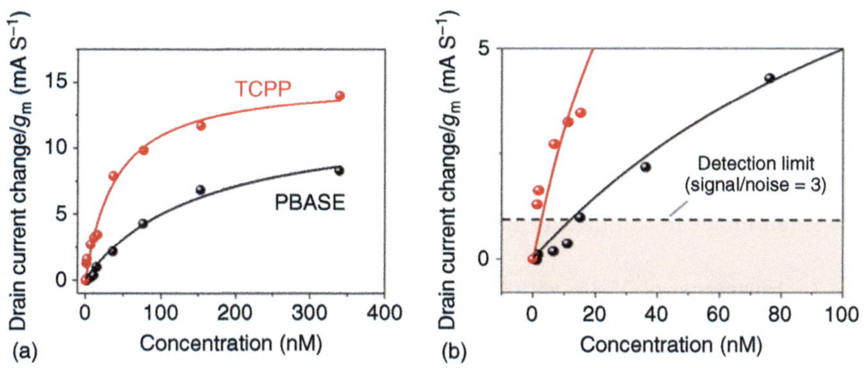

Figure 18.6 Electrical responses of the TCPP-GFETs and PBASE-GFETs as a function of IgE concentrations. (b) A magnified view of the low concentration region in (a). Source: Reproduced from Kawata et al. [18]/with permission from IOP Publishing.

in the Langmuir kinetics model implies an increase in the density of the receptor via the TCPP linker. (iii) The TCPP-GFETs produced a smaller LOD, where the signal exceeded the background noise level (signal-to-noise ratio of 3), indicating that the sensitivity was enhanced by TCPP modification. The LOD for TCPP-GFETs was 2.2 nM, which was equivalent to the IgE concentration in the serum of healthy normal subjects. Overall, the results indicate that our GFETs can potentially be applied to practical applications.

18.3.2 Ultrasensitive Detection of Small Antigens by Open-Sandwich Immunoassay on GFETs

In this section, we introduce OS-IAs to improve the sensitivity of GFET biosensors. We have demonstrated that small antigens were detected by OS-IAs on GFETs with 100-fold higher sensitivity and wider antigen detection range compared to OS-ELISA [11].

18.3.2.1 Principle of Open-Sandwich (OS) Immunoassay

OS-IA is a relatively new immunoassay, which employs antigen dependency in the interaction between the separated heavy-chain and light-chain variable regions (V_H and V_L) of an antibody in the presence of an antigen. The interaction is enhanced when the two antibody variable region fragments bind to the antigen simultaneously, resulting in their association, whereas they are prone to dissociate in the absence of the antigen. Since its first report in 1996 [31], the assay principle has been further studied and successfully applied to the detection of various antigens, especially small molecules, including pesticides, peptides, medicines, hormones, and oxidized lipids [32–36]. The OS-IA can be conducted with the two fragments derived from an antibody and can be carried out with less handling and in a shorter time compared to the sandwich immunoassay (Figure 18.7). However, the most remarkable feature of OS-IA is that it detects antigens noncompetitively (even if their molecular weight is less than 1000) with higher sensitivity and broader detection range compared to those of competitive immunoassays [37]. This differentiates OS-IA as a unique class of immunoassays. Because it can detect peptides and their modifications, the OS-IA principle has been applied for the detection of proteins as well, with specific epitopes such as the furin cleavage site of hemagglutinin derived from the highly pathogenic avian influenza virus H5N1, allowing its application in disease diagnostics [38]. Moreover, in combination with phage display technology, an effective and rapid method for screening antibodies suitable for OS-IA has been developed [39]. In this section, the application of OS-IA to graphene-based detection of small-molecule antigens is introduced.

18.3.2.2 Advantages of OS-IAs with GFETs

Because of the Debye length, smaller antibodies are preferred, and target antigens are more easily detected in GFET biosensors. One potential advantage of performing OS-IAs with GFETs instead of the conventional competitive immunoassay for small molecules is that the former uses only the variable region fragments of an

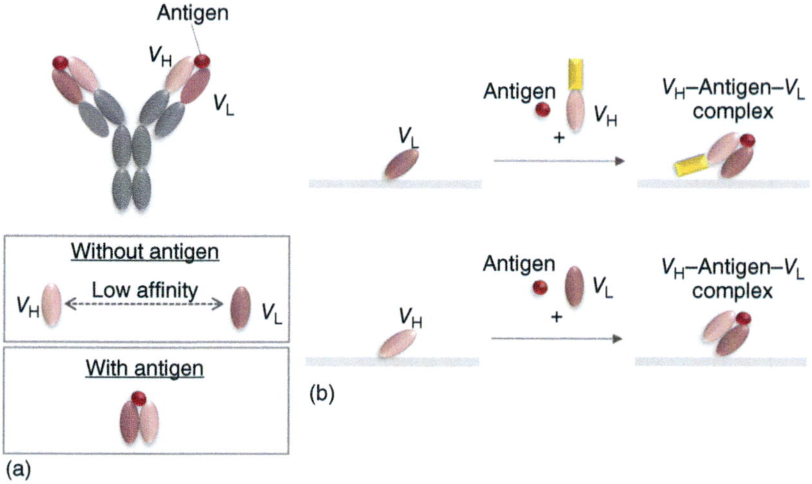

Figure 18.7 OS immunoassay (a) Principle of OS immunoassay. (b) The procedure of OS-ELISA. Source: Reproduced from Kawata et al. [18]/with permission from American Chemical Society.

antibody, which are much smaller than the entire antibody. However, the detection of small molecules has been challenging because of their small charge. Therefore, V_L fragments derived from an antibody were immobilized on graphene and the V_H fragments were fused to the tagged charged protein. The current change in graphene due to the binding of V_H is expected to be higher than in the case of antigen alone because of the stronger charge of V_H or that of the tagged charged protein-bound V_H.

18.3.2.3 Antibody Fragments and Device Fabrication

The recombinant variable region fragments of an anti-bone Gla protein (BGP or osteocalcin) antibody were prepared and used for OS-IAs to detect BGP-C7, a C-terminal 7-mer peptide of human osteocalcin (BGP), which is a minor bone protein used as a bone disease marker. Recently, BGP has attracted attention for its unprecedented functions in glucose homeostasis [40], brain development [41], and male fertility [42], attracting substantially more attention. First, the antigen-dependent interactions between the prepared thioredoxin (Trx) and biotin acceptor peptide (BAP)-fused V_H and maltose-binding protein (MBP)-fused V_L were successfully confirmed using OS-ELISA. Next, to enhance the current change of the GFET by V_H binding, V_H was fused to an APP tag derived from an intrinsically disordered region [43] of the murine amyloid precursor protein [44], which contains many acidic amino acids (e.g. aspartic acid [D] and glutamic acid [E]) to attain a higher negative charge (Figure 18.8a).

Graphene was synthesized on copper foils by chemical vapor deposition and transferred onto Si/SiO_2 substrates with Ti/Au electrodes. PBASE was modified on graphene and MBP-V_L was reacted with PBASE to immobilize it. Figure 18.8b shows the schematic of the GFET biosensor device.

AEESDSVDSADAEEDDSDVWWGGADTDYADGGEDKVVEVAEEEEVADVEEEEA
DDDEDVEDGDEVEEEAEEPYEEATERTTSTATTTTTTTESVEEV

(a)

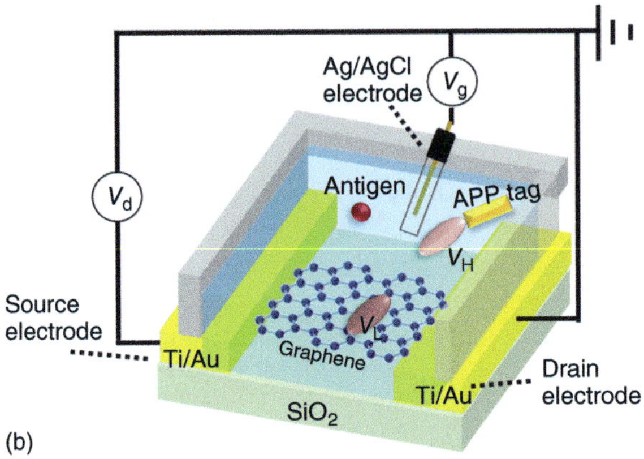

(b)

Figure 18.8 (a) Amino acid sequence of APP tag. (b) Structure of a GFET biosensor. Source: Reproduced from Kawata et al. [18]/with permission from American Chemical Society.

18.3.2.4 OS-IAs on GFETs

Figure 18.9a shows the $\Delta I_D/g_m$ value at different concentrations of BGP-C7 in 10 mM phosphate buffer (PB, pH 7.0) and 1 μg ml^{-1} APP-V_H solution at $V_D = -0.1$ V and $V_G = 0$ V. The transconductance g_m was evaluated based on the transport characteristics and the charge carriers are holes at $V_G = 0$ V. $\Delta I_D/g_m$ increased depending on the concentration of BGP-C7 up to 100 ng ml^{-1}, which indicates that negative ions were adsorbed on graphene surfaces. Because BGP-C7 and V_H bind V_L on graphene surfaces, $\Delta I_D/g_m$ was changed by negative ions in the disordered APP tag-fused on V_H. Figure 18.9b shows $\Delta I_D/g_m$ value at different BGP-C10dV concentrations. BGP-C10dV has a sequence very similar to that of BGP-C7, but does not bind to this antibody, which specifically recognizes the C-terminal valine residue of BGP. $\Delta I_D/g_m$ did not increase by the addition of BGP-C10dV. This result indicates that GFETs do not detect BGP-C10dV, which implies that GFETs can detect BGP-C7 selectively. The LOD of BGP-C7 was dramatically improved to 10 fg ml^{-1} (Figure 18.9c). The sensitivity in terms of the LOD was more than 100-fold higher than that of OS-ELISA using this antibody.

18.3.2.5 OS-IAs on GFETs in Human Serum

To investigate the possible use of GFETs in clinical diagnostics, the samples were measured in PB with 0.5% human serum (Figure 18.10a). We found that the current increased depending on the concentration of BGP-C7 up to 1000 ng ml^{-1}, although the current increase upon addition of BGP-C10dV was very small. Moreover, the current did not increase in the absence of APP-V_H and MBP-V_L-unmodified GFETs.

Altogether, these results indicate that MBP-V_L-modified GFETs detected BGP-C7 selectively. The LOD of GFETs for BGP-C7 in 0.5% serum was approximately 100 fg ml^{-1}, as shown in Figure 18.10b. This concentration translates to 20 pg ml^{-1} of

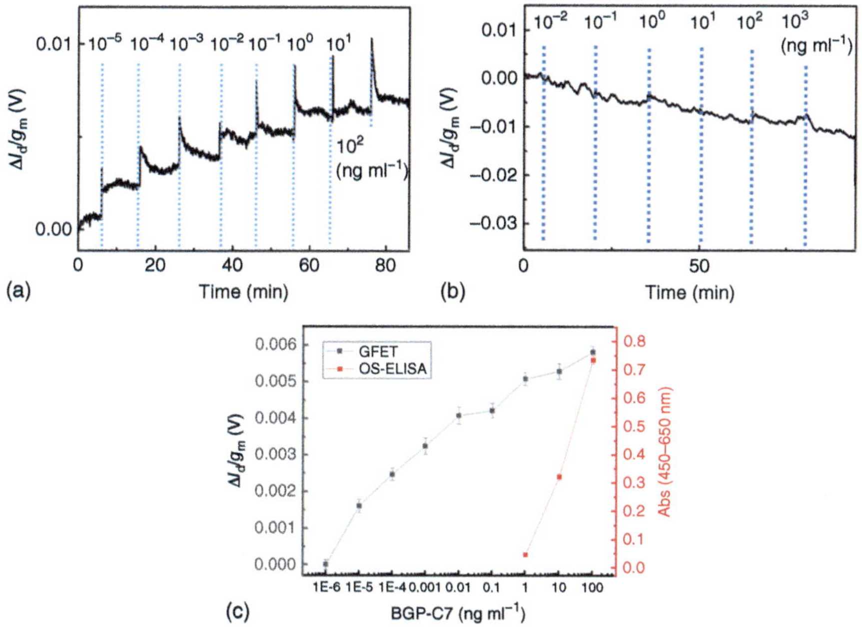

Figure 18.9 Application of APP-VH: (a, b) time course of averaged $\Delta I_D/g_m$ at V_D of −0.1 V, V_G of 0 V for V_H modified GFETs, the various concentrations of BGP-C7 (a) and BGP-C10dv (b) and constant concentration of V_L of 1 mg ml^{-1}. Source: Reproduced from Kawata et al. [18]/ with permission from American Chemical Society. (c) Dose-dependency plots of GFETs and OS-ELISA. Error bar, ±1 SD. Source: Adapted from Kanai et.al. [11].

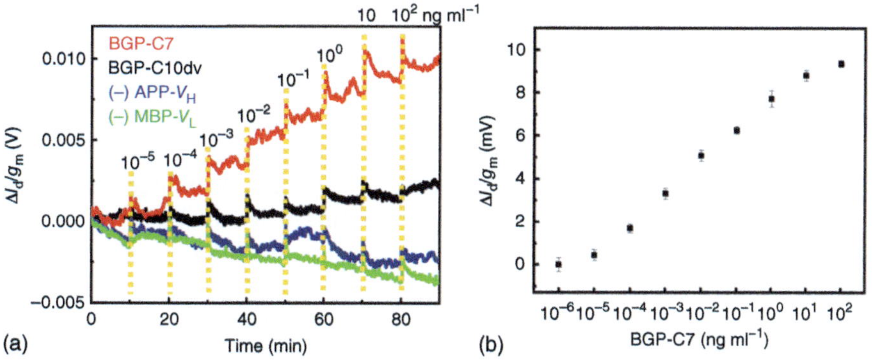

Figure 18.10 (a) Time course of average $\Delta I_D/g_m$ at V_D of 0.1 V, V_G of 0 V for V_L modified GFETs, various concentrations of BGP-C7 (red) or BGP-C10dV (black) with a constant APP-VH concentration of 1 μg ml^{-1}; 0.5% human serum were applied. As a control, BGP-C7 alone (blue) was applied to V_L modified GFET, whereas BGP-C7 and APP-V_H were applied to intact GFET without V_L (green). (b) Dose-dependency response plot. Error bars represent ±1 SD. Source: Reproduced from Kawata et al. [18]/with permission from American Chemical Society.

BGP-C7 (mw: 894) in 100% serum. The normal ranges of osteocalcin (mw: about 6 k) in Japan for healthy women are in the range 7.8–30.8 and 14.2–54.8 ng ml^{-1} before and after menopause, respectively, and 8.4–33.1 ng ml^{-1} for men. Therefore, the obtained LOD (130 pg ml^{-1}) and the working range are sufficient to cover all these ranges. The high sensitivity and wide dynamic range of GFET sensors suggest their potential applications in clinical diagnostics.

18.3.3 Real-Time Measurement of Enzyme Reaction in Microdroplets Using GFETs and Its Application to Pathogen Detection

18.3.3.1 Introduction

Debye screening is a critical issue for FET biosensors [10, 45, 46]. It is a phenomenon in which the surface charge of a target is neutralized by counterions in aqueous solutions. At physiological salt concentrations, such as *in vivo*, the charge of the target molecules is neutralized within 1 nm of the surface. Thus, it is generally difficult to detect molecules with whole IgG antibodies because their size is approximately 15 nm [47, 48]. However, lowering salt concentrations is also problematic because reaction efficiency, such as intermolecular bonding, may be degraded [49]. This has been a barrier to the commercial implementation of FET biosensors. Here, we demonstrate detection by measuring enzyme reactions that occur with target molecules rather than the charge of the target molecules [9].

Helicobacter pylori (*H. pylori* [HP]) was selected as a model target for the proof of concept. It has been recently discovered that most gastric cancers are caused by *H. pylori* infection, which makes early diagnosis and treatment increasingly important [50–54]. To survive and infect the highly acidic surface of the stomach, *H. pylori* uses its urease (a type of catalyst enzyme made of protein) to produce ammonia that neutralizes stomach acid. The ammonia-produced reaction is utilized in medical practice to detect and diagnose *H. pylori* infection [55, 56]. Our goal is to detect the enzyme reaction using our GFETs biosensors. *H. pylori* was captured by antibodies immobilized on graphene; however, under physiological conditions, the surface charge of the bacteria was shielded by ions in solution, which prevented charge detection by conventional GFETs. If ammonia produced via the enzymatic reaction can be detected by GFETs, then the presence of *H. pylori* can be indirectly, but stably, detected. There have been reports demonstrating that ammonia gas can be detected with graphene [57–60]; thus, we expected ammonia in the solution to be detected.

It should also be noted that the number of *H. pylori* captured by the antibody was very small. Consequently, the ammonia produced was very small and quickly diffused into the bulk solution. Therefore, ammonia cannot be detected using conventional GFETs because of its low concentration. Hence, we introduced a novel technique; we fabricated a microwell on a GFET to hold a microdroplet. The holes were micrometers in size and femtoliters (10^{-15} l) in volume. If *H. pylori.* is captured, the enzyme reaction occurs in the droplet, and ammonia is concentrated so that it can be detected. By making the droplet volume identical, this approach offers an additional advantage over conventional GFETs. The reaction rate within the droplet can be calculated, and the target concentration can be quantified from the non-uniform current changes of the GFETs.

18.3.3.2 Measurement Mechanism and Model Measurement System

Prior to encapsulating the reaction system, the possibility of measuring the reaction product (ammonia) in the bulk solution was verified (Figure 18.11a). The drain current of the GFETs followed the Langmuir kinetics of ammonia adsorption, and it was found that ammonia in the order of mM can be quantified.

The reaction rate of the enzyme can be described using the Michaelis–Menten equation. If ammonia produced from time to time is adsorbed according to Langmuir kinetics under the maximum reaction rate of urease in excess substrate, the time evolution of the drain current is given by

$$\Delta I = \Delta I_{\max} t \bigg/ \left(\frac{K_{\mathrm{D}}}{2C_{\mathrm{urease}} v_{\max}} + t \right) \tag{18.1}$$

where ΔI, $\Delta I_{\max}$, t, K_{D}, C_{urease}, and $v_{\max}$ are the change in the drain current, maximum change in the drain current, reaction time, dissociation constant of ammonia from the graphene surface, urease concentration, and maximum reaction rate, respectively. The measurement results obtained using the GFETs are well explained by this equation (Figure 18.11b).

Subsequently, microwells were fabricated on GFETs using a fluoropolymer (Figure 18.12). Fluoropolymer microwells have been reported to be effective in holding very small volumes in the femtoliter (fl) to atoliter (al) scale [61, 62]. Encapsulation of enzyme solutions in microwells has also been used for ultrasensitive enzymatic detection at the single-molecule level [63–65]. The response to the

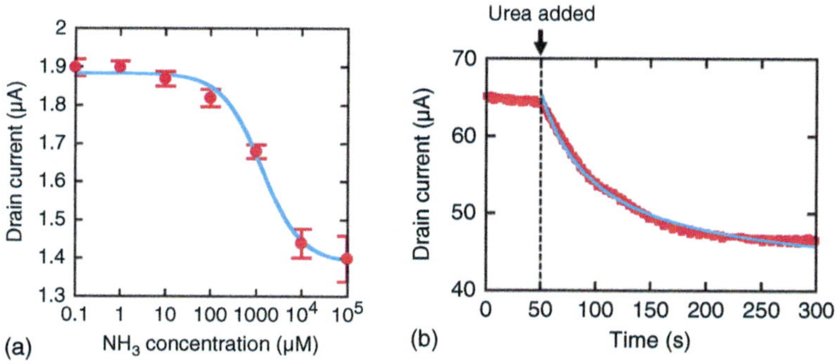

Figure 18.11 (a) Response of GFETs to ammonia solution. The blue line represents the fitting curve given by the Langmuir adsorption isotherm. Error bars show three standard deviations of the drain current fluctuation in the 300 seconds measurement for each data point. (b) Response of GFETs to the urease reaction in bulk solution. The blue line represents the fitting curve given by Eq. (18.1). Source: Reproduced from Ono et al. [9]/with permission from American Chemical Society.

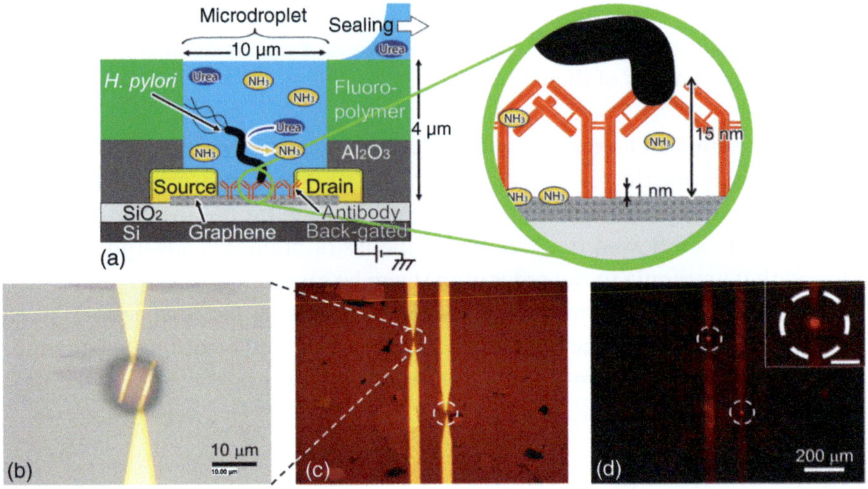

Figure 18.12 Device overview. (a) Schematic of a GFET with a microwell. The target, *H. Pylori*, is captured away from the graphene by the antibody and the ammonia produced by the bacteria is concentrated in the microdroplets for detection. (b, c) Bright-field images of the device. (d) Fluorescent image of the device encapsulating Rhodamine 6G. The inset shows droplet formation. Source: Reproduced from Ono et al. [9]/with permission from American Chemical Society.

reaction in the microwells was similar to that in the bulk solution, indicating that GFETs can be used in reaction systems even though the reaction volumes were nine orders of magnitude (ml–fl) smaller compared to the bulk solution (Figure 18.13a).

In addition, measurements of urease trapped on graphene rather than the amount diffused in solution were performed (Figure 18.13b). The urease was immobilized

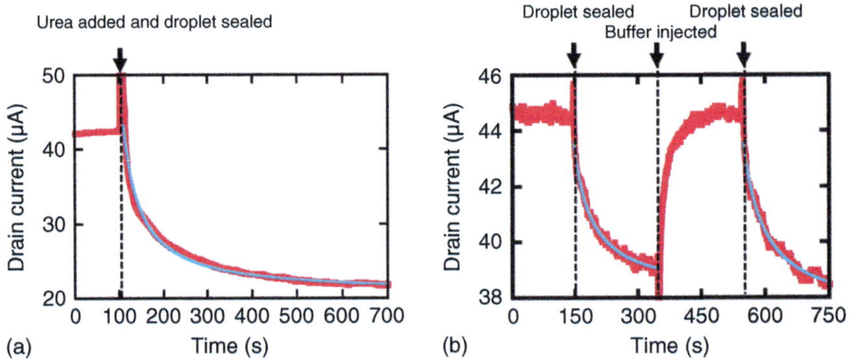

Figure 18.13 (a) Response of a GFET to urease reaction in microdroplet. The blue line represents the fitting curve given by Eq. (18.1). (b) Response of a GFET by reaction of urease immobilized on graphene in microdroplets. The blue lines represent the fitting curves given by Eq. (18.1). Source: Reproduced from Ono et al. [9]/with permission from American Chemical Society.

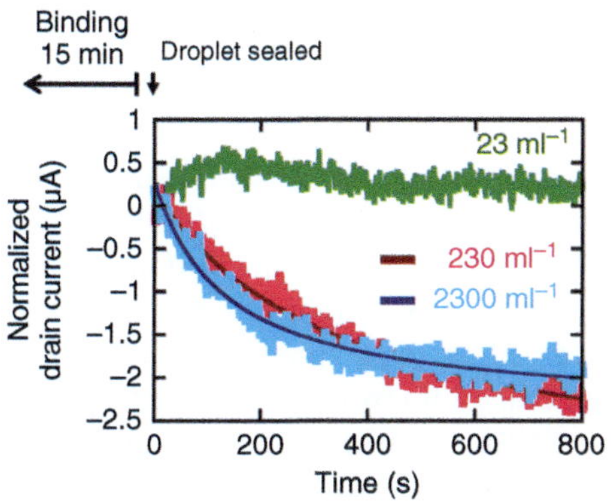

Figure 18.14 Detection of *H. pylori* by GFETs equipped with microwells. *H. pylori* was bound on graphene for 15 minutes and then the droplets were sealed. Fitting curves by Eq. (18.1) to the drain current values are shown in dark blue and red for 2.3×10^3 and 2.3×10^2 bacterial cells ml^{-1} solution, respectively. Source: Reproduced from Ono et al. [9]/ with permission from American Chemical Society.

on graphene via avidin–biotin reactions, and subsequently, microdroplets were formed. Before encapsulation, ammonia quickly diffused from the graphene surface into the bulk solution and became low in concentration. However, after encapsulation, ammonia became highly concentrated, making it measurable by GFETs. When the encapsulation was broken, ammonia diffused into the bulk, and upon further encapsulation, the same changes were measured as during the initial encapsulation. This method detected urease at concentrations as low as 4 pM.

Finally, we performed *H. pylori* measurements using GFETs. The response of trapped *H. pylori* was measured at a low concentration of $230\,ml^{-1}$ for the bacterial solution. The number of captured bacteria was estimated from the fitting curve. A minimum of 0.04 bacteria (fragments of bacteria) was detected. This indicates the extremely high sensitivity of the proposed method. The entire measurement was completed within 30 minutes (Figure 18.14).

18.4 Practical Issues: Baseline Drift and Inspection Methods

In this section, we shall focus on two practical issues: (i) baseline drift and (ii) inspection methods. Baseline drift is commonly observed in the response of FET-based sensors, particularly GFET biosensors, owing to their high sensitivity. Baseline drift makes it difficult to accurately estimate the concentrations of the target molecules. In Section 18.4.1, we introduce two approaches (hardware-based and software-based) for drift suppression and compensation. Both approaches are useful for accurate estimation of target concentrations in practical applications.

18.4.1 Drift Suppression and Compensation of GFET Biosensors

18.4.1.1 Drift Suppression in GFETs by Cation Doping

Since the introduction of ion-sensitive field-effect transistors (ISFETs) in 1970 [66], sensor or baseline drift has been an inherent drawback in ISFETs and GFETs. Some studies have shown that this drift persists for more than one hour [67–69]. Various models have been proposed and studied extensively to describe the drift in ISFETs, which includes the diffusion of ions from electrolyte solutions into the oxide layers [70, 71]. However, in most solution-gated GFETs, the graphene channels are in direct contact with the electrolyte solution, that is, no oxide covers the graphene channels. Therefore, the established theories on ISFETs cannot be applied to SG-GFETs; further studies are required to understand the drift in GFETs.

One possible reason for this drift could be related to the p-doping in GFETs being countered by cations in the solution, as the cations can permeate into the polymer residue and/or between graphene and SiO_2 substrates. Thus, we propose a method to suppress the drift, in which sufficient cations are doped to counter the p-doping of GFETs prior to measurements [72].

Figure 18.15b depicts a pretreatment in which GFETs were immersed in a 15 mM sodium chloride (NaCl) aqueous solution for 25 hours, whereas nontreated samples were shown in Figure 18.15a. Figure 18.15c shows the time course of the average CNP values from GFETs with and without pretreatment. It is shown that for GFETs with pretreatment, CNP values were stable, but these values changed gradually for GFETs without pretreatment because, as CNP rapidly changed, the rate of change decreased over time. We believe that this rate correlates with ion diffusion into the GFET. As the ions accumulate in the GFET, their diffusion rate decreases. As such, the rate of change in CNP is assumed to decrease. The CNP for the sample without pretreatment drifted by 49 mV in the first hour. However, the drift was reduced to

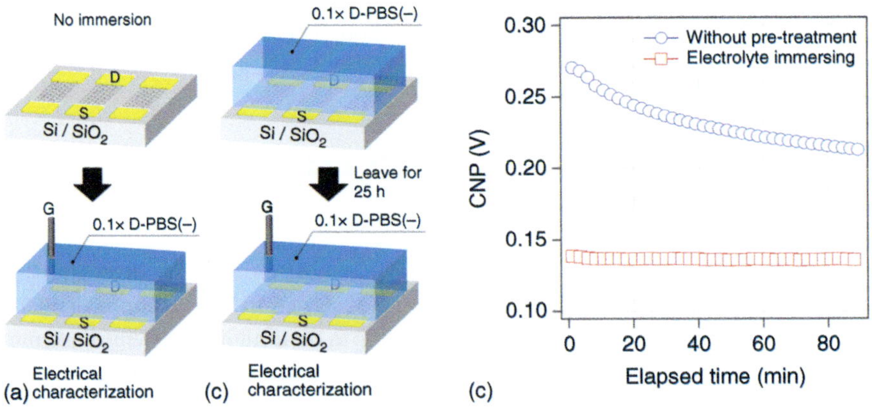

Figure 18.15 Schematic illustration of pretreatments and electrical evaluation. (a) No immersion (b) immersion in 0.1 × D-PBS(–). (c) Time-course of the calculated average CNP of the GFETs without the pre-treatment (blue circle) and with the electrolyte immersing (red square). Source: Reproduced from Miyakawa et al. [72], © MDPI/Public Domain CC BY 4.0.

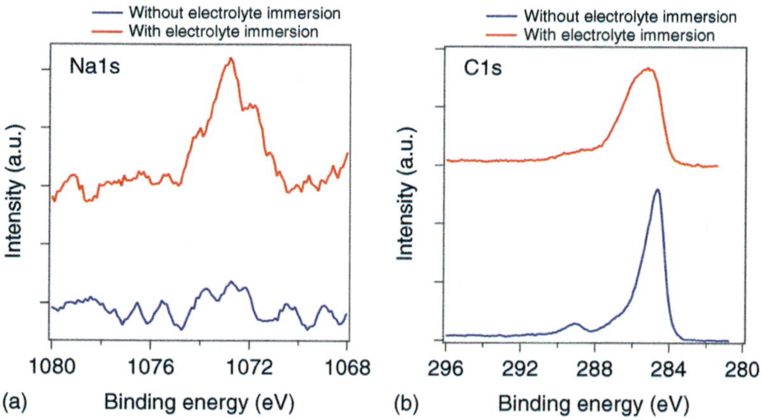

Figure 18.16 XPS analysis results of graphene with (red line) and without (blue line) electrolyte immersion. (a) Na1s peak and (b) C1s peak. Source: Reproduced from Miyakawa et al. [72], © MDPI/Public Domain CC BY 4.0.

2 mV for the pretreated sample. The results clearly indicate that the electrolyte immersion pretreatment was effective in suppressing the CNP drift by 96%.

Based on the pretreatment effect on the CNP drift, it was expected that cations would permeate the GFETs. To determine whether cations accumulated around graphene, X-ray photoelectron spectroscopy (XPS) was performed on graphene before and after immersion in the electrolyte solution. Two GFET substrates were prepared as samples for XPS evaluation. After GFETs were fabricated, one substrate was immersed in 0.1×D-PBS(−) for 25 hours, whereas the other was not. Finally, both substrates were rinsed with deionized water and dried using N_2 gas flow and heating on a hot plate at 120 °C. XPS analysis results for the Na1s and C1s are shown in Figure 18.16. Figure 18.16a shows that the sodium-derived signals were detected only from GFETs immersed in D-PBS(−), which indicates that they absorbed cations during immersion. The relative atomic concentration of Na to carbon was calculated at 3%, considering the peak areas of the Na1s and C1s. However, from the XPS analysis, it was difficult to understand where sodium accumulated because of the low spatial resolution and low intensity of Na1s. Some reports have shown that water molecules penetrate between graphene and substrates [73, 74] and into the PMMA resin [75]. Therefore, we expected that sodium would be located in these regions, together with water molecules, because positively charged sodium ions may be adsorbed on the negatively charged substrate surface and PMMA. In addition, cation–π interactions should facilitate the adsorption of Na^+ onto the surface of graphene [76, 77]. Furthermore, based on these results, Na^+ as well as other cations, such as K^+, should be adsorbed.

Figure 18.16b shows the peak of the C1s. The sp^2 C–C bond in graphene is assigned at approximately 285 eV. Other tails appearing on the high-energy side correspond to the binding energies of species in PMMA or functional groups on graphene [78]. For the GFET without electrolyte immersion, the graphene-derived 285 eV peak was more prominent than the PMMA-derived 289 eV peak, indicating that the GFET surface was less contaminated with PMMA. The GFET with electrolyte immersion has a wider C1s peak, especially on the high-energy side, compared to those without electrolyte immersion. This result implies that the C–C bonds of graphene partially become C–O bonds during electrolyte immersion. In GFETs, graphene and gold electrodes were electrically connected to the electrolyte solution. Under such circumstances, a galvanic current may flow through graphene, gold electrodes, and electrolytes, which occurs when two different metals are present in an electrolyte solution. This current possibly induces a redox reaction on the graphene surface [79]. Consequently, graphene may have been partially oxidized. This graphene oxidation could have also contributed to the change in the electrical properties, although the transfer curves were not significantly changed by the pretreatments, in addition to the CNP shift.

18.4.1.2 Drift Compensation by State-Space Modeling

In this section, we propose state-space models (SSMs) to describe the time-series data of a GFET biosensor and to separate the response against target molecules from the baseline drift [69]. State-space modeling is a framework established to understand stochastic and deterministic dynamical systems, referred to as states, which are observed through a stochastic process [80]. SSMs have been widely used in time-series analyses, such as gravimetric chemical sensors [81], to understand systems that can generate time-series data.

Figure 18.17a shows a schematic of the measurement system, including the GFET sensor. A graphene film was first grown on Cu foil by chemical vapor deposition and then transferred onto a Si/SiO_2 substrate. A source/drain electrode was formed with 10 nm Ti and 90 nm Au. Finally, a graphene channel was formed via oxygen plasma etching. The obtained GFET was immersed in 15 mM D-PBS (Nacalai Tesque Inc.). A bias voltage ($V_{DS} = 0.1$ V) was applied between the source and the drain, and a top-gate voltage (V_{GS}) was applied through the solution using an Ag/AgCl electrode. While sweeping V_{GS}, the drain current (I_{DS}) was measured using a semiconductor parameter analyzer (Keysight Technologies, B1500A). From the I_{DS}–V_{GS} plot, CNP was calculated. Figure 18.17c shows the time-series data for CNP. While monitoring

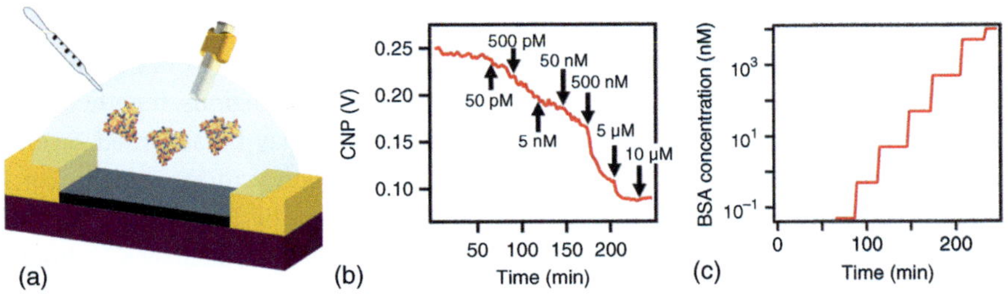

Figure 18.17 (a) Schematic of a GFET sensor. (b) Time-series data of the CNP against BSA. The arrows indicate the time at which BSA is introduced to the sensor. (c) Time-series data of the BSA concentration during the measurement. The concentration during the time interval 0–65 minutes is not plotted in the semi-logarithm graph because the concentration was 0 nM during this time. Source: Reproduced from Ushiba et al. [69]/with permission from IOP Publishing.

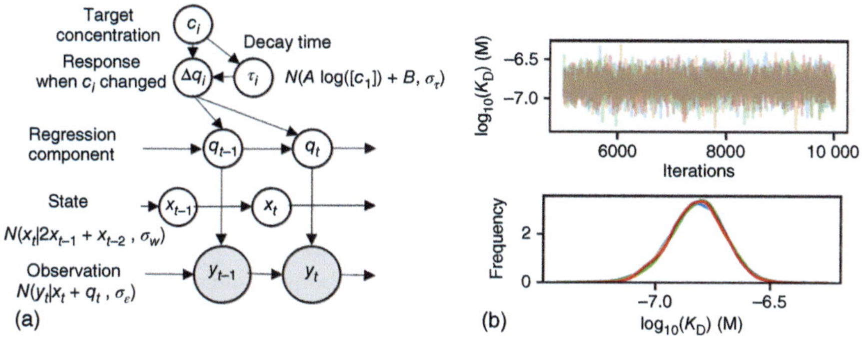

Figure 18.18 (a) Graphical model of the developed SSM. (b) Trace plots of four chains for K_D. (c) Posterior distribution of K_D calculated from (b). The colors represent each chain. Source: Reproduced from Ushiba et al. [69]/with permission from IOP Publishing.

the CNP, BSA molecules (Sigma-Aldrich), as a sensing target, were intermittently exposed to the sensor at various concentrations of 50 pM, 500 pM, 5 nM, 50 nM, 500 nM, 5 μM, and 10 μM in D-PBS solution (Figure 18.17c). The BSA molecules were physisorbed onto graphene. In the dynamic response of the sensor, shown in Figure 18.17b, the CNP varied with the BSA concentration. In addition, there was an apparent baseline drift as the CNP sequentially varied even in absence of BSA in the solution.

To model the time-series data obtained in Figure 18.17c, we propose an SSM, as shown in Figure 18.18a, described by Eqs. (18.2)–(18.6). We define this model as model 1.

$$y_t = x_t + q_t + \varepsilon_t, \quad y_t \sim N(0, \sigma_\varepsilon) \tag{18.2}$$

$$x_t - x_{t-1} = x_{t-1} - x_{t-2} + w_t, \quad w_t \sim N(0, \sigma_w) \tag{18.3}$$

$$q_t = \sum_{i=1}^{N} \Delta q_i \tag{18.4}$$

$$\Delta q_t = \begin{cases} 0(t < t_i) \\ (Q_i - Q_{i-1})\left\{1 - \exp\left(-t - {t_i}/{\tau_i}\right)\right\}(t \geq t_i) \end{cases} \tag{18.5}$$

$$Q_i = \frac{ac_i}{K_D + c_i} \tag{18.6}$$

According to Eq. (18.2), which is called the observation equation, the observed CNP (y_t) is the summation of the baseline (x_t), regression component of BSA (q_t), and observation noise [$\varepsilon_t \sim N(0, \sigma_\varepsilon)$]. Subscript t (=1, 2, 3, . . ., 104) corresponds to the data index. In Eq. (18.3), which is called the state equation, the unobserved baseline (x_t) is assumed to follow a quadratic trend with system noise [$w_t \sim N(0,$

σ_w)]. The regression component (q_t) is described based on two concepts: the first is that the signal reaches a plateau, as defined by the Langmuir model in Eq. (18.6), where K_D, c_i, and a are the dissociation constant, BSA concentration, and the coefficient, respectively. Subscript i (=0, 1, 2, . . ., 7) is the sample index, which increases when BSA is introduced to the sensor. This assumption, represented by Eq. (18.6), agrees with experimental data [3, 82], confirming that the sensor response to BSA follows the Langmuir equation. The second concept is that the signal varies following exponential decay after BSA is introduced (Eq. (18.5)). According to previous studies, the response time of FET-based biosensors is in the order of seconds to minutes, even though the target concentration ranges in the order of fM to μM. This implies that the decay time τ_i is not directly proportional to concentration (c_i). Therefore, in Eq. (18.7), we assume that the mean of τ_i is proportional to the logarithm of c_i, not directly proportional to c_i.

We estimated the states and parameters of Model 1 to fit the dynamic response shown in Figure 18.17c. As the regression component (q_t) in our SSMs is highly nonlinear, Kalman filter algorithms cannot be applied to subsequent parameter estimation. Therefore, we used Markov chain Monte Carlo (MCMC) methods [83] implemented in Python and Stan's probabilistic programming languages [84] to estimate the state and parameters of Eqs. (18.2)–(18.7). In the computation, four chains were run, with 10 000 iterations for each chain. The first 5000 iterations were discarded as burn-in, and the last 5000 iterations in each chain were used for subsequent inferences. Figure 18.18b shows the trace plots for K_D. The posterior distribution is well sampled because there is no serial correlation, and the chains explore the sample space many times. The other model parameters have similar trace plots. The *Rhat* value, which is an indicator of model convergence, was lower than 1.1 for all parameters (Table 18.1). This result indicated that the model converged. The histogram of the trace plots shown in Figure 18.18b corresponds to the posterior distribution of the dissociation constant K_D (Figure 18.18c). The posterior mean and the 95% credible interval for K_D were calculated to be 151, 85, and 257 nM, respectively. These values are comparable to those reported previously [3]. Table 18.1 lists some of the representative parameters. Using the proposed SSM and

Table 18.1 Summary of representative parameters for the SSM estimated using the MCMC method.

	Mean	2.5%	97.5%	Rhat
K_D (nM)	151	85	257	1.0
a (mV)	−81	−120	−53	1.0
σ_ε (mV)	2.2	1.9	2.6	1.0
τ_6 (min)	6.35	1.46	13.83	1.0

The posterior mean, the 95% credible interval, and *Rhat* are shown for K_D, a, σ_ε, and τ_6. τ_6 corresponds to the decay time when the BSA concentration increases from 500 nM to 5 μM.
Source: Reproduced from Ushiba et al. [69]/with permission from IOP Publishing.

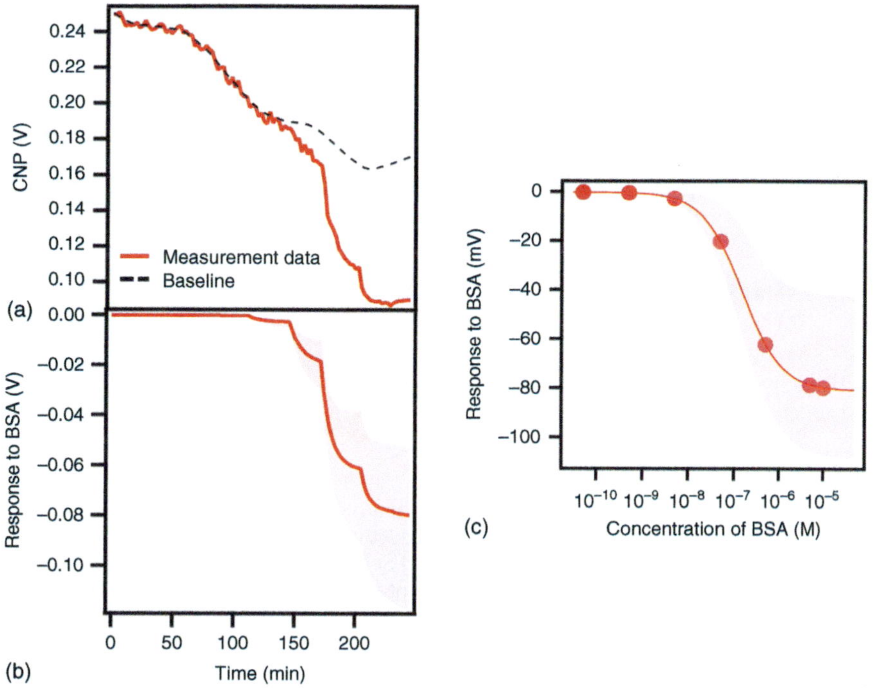

Figure 18.19 (a) Measurement data of CNP (red) and baseline (x_t) estimated using the SSM (black). (b) Dynamic response to BSA. The shaded area indicates the 95% Bayesian confidence interval. (c) GFET sensor response against BSA as a function of the concentration (c_i) calculated from Model 1. Source: Reproduced from Ushiba et al. [69]/with permission from IOP Publishing.

estimated parameters, we divided the time-series data, shown in Figure 18.17c, into the baseline (x_t, Figure 18.19a) and a signal related to the BSA (q_t, Figure 18.19b). Figure 18.19c shows the sensor response to BSA as a function of concentration. It should be noted that the effect of the baseline drift is excluded in Figure 18.19c, which indicates that the dynamic response in Figure 18.17c, in which the BSA concentration ranges from 0 to 5 nM, was mainly attributed to the baseline drift.

In Section 18.4.1, we propose two methods to suppress and compensate for drift. Both approaches are useful for accurate estimation of target concentrations in practical applications. The state-space modeling approach can be applied to GFET biosensors as well as to other FET-based sensors, and enables an accurate analysis of the sensor response. Therefore, these studies pave the way for further applications based on FET-based sensors.

18.4.2 Deep-Learning-Based Optical Inspection of GFETs

To implement cutting-edge technologies in society, mass production of GFETs is indispensable. Although numerous studies have reported the successful production of GFET arrays, the uniform transfer characteristics over the array platform

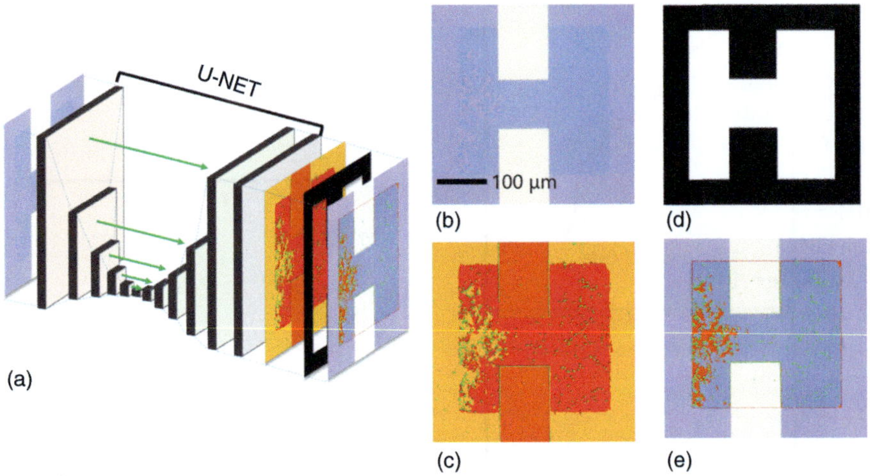

Figure 18.20 (a) Schematic illustration of a deep-learning-based semantic segmentation pipeline with U-Net architecture, in which the encoder part is in orange, and the decoder part is in green. (b) Raw optical image of a GFET. (c) Semantic segmentation of the optical image in (b). Graphene, electrode, substrate, and contaminants are flagged in red, red-orange, orange, and green, respectively. (d) Mask used for the graphene area extraction. (e) Overlay of the raw optical and segmented images, where holes and contaminants are in red and green, respectively. Scale bar is 100 μm. Source: Reproduced from Ushiba et al. [69]/ with permission from IOP Publishing.

constitute a complication that remains to be addressed. Therefore, establishing a fast screening method for mass production of graphene devices is necessary. Optical microscopy is widely used to inspect GFETs, but the quantitative evaluation of their optical images is challenging because they are not classified. Thus, a more appropriate and fast screening method remains to be developed. In this context, we introduce a deep-learning (DL)-based semantic image segmentation algorithm [85].

Figure 18.20a shows a schematic diagram of our four-step DL-based image segmentation system: (i) the optical images of GFETs were captured using a motorized optical microscope (Figure 18.20b); (ii) the captured optical images were input into a customized software pipeline based on the U-Net architecture comprising two parts: a contracting path and a symmetric expanding path [86]. The neural network architecture outputs pixel-segmented images, where each pixel is categorized into four classes: graphene (red), electrode (red-orange), substrate (orange), and contaminants (green) (Figure 18.20c); (iii) a mask was prepared to extract the expected graphene areas (Figure 18.20d); and (iv) the overlay image of the raw optical and segmented images were output, where holes and contaminants are depicted in red and green, respectively (Figure 18.20e).

To train the neural network, we prepared 808 optical images and their corresponding segmentation maps using the Image Labeler application in MATLAB (MathWorks, Inc.). The size of the dataset was sufficiently large to train the network because U-Net requires relatively few training images. The 808 images were divided into training, validation, and testing sets with 566, 162, and 80 images, respectively. Data augmentation was applied to the 566 training images, including horizontal and vertical flips and 180° rotation, thus generating 2264 training images. The encoder and decoder weights were initialized using the technique described by He et al. [87]. To train all the variants, we employed Adam [88], which is an algorithm for the first-order gradient-based optimization of stochastic objective functions, with a fixed learning rate, β_1, β_2, and ε of 0.001, 0.9, 0.999, and 10^{-7}, respectively. The batch size and epoch number were set to 4 and 40, respectively. Prior to each epoch, the training set was shuffled, and each mini-batch was picked subsequently to ensure that each image was used only once per epoch. To evaluate segmentation performance, we used the Dice coefficient, which ranged from 0 to 1; the latter corresponded to perfect segmentation. The Dice coefficient curves for both the training and validation datasets over the training epochs converged. The training procedure was completed after 1 and 40 minutes. Subsequently, we conducted a holdout validation to select the highest-performing model on the validation dataset. To evaluate the performance of our model, we calculated the evaluation metrics of precision p (also referred to as the positive predictive value), recall r (also known as sensitivity), and F_1-score, according to Eqs. (18.7)–(18.9) [89]:

$$p = \mathrm{TP} / (\mathrm{TP} + \mathrm{FP}) \tag{18.7}$$

$$r = \mathrm{TP} / (\mathrm{TP} + \mathrm{FN}) \tag{18.8}$$

$$F_1 = 2pr / (p + r) \tag{18.9}$$

where the TP, FP, and FN labels correspond to true positives, false positives, and false negatives, respectively. The results for each label are summarized in Table 18.2, which indicated that our model successfully classified pixels, while achieving the F_1-scores above 80%; this is a practical value for screening devices. The F_1-score of the contaminants was lower than that of other categories, which was attributed to the complications arising from the semantic boundary prediction process [90]. In general, semantic boundary prediction, which identifies pixels that belong to object boundaries, can be a challenging and laborious procedure because the precise detection and annotation of object boundaries is a difficult task, frequently resulting in severe label noise. The objects labeled as contaminants were small and contained a relatively large number of boundary pixels, which induced a decrease in the

Table 18.2 Evaluation metrics of the precision p, recall r, and F_1-score for each category.

	Graphene	Electrode	Substrate	Contaminants
p (%)	97.15	99.72	99.20	76.90
r (%)	99.59	95.63	98.15	88.60
F_1-score (%)	98.35	97.62	98.67	81.78

Source: Reproduced from Ushiba et al. [69]/with permission from IOP Publishing.

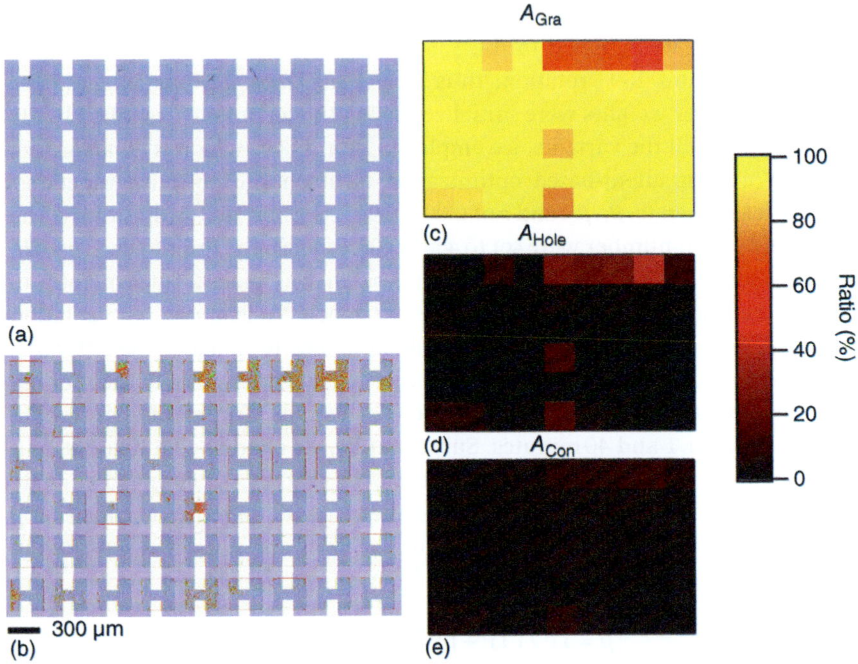

Figure 18.21 (a) Rearranged raw optical images of a GFET array. (b) Rearranged overlay images. The scale bar for (a) and (b) is 300 μm. (c–e) 2D maps of A_{Gra} (c), A_{Hole} (d), and A_{Con} (e). Source: Reproduced from Ushiba et al. [69]/with permission from IOP Publishing.

F_1-score. However, the F_1-score can be improved if the network is trained using higher-resolution images. To quantitatively evaluate the images, the respective ratios of the graphene films (A_{Gra}), holes (A_{Hole}), and contaminants (A_{Con}) to the expected graphene area were introduced. A device with no holes or contaminants yielded an A_{Gra} of 100%. In the overlay image in Figure 18.20e, A_{Gra}, A_{Hole}, and A_{Con} ratios were calculated as 81.2%, 12.5%, and 6.3%, respectively. These results clearly verified the quantitative evaluation of the optical image of the GFET. The time interval required until a segmented image and corresponding overlay image were generated from a single image was 10 seconds with computer specifications: Intel (R) Xeon (R) gold 6126 CPU @ 2.60 GHz, 32 GB of memory.

To perform fast screening, we used a motorized optical microscope and captured optical images of GFETs arranged in an array of six rows and nine columns, for a total of 54 images per chip. Subsequently, all images were individually input into the network, producing an equal number of overlay images from the corresponding raw and segmented images, as shown in Figure 18.20e. The images were then rearranged to the corresponding positions in a six rows/nine columns array (Figure 18.21a,b). It was difficult to evaluate the chip quality from a brief visual observation of the raw optical images shown in Figure 18.21a. Conversely, in Figure 18.21b, it is easily discernible which GFETs were damaged and/or contaminated via visual observation; for example, the top-row GFETs were relatively

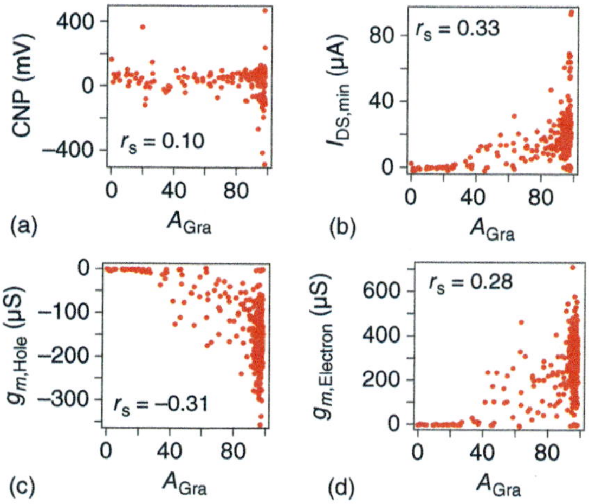

Figure 18.22 Scatter plots between A_{Gra} and CNP (a), $I_{DS,min}$ (b), $g_{m,Hole}$ (c) and $g_{m,Electron}$ (d). Spearman's rank correlation coefficient r_s was also calculated. Source: Reproduced from Ushiba et al. [69]/with permission from IOP Publishing.

damaged. Quantitative information regarding A_{Gra}, A_{Hole}, and A_{Con} ratios for the entire chip was also obtained as shown in Figure 18.21c–e, with their average and standard deviations calculated as $91.6 \pm 0.1\%$, $4.7 \pm 7.5\%$, and $3.7 \pm 1.9\%$, respectively. Furthermore, the 2D maps indicated quality variation over the chips, enabling quantitative quality comparison between them.

To investigate the correlation between the optical images and transfer characteristics of GFETs, we captured more than 500 optical images and subsequently produced their corresponding segmented images. The transfer characteristics of the corresponding GFETs were also measured. Figure 18.22a–d display the scatter plots of A_{Gra} in relation to CNP, $I_{DS,min}$, $g_{m,Hole}$, and $g_{m,Electron}$, respectively. Spearman's rank correlation coefficient, r_s, which is a nonparametric measure of rank correlation, was also calculated for each plot. There was no apparent correlation between A_{Gra} and CNP values, as r_s was 0.1, which was between −0.2 and 0.2 (Figure 18.22a). In contrast, A_{Gra} showed correlation with $I_{DS,min}$, $g_{m,Hole}$, and $g_{m,Electron}$, and the corresponding r_s values were 0.33, −0.31, and 0.28, respectively. This was attributed to the larger holes, implying that both A_{Gra} and the effective channel width decreased, inducing a decrease in conductance by extension, that is, a decrease in $I_{DS,min}$, $g_{m,Hole}$, and $g_{m,Electron}$. Although A_{Gra} had almost reached 100%, a significant variation in $I_{DS,min}$, $g_{m,Hole}$, and $g_{m,Electron}$ trends still remained, presumably due to the variation in the contact resistance between the electrode and the graphene films [91], which severely impacted the conductance as well. Because the optical images contained no information regarding the contact resistance, predicting the transfer characteristics from these images alone was not feasible. However, the correlation between the optical images and the conductance enables the screening of GFETs based on segmented images. In conclusion, a DL-based semantic segmentation algorithm was implemented to quantitatively evaluate the optical images of GFETs.

Our developed U-Net-based network generated semantically segmented images where every pixel was labeled as graphene, electrode, substrate, or contaminant, achieving F_1-scores exceeding 80%. Combined with a motorized optical microscope, the chip was effectively scanned, and all images were promptly and quantitatively evaluated, which would be impossible to perform with the naked eye. Our proposed strategy may also lead to a better understanding of the origins of process variation, as it can statistically analyze which position in the array is frequently damaged or contaminated. Furthermore, as long as the network is provided with training images, the algorithm can yield results even if the microscopic conditions, such as illumination and color balance, change. Hence, our approach paves the way for the large-scale and stable manufacturing of GFET arrays.

18.5 Conclusion

In this chapter, the application of graphene-based FET biosensors for selective high-sensitivity detection has been thoroughly discussed. We first presented the functionalizing GFETs with sialoglycans, which mimic cell surfaces where the influenza virus infection response occurs. The capability of GFETs in distinguishing species-specific infectivity with high sensitivity is demonstrated. This phenotype-based sensing method has the potential to distinguish between variant avian influenza viruses faster than conventional methods such as general-purpose PCR. Thus, we believe that our device will be useful as a borderline measure to prevent the spread of variant viruses. We then presented and discussed several approaches to further enhance the sensitivity and expand the potential applications of GFET biosensors. Furthermore, we addressed the practical issues of baseline drift and inspection methods by proposing two main approaches, hardware- and software-based, to tackle the drift problem. Both approaches were shown to be efficient and could be applied to further improve inspection methods. Finally, we presented a new deep-learning-based method, which we developed, to implement automated screening based on algorithms trained with sets of optical images. We showed that our approach creates new means for large-scale and stable manufacturing of GFET arrays.

References

1 Maehashi, K., Matsumoto, K., Kerman, K. et al. (2004). Ultrasensitive detection of DNA hybridization using carbon nanotube field-effect transistors. *Jpn. J. Appl. Phys.* 43: 1558–1560. https://doi.org/10.1143/JJAP.43.L1558.

2 Maehashi, K., Katsura, T., Kerman, K. et al. (2007). Label-free protein biosensors based on aptamer-modified carbon nanotube field-effect transistors. *Anal. Chem.* 79: 782–787.

3 Ohno, Y., Maehashi, K., Yamashiro, Y., and Matsumoto, K. (2009). Electrolyte-gated graphene field-effect transistors for detecting pH and protein adsorption. *Nano Lett.* 9: 3318–3322. https://doi.org/10.1021/nl901596m.

4 Hwang, E.H., Adam, S., and Das Sarma, S. (2007). Carrier transport in two-dimensional graphene layers. *Phys. Rev. Lett.* 98: 186806.

5 Bolotin, K., Sikes, K., Jiang, Z. et al. (2008). Ultrahigh electron mobility in suspended graphene. *Solid State Commun.* 146: 351–355.

6 Ushiba, S., Ono, T., Kanai, Y. et al. (2018). Graphene as an imaging platform for charged molecules. *ACS Omega* 3: 3137–3142. https://doi.org/10.1021/acsomega.7b02008.

7 Ohno, Y., Maehashi, K., and Matsumoto, K. (2010). Label-free biosensors based on aptamer-modified graphene field-effect transistors. *J. Am. Chem. Soc.* 132: 18012–18013.

8 Shimatani, M., Ogawa, S., Fujisawa, D. et al. (2016). Giant Dirac point shift of graphene phototransistors by doped silicon substrate current. *AIP Adv.* 6: 035113: https://doi.org/10.1063/1.4944622.

9 Ono, T., Kanai, Y., Inoue, K. et al. (2019). Electrical biosensing at physiological ionic strength using graphene field-effect transistor in femtoliter microdroplet. *Nano Lett.* 19: 4004–4009.

10 Okamoto, S., Ohno, Y., Maehashi, K. et al. (2012). Immunosensors based on graphene field-effect transistors fabricated using antigen-binding fragments. *Jpn. J. Appl. Phys.* 51: 06FD08.

11 Kanai, Y., Ohmuro-Matsuyama, Y., Tanioku, M. et al. (2020). Graphene field-effect transistor-based immunosensor for ultrasensitive non-competitive detection of small antigens. *ACS Sens.* 5: 24–28. https://doi.org/10.1021/acssensors.9b02137.

12 Ono, T., Oe, T., Kanai, Y. et al. (2017). Glycan-functionalized graphene FETs toward selective detection of human-infectious avian influenza virus. *Jpn. J. Appl. Phys.* 56: 030302.

13 Watanabe, Y., Ibrahim, M.S., Ellakany, H.F. et al. (2011). Acquisition of human-type receptor-binding specificity by new H5N1 influenza virus sublineages during their emergence in birds in Egypt. *PLoS Pathog.* 7 (5): https://doi.org/10.1371/journal.ppat.1002068.

14 Suzuki, Y., Nakao, T., Ito, T. et al. (1992). Structural determination of ganglio- sides that bind to influenza A, B, and C viruses by an improved binding assay: Strain-specific receptor epitopes in sialo-sugar chains. *Virology* 189: 121.

15 Imai, M., Watanabe, T., Hatta, M. et al. (2012). Experimental adaptation of an influenza H5 haemagglutinin confers respiratory droplet transmission to a reassortant H5 HA/H1N1 virus in ferrets. *Nature* 486: 420.

16 Ohno, Y., Maehashi, K., and Matsumoto, K. (2010). Chemical and biological sensing applications based on graphene field-effect transistors. *Biosens. Bioelectron.* 26: 1727–1730. https://doi.org/10.1016/j.bios.2010.08.001.

17 Ohno, Y., Maehashi, K., Inoue, K., and Matsumoto, K. (2011). Label-free aptamer-based immunoglobulin sensors using graphene field-effect transistors. *Jpn. J. Appl. Phys.* 50: 070120: https://doi.org/10.1143/JJAP.50.070120.

18 Kawata, T., Ono, T., Kanai, Y. et al. (2018). Improved sensitivity of a graphene FET biosensor using porphyrin linkers. *Jpn. J. Appl. Phys.* 57: 065103.

19 Murakami, H., Nomura, T., and Nakashima, N. (2003). Non-covalent porphyrin-functionalized single-walled carbon nanotubes in solution and the formation of porphyrin–nanotube nanocomposites. *Chem. Phys. Lett.* 378: 481.

20 Penza, M., Rossi, R., Alvisi, M. et al. (2010). *Sens. Actuators, B* 144: 387.

21 Bassiouk, M., Basiuk, V.A., Basiuk, E.V. et al. (2013). Noncovalent functionalization of single-walled carbon nanotubes with porphyrins. *Appl. Surf. Sci.* 275: 168.

22 Chen, R.J., Zhang, Y., and Wang, D. (2001). Noncovalent sidewall functionalization of single-walled carbon nanotubes for protein immobilization. *J. Am. Chem. Soc.* 123: 3838–3839.

23 Li, T., Yang, M., and Li, H. (2011). Label-free electrochemical detection of cancer marker based on graphene–cobalt hexacyanoferrate nanocomposite. *Electroanal. Chem.* 655: 50–55.

24 Guo, S.-R., Lin, J., Penchev, M. et al. (2011). Label free DNA detection using large area graphene based field effect transistor biosensors. *J. Nanosci. Nanotechnol.* 11: 5258.

25 Winter, W.E., Hardt, N.S., and Fuhrman, S. (2000). Immunoglobulin E: importance in parasitic infections and hypersensitivity responses. *Arch. Pathol. Lab. Med.* 124: 1382.

26 Maehashi, K., Matsumoto, K., Takamura, Y., and Tamiya, E. (2009). Aptamer-based label-free immunosensors using carbon nanotube field-effect transistors. *Electroanalysis* 21: 1285.

27 Liss, M., Petersen, B., Wolf, H., and Prohaska, E. (2002). An aptamer-based quartz crystal protein biosensor. *Anal. Chem.* 74: 4488.

28 Mendonsa, S.D. and Bowser, M.T. (2004). In vitro selection of high-affinity DNA ligands for human IgE using capillary electrophoresis. *Anal. Chem.* 76: 5387.

29 Katilius, E., Katiliene, Z., and Woodbury, N.W. (2006). Signaling aptamers created using fluorescent nucleotide analogues. *Anal. Chem.* 78: 6484.

30 Turgeon, R.T., Fonslow, B.R., Jing, M., and Bowser, M.T. (2010). Measuring aptamer equilibria using gradient micro free flow electrophoresis. *Anal. Chem.* 82: 3636.

31 Ueda, H., Tsumoto, K., Kubota, K. et al. (1996). Open sandwich ELISA: a novel immunoassay based on interchain interaction of antibody variable region. *Nat. Biotechnol.* 14: 1714–1718. https://doi.org/10.1038/nbt1296-1714.

32 Lim, S.L., Ichinose, H., Shinoda, T., and Ueda, H. (2007). Noncompetitive detection of low molecular weight peptides by open sandwich immunoassay. *Anal. Chem.* 79: 6193–6200. https://doi.org/10.1021/ac070653z.

33 Ihara, M., Suzuki, T., Kobayashi, N. et al. (2009). Open-sandwich enzyme immunoassay for one-step non-competitive detection of corticosteroid 11 -deoxycortisol. *Anal. Chem.* 81: 8298–8304. https://doi.org/10.1021/ac900700a.

34 Sakata, T., Ihara, M., Makino, I. et al. (2009). Open sandwich-based immuno-transistor for label-free and noncompetitive detection of low molecular weight antigens. *Anal. Chem.* 81: 7532–7537. https://doi.org/10.1021/ac900457m.

35 Islam, K.N., Ihara, M., Dong, J. et al. (2011). Direct construction of an open-sandwich enzyme immunoassay for one-step noncompetitive detection of thyroid hormone T4. *Anal. Chem.* 83: 1008–1014. https://doi.org/10.1021/ac102801r.

36 Dong, J., Shichiri, M., Chung, C.-I. et al. (2017). An open sandwich immunoassay for detection of 13(*R*,*S*)-hydroxy-9(*E*),11(*E*)-octadecadienoic acid. *Analyst* 142: 787–793. https://doi.org/10.1039/C6AN02437H.

37 Dong, J. and Ueda, H. (2017). Open sandwich immunoassay and its applications. In: *Advances in Medicine and Biology* (ed. L.V. Berhardt). New York: Nova Science Publishers Inc.

38 Dong, J., Sakurai, A., Nomura, N. et al. (2013). Isolation of recombinant phage antibodies targeting the hemagglutinin cleavage site of highly pathogenic avian influenza virus. *PLoS One* 8: e61158: https://doi.org/10.1371/journal.pone.0061158.

39 Dong, J., Ihara, M., and Ueda, H. (2009). Antibody Fab display system that can perform open sandwich ELISA. *Anal. Biochem.* 386: 36–44. https://doi.org/10.1016/j.ab.2008.11.045.

40 Lee, N.K., Sowa, H., Hinoi, E. et al. (2007). Endocrine regulation of energy metabolism by the skeleton. *Cell* 130: 456–469. https://doi.org/10.1016/j.cell.2007.05.047.

41 Oury, F., Khrimian, L., Denny, C.A. et al. (2013). Maternal and offspring pools of osteocalcin influence brain development and functions. *Cell* 155: 228–241. https://doi.org/10.1016/j.cell.2013.08.042.

42 Oury, F., Sumara, G., Sumara, O. et al. (2011). Endocrine regulation of male fertility by the skeleton. *Cell* 144: 796–809. https://doi.org/10.1016/j.cell.2011.02.004.

43 Sangawa, T., Tabata, S., Suzuki, K. et al. (2013). A multipurpose fusion tag derived from an unstructured and hyperacidic region of the amyloid precursor protein. *Protein Sci.* 22: 840–850. https://doi.org/10.1002/pro.2254.

44 Wongso, D., Dong, J., Ueda, H., and Kitaguchi, T. (2017). Flashbody: a next generation fluobody with fluorescence intensity enhanced by antigen binding. *Anal. Chem.* 89: 6719–6725. https://doi.org/10.1021/acs.analchem.7b00959.

45 Gao, N., Gao, T., Yang, X. et al. (2016). Specific detection of biomolecules in physiological solutions using graphene transistor biosensors. *Proc. Natl. Acad. Sci. U.S.A.* 113: 14633–14638.

46 Nakatsuka, N., Yang, K.-A., Abendroth, J.M. et al. (2018). Aptamer-FETs overcame Debye length limitations for small-molecule sensing. *Science* 362: 319–324. https://doi.org/10.1126/science.aao675.

47 Vacic, A., Criscione, J.M., Rajan, N.K. et al. (2011). Determination of the molecular configuration by Debye length modulation. *J. Am. Chem. Soc.* 133: 13886–13889.

48 Harris, L.J., Skaletsky, E., and McPherson, A. (1998). Crystallographic structure of an intact IgG1 monoclonal antibody. *J. Mol. Biol.* 275: 861–872.

49 Oka, Y., Ushiba, S., Miyakawa, N. et al. (2022). Ionic strength-sensitive and pH-insensitive interactions between C-reactive protein (CRP) and an anti-CRP antibody. *Biophys. Physicobiol.* 19: 1–9. https://doi.org/10.2142/biophysico.bppb-v19.0003.

50 Warren, J.R. and Marshall, B. (1983). Unidentified curved bacilli on gastric epithelium in active chronic gastritis. *Lancet* 321: 1273–1275.

51 World Health Organisation (1994). Infection with Helicobacter pylori. In: *IARC Monographs on the Evaluation of Carcinogenic Risks to Humans*, vol. 61, 177–241. Lyon, France: International Agency for Research on Cancer.

52 Uemura, N., Okamoto, S., Yamamoto, S. et al. (2001). Helicobacter pylori infection and the development of gastric cancer. *N. Engl. J. Med.* 345: 784–789.

53 Plummer, M., Franceschi, S., Vignat, J. et al. (2015). Global burden of gastric cancer attributable to Helicobacter pylori. *Int. J. Cancer* 136: 487–490.

54 Hayashi, T., Senda, M., Suzuki, N. et al. (2017). Differential mechanisms for SHP2 binding and activation are exploited by geographically distinct helicobacter pylori CagA oncoproteins. *Cell Rep.* 20: 2876–2890.

55 McNulty, C.A.M. and Wise, R. (1985). Rapid diagnosis of *Campylobacter*-associated gastritis. *Lancet* 325: 1443–1444.

56 Mobley, H.L.T. and Island, M.D. (1995). Molecular biology of microbial urease. *Microbiol. Rev.* 59: 451–480.

57 Schedin, F., Geim, A.K., Morozov, S.V. et al. (2007). *Nat. Mater.* 6: 652–655.

58 Romero, H.E., Joshi, P., Gupta, A.K. et al. (2009). Adsorption of ammonia on graphene. *Nanotechnology* 20: 245501.

59 Gautam, M. and Jayatissa, A.H.J. (2012). Graphene based field effect transistor for the detection of ammonia. *Appl. Phys.* 112: 064304.

60 Inaba, A., Yoo, K., Takei, Y. et al. (2014). Ammonia gas sensing using a graphene field–effect transistor gated by ionic liquid. *Sens. Actuators, B* 195: 15–21.

61 Kim, S.H., Iwai, S., Araki, S. et al. (2012). Large-scale femtoliter droplet array for digital counting of single biomolecules. *Lab Chip* 12: 4986–4991.

62 Ono, T., Ichiki, T., and Noji, H. (2018). Digital enzyme assay using attoliter droplet array. *Analyst* 143: 4923–4929.

63 Rondelez, Y., Tresset, G., Tabata, K.V. et al. (2005). Microfabricated arrays of femtoliter chambers allow single molecule enzymology. *Nat. Biotechnol.* 23: 361–365.

64 Rissin, D.M., Kan, C.W., Campbell, T.G. et al. (2010). Single-molecule enzyme-linked immunosorbent assay detects serum proteins at subfemtomolar concentrations. *Nat. Biotechnol.* 28: 595–599.

65 Sakakihara, S., Araki, S., Iino, R., and Noji, H. (2010). A single-molecule enzymatic assay in a directly accessible femtoliter droplet array. *Lab Chip* 10: 3355–3362.

66 Bergveld, P. (1970). Development of ion-sensitive solid-state device for neurophysiological measurements. *IEEE Trans. Biomed. Eng.* BME-17: 70–71.

67 Woias, P., Meixner, L., and Fröstl, P. (1998). Slow-pH response of silicon nitride ISFET sensors. *Sens. Actuators, B* 48: 501–504.

68 Jamasb, S., Collins, S., and Smith, R.L. (1998). Physical model for drift in pH ISFETs. *Sens. Actuators, B* 49: 146–155.

69 Ushiba, S., Okino, T., Miyakawa, N. et al. (2020). State-space modeling for dynamic response of graphene FET biosensors. *Jpn. J. Appl. Phys.* 59: SGGH04.

70 Bousse, L. and Bergveld, P. (1984). The role of buried OH sites in the response mechanism of inorganic-gate pH-sensitive ISFETs. *Sens. Actuators* 6: 65–78.

71 Zhang, J., Zhao, D., Yang, H. et al. (2019). Dual-mechanism model to describe slow response of ISFETs. *IEEE Sens. J.* 19: 7471–7478.

72 Miyakawa, N., Shinagawa, A., Kajiwara, Y. et al. (2021). Drift suppression of solution-gated graphene field-effect transistors by cation doping for sensing platforms. *Sensors* 21: 7455. https://doi.org/10.3390/s21227455.

73 Kim, J.S., Choi, J.S., Lee, M.J. et al. (2013). Between Scylla and Carybdis: hydrophobic graphene-guided water diffusion on hydrophilic substrates. *Sci. Rep.* 3: 2309.

74 Xu, K., Cao, P., and Heath, J.R. (2010). Graphene visualized the first water layer on mica under ambient conditions. *Science* 329: 1188–1191.

75 Linossier, I., Gaillard, F., Romand, M., and Feller, J.F. (1997). Measuring water diffusion in polymer films on the substrate by internal reflection Fourier transform infrared spectroscopy. *J. Appl. Polym. Sci.* 66: 2465–2473.

76 Jia, X., Hu, M., Soundarapandian, K. et al. (2019). Kinetic ionic permeation and interfacial doping of the supported graphene. *Nano Lett.* 19: 9029–9036.

77 Chen, J.H., Jang, C., Adam, S. et al. (2008). Charged-impurity scattering in graphene. *Nat. Phys.* 4: 377.

78 Lin, Y.C., Lu, C.C., Yeh, C.H. et al. (2012). Graphene annealing: how clean can it be? *Nano Lett.* 12: 414.

79 Rao, K.S., Senthilnathan, J., Liu, Y.F., and Yoshimura, M. (2014). Role of peroxide ions in formation of graphene nanosheets by electrochemical exfoliation of graphite. *Sci. Rep.* 4: 4237.

80 Patterson, T.A., Thomas, L., Wilcox, C. et al. (2008). State-space models of individual animal movement. *Cell* 23: 87.

81 Thalmayr, F. and Fischerauer, G. (2010). Sensor signal processing for gravimetric chemical sensors based on a state-space model. *Sens. Actuators, B* 144: 27–36. https://doi.org/10.1016/j.snb.2009.10.014.

82 Seredych, M., Mikhalovska, L., Mikhalovsky, S., and Gogotsi, Y. (2018). Adsorption of bovine serum albumin on carbon-based materials. *J. Carbon Res.* 4: 3.

83 Brooks, S., Gelman, A., Jones, G.L., and Meng, X.-L. (2011). *Handbook of Markov Chain Monte Carlo*. Boca Raton, FL: CRC Press.

84 Carpenter, B., Gelman, A., Hoffman, M.D. et al. (2017). Stan: a probabilistic programming language. *J. Stat. Software* 76: 1.

85 Ushiba, S., Miyakawa, N., Ito, N. et al. (2021). Deep-learning-based semantic image segmentation of graphene field-effect transistors. *Appl. Phys Express* 14: 036504. https://doi.org/10.35848/1882-0786/abe3db.

86 Ronneberger, O., Fischer, P., and Brox, T. (2015). U-Net: convolutional networks for biomedical image segmentation. arXiv: 1505.04597. https://doi.org/10.48550/arXiv.1505.04597.

87 He, K., Zhang, X., Ren, S., and Sun, J. (2015). Delving deep into rectifiers: surpassing human-level performance on ImageNet classification. arXiv: 1502.01852. https://doi.org/10.48550/arXiv.1502.01852.

88 Kingma, D.P. and Ba, J.L. (2014). ADAM: a method for stochastic optimization. arXiv: 1412.6980. https://doi.org/10.48550/arXiv.1412.6980.

89 Minaee, S., Boykov, Y., Porikli, F., et al. (2020). Image segmentation using deep learning: a survey. arXiv: 2001.05566. https://doi.org/10.48550/arXiv.2001.05566.

90 Acuna, D., Kar, A., and Fidler, S. (2019). Devil is in the edges: learning semantic boundaries from noisy annotations. arXiv: 1904.07934. https://doi.org/10.48550/arXiv.1904.07934.

91 Giubileo, F. and Di, A. (2017). The role of contact resistance in graphene field-effect devices. *Prog. Surf. Sci.* 92: 143.

19

High-Performance Detection of Extracellular Vesicles Using Graphene Field-Effect Transistor Biosensor

Ding Wu[1,2], Yi Yu[1], Zhi-Yong Zhang[3], and Guo-Jun Zhang[1]

[1] *Hubei University of Chinese Medicine, School of Laboratory Medicine, 16 Huangjia Lake West Road, Wuhan, 430065, China*
[2] *Zhongnan Hospital of Wuhan University, Department of Laboratory Medicine, 169 Donghu Road, Wuhan, 430071, China*
[3] *Peking University, Key Laboratory for the Physics and Chemistry of Nanodevices, Department of Electronics, 5 Yiheyuan Road, Beijing, 100871, China*

Extracellular vesicles (EVs), which are nanoscale lipid bilayer vesicles released by almost all cells, have been recognized as completely new and promising biomarkers for next-generation liquid biopsy of tumors and have been studied extensively all over the world in recent years. Thus, high-performance detection of EVs is of great significance for the early diagnosis of various tumors. The selective recognition of biomolecules through the combination of nanomaterials and biosensing elements has promoted the development of nanobiosensors. Compared with the traditional detection technology, graphene field-effect transistor (GFET), as a new type of biosensing platform developed in recent years, hold the following important advantages, including high sensitivity and selectivity, fast analysis speed, label-free detection and cost-effectiveness. Therefore, GFET is promising to be a powerful tool for EVs-based tumor diagnosis in future. This chapter introduces the definition, classification, function, and clinical significance of EVs and focuses on the working principle and application of GFET biosensors for EVs detection.

19.1 What is Extracellular Vesicles

EVs are membranous vesicles secreted by a variety of living cells into the extracellular environment and contain abundant macromolecular substances carried from parent cells, including proteins, nucleic acids, and other biological molecules [1–5]. The size of EVs ranges from 30 to 1000 nm and can be classified according to their biogenesis, size, and biophysical properties [6–8]. The classic description of EVs is based on the size of vesicles. The internal vesicles produced in the multivesicular

Graphene Field-Effect Transistors: Advanced Bioelectronic Devices for Sensing Applications, First Edition. Edited by Omar Azzaroni and Wolfgang Knoll.

endosome compartment are called exosomes (EXOs), which are secreted by the fusion of the multivesicular body membrane and the plasma membrane. EXOs are usually less than 150 nm in diameter and contain proteins related to endosomes. The general consensus is that certain EVs released directly from the plasma membrane of the cells are usually called microvesicles (MVs). MVs are larger (approximately 10–1000 nm) plasma membrane-derived vesicles. Apoptotic bodies originate from apoptosis and are also large vesicles (>1000 nm). The above definition has not yet reached a consensus. It is known that EV populations released from cells and found in circulating biological fluids are heterogeneous in source, size, and composition. EVs exist in almost all human body fluids, including blood, urine, and saliva, and can also be collected from cell culture supernatants [9–11].

19.2 The Clinical Significance of Extracellular Vesicles

EVs, as a new type of intracellular signal transduction carrier and extracellular communication messenger, are almost involved in tumorigenesis, development, metastasis, multidrug resistance, and other links [12–15]. Furthermore, EVs also play a vital role in neurodegenerative diseases [16, 17], metabolic homeostasis [18, 19], angiogenesis [20, 21], inflammation [22, 23], neuronal plasticity [24], migration [25], nutritional support [26], pathogen infection [27], etc. Therefore, EVs are recognized as one of the three major liquid biopsy biomarkers along with circulating tumor DNA (ctDNA) and circulating tumor cells (CTCs) and have become an ideal candidate biomarker and research hotspot for diagnosis and treatment of various diseases, especially cancer [28, 29]. Furthermore, the superior cargo transport capability, biocompatibility, and natural nanoscale dimensions of EVs have led to the exploration of EVs as a vesicle for drug delivery [30, 31]. Due to their low immunogenicity, EVs can be repeatedly administered, which overcomes the barriers of repeated administration of viral vectors [32]. As reported, EV is capable of crossing the blood-brain barrier in mice [33], which makes targeted treatment of brain diseases possible. In conclusion, EVs not only have been recognized as the next-generation tumor liquid biopsy biomarkers but also can be applied as natural therapeutic drug carriers to treat various diseases.

19.3 Introduction to Graphene Field-Effect Transistor Biosensor

There are many traditional EVs detection technologies, such as flow cytometry (FCM), enzyme-linked immunosorbent assay (ELISA), nanoparticle tracking analysis (NTA), dynamic light scattering (DLS), tunable resistance pulse sensing (TRPS), etc. [34–36]. However, these technologies often suffer from disadvantages,

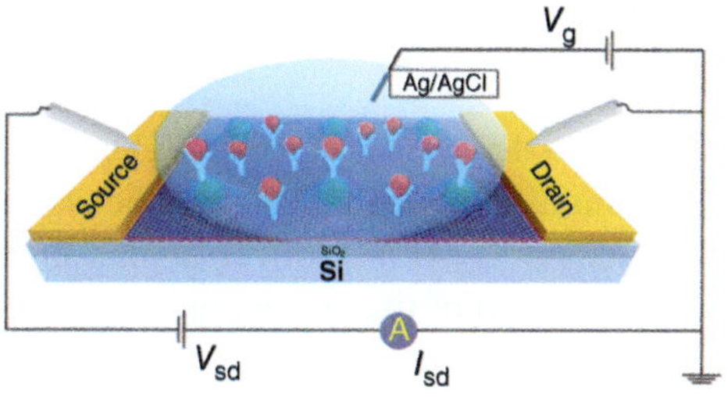

Figure 19.1 Basic structure of GFET. Source: Reproduced from Zhang et.al. [38]/with permission from American Chemical Society.

such as low sensitivity, target labeling, long detection times, high detection costs, etc. With the continuous development of nanotechnology and semiconductor technology, the fabrication craft of GFET biosensors has become increasingly mature. GFET biosensor is usually composed of three electrodes: source electrode (S), drain electrode (D), and gate electrode (G) [37]. Commonly, graphene is fixed on the channel surface between S and D electrodes, so as to form a conductive channel and serve as the sensing element of the device (Figure 19.1). Before detecting a target, probe that can specifically bind to the target is usually immobilized on the surface of graphene through different methods. During the detection process, the specific binding event between the target and the probe causes rapid charge density changes on the graphene surface, which in turn leads to conductance changes of the device. Finally, the quantitative detection of target content can be achieved by detecting electrical signal changes in the GFET biosensor before and after the reaction. Benefiting from the natural advantages of graphene, such as its large active detection area, high electron mobility, and low preparation cost, GFET biosensors exhibit excellent analytical performance and have great value in clinical practice. Compared with other biosensors, GFET biosensors hold much superiority, such as high sensitivity, good specificity, short detection time, small device size and low detection cost. Currently, GFET biosensors have been applied for the detection of various substances, such as nucleic acids [39–45], proteins [38, 46–50], bacteria [51–54], viruses [55–61], and small molecules [62–66]. Thus, GFET biosensor is promising to be a point-of-care tool for the detection EVs.

19.4 GFET Biosensor for High-Performance Detection of Extracellular Vesicles

To date, there have been many reports about using GFET for the detection of EVs. The following content shows the relevant research results of Zhang's group using GFET biosensors to detect EVs.

19.4.1 Detection of the Overall Level of Microvesicles Using GFET Biosensor

Wu et al. [67] reported a streptavidin (SA)-functionalized GFET for the highly sensitive and specific detection of MVs derived from different cells. Unlike other EVs subtypes (such as EXOs), the surface of MVs lacks well-defined universal markers, so it is challenging to realize the quantitative detection of the entire population of MVs. To artificially label the surface of MVs with "Biotin," 1,2-distearoyl-*sn*-glycero-3-phosphoethanolamine-*N*-[biotinyl(polyethylene glycol)-2000)] (DSPE-PEG-Biotin) is added to the cell culture medium when culturing different cells, as illustrated in Figure 19.2a. For specific detection of Biotin-MVs, the SA probe is modified and placed on the channel surface of the GFET. The SA probe on the graphene surface can specifically capture Biotin-MVs (negatively charged), so the negative charge on the surface of Biotin-MVs causes an immediate surface charge

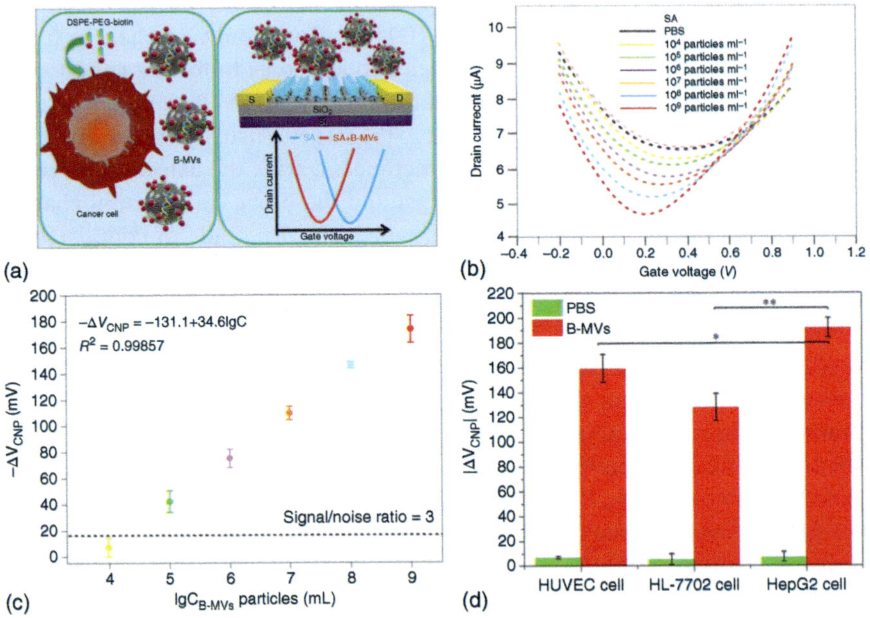

Figure 19.2 (a) Microvesicles detection by GFET biosensor based on a membrane biotinylation strategy. (b) Transfer characteristics and (c) shift of V_{CNP} of the SA-immobilized GFET biosensors treated with the B-MVs at a series of concentrations (10^4 particles ml^{-1} − 10^9 particles ml^{-1}). (d) The shift of V_{CNP} of the SA-immobilized FET biosensors for detecting B-MVs derived from HUVEC, HL-7702 and HepG2 cells, respectively. Error bars represent standard deviations of measurements ($n = 3$). Differences between B-MVs derived from different cells were statistically significant (Welch *t*-test). **: $p < 0.01$, *: $p < 0.05$. Source: Reproduced from Wu et al. [67]/with permission from the Royal Society of Chemistry.

change of the graphene. Therefore, the specific detection of Biotin-MVs can be achieved by monitoring the electrical signal changes (V_{CNP}) of GFET chips. Due to the high affinity and multivalent (1 : 4) combination between SA and biotin, the sensitivity of the SA-functionalized GFET biosensor for Biotin-MVs detection is greatly improved. The limit of detection is as low as 20 particles μl^{-1} (Figure 19.2b,c). The GFET biosensor cannot only identify Biotin-MVs with high specificity but also detect Biotin-MVs derived from different cell lines (Figure 19.2d) and Biotin-MVs in clinical samples. The development of this platform provides a new and reliable platform for the MVs detection in various biological samples. Though the high-performance detection of MVs is realized through the above research, the SA-modified GFET biosensor can only be applied to detect the overall level of MVs, and its significance for tumor diagnosis is still limited.

19.4.2 Specific Detection of Hepatocellular Carcinoma-Derived Microvesicles Using Dual-Aptamer Modified GFET Biosensor

Inspired by previously reported detection strategies for HepG2 cells, Wu et al. [68] further demonstrated an *in situ* gold nanoparticles (AuNPs) deposited, dual aptamer-functionalized reduced graphene oxide (RGO) FET (AAP-GFET) nanosensor for the specific detection of MVs derived from HepG2 cells (HepG2-MVs), as shown in Figure 19.3a. The AAP-GFET nanosensor is capable of detecting HepG2-MVs with high selectivity and sensitivity (84 particles μl^{-1}, Figure 19.3b,c) through a three-step process. Firstly, HepG2-MVs were purified from clinical blood samples and/or cell culture supernatants using differential centrifugation (step 1). Subsequently, AuNP was in situ deposited onto the surface of RGO sheets according to previously reported literature [69] and the thiolated HepG2 cell-specific TLS11a aptamer and thiolated EpCAM aptamer were then simultaneously functionalized onto AuNPs through the Au—S bond. Therefore, the AuNP-deposited, dual-aptamer-modified RGO-FET nanosensor was finally constructed (step 2). Due to the specific recognition between HepG2-MVs and aptamers, HepG2-MVs specifically bind onto the channel surface of the AAP-GFET nanosensor. This combination accumulates negative charges onto the surface of the RGO (HepG2-MVs were negatively charged proven by zeta potential detection), causing the Dirac point of the transfer characteristic curve to change immediately. Finally, the quantitative detection of HepG2-MVs was achieved by calculating the amount of Dirac point change before and after the reaction (step 3). This dual aptamer-based recognition strategy and the RGO-FET biosensor as an excellent charge conduction platform finally realized the high sensitivity and high specificity detection of HepG2-MVs. Most importantly, this well-designed nanosensor was able to distinguish hepatocellular carcinoma (HCC) patients from healthy individuals, showing its great application potential in the early diagnosis of HCC (Figure 19.3d).

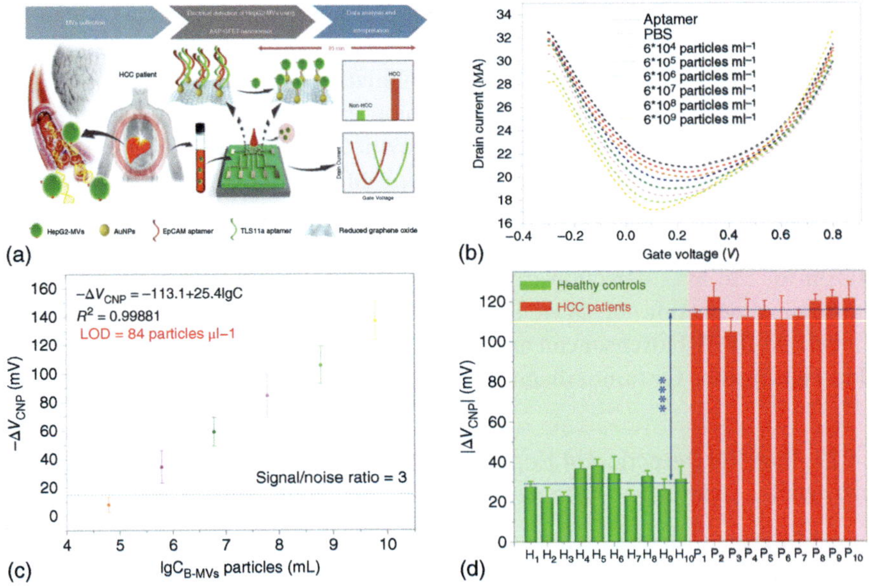

Figure 19.3 (a) Dual-aptamer modified GFET biosensor for label-free and specific detection of hepatocellular carcinoma-derived microvesicles. (b) Transfer curves of the AAP-GFET nanosensor incubated with varying concentrations of HepG2-MVs (6×10^4 particles ml^{-1} – 6×10^9 particles ml^{-1}). (c) Logarithmic plot of the shift of V_{CNP} ($-\Delta V_{CNP}$) vs. different concentrations of HepG2-MVs. The dashed line represents the threefold noise level (~12 mV) from the blank control test. (d) Clinical blood samples analysis using the AAP-GFET nanosensor. The clinical samples involved 10 HCC patients and 10 healthy controls. The dotted line shows the mean value of HepG2-MVs signals ($|\Delta V_{CNP}|$) from 10 healthy controls or 10 HCC patients detected by the AAP-GFET nanosensor. ****: $p < 0.0001$ (Welch t-test). Error bars represent standard deviations of measurements (n = 3). Source: Reproduced from Wu et.al. [68]/with permission from American Chemical Society.

19.4.3 Label-Free Detection of Cancerous Exosomes Using GFET Biosensor

Compared with the detection of MVs, the research on the detection of EXOs is more popular and extensive, since the separation technology for EXOs is more mature, and its significance for clinical diagnosis and treatment is likely more profound. As diagrammed in Figure 19.4a, Yu et al. [70] fabricated an anti-CD63 antibody-decorated RGO-FET biosensor for the sensitive detection of EXOs. CD63 molecule is a universal and highly expressed protein marker on the surface of EXOs, so its specific antibody was selected as the detection probe. After the successful fabrication of RGO-FET, anti-CD63 antibody was modified onto the RGO using the 1-pyrenebutanoic acid succinimidyl ester (PASE) as a crosslinker. Since EXO is negatively charged, the EXOs captured by the antibody continuously accumulate negative charges onto the surface of the RGO, leading to their n-doping of the RGO and changing the Dirac point in the transfer characteristic immediately. As a result, the identification and quantitative detection of EXOs was finally realized. This GFET

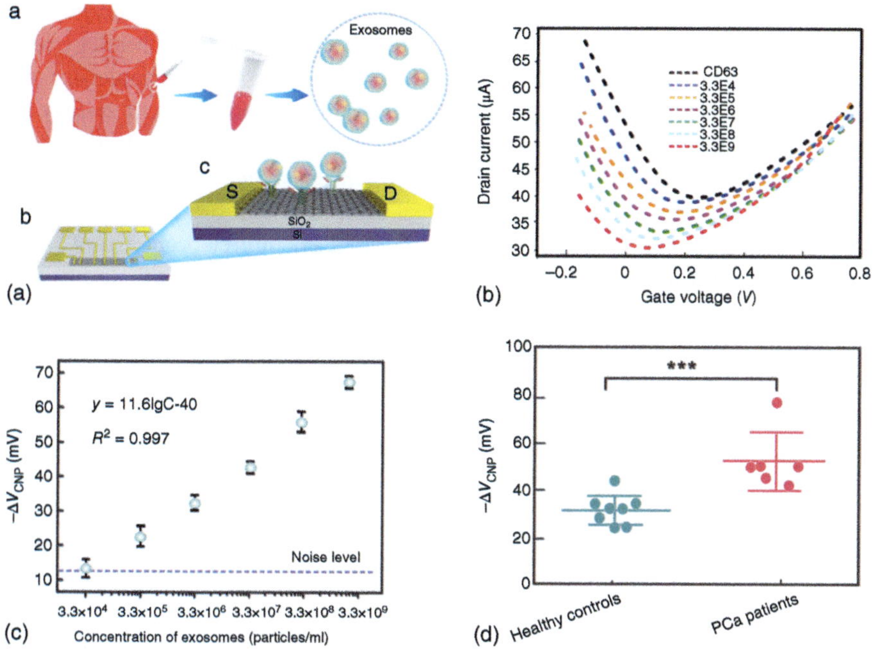

Figure 19.4 (a) Label-free and electrical detection of exosomes with a CD63 antibody-functionalized GFET biosensor. (b) Transfer curves of the GFET biosensor interacted with varying concentrations of exosomes. (c) The shift of V_{CNP} of the GFET biosensor reacted with a series of concentrations of exosomes. Dashed line represents the signal/noise ratio of 3 from the blank control test. (d) The expression level of exosomes in the serum obtained from healthy volunteers and cancer patients, ***: $p < 0.001$ (Welch *t*-test). Source: Reproduced from Yu et.al. [70]/with permission from American Chemical Society.

biosensor was capable of detecting EXOs with a limit of detection as low as 33 particles μl^{-1} (Figure 19.4b,c). Moreover, the platform was applied for the detection of EXOs separated from clinical serum samples and successfully distinguished prostate cancer patients from healthy persons, indicating its great potential for clinical application (Figure 19.4d). This EXOs detection platform demonstrated lots of advantages, such as label-free, short detection time (30 minutes), high sensitivity, excellent selectivity and low sample consumption (10 μl).

19.5 Some Prospects for Graphene Field-Effect Transistor Biosensor

This chapter discusses the application of GFET biosensors for the detection of EVs. These GFET biosensors applied for the detection of EVs usually hold the characteristics of high sensitivity, good specificity, short detection time, label-free detection, etc., representing their broad application prospects in clinical diagnosis and treatment. With the increasing development of nanosensors, it is believed that GFET will play an increasingly important role in disease diagnosis, monitoring, prognostic judgment, medical research, etc. by integrating with microfluidic technology and AC readout. On the other hand, GFET nanosensors are also facing unprecedented challenges, such as: (i) due to the immature fabrication technology, it is easy to cause problems such as the uniformity and poor stability of the sensor; (ii) the need for complex micro-processing technology causes the problem of high fabrication cost; and (iii) the composition of actual clinical samples is complex, so these GFET biosensors that only stay in the research stage often encounter problems of poor signal stability and high background signal when detecting clinical samples. In the future, GFET biosensors may break through these bottlenecks in several directions: (i) discovering new fabrication processes through technological innovation, especially the research and development of new processes for sensor surface functionalization that make GFET biosensors with low cost, high uniformity, and good stability possible. (ii) Increasing the research on the factors affecting the surface blocking effect of functionalized GFET biosensors so as to reduce the binding of nonspecific substances on the sensor surface. (iii) Intensifying research on new methods for reducing background signals in clinical sample detection so that GFET biosensors can be used in clinical practice as soon as possible, leading technological innovation, and bringing the gospel of a new era to the majority of patients.

References

1 Robbins, P.D. and Morelli, A.E. (2014). Regulation of immune responses by extracellular vesicles. *Nat. Rev. Immunol.* 14 (3): 195–208.

2 Tkach, M. and Thery, C. (2016). Communication by extracellular vesicles: where we are and where we need to go. *Cell* 164 (6): 1226–1232.

3 Raposo, G. and Stahl, P.D. (2019). Extracellular vesicles: a new communication paradigm? *Nat. Rev. Mol. Cell Biol.* 20 (9): 509–510.

4 van der Pol, E., Boing, A.N., Harrison, P. et al. (2012). Classification, functions, and clinical relevance of extracellular vesicles. *Pharmacol. Rev.* 64 (3): 676–705.

5 van Niel, G., D'Angelo, G., and Raposo, G. (2018). Shedding light on the cell biology of extracellular vesicles. *Nat. Rev. Mol. Cell Biol.* 19 (4): 213–228.

6 Raposo, G. and Stoorvogel, W. (2013). Extracellular vesicles: exosomes, microvesicles, and friends. *J. Cell Biol.* 200 (4): 373–383.

7 Thery, C., Witwer, K.W., Aikawa, E. et al. (2018). Minimal information for studies of extracellular vesicles 2018 (MISEV2018): a position statement of the International Society for Extracellular Vesicles and update of the MISEV2014 guidelines. *J. Extracell. Vesicles* 7 (1): 1535750.

8 Kalra, H., Drummen, G.P.C., and Mathivanan, S. (2016). Focus on extracellular vesicles: introducing the next small big thing. *Int. J. Mol. Sci.* 17 (2): 170.

9 Gardiner, C., Di Vizio, D., Sahoo, S. et al. (2016). Techniques used for the isolation and characterization of extracellular vesicles: results of a worldwide survey. *J. Extracell. Vesicles* 5 (1): 32945.

10 van der Pol, E., Boing, A.N., Gool, E.L., and Nieuwland, R. (2016). Recent developments in the nomenclature, presence, isolation, detection and clinical impact of extracellular vesicles. *J. Thromb. Haemost.* 14 (1): 48–56.

11 Zaborowski, M.P., Balaj, L., Breakefield, X.O., and Lai, C.P. (2015). Extracellular vesicles: composition, biological relevance, and methods of study. *Bioscience* 65 (8): 783–797.

12 Shoucair, I., Mello, F.W., Jabalee, J. et al. (2020). The role of cancer-associated fibroblasts and extracellular vesicles in tumorigenesis. *Int. J. Mol. Sci.* 21 (18): 6837.

13 Becker, A., Thakur, B.K., Weiss, J.M. et al. (2016). Extracellular vesicles in cancer: cell-to-cell mediators of metastasis. *Cancer Cell* 30 (6): 836–848.

14 Wu, X., Zhou, Z., Xu, S. et al. (2020). Extracellular vesicle packaged LMP1-activated fibroblasts promote tumor progression via autophagy and stroma-tumor metabolism coupling. *Cancer Lett.* 478: 93–106.

15 Bouvy, C., Wannez, A., Laloy, J. et al. (2017). Transfer of multidrug resistance among acute myeloid leukemia cells via extracellular vesicles and their microRNA cargo. *Leukemia Res.* 62: 70–76.

16 Croese, T. and Furlan, R. (2018). Extracellular vesicles in neurodegenerative diseases. *Mol. Aspects Med.* 60: 52–61.

17 Thompson, A.G., Gray, E., Heman-Ackah, S.M. et al. (2016). Extracellular vesicles in neurodegenerative disease – pathogenesis to biomarkers. *Nat. Rev. Neurol.* 12 (6): 346–357.

18 Li, C.J., Fang, Q.H., Liu, M.L., and Lin, J.N. (2020). Current understanding of the role of adipose-derived extracellular vesicles in metabolic homeostasis and diseases: communication from the distance between cells/tissues. *Theranostics* 10 (16): 7422–7435.

19 Martinez, M.C. and Andriantsitohaina, R. (2017). Extracellular vesicles in metabolic syndrome. *Circ. Res.* 120 (10): 1674–1686.

20 Todorova, D., Simoncini, S., Lacroix, R. et al. (2017). Extracellular vesicles in angiogenesis. *Circ. Res.* 120 (10): 1658–1673.

21 Kholia, S., Ranghino, A., Garnieri, P. et al. (2016). Extracellular vesicles as new players in angiogenesis. *Vasc. Pharmacol.* 86: 64–70.

22 Buzas, E.I., Gyorgy, B., Nagy, G. et al. (2014). Emerging role of extracellular vesicles in inflammatory diseases. *Nat. Rev. Rheumatol.* 10 (6): 356–364.

23 Robbins, P.D., Dorronsoro, A., and Booker, C.N. (2016). Regulation of chronic inflammatory and immune processes by extracellular vesicles. *J. Clin. Invest.* 126 (4): 1173–1180.

24 Lemaire, Q., Raffo-Romero, A., Arab, T. et al. (2019). Isolation of microglia-derived extracellular vesicles: towards miRNA signatures and neuroprotection. *J. Nanobiotechnol.* 17 (1): 1–21.

25 Nguyen, M.A., Karunakaran, D., Geoffrion, M. et al. (2018). Extracellular vesicles secreted by atherogenic macrophages transfer microRNA to inhibit cell migration. *Arterioscler. Thromb. Vasc. Biol.* 38 (1): 49–63.

26 Zempleni, J., Aguilar-Lozano, A., Sadri, M. et al. (2017). Biological activities of extracellular vesicles and their cargos from bovine and human milk in humans and implications for infants. *J. Nutr.* 147 (1): 3–10.

27 Schorey, J.S. and Harding, C.V. (2016). Extracellular vesicles and infectious diseases: new complexity to an old story. *J. Clin. Invest.* 126 (4): 1181–1189.

28 Pang, B.R., Zhu, Y., Ni, J. et al. (2020). Extracellular vesicles: the next generation of biomarkers for liquid biopsy-based prostate cancer diagnosis. *Theranostics* 10 (5): 2309–2326.

29 Campos, C.D.M., Jackson, J.M., Witek, M.A., and Soper, S.A. (2018). Molecular profiling of liquid biopsy samples for precision medicine. *Cancer J.* 24 (2): 93–103.

30 Vader, P., Mol, E.A., Pasterkamp, G., and Schiffelers, R.M. (2016). Extracellular vesicles for drug delivery. *Adv. Drug Delivery Rev.* 106: 148–156.

31 van der Meel, R., Fens, M., Vader, P. et al. (2014). Extracellular vesicles as drug delivery systems: lessons from the liposome field. *J. Controlled Release* 195: 72–85.

32 Tang, T.T., Wang, B., Lv, L.L., and Liu, B.C. (2020). Extracellular vesicle-based nanotherapeutics: emerging frontiers in anti-inflammatory therapy. *Theranostics* 10 (18): 8111–8129.

33 Saint-Pol, J., Gosselet, F., Duban-Deweer, S. et al. (2020). Targeting and crossing the blood-brain barrier with extracellular vesicles. *Cells* 9 (4): 851.

34 Morales-Kastresana, A., Telford, B., Musich, T.A. et al. (2017). Labeling extracellular vesicles for nanoscale flow cytometry. *Sci. Rep.* 7 (1): 1878.

35 Serrano-Pertierra, E., Oliveira-Rodriguez, M., Matos, M. et al. (2020). Extracellular vesicles: current analytical techniques for detection and quantification. *Biomolecules* 10 (6): 824.

36 Szatanek, R., Baj-Krzyworzeka, M., Zimoch, J. et al. (2017). The methods of choice for extracellular vesicles (EVs) characterization. *Int. J. Mol. Sci.* 18 (6): 1153.

37 Zhan, B.B., Li, C., Yang, J. et al. (2014). Graphene field-effect transistor and its application for electronic sensing. *Small* 10 (20): 4042–4065.

38 Zhang, C., Xu, J.Q., Li, Y.T. et al. (2016). Photocatalysis-induced renewable field-effect transistor for protein detection. *Anal. Chem.* 88 (7): 4048–4054.

39 Hajian, R., Balderston, S., Tran, T. et al. (2019). Detection of unamplified target genes via CRISPR-Cas9 immobilized on a graphene field-effect transistor. *Nat. Biomed. Eng.* 3 (6): 427–437.

40 Cai, B.J., Huang, L., Zhang, H. et al. (2015). Gold nanoparticles-decorated graphene field-effect transistor biosensor for femtomolar microRNA detection. *Biosens. Bioelectron.* 74: 329–334.

41 Ganguli, A., Faramarzi, V., Mostafa, A. et al. (2020). High sensitivity graphene field effect transistor-based detection of DNA amplification. *Adv. Funct. Mater.* 30 (28): 2001031.

42 Gao, Z.L., Xia, H., Zauberman, J. et al. (2018). Detection of Sub-fM DNA with target recycling and self-assembly amplification on graphene field-effect biosensors. *Nano Lett.* 18 (6): 3509–3515.

43 Balderston, S., Taulbee, J.J., Celaya, E. et al. (2021). Discrimination of single-point mutations in unamplified genomic DNA via Cas9 immobilized on a graphene field-effect transistor. *Nat. Biomed. Eng.* 5 (7): 713–725.

44 Campos, R., Borme, J., Guerreiro, J.R. et al. (2019). Attomolar label-free detection of DNA hybridization with electrolyte-gated graphene field-effect transistors. *ACS Sens.* 4 (2): 286–293.

45 Zheng, C., Huang, L., Zhang, H. et al. (2015). Fabrication of ultrasensitive field-effect transistor DNA biosensors by a directional transfer technique based on CVD-grown graphene. *ACS Appl. Mater. Interfaces* 7 (31): 16953–16959.

46 Lei, Y.M., Xiao, M.M., Li, Y.T. et al. (2017). Detection of heart failure-related biomarker in whole blood with graphene field effect transistor biosensor. *Biosens. Bioelectron.* 91: 1–7.

47 Kanai, Y., Ohmuro-Matsuyama, Y., Tanioku, M. et al. (2020). Graphene field effect transistor-based immunosensor for ultrasensitive noncompetitive detection of small antigens. *ACS Sens.* 5 (1): 24–28.

48 Kim, D.H., Oh, H.G., Park, W.H. et al. (2018). Detection of alpha-fetoprotein in hepatocellular carcinoma patient plasma with graphene field-effect transistor. *Sensors* 18 (11): 4032.

49 Hao, Z., Zhu, Y.B., Wang, X.J. et al. (2017). Real-time monitoring of insulin using a graphene field-effect transistor aptameric nanosensor. *ACS Appl. Mater. Interfaces* 9 (33): 27504–27511.

50 Hinnemo, M., Makaraviciute, A., Ahlberg, P. et al. (2018). Protein sensing beyond the debye length using graphene field-effect transistors. *IEEE Sens. J.* 18 (16): 6497–6503.

51 Kim, K.H., Park, S.J., Park, C.S. et al. (2020). High-performance portable graphene field-effect transistor device for detecting Gram-positive and -negative bacteria. *Biosens. Bioelectron.* 167: 112514.

52 Thakur, B., Zhou, G.H., Chang, J.B. et al. (2018). Rapid detection of single *E.coli* bacteria using a graphene-based field-effect transistor device. *Biosens. Bioelectron.* 110: 16–22.

53 Wu, G.F., Dai, Z.W., Tang, X. et al. (2017). Graphene field-effect transistors for the sensitive and selective detection of *Escherichia coli* using pyrene-tagged DNA aptamer. *Adv. Healthcare Mater.* 6 (19): 1700736.

54 Pourasl, A.H., Ahmadi, M.T., Rahmani, M., and Ismail, R. (2017). Graphene based biosensor model for *Escherichia Coli* bacteria detection. *J. Nanosci. Nanotechnol.* 17 (1): 601–605.

55 Jin, X., Zhang, H., Li, Y.T. et al. (2019). A field effect transistor modified with reduced graphene oxide for immunodetection of Ebola virus. *Microchim. Acta* 186 (4): 1–9.

56 Seo, G., Lee, G., Kim, M.J. et al. (2020). Rapid detection of COVID-19 causative virus (SARS-CoV-2) in human nasopharyngeal swab specimens using field-effect transistor-based biosensor. *ACS Nano* 14 (4): 5135–5142.

57 Li, J.H., Wu, D., Yu, Y. et al. (2021). Rapid and unamplified identification of COVID-19 with morpholino-modified graphene field-effect transistor nanosensor. *Biosens. Bioelectron.* 183: 113206.

58 Li, Y.X., Peng, Z.K., Holl, N.J. et al. (2021). MXene-graphene field-effect transistor sensing of influenza virus and SARS-CoV-2. *ACS Omega* 6 (10): 6643–6653.

59 Kwon, J., Lee, Y., Lee, T., and Ahn, J.H. (2020). Aptamer-based field-effect transistor for detection of avian influenza virus in chicken serum. *Anal. Chem.* 92 (7): 5524–5531.

60 Islam, S., Shukla, S., Bajpai, V.K. et al. (2019). A smart nanosensor for the detection of human immunodeficiency virus and associated cardiovascular and arthritis diseases using functionalized graphene-based transistors. *Biosens. Bioelectron.* 126: 792–799.

61 Aspermair, P., Mishyn, V., Bintinger, J. et al. (2021). Reduced graphene oxide-based field effect transistors for the detection of E7 protein of human papillomavirus in saliva. *Anal. Bioanal. Chem.* 413 (3): 779–787.

62 Xie, H., Li, Y.T., Lei, Y.M. et al. (2016). Real-time monitoring of nitric oxide at single-cell level with porphyrin-functionalized graphene field-effect transistor biosensor. *Anal. Chem.* 88 (22): 11115–11122.

63 Wang, S.Y., Hossain, M.Z., Shinozuka, K. et al. (2020). Graphene field-effect transistor biosensor for detection of biotin with ultrahigh sensitivity and specificity. *Biosens. Bioelectron.* 165: 112363.

64 Wang, C., Kim, J., Zhu, Y.B. et al. (2015). An aptameric graphene nanosensor for label-free detection of small-molecule biomarkers. *Biosens. Bioelectron.* 71: 222–229.

65 Danielson, E., Dindo, M., Porkovich, A.J. et al. (2020). Non-enzymatic and highly sensitive lactose detection utilizing graphene field-effect transistors. *Biosens. Bioelectron.* 165: 112419.

66 Wang, Z., Yi, K.Y., Lin, Q.Y. et al. (2019). Free radical sensors based on inner-cutting graphene field-effect transistors. *Nat. Commun.* 10 (1): 1544.

67 Wu, D., Zhang, H., Jin, D. et al. (2019). Microvesicle detection by a reduced graphene oxide field-effect transistor biosensor based on a membrane biotinylation strategy. *Analyst* 144 (20): 6055–6063.

68 Wu, D., Yu, Y., Jin, D. et al. (2020). Dual-aptamer modified graphene field-effect transistor nanosensor for label-free and specific detection of hepatocellular carcinoma-derived microvesicles. *Anal. Chem.* 92 (5): 4006–4015.

69 Dong, X.C., Shi, Y.M., Huang, W. et al. (2010). Electrical detection of DNA hybridization with single-base specificity using transistors based on CVD-grown graphene sheets. *Adv. Mater.* 22 (14): 1649–1653.

70 Yu, Y., Li, Y.T., Jin, D. et al. (2019). Electrical and label-free quantification of exosomes with a reduced graphene oxide field effect transistor biosensor. *Anal. Chem.* 91 (16): 10679–10686.

Index

a

acetylcholine (Ach) 61, 136, 322–327, 332
acetylcholinesterase (AChE) 61, 136, 137, 323–326, 328, 332
affinity biosensors 45, 61–63
3-amino-benzylamine (ABA) monomers 137, 318–332
aminophenylboronic acid (APBA) 136
aniline (ANI) 247, 319, 325
anti-BNP antibody 305, 306, 309
aptamers 61, 63, 80, 82, 150, 191, 198, 219, 223, 225, 252, 356, 363, 364, 374, 409
aptasensor 61, 91, 117, 118
atomic force microscopy (AFM) 32, 33, 104, 189, 193, 194, 209, 210, 213, 214, 221, 222, 351, 375, 376
Avian influenza virus infection 374

b

baseline drift 374, 387–398
bilayer graphene (BLG) 107, 155, 169–181, 221
bioanalytical tools 251
biocompatibility 26, 59, 60, 88, 126, 127, 138, 141, 160, 162–163, 285, 286, 303, 340, 367, 406
bioelectronic noses 106, 107
bioelectronics
  using bioprobe-modified graphene 112–116
  wearable 358, 360
biological recognition element (BRE) 339
biosensors
  classification 339
  for environmental component monitoring 116–120
biostability 160, 162–163
blood glucose biosensors 71
bottom-up syntheses 105
bovine serum albumin (BSA) 35, 36, 82, 128, 130, 131, 305, 309, 376, 390–393
B-type natriuretic peptide 301

c

carbon-based nanomaterials 103
carbon nanotube (CNT) 214–215, 231–233
cation doping 388–390
cellulosome complex 198
charge carrier mobility 36, 75, 159, 169, 174, 215, 216, 238–240, 272, 320, 347
charge neutrality point (CNP/V_{CNP}) 58, 73–75, 78, 83, 177, 178, 192, 195, 343, 344, 346, 375

Graphene Field-Effect Transistors: Advanced Bioelectronic Devices for Sensing Applications,
First Edition. Edited by Omar Azzaroni and Wolfgang Knoll.

chemical vapor deposition (CVD) 2, 43, 76, 105, 125, 207
  carrier mobility 208
  diazonium 217
  for FET 214–224
  growth 210
  pyrene chemistry 220–224
clustered regularly interspaced short palindromic repeats (CRISPR) 62, 79, 83, 115, 116, 253, 254
computer chips 154
conducting polymer nanotubes (CPNTs) 111
conducting polymers (CPs)
  films 186, 317–333
conductivity, BLG 178
continuous glucose monitor (CGM) 150, 161
coronavirus disease 2019 (COVID-19) 62, 80, 87, 114, 128, 142, 150, 301, 302, 360
cortisol 162, 342, 343, 354, 355, 364
cut-off frequency 11, 14, 16
cutting-edge technologies 374, 393

d

Debye–Hückel model 252
Debye length 251, 380
  GFET 81–82
  screening 63–64
Debye screening 384
  effect 87, 362–364
  length ($\lambda_{D)}$ 64, 81, 254, 354, 362, 364–365, 367
deep-learning-based optical inspection 393–398
density functional theory (DFT) 196–198, 200–202, 209
deoxyribonucleic acid (DNA)
  biosensors 83–85
  methylation 84
device fabrication, GFET biosensors 349–350
device-to-device variability 362, 366, 367
diabetes mellitus 327
1,5-diamoninaphalene (DAN) 107, 110
diazonium chemistry 217–220
Dice coefficient 395
Dirac point, graphene 173–174
Dirac voltage (V_{Dirac}) 11, 188, 189, 344, 346, 347, 350, 351, 353, 354, 358–360, 363, 365
Donnan equilibrium 253, 254, 257
Donnan potential 253–255, 257, 265
dopamine (DA) 51, 78, 82, 135, 138, 151, 152, 209, 260, 325, 342, 355, 356
Duchenne muscular dystrophy (DMD) 83

e

elaborate theoretical model 245
electrical double layer (EDL/EDL1) 6, 44, 73, 127, 154, 174, 218, 232, 251, 346, 363
electroactive molecule 151
electrochemical double-layer (EDL) 216
electrochemical techniques 51
electrolyte-gated FETs 44, 72, 232
electrolyte-gated gFETs (EG-gFETs)
  biomolecule sensing 251
  characterization 73
  principle 72–75
electron beam (E-beam) evaporator 105
electronic biosensors 71–93
  disease diagnosis 71
electronic sensors 43, 106, 125, 285
electrosynthesized polymer 318
enable nearfield communication (NFC) 162
enzymatic cascades 293, 294, 297
enzyme immobilization 293, 297, 322, 323
enzyme-linked immunosorbent assay (ELISA) 71, 113, 254, 302, 381–383, 406
exosomes (EXOs) 61, 406, 410–411
extracellular vesicles (EVs) 405
  clinical significance 406
  description of 405–6

f

fabrication process
  developments 78
  of G-FET sensor 304
  graphene transfer over
    substrates 76–77
  synthesis 75–76
Faradaic effect 243–245
fatigue biosensors 353–357
Fe(III) meso-tetra(4-carboxyphenyl)
    porphyrin (FeTCP) 138
Fermi energy 177, 178, 289
field-effect transistors (FETs) 9, 339
  biosensors 373–398
  graphene 43, 72, 151–152, 185
figures of merit (FOMs) 5, 14
fitness trackers 149, 161
flexible electronics 11, 53, 340
fluid-sensor interface 160
fluorescence microscopy 189, 190
fluorescent dyes 189
food safety 106, 118, 285
full width half maximum (FWHM)
    175

g

gas detection 125, 138–140, 276
gas sensing technology 271
  graphene 272, 274–275
  graphene FET 272, 274–275
  hybrid graphene FET 277–280
glassy carbon (GCE) 51, 152
glucose detection 80, 90, 136, 327–332
glucose oxidase (GOx) 80,
    327–328, 330–333
Grahame equation 252
Grand Canonical MC (GCMC)
    method 201
graphene 2, 343
  capacitors 9
  characterize molecular
    functionalization 186–196
  frequency multipliers 18
  functionalization 350–353
  Hall structures 156–159
  inductors 5
  membranes 159
  mixers 18
  oscillators 19
  production 348–349
  quantum capacitance 8
  synthesis 104
  transistors 153
  tunable capacitor 7–8
graphene-based field effect
    biosensors (GFET)
  fabrication of 170–173
  fundamental sensing
    parameters 173–174
  with pyrene derivatives 174–176
  quadratic fit analysis 177–181
graphene-based field effect transistors
    (gFETs) 10, 186, 251, 285, 286
graphene binding peptides
    (GrBP) 189, 191
graphene field-effect transistor (GFET)
  antibody-antigen interactions
    85–87
  biosensor 411–412
  biosensor fabrication 348–353
  challenges 92–93
  Debye length 81–82
  enzymatic biosensors 87–90
  functionalization and
    passivation 78–83
  nucleic acids 83–85
  platform 106
  probe molecules 80–81
  sensing mechanism 346–348
  of small ions 90–92
  structure 345–346
graphene field-effect transistor (G-FET)
    biosensors
  deep-learning-based optical
    inspection 393–398
  drift suppression and compensation of
    cation doping 388–390
    state-space models (SSMs)
      390–393
  influenza viruses 374

graphene field-effect transistor (G-FET) biosensors (*Continued*)
mimicking cell surfaces 374–377
real-time measurement, enzyme reaction in microdroplets
Helicobacter pylori (*H. pylori*, [HP]) 384–385
measurement mechanism and model measurement system 385–387
sialoglycan 374
sialoglycan-functionalized
evaluation of 375–377
fabrication 375
signal enhancement, G-FETs 377–387
graphene oxide (GO) 25
biosensors 25–26, 34–38
chemical reduction 48–49
cyclic voltammograms 52
dispersion 46
electrochemical reduction 51–52
modified Hummers method 27–29
production cost 47–52
properties 45
reduced growth 29–34
structure 46
synthesis 26, 47
thermal reduction 49–51
graphene quantum dots (GQDs) 50
graphene topographical quality 212–214
graphite, chemical exfoliation 44

h

Hall effect 103, 156–158, 273
Hall-structure-based sensors 152, 156–158
health sensing 149–151, 160
health sensor 149–163
heart failure (HF) 301–312
Helicobacter pylori (*H. pylori*, [HP]) 384
human immunoglobulin E (IgE) 378
human olfactory receptor 2AG1 (hOR2AG1) 107
human umbilical vein endothelial cells (HUVECs) 138
Hummer's method 26–27, 44, 126
hydrocarbon source 29
hydrothermal process 49, 50

i

immobilization process 80, 117, 323
implantable biosensors 78
inductively coupled plasma mass spectrometry (ICP-MS) 132
inflammation 341–343, 360, 406
interdigitated electrodes 51, 52, 56, 57, 78, 210, 218, 225
interface-focused analytical approaches 375
interleukin-6 (IL-6) 35, 254, 342, 343
ion-exclusion 253–255
ionic etchants 76
ionophore membranes 80, 90
ion-selective-electrodes (ISEs) 241, 242
ion-selective field effect transistor (ISFET) 72, 241, 242, 388

j

Japan Science and Technology Agency (JST) 374

k

Kelvin Force Microscopy (KFM) 319
kinetic inductance 3–5

l

lab-on-a-graphene-FET 373
Langmuir–Blodgett (LB) method 104, 134
Langmuir equation 392
L-arginine sensor 293
lateral flow assay (LFA) 71, 115
Launder's approach 5
layer-by-layer (LbL) assembly
enzyme-based gFET sensors 287–296
of polyethylenimine and urease 288–292

LC tank oscillators 19
leachables 162, 163
Lerf-Klinowski's GO model 30
limit of detection (LOD) 108, 152, 219, 297, 325, 331, 360, 376, 409, 411
liquid
 electrochemical perspective 241–245
 field-effect operation 232–233
 graphene FET 235–236
 measurement modes 236–237
 sensing mechanism 238–240
liquid-phase exfoliation (LPE) 2, 104
living-cell biosensor 137–138
low gate voltages 36, 325

m

Maackia amurensis (MAM) 375–377
magainin I 113, 114, 259
material flexibility 358
MATLAB 394
maximum available power gain 15
mechanically exfoliation 125
medical imaging compatibility 160, 163
metal-containing systems 201
metal-ion detection 131–134
metal–oxide–semiconductor field-effect transistor (MOSFET) 72, 126, 232, 354
microelectromechanical system (MEMS) 105–107, 115
microvesicles (MVs) 406
 detection 408–409
 dual-aptamer modified GFET biosensor 409–410
 label-free detection 410–411
molecular biomarkers 341–343
monolayer graphene 72, 76, 91, 105, 155, 170, 171, 175–177, 181, 210, 213, 221, 340, 350, 366
monolayer graphene fabrication technology 105
Monte Carlo simulations 200, 201, 392
Moore's Law 1
multilayer graphene (MLG) inductor 3, 76, 214, 220, 348

n

nanofluidics 158, 159, 287
neuropeptide Y (NPY) sensing 342
nucleic acids 246
 detection 134–135
 sequences 79

o

observation equation 391
olfactory receptor-conjugated graphene 106–112
open-sandwich immunoassay (OS-IA) 374
 advantages 380–381
 antibody fragments and device fabrication 381–382
 BGP-C7 selectively 382
 in human serum 382–384
 principle of 380
organic thin-film transistors (OFETs) 207

p

p-doping 216, 219, 291, 306, 308, 323, 328, 344, 347, 388
peptide nucleic acids (PNA) 55, 79, 83, 135
photolithography (PL) 56, 77, 105, 110, 172, 304, 350, 354
platinum nanoparticles, decoration of 304–305
p-n junctions 16
poly(3-amino-benzylamine-co-aniline) (PABA) 137
 functionalization 318–322
 nanofilms 318–322
polyethylene glycol (PEG) 82, 130, 253, 408
polyethylene terephthalate (PET) film 53, 107, 129, 137
polyethylenimine (PEI) 59, 136, 288–292

polymethyl methacrylate (PMMA) 76, 105, 172–173, 210–211, 213–214, 235, 349, 389–390
polypyrrole (PPy) 111, 135, 186, 318
potentiometric sensors 241, 303
projected-to-atoms density of states (PDOS) 200
pyrenebutanoic acid succinimidyl ester (PASE) 81, 83, 86, 127, 132, 135, 221, 222, 262, 375, 410
1-pyrenebutanoic acid succinimidyl ester (PBASE) 127, 221, 262, 375, 377, 410
pyrene-maleimide 174, 175, 179, 180

q

quantum capacitance 6, 8, 9, 44, 73, 235, 346

r

Raman spectroscopy 50, 175, 181, 189, 190, 209, 212, 213, 217, 351
reactive-ion etching (RIE) processes 105, 110, 172, 349, 354
reactive signal amplifiers 287, 291, 293
reduced graphene oxide (rGO) 128
- chemical reduction 48–49
- field-effect transistors 286–287
- functional groups 44
- gas detection 138–140
- living-cell biosensor 137–138
- metal-ion detection 131–134
- nucleic acid detection 134–135
- production cost 47
- properties 126
- protein detection 128–131
- small biomolecular biosensor 135–137
- structure 46
- thermal reduction 49–51

rGO SG-FET 53
- affinity biosensors 61–62
- applications 57–64
- challenges 64–65
- Debye length screening 63–64
- enzymatic biosensors 60–61
- functionalization 59
- manufucturing strategies 53–57

ring oscillators 19
roll-to-roll (R2R) transfer 76, 78
root-mean-square deviation (RMSD) 198

s

Sambucus sieboldiana (SSA) 375–377
scanning electron microscopy (SEM) 27, 30, 189, 209, 211, 212, 288, 306, 319, 350, 351, 365
scanning tunnelling microscopy (STM) 188, 189
Scotch tape method 75, 170, 348
self-assembled monolayer (SAM) 88, 91
self-assembly 185
- of (bio)molecular films 186

sensing performance
- advantages, limitations 311–312
- clinical application 311
- selectivity 309
- sensitivity 309–310
- stability and reproducibility 308–309

short-channel effects (SCEs) 185
sialoglycan 374–377
sialoglycan-functionalized G-FETs 374
- evaluation of 375–377
- fabrication of 375–376

sialylglycopeptide (SGP) 375
signal enhancement, G-FETs
- increasing receptor density
  - case of linkers 377–378
  - of linker-based performance enhancement 378
  - performance enhancement by linkers 378–380
- open-sandwich immunoassay
  - advantages 380–381
  - antibody fragments and device fabrication 381–382
  - BGP-C7 selectively 382
  - in human serum 382–384
  - principle of 380

sodium borohydride ($NaBH_4$) 48, 56
solution-gated FETs (SG-FETs) 44, 45, 52–60, 62–64
solution-processable graphene 125, 126
sonication 46, 47, 104, 159
S parameters 13–15
spectroscopic ellipsometry 191
stability factor 15
state equation 391
state-space models (SSMs) 390–393
steroid 343
"stiction" 186, 187
stress, chronic 341–342
stress hormone 343, 354
surface binding process 192
Surface-Enhanced Raman Scattering (SERS) 50, 72, 118
surface functionalization 126, 132, 137, 141, 187, 196, 287, 305, 333, 352, 412
surface modification
  Donnan potential 253–255
  ion-exclusion 253–255
  kinetic cost 262–264
  with lipid layers 258–259
  with mesoporous materials 260–261
  with polymer films 255–258

t

tetrakis(4-carboxyphenyl) porphyrin (TCPP) 377–380
thermal reduction methods 49
thin-film transistors (TFTs)
  principle 126–127
  rGO 128–140
thioglycolic acid (TGA) 131, 133, 134, 305
traditional analytical methods 285
traditional immobilization techniques 286
transition-metal dichalcogenides (TMDs) 2, 3, 142
tripodal pyrene binders 221
tumor necrosis factor-alpha (TNF-α) 342, 343
2D materials
  diodes 15–18
  modelling 17–18
  p-n junctions 16
  transistors 9–15

u

ultrasensitive sensing 292–296

v

vancomycin 113, 114
volatile organic compounds (VOCs) 111, 112, 271, 279, 281

w

wearable bioelectronics 358, 360
wearable sensors 53, 149, 160, 358
wet transfer technique 76, 105
whole blood, immunodetection in 305
whole-gene sequencing 115

x

X-enes 2
X-ray photoelectron spectroscopy (XPS) 189, 209, 213, 307, 389